5th Edition

Competency in College Mathematics

Jack Gill

Robert Blitzer
Miami-Dade
Community College

H&H Publishing Company
1231 Kapp Drive
Clearwater, FL 34625
(813) 442-7760

Competency in College Mathematics, 5th Edition
Jack Gill, Robert Blitzer

Copyright 1992
H&H Publishing Company, Inc.
1231 Kapp Drive
Clearwater, FL 34625
telephone (813) 442-7760

Editorial, Production, and Marketing
    Robert Hackworth, Karen Davis, Thomas Howland

Operations
    Thomas Howland, Michael Ealy, Sally Marston,
    Janie Howland, Priscilla Trimmier

Computers
    Apple Computer, Inc.

ISBN 0-943202-39-6

Library of Congress card catalog number 92-071295

Printing is the lowest number: 20 19 18 17 16 15 14 13 12 11

Preface

The challenge of the times for mathematics educators is to meet new demands for content and quality.

We have accepted this challenge and are keeping pace with the dynamic changes in the teaching of mathematics. This fifth edition affirms our commitment to the development of timely, quality materials to meet the changing needs of our students, professors, and institutions. It also contains the latest information, as of summer 1992, on the mathematics content that every college student is expected to know. **Competency in College Mathematics, 5th Ed.**, is designed to teach those college students to attain higher performance levels with regard to required competencies in mathematics.

Many states are now mandating that students achieve a mastery of certain mathematics competencies. As these mandates have developed there is an increasing consensus toward a mastery of many skills and concepts associated with courses in traditional liberal arts mathematics and finite mathematics. This text meets the political mandates of our times, but also maintains the solid academic integrity of our profession.

As suggested above, the content of **Competency in College Mathematics** reflects the development of curricula which emphasize elements of Liberal Arts mathematics and finite mathematics. The text, then, could be highly recommended for any of three courses:

- •1 A course specifically designed to meet competency requirements in higher education.
- •2 A traditional liberal arts (general education) course.
- •3 A finite mathematics course.

We would like to draw your attention to the format of this text. First, the objectives of each chapter may be found by studying the detailed Table of Contents. Second, there are a number of exercises in each problem set which review ideas discussed in previous sections of the chapter—especially those relating to the mastery of particular competencies. We believe that continuous review is necessary if students are to assimilate and retain many of the concepts presented. Thus, a student is not permitted the opportunity to forget ideas discussed earlier in the text. Third, a student is given the opportunity to sharpen basic arithmetic skills in most problem sets throughout the text.

Fourth, problem sets occur after just a few concepts are introduced as opposed to introducing several ideas followed by a large set of problems. This helps the student to master concepts sequentially and in relatively small increments. Answers to all problems have been included at the end of the text to insure maximum feedback.

Ideally a student should possess a sound mastery of arithmetic by the time he or she reads this text. A student should also possess some manipulative skills in elementary algebra. **Competency in College Mathematics, 5th Ed.**, really is not appropriate for a student who has little or no background in mathematics. Prerequisite courses should be taken.

The major changes in this fifth edition are highlighted in the following list.

Chapter 1

Problems are added where students must identify the correct statement of relationship among sets and set membership.

Chapter 2

a. Almost all English examples are changed to bridge the gap between student's intuitive understanding of logic and formal symbolic logic.
b. More emphasis is placed (both intuitively and formally) on equivalences, nonequivalences, and negations of statements.
c. Arithmetic verbal problems are added to the review competencies.
d. A subsection is added (5.3) on using valid forms of arguments to deduce conclusions that result in valid arguments.

e. The section on Euler circles and arguments is expanded.

f. A section (7) is added on logic in everyday life. This section requires students to use more informal methods of reasoning to deduce logical conclusions.

g. The chapter self-test is rewritten and expanded to reflect changes in the chapter.

Chapter 3

Problems are added that go beyond students merely recognizing the names of properties of operations. These problems require students to use the properties correctly and recognize when they are not used correctly.

Chapter 4

a. Problems on base 10 numerals are expanded.

b. More detail on exponential simplification is presented in an expanded section.

Chapter 5

a. Problems are added which ask students to use applicable properties to select equivalent equations and inequalities.

b. The section on algebraic word problems is expanded to contain percent problems, consecutive integer problems, and problems on the digits of a two-digit number.

c. A section (7) is added on solving a system of two linear equations in two variables by substitution and elimination methods.

d. The chapter self-test is rewritten and expanded to reflect changes in the chapter.

Chapter 6

a. Notations for parallel and perpendicular lines are included.

b. The new chapter self-test now contains 50 items with increased emphasis on critical thinking, the ability to reason based upon geometric figures, and the ability to generalize knowledge. These unique multiple-choice items were class tested with thousands of students at Miami-Dade Community College South Campus over the past five years.

Chapter 7

More emphasis is placed on real-world problems

that utilize the fundamental counting principle and basic concepts of probability. The chapter self-test is improved and expanded.

Chapter 8

a. Problems on interpreting real-world data involving frequency and cumulative frequency tables are added.

b. Problems and graphs, including plot diagrams, requiring students to infer relationships and make accurate predictions from studying statistical data are now included.

c. There are more illustrative examples and problems dealing with properties and interrelations among the mean, median, and mode in a variety of distributions.

d. The chapter self-test is improved and expanded.

A new feature is the inclusion of an appendix containing 130 multiple-choice items that review the competencies presented throughout the book. Also included are answers and brief explanations. This cumulative all-inclusive review should be useful to students preparing for a final course examination as well as for state-mandated competency examinations.

We are indebted to the students and professors who have used the earlier editions and given us the helpful suggestions which have improved this fifth edition. Special thanks are due Professors Dale Grussing, J. Louis Nanney, Kay Waterman, and Kenneth Goldstein of Miami-Dade Community College who made many constructive comments during the preparation of the original manuscript and through the first and second editions.

We are also indebted to the staff of H&H Publishing Company, Inc. which has improved the technical quality of this book, delivered this edition in a timely manner, while maintaining a reasonable book price for our students. Special thanks go to Tom Howland and, above all, to our ever-efficient and helpful editor, Karen H. Davis.

Jack C. Gill   Robert Blitzer
May, 1992   Miami-Dade Community College

Contents

one

Set
Theory

In this chapter a concept will be studied that has become extremely important in modern mathematics. The idea of a set was once viewed as a very difficult and sophisticated concept. It was studied only at the graduate level of colleges and universities. We have now come to realize that simple set ideas can even be introduced in the kindergarten.

Georg Cantor, a German mathematician, contributed to the development of the theory of sets in the nineteenth century; however, it was not until the middle of this century that this concept came to be appreciated as a basic unifying principle in mathematics. The set concept connects many mathematical ideas. It also provides a very precise way of describing and communicating these ideas. At the most basic level the set idea is used to develop the abstract notion of "number" itself and to develop the ideas of addition and multiplication. At the college level, the set concept finds its way into courses in finite mathematics, statistics, probability and algebra.

It is possible that you may already be familiar with certain aspects of set theory. If so, this chapter can serve to refine your knowledge.

Section 1   Set Notation

A **set** is said to be a collection or group of objects. The objects in the set are called the **elements** or **members** of the set. There are many examples of sets that we encounter every day. We might speak about a class of students or a herd of cattle or a flock of sheep.

We will not concern ourselves with these general types of sets. Rather, we will confine our interest to sets in which it is clear whether or not some element is in them. This consideration leads us to state the following.

> **Definition**
> A set is **well-defined** if it is possible to determine whether or not a given element belongs to it.

Examples

1. The set of letters in the English alphabet is a well-defined set since it is possible to determine whether or not a given element belongs to it.

2. The set of names of the months of the year is a well-defined set. We can easily determine whether or not an element belongs to the set. It is possible to identify all the elements.

3. The set of good baseball players is not a well-defined set. It would be difficult to determine whether or not a given element belongs to it. Would everyone agree as to what makes a "good" baseball player?

Problem Set 1.1

Which of the following sets would be well-defined?

1. The set of teams in the American Football Conference
2. The set of funny-looking teachers
3. The set of months of the year that have 30 days
4. The set of interesting people
5. The set of large community colleges

Review Competencies

6. $\frac{3}{5} = \frac{?}{20}$

7. Express $5\frac{3}{7}$ as an improper fraction.

We use capital letters (A,B,C, etc.) to indicate sets. All the elements in the set are enclosed in braces and separated by commas. Consider the set A with the elements x, y, and z. If we wished to express this set A in set notation, we write:

$$A = \{x,y,z\}$$

If B is the set of all letters of the English alphabet, we could write:

$$B = \{a,b,c,\ldots,z\}$$

We use three dots to represent the elements that are not listed but which are still understood to be in the set. The set B has a limited number of elements (26 in fact) and is called a **finite** set.

The set of **whole numbers** is represented as {0,1,2,3,...}. This set contains an unlimited number of elements and is said to be an **infinite** set.

The set of **counting numbers** (or natural numbers) is represented as {1,2,3,...}. The set of counting numbers is also an infinite set.

Problem Set 1.2

1. What is incorrect about each of the following?   Review Competencies
   a. A = 2,4,6
   b. B = 2,...
   c. C = (1,3,5)
   d. D = {1,2,3
   e. E = {abc}

   3. Which of the following sets is well-defined?
      a. The set of large numbers
      b. The set of whole numbers

2. Which of the following sets is infinite?
   a. A = {2,4,6,...,100}
   b. B = {5,10,15,...}
   c. C = {1,$\frac{1}{2}$,$\frac{1}{4}$,$\frac{1}{8}$,...}

   4. Express $\frac{23}{6}$ as a mixed number.

Notation

The symbol $\in$ is used to indicate that an object is an element of a set, and the symbol $\notin$ is used to indicate that an object is not an element of a set.

Examples

1. It is correct to write: a $\in$ {a,b,c,d} because the letter a is certainly an element of this set.

2. It is correct to write: e $\notin$ {a,b,c,d} since element e is not an element of this set.

3. It is not correct to write: {b} $\in$ {a,b,c,d}. However, it is correct to write: b $\in$ {a,b,c,d}.

4. It is correct to write {b} $\in$ {{a}, {b}, {c}, {d}}.

Problem Set 1.3

1. Let J represent "the set of months of the year that begin with the letter J," and let A represent "the set of all months of the year." Indicate whether the following statements are true:
   a. June ∈ J   b. July ∈ A

2. Suppose X represents "the set of all the states in the United States." Determine the truth or falsity of each statement.
   a. Florida ∈ X T   b. Puerto Rico ∉ X
   c. Lower Slobbovia ∈ X T

3. Which of the following is true?
   a. 4 ∈ {1,6,4,9} T   b. {3} ∈ {3,4,5,6} T
   c. 2 ∉ {1,3,5,7}

Review Competencies

4. Is the following set well-defined?
   The set of community colleges in Florida

5. Determine if the following set is infinite or finite. $A = \{1, \frac{1}{2}, \frac{1}{3}, \ldots\}$

6. Add: $\frac{5}{6} + \frac{1}{4}$

Let us now consider an important concept of set theory: equality of sets.

Definition
Sets that possess the same exact elements regardless of the order in which they appear are called **equal** sets or **identical** sets.

Examples

1. If A = {x,y,z} and B = {z,x,y}, then set A is equal or identical to set B because they both contain the same exact elements. We say A = B.

2. If C = {x,x,y,z} and D = {x,y,z}, then set C is still equal to set D. Repeating an element more than once in a set does not affect its membership. Set C only contains the distinct elements x,y, and z.

Problem Set 1.4

Indicate whether or not the following sets are equal sets.

1. {r,a,t}; {t,a,r} T

2. {m,i,a,m,i }; {m,i,a} F

3. {2,4,6,8}; {2,4,6,8,0} F

4. True or false? $4 \notin \{3,5,7,9\}$

5. Infinite or finite? {1,2,3,...,1000}

6. True or false? $\{4\} \in \{1,2,3,4\}$

7. Subtract: $\frac{7}{9} - \frac{3}{5}$

Some sets are defined by listing their elements or by describing their elements in words. It is important that the description be as precise as possible, although it is often possible to supply more than one correct description of a set.

Examples

1. The set {1,2,3,4,5,6} may be described as
   "The set of counting (natural) numbers from 1 to 6 inclusive" or
   "The set of counting numbers less than 7."
   This set is not described precisely as
   "The set of counting (natural) numbers."

2. The set {2,4,6,8} may be described as
   "The set of even counting numbers less than 10."
   The set is not described precisely as
   "The set of even counting numbers."

3. The set {11,13,15,...} may be described as
   "The set of odd counting (natural) numbers greater than 10" or
   "The set of odd counting (natural) numbers greater than or equal to 11."

Problem Set 1.5

Supply a description for each of the following sets.

1. {Monday, Tuesday, Wednesday, Thursday, Friday, Saturday, Sunday} Set of the week Days.

2. {Florida, Alabama, Mississippi, Louisiana, Texas} Set of the south west state

3. {21,22,23,...}   4. {4,8,12,...}
   natural # greater than 20

5. {1,2,3,...,50} Set of natural # greater than 0 and less than 51

6. Are the following sets equal?
   {1,2,3,4,5}, {3,1,5,4,x}

7. True or false? $0 \notin \{1,2,3,4\}$

8. Add: $5\frac{1}{3} + 3\frac{1}{4}$

Chapter 1

Sometimes it is tedious to list all the elements of large sets. Thus, another notation is used. We could describe the set of books in the library of Congress by using a letter, say x, to obtain:

{x | x is a book in the Library of Congress}

We read this as:

"The set of all elements x such that x is a book in the Library of Congress."

The vertical line is read "such that." The remaining notation defines the conditions necessary for a candidate to be a member of the set. By substituting specific books for the letter x we are able to build a set. We refer to this as the **set-builder notation.**

As another example: {x | x is an even whole number greater than 3 and less than 200} implies the following set: {4,6,8,...,198}

Problem Set 1.6

1. List the elements in the following sets.
    a. {x | x is a whole number greater than 10}
    b. {x | x is a whole number between 4 and 20}

2. Is the following set finite or infinite?
    {x | x is a fraction between 1 and 2}

3. Is the following set a well-defined set?
    {x | x is a smart student}

4. Are these sets equal?
    {x | x is a lower-case letter in the English alphabet} and {a,b,c,...,z}

Review Competencies

5. Are the following sets equal?
    {0,2,4,6} and {2,4,6}

6. Write a description for the set: {7,14,21,...}

7. True or false? $0 \in \{0,1,2,3,...\}$

8. Subtract: $5 - 1\frac{3}{7}$

Section 2   Subsets and Complements

In this section we discuss some special sets. To do this we must first recognize that no matter what problem in set theory is under consideration it is necessary to establish a set that meets the specific needs of that particular problem.

Definition
The set that contains all the possible elements under consideration in a problem is called the **universal set.** This set is indicated by the letter U .

6

Examples

1. In a discussion about states in the United States, the universal set would be the entire set of 50 states. Thus, set U would contain 50 elements.

2. If a universal set U is given to be the set of elements {2,4,6,8,10}, we can only work with these particular elements in our discussion. The universal set restricts us to using any or all of the elements in the universal set.

The universal set will vary from one problem to another, but it must be specific for each problem under consideration.

Problem Set 2.1

1. What is the universal set in a discussion about months of the year whose names begin with the letter "J." *the months of the year*

2. What is the universal set in a discussion about vowels in the English language? *a e i o u.*

Review Competencies

3. Write a description for the set {5,10,15,...}.

4. List the elements in the set {x | x is a whole number between 2 and 8}. *F*

5. Is the following set well-defined? {x | x is a big person} *F*

6. Subtract: $7\frac{2}{5} - 3\frac{3}{4}$
   $\frac{57}{5} - \frac{15}{4} = \frac{148-75}{20} = \frac{73}{20} = 3\frac{13}{20}$

We often encounter situations in which all the elements of one set are also elements of another set. This leads us to the following.

> ### Definition
> Set A is said to be a **subset** of set B if every element in set A is also in set B. Symbolically, this is expressed as: $A \subseteq B$. $A \nsubseteq B$ means that set A is not a subset of set B.

Examples

1. If set A = {1,2,3} and set B = {1,2,3,4,5,6}, we see that every element in set A is also an element of set B. Thus we can write: $A \subseteq B$ Notice that set B is not a subset of set A. We write: $B \nsubseteq A$.

2. If set X = {a,b,c} and set Y = {a,b,c}, we see that every element in set X is also an element in set Y. Thus, we write $X \subseteq Y$. In fact, X = Y. Notice that when X = Y, both $X \subseteq Y$ and $Y \subseteq X$ are true. This example illustrates that a set is a subset of itself.

Problem Set 2.2

1. Find a subset of M = {1,2,3,4,5,6,7,8}
    that contains:
    a. all the odd numbers in M
    b. the numbers in M whose triples are in M
    c. the numbers in M that are divisible by 2

2. Indicate which of the following is true or false.
    a. {1,3} ⊆ {2,4,9,3} F
    b. {0,1,5} ⊄ {1,5,7,9} T
    c. {4,6} ⊆ {x | x is a whole number}
    d. {1,2,3,4} ⊆ {1,2,3} F

3. Is the following set well-defined?
    The set of universities in Florida

4. Is the set {1, $\frac{1}{3}$, $\frac{1}{9}$,...} infinite or finite?

5. True or false? 3 ∈ {2,4,6,8}

6. Multiply: $\frac{5}{9} \cdot \frac{4}{7} \cdot \frac{3}{10}$

We now consider special kinds of subsets.

Definition
Set A is said to be a **proper subset** of set B if
every element in set A is also in set B, and there
is at least one element in set B that is not in set A.
Symbolically, this is expressed as A ⊂ B.

If set A is a proper subset of set B, then set A must contain **fewer** elements
than set B. Notice the similarity between A ⊂ B in set theory and A < B
(A is less than B) in algebra.

Examples

1. If set A = {1,2} and set B = {1,2,3}, we can say that set A
    is a proper subset of set B because there is one element in B, 3,
    that is **not** in set A. Furthermore, every element in set A is also in
    set B. Symbolically, we write: A ⊂ B.

2. If set C = {2,4,6,8} and set D = {1,2,3}, then set D is neither a
    proper subset nor a subset of set C. Symbolically, we write D ⊄ C.
    D ⊄ C means D is not a proper subset of C.

3. If E = {3,4,5} and F = {4,5,3}, then E ⊄ F, but E ⊆ F.

We now discuss a most important set.

> ### Definition
> The set that contains no elements at all is called the **null set** or the **empty set**. This may be indicated by $\emptyset$ or { }.

The null set is not represented by {$\emptyset$}. This is a set containing one element, the Greek letter phi (pronounced fi). There is one and only one null set, but many ways to describe it. Two ways of describing the null set are shown by the examples below.

Examples

1. The set of counting numbers between 1 and 2 is the null set because there are no counting numbers between 1 and 2.

2. The set of states of the United States that touch both Canada and Mexico is the null set because there are no states that satisfy this condition.

Some interesting properties of the null set are the facts that the **null set is a subset of every set**, and the **null set is a proper subset of every set except itself**. Symbolically, this is expressed as: $\emptyset \subset A$ and $\emptyset \subseteq A$ if A is not the empty set. When finding all the subsets that can be formed from a given set, always list the given set itself and the null set. The set itself will be the only set that is not a proper subset. The null set will be the only set with no elements.

Example

The list of all subsets of {a,b,c} must include the three-element set {a,b,c} and the empty set, $\emptyset$. Other subsets will be all possible two-element and one-element sets.

The set itself is a subset	Two element subsets	One element subsets	Empty set
{a,b,c}	{a,b}	{a}	$\emptyset$
	{a,c}	{b}	
	{b,c}	{c}	

Notice that there are 8 subsets that can be formed from the set {a,b,c}. Seven of these subsets are proper subsets.

The **number of subsets** that can be formed from a given set is given by the expression 2^n in which n represents the number of elements in the given set.

Chapter 1

The letter n is an **exponent**. Exponents will be discussed more fully in a later chapter. However, for the moment you might realize the following:

$$2^0 = 1, \quad 2^1 = 2, \qquad 2^2 = 2 \cdot 2 = 4, \qquad 2^3 = 2 \cdot 2 \cdot 2 = 8, \text{ etc.}$$

The **number of proper subsets** that can be formed from a given set of n elements is represented by the expression $2^n - 1$. The given set itself is the only set that is not a proper subset.

Examples

1. The number of subsets of {a,b,c} is given by the expression
$2^3 = 2 \cdot 2 \cdot 2 = 8$.

   The number of proper subsets is given by $2^3 - 1 = 8 - 1 = 7$. These results can be verified by the previous example.

2. The number of subsets of the set {a,b,c,d,e,f} is given by the expression
$2^6 = 2 \cdot 2 \cdot 2 \cdot 2 \cdot 2 \cdot 2 = 64$. It is a tedious task to verify this result by

   listing all 64 subsets. The number of proper subsets, excluding the original set from our 64 subsets, is $64 - 1 = 63$.

3. The number of subsets of {1,Ø} is $2^2 = 4$. The subsets are {1,Ø},{1},

   {Ø}, and Ø.

Problem Set 2.3

1. If the universal set is U = {0,1,2,3,4}, which of the following are subsets of U?
   a. Ø F
   b. {0,2} T
   c. {4,3,2,1,0} T
   d. {2,Ø} F

2. What set represents the set of all the elements which are not in the universal set?

   proper subset

3. Write all the subsets that can be formed from each of the following.
   a. {x,y}
   b. {x,y,z}
   c. {0}
   d. {x,y,Ø}
   e. Ø

4. Use the results from problem #3 to complete the following.

   a.
Number of elements in a set	0	1	2	3
Number of subsets that can be formed				

   b. How many subsets can be formed from a set containing 4 elements? 5 elements?

5. How many proper subsets can be formed from a set which has:
   a. no elements? 0
   b. one element? 1
   c. two elements? 1
   d. three elements? 1
   e. four elements? 1
   f. five elements? 1

6. Is there a difference between Ø and {Ø}?
   No

7. If A = {2,3,8,10}, which of the following are true?
    a. 3 ⊂ A
    b. 2 ∉ A
    c. Ø ⊂ A
    d. Ø ∈ A
    e. A ⊄ A

8. Can it be said that Ø ∈ Ø? F

9. Can it be said that 0 ∈ Ø? F

10. Is the following set well-defined or not well-defined?
    {x | x is an elephant that flies}

11. Are the following sets equal?
    {0,1,3,5}; {1,3,5}

12. Infinite or finite?
    {x | x is a fraction between 0 and 1}

13. True or false? {2,4} ⊆ {1,2,3,4 }

14. Multiply: $4\frac{2}{3} \cdot 3\frac{1}{4}$

We are often interested in the elements that are not members of a given set. We therefore introduce an important set operation in the following discussion.

> ## Definition
> The **complement** of set A is the set of elements that are not in set A but are still in the universal set. The set is denoted as A' (read "A prime"or "A complement").

Problem 1   If U = {1,2,3,...,10} and A = {1,4,5,8,9},
find the complement of A, which is symbolized as A'.

Solution   The complement of A, A', is the set which contains elements of U that are not in A. Check each element of U. Those that are not in A must be in A'. Thus, A' = {2,3,6,7,10}.

Problem 2   If U = {0,1,2,3,...} and B = {0,1,2,...,100}, find the complement of B.

Solution   Since the universal set contains all the whole numbers, and set B contains only those whole numbers which are 100 or less, the complement of B must contain all the whole numbers greater than 100.
    B' = {101,102,103,...}

Chapter 1

Problem 3 If U = {1,2,3,4,5} and A = {1,3,5}, find (A')'.

Solution First find A'. A' = {2,4}. Now find the complement of this set A' or (A')'. Thus, (A')' = {1,3,5} or A. In set notation this says (A')' = A. In words this says that the complement of the complement of a set is the set itself.

Problem Set 2.4

1. If U = {F,I,N,K}, find each of the following:
   a. A = {F,I,N}; A' = ? (K)
   b. B = {I,K}; B' = ? (F,N)
   c. X = {F,I,N,K}; X' = ? ∅
   d. C = {F,N,K}; C' = ? (I)
   e. Y = ∅; Y' = ? (F,I,N,K)
   f. D = {I,N}; (D')' = ? (F,K)

2. Suppose that U represents the set of numbers {1,2,3,...}. Suppose also that A represents the set of odd numbers {1,3,5,...}. How could A' be represented? (2,4,...)

3. What is the complement of the universal set? null ∅ What is the complement of the null set? Universal set

4. Why do we say that the null set can be a proper subset of any set other than the null set itself? Because it represents 0.

5. If set A = {all humans in Florida universities}, and U = {all humans}, express set A' in words. (all humans that don't go to FL Universities)

Review Competencies

6. How many subsets and proper subsets can be formed from a set that contains two elements?

7. True or false? ∅ ∈ ∅

8. True or false? {2} ⊂ {1,2,3}

9. Divide: $\frac{6}{7} \div \frac{5}{6}$

Section 3   Equivalent Sets and Set Cardinality

In this section we will see that statements can be made about sets relative to the number of elements they contain. We must first discuss a way of demonstrating whether or not sets have the same number of elements.

> **Definition**
> A **one-to-one correspondence** between two sets is a relationship such that for every element in one set there may be found exactly one element in the other set to pair with it.

Examples

1. Consider the following two sets: A = {1,2,3} and B = {a,b,c}.

A = { 1, 2, 3 }

B = { a, b, c }

A one-to-one correspondence is said to exist between set A and set B because each element in set A can be paired or matched with one and only one element in set B.

2. Consider the following two sets: A = {w,x,y,z} and B = {1,2,3}.

A = { w, x, y, z }

B = { 1, 2, 3 }

A one-to-one correspondence does not exist between set A and set B because not every element in set A can be paired with some element in set B.

Problem Set 3.1

1. Identify whether or not the following pairs of sets possess a one-to-one correspondence:
   a. {fathers} -- {sons} T
   b. {U. S. States} -- {U. S. Senators} T
   c. {brothers} -- {sisters} T
   d. {odd whole numbers} -- {even whole numbers} T
   e. {one-digit whole numbers} -- {two-digit whole numbers} ←

2. How could one determine which of the following two sets has more elements without resorting to counting the elements?
   {F,L,O,R,I,D,A} or {C,L,O,D}   1st one

3. Do the following sets possess a one-to-one correspondence?
   {0,1,2,...} and {1,2,3,...}   Yes.

Review Competencies

4. If U = {F,L,O,R,I,D,A} and A = {L,I,D,A}, find A'.

5. What is the complement of the null set?

6. If U = {1,2,3,4} and A = {1,3}, find (A')'.

7. List the elements in the set:
   {x | x is a whole number greater than three}.

8. Divide: $4\frac{1}{4} \div 2\frac{3}{5}$.

In the discussion that follows we will see that we can describe sets that can be placed in a one-to-one correspondence.

> **Definition**
>
> Finite sets that have the same number of distinct elements are called **equivalent sets**.

Examples

1. If set A = {1,2,3} and set B = {x,y,z}, then set A is equivalent to set B because they both contain the same number of distinct elements (three).

2. If set C = {p,q,r} and set D = {1,2,3,3}, then set C is considered equivalent to set D because they both contain three distinct elements.

Problem Set 3.2

1. Are two equivalent sets always equal?

2. Are two equal sets always equivalent?

3. Are these two sets equivalent?
    {J,E,R,K} and {R,E,J,K}

4. Are these two sets equivalent?
    {S,E,T,S} and {S,E,X}

5. Are these two sets equivalent? {0} and Ø?

Review Competencies

6. List the subsets that can be formed from the set {2}.

7. What is the complement of the universal set?

8. Does the following pair of sets possess a one-to-one correspondence?
    {mothers}   {daughters}

9. A family spends $\frac{1}{4}$ of its income for food and $\frac{1}{3}$ for rent. What fractional part of the income does the family still have?

Since the number of elements in a finite set is a counting number, we now proceed to define that number.

> **Definition**
>
> The number that indicates how many elements are present in a finite set is called the **cardinal number** of the set. The notation "n(A)" is used to refer to "the number of elements in set A."

Examples

1. If set A = {a,b,c,d,e}, then set A is said to contain 5 elements. This number 5 is the cardinal number of set A. Symbolically, this can be expressed as: n(A) = 5. Notice that this set of elements can be placed in a one-to-one correspondence with the set of counting numbers {1,2,3,4,5}.

2. If A = Ø, then the cardinal number of A is 0. Thus, n(Ø) = 0. It should be obvious that two sets A and B are equivalent if sets A and B have the same cardinal number.

Problem Set 3.3

1. Indicate the cardinal number of each of the following sets:
   a. {2,4,6,8,10} 5
   b. {3,10,20} 3
   c. {a} 1
   d. {w,x,y,z} 4
   e. {4,Ø} 1
   f. {1,2,3,...,20}
   g. Ø 0
   h. {0} 1

2. Criticize the following student's reasoning:
   The cardinal number of the set {1,2,3,4} is 4 so the cardinal number of a set should always be the last number listed in the set. *Cardinal # is the number of elements that the set has.*

3. If set A is equivalent to set B, can it be asserted that n(A) = n(B)? *Yes*

4. Find n(A) if A = {x | x is a state in the United States}. *50.*

5. Find n(A) if A = {x | x is a city that lies in both Florida and Georgia}. *Ø*

6. Let A = {1,2,3,...} and B = {2,4,6,...}.
   a. Why is the cardinal number of set A equal to the cardinal number of set B?
   b. Why is set B a subset of set A? (Notice the strange statement that now emerges for infinite sets: The part, the subset, has the same cardinal number as the whole set.)
   c. We can pair 1, in set A, with 2, in set B; 2, in set A, with 4, in set B; etc. Following this pattern, we can pair n in set A with _____ in set B.

Review Competencies

7. Are the following sets equivalent?
   {a,b,c} and {b,a,c}

8. If U = {1,2,3,4} and A = Ø, what is (A')'?

9. Is the following true? {1,2,3} ⊆ {1,2}

10. Express $\frac{5}{8}$ in decimal form (to three places).

Section 4   Intersection and Union of Sets

Some additional set operations will be investigated in this section. It is extremely important that you understand them.

It is often necessary to determine the elements that two given sets have in common. The set that contains these common elements is discussed below.

> ### Definition
> The **intersection** of two sets A and B, denoted by $A \cap B$, is the set that consists of the elements that are in **both** set A and set B.

It will be helpful to remember that the word "and" in the definition above will be associated with the intersection operation.

Examples

1. If set $A = \{1,2,3\}$ and set $B = \{2,3,4\}$, then $A \cap B = \{2,3\}$ because the elements 2 and 3 are in both sets. Also, $n(A \cap B) = 2$.

2. If set $A = \{1,2\}$, set $B = \{2,3\}$, and set $C = \{2,4\}$, then $(A \cap B) \cap C$ is found by first determining what is contained in the parentheses, namely $A \cap B$, or $\{2\}$. Then, the intersection of this set, $\{2\}$, and set C is determined. This final intersection is $\{2\}$.

$$(A \cap B) \cap C = \{2\}.$$

Using set-builder notation it is possible to define the intersection of set A and set B as: $A \cap B = \{x \mid x \in A \text{ and } x \in B\}$.

> ### Definition
> Two sets that have no elements in common are called **disjoint sets.** Their intersection is the null set.

Example

If set $A = \{1,2\}$ and set $B = \{a,b\}$, then set A and set B are disjoint sets because there are no elements that are common to both sets.
$$A \cap B = \emptyset$$

Problem Set 4.1

1. If set A = {2,4,6}, set B = {1,2,3} and
   set C = {3,4,6}, find:
   a. A∩B   2   b. B∩C   3
   c. (A∩B)∩C   2   d. A∩(B∩C)   ∅
   e. A∩∅   0   f. n(A∩C)   3
   g. n(B∩∅)   2(4,6)

2. Find A∩B if:
   a. A = {2,4,6,8} and B = {1,5,6,7} = 6
   b. A = {2,4,6,...} and B = {1,3,5,...} = ∅
   c. A = {3,6,9,...} and B = ∅   = 0

3. Is the following true? ∅∩∅ = ∅   yes

4. {x l x is an even whole number} ∩ {x l x is
   an odd whole number}= ?

5. Consider the following sets:
   U = {all citizens}
   A = {all citizens of Florida}
   B = {all employed citizens of Florida}
   C = {all registered Democrats in Florida}
   Express each of the following sets in words.
   a. A∩B   b. B∩C'
   c. (A∩B)∩C

Review Competencies

6. What is the cardinal number of the following
   set: {0,∅}?

7. Are two equivalent sets always equal?

8. Find n(A) if A = {x l x is a whole number
   between 2 and 8}.

9. List the subsets that can be formed from the
   set {1,∅}.

10. Subtract: 6 – 0.563

We will now see that it is important to be able to unite two sets in order to form
a new set. This new set of elements is discussed below.

> ### Definition
> The **union** of two sets A and B, denoted by A ∪ B,
> is the set that consists of all the elements in set A
> or in set B or in both set A and set B.

The union of two sets is the set of elements whose members are elements of
at least one of the given sets. We form the union of set A and set B by listing
all the elements of set A and then including any elements of set B that have
not already been listed. It will be helpful to remember that the word "or" is
a word that will be associated with the union operation.

Examples

1. If set A = {a,b} and set B = {c,d}, then A ∪ B = {a,b,c,d} because these are the elements that are in set A or in set B.

2. If set A = {1,2,3} and set B = {2,3,4}, then A ∪ B = {1,2,3,4} because these are the elements that are in set A or in set B or in both set A and set B.

3. If set A ={x,y} and set B = ∅, then A ∪ B = {x,y}.
   A ∪ B is not {x,y,∅}.

Using set-builder notation it is possible to define the union of set A and set B as: A ∪ B = {x | x ∈ A or x ∈ B}.

Problem Set 4.2

1. If set A = {2,4,6}, set B = {1,2,3}, and set C = {3,4,6}, find:
   a. A ∪ B *(1,2,3,4,6)*   b. B ∪ C *(1,2,3,4,6)*
   c. (A ∪ B) ∪ C *(1,2,3,4,6) ∪ (3,4,6)*
   (Hint: First find the set that is contained in the parentheses.)
   d. A ∪ (B ∪ C) *(1,2,3,4,6)*   e. n(A ∪ C) *(2,3,4,6)*
   f. n(A ∪ ∅) *(2,4,6)*

2. Find A ∪ B if:
   a. A = {1,5,6,7} and B = {2,4,6,8} *(1,2,4,5,6,7,8)*
   b. A = {1,3,5, ...} and B = {2,4,6, ...} *(1,2,3,4,5,6 ...)*
   c. A = ∅ and B = {3,6,9, ...} *(3,6,9 ...)*

3. Which of the following is/are true?
   a. ∅ ∪ ∅ = ∅   T   b. A ∪ ∅ = A   T

4. Consider the following sets:
   A = {Miami Dolphins fans}
   B = {Tampa Bay Buccaneers fans}
   C = {Dallas Cowboys fans}
   Express each of the following sets in words.
   a. A ∪ B   b. (A ∪ B) ∪ C

Review Competencies

5. Are two equal sets always equivalents?

6. If U = {a,b,c,d} and A = {a,b,c,d}, find A'.

7. If set A = {6,12,18, ...}, what is the next element in set A after 18?

8. Are the following sets equivalent?
   ∅ and {∅}

9. Find A ∩ B and n(A ∩ B) if A = {1,5,7,9} and B = {5,6,9,10}.

10. Is the following true? ∅ ∈ {1,2,3}

11. Divide: 1.8 ÷ 0.003

Miami Dolphins and
18 *Tampa Bay Buccaneers fans*

We now consider the solution of more complex problems that involve the set operations of union, intersection and complement. Be sure that you don't confuse $\cup$ with $\cap$, and don't confuse $\cup$ with the symbol for the universal set U.

Problem 1   If U = {1,2,3,4,5,6,7,8,9,10} and if X = {2,4,6} and
Y = {4,6,10}, find X' $\cap$ Y'.

Solution

step l
First find the elements in X': X' = {1,3,5,7,8,9,10}.

step 2
Find the elements in Y': Y' = {1,2,3,5,7,8,9}.

step 3
Finally, find X' $\cap$ Y': X' $\cap$ Y' = {1,3,5,7,8,9}.

Problem Set 4.3

1. Let U = {1,2,3,4,5,6,7,8,9,10}. Determine X',
    Y', X' $\cap$ Y, X $\cup$ Y', X' $\cap$ Y', and X' $\cup$ Y' for
    each of the following.
    a.   X = {3,5,7}; Y = {1,6,7}
    b.   X = {2,4,6,8}; Y = {1,2,3,4}

2. If U is the universal set and $\emptyset$ is the null set
    and A is any set whatever, find:
    a.   A $\cap$ A'   b.   A $\cup$ A'
    c.   U'   d. $\emptyset$'
    e.   A $\cap$ U   f.   A $\cup$ U
    g.   A $\cap$ $\emptyset$ h.   A $\cup$ $\emptyset$

Review Competencies

3. Find A $\cup$ B and n(A $\cup$ B) if A = {1,2,3,...,10}
    and B = $\emptyset$.

4. If U = {1,2,3,4,5,6}, A = {1,3,5,6},
    B = {3,4,5,6} and C = {1,2,3,4},
    find (A $\cup$ B) $\cap$ C.

5. Express 45% as a fraction in lowest terms.

The next two problems involve set operations and the use of parentheses.

Problem 2   If U = {1,2,3,4,5,6,7}, set A = {1,4,6,7}, and set B = {2,4,5,6},
find (A $\cup$ B')'.

Solution

step l
Begin by working **inside** the parentheses. Find B': B' = {1,3,7}.

step 2
Find A $\cup$ B': A $\cup$ B' = {1,3,4,6,7}

step 3
Finally, find the complement of the set, {1,3,4,6,7}. (A $\cup$ B')' = {2,5}.

Problem 3   If U = {1,2,3,...,10} and A = {1,3,7,9} and B = {3,7,8,10},
find (A ∪ B)', A' ∩ B', (A ∩ B)', and A' ∪ B'.

Solution

step 1
To find (A ∪ B)', first find
the elements in A ∪ B.   A ∪ B = {1,3,7,8,9,10}

step 2
To find (A ∪ B)', find the
complement of A ∪ B.   (A ∪ B)' = {2,4,5,6}

step 3
Find A' and B'.   A' = {2,4,5,6,8,10} and B' = {1,2,4,5,6,9}

step 4
Now find A' ∩ B'.   A' ∩ B' = {2,4,5,6}

Thus, (A ∪ B)' = A' ∩ B' = {2,4,5,6}. This statement says that **the
complement of a union of two sets is equal to the intersection
of the complements of the two sets.**

step 5
To find (A ∩ B)', first find
the elements in A ∩ B.   A ∩ B = {3,7}

step 6
To find (A ∩ B)', find the
complement of A ∩ B.   (A ∩ B)' = {1,2,4,5,6,8,9,10}

step 7
From Step 3, A' = {2,4,5,6,8,10} and B' = {1,2,4,5,6,9}.

step 8
Finally, find A' ∪ B': A' ∪ B' = {1,2,4,5,6,8,9,10}.
Therefore, (A ∩ B)' = A' ∪ B' = {1,2,4,5,6,8,9,10}.

This statement says that **the complement of the intersection of two
sets is equal to the union of the complements of the two sets.**

The statements verified in the previous problem are known as **De Morgan's
Laws,** named for the British logician Augustus De Morgan.

De Morgan's Laws

$$(A \cup B)' = A' \cap B'$$
$$(A \cap B)' = A' \cup B'$$

These two laws will form the basis of an important discussion in another chapter on logic.

Problem Set 4.4

1. Let $U = \{2,4,6,8,10,12,14,16,18,20\}$,
   $A = \{4,10,14,16,18\}$
   and $B = \{4,6,10,12,18\}$. Find:
   a. $A \cap B'$ b. $A' \cup B$
   c. $A' \cup B'$ d. $(A \cap B)'$
   e. $(A \cup B)'$ f. $(A \cup B')'$
   g. $(A' \cap B)'$ h. $(A' \cap B')'$
   i. $(A' \cup B')'$ j. $n(B')$
   k. $n(A' \cap B')$

2. What is $A \cup B$ and $A \cap B$ if:
   a. A and B are equal sets?
   b. A and B are disjoint sets?

3. If A and B are two sets, which of the following statements are true?
   a. $B \subseteq (A \cup B)$ b. $B \subseteq (A \cap B)$
   c. $(A \cup B) \subseteq B$ d. $(A \cap B) \subseteq A$

4. If $U = \{1,2,3,4,5,6,7\}$, $A = \{1,4,5,7\}$, and $B = \{1,2,6,7\}$, verify **De Morgan's Laws** stated below.
   a. $(A \cup B)' = A' \cap B'$
   b. $(A \cap B)' = A' \cup B'$

5. Using **De Morgan's Laws** and the facts that $(A')' = A$ and $(B')' = B$, determine which of the following statements are true.
   a. $(A \cup B')' = A' \cap B$
   b. $(A' \cap B')' = A \cup B$
   c. $(A' \cup B)' = A \cup B'$

6. Let $A = \{1,2,3,4,5,6\}$, $B = \{2,4,5,6,7,8\}$, and $C = \{1,3,5,6,7,8\}$. Verify the following.
   a. $A \cup B = B \cup A$ **(The Commutative Property for set union)**
   b. $A \cap B = B \cap A$ **(The Commutative Property for set intersection)**
   c. $(A \cup B) \cup C = A \cup (B \cup C)$ **(The Associative Property for set union)**
   d. $(A \cap B) \cap C = A \cap (B \cap C)$ **(The Associative Property for set intersection)**
   e. $A \cup (B \cap C) = (A \cup B) \cap (A \cup C)$ **(The Distributive Property for set union over set intersection)**
   f. $A \cap (B \cup C) = (A \cap B) \cup (A \cap C)$ **(The Distributive Property for set intersection over set union)**

Review Competencies

7. Consider the following sets:
   $U = \{$all visitors to a Florida Attraction$\}$
   $A = \{$all visitors to Disney World$\}$
   $B = \{$all visitors to Everglades National Park$\}$
   $C = \{$all visitors to Busch Gardens$\}$
   Express the following set in words:
   $(A \cap B) \cap C'$

8. If $U = \{1,2,3,...,10\}$, $A = \{2,4,6,8\}$, and $B = \{1,2,3,4\}$, find $A \cap B'$.

9. Express $\frac{11}{25}$ as a percent.

Section 5   Venn Diagrams

It is possible to obtain a more thorough understanding of sets and their operations by considering graphic illustrations that allow visual analyses. **Venn diagrams,** named for the nineteenth century British mathematician John Venn, are used to show the visual relationships among sets.

There are six basic ideas that are used to construct Venn diagrams:

1. In a Venn diagram the **universal set** U will always be represented by an area inside a rectangle. The figure at the right represents the universal set U.

2. Any **subset** of U must be completely within the rectangle representing U. In the figure at the right, the set A is represented by the area inside the circle. Set A is a subset of U because the circle is inside the rectangle. Any subset of A must be a figure inside the circle of A.

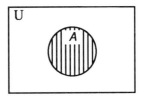

3. The **complement** of set A with respect to U is shown in the figure at the right. The area outside the circle, but within the rectangle, represents the set of elements in U that are not in set A. Thus, the shaded area represents the complement of A which is A'.

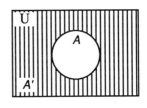

4. Two sets that have no elements in common are **disjoint sets**. Two disjoint sets A and B are shown in the figure at the right. Disjoint sets appear as circles which do not overlap.

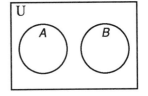

5. If two sets have some common elements, then the circles representing them must overlap. The figure at the right shows a shaded area where sets A and B overlap. This shaded area represents the **intersection** of sets A and B.

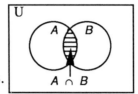

6. The **union** of sets A and B includes elements in A or in B or in both A and B. The shaded area of the figure at the right includes all of the two circles. This shaded area is the union of the two sets.

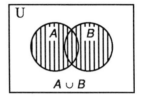

We will illustrate these six basic ideas of Venn diagrams in the following problems. Two techniques will be demonstrated: the **shading technique** and the **numbered regions** technique.

Problem 1   Draw a Venn diagram and determine the region represented by A' ∩ B in which A and B are overlapping sets. Use the shading technique.

Solution   **step 1**
Because A and B are overlapping sets they can be represented by circles like those shown at the right. Note: The circles will always overlap unless the directions state that A and B are disjoint sets or that there is a subset relationship between A and B.

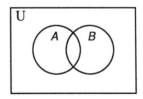

step 2
The area representing A' is shaded. A' is the region of U that is outside circle A. This region is shaded using vertical lines.

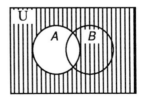

23

step 3
The area representing B is shaded.
This is the region inside circle B.
Horizontal lines are used to shade set B.

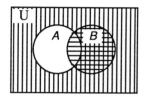

step 4
The intersection of set A' and set B
is the region where the two shaded
areas overlap. This cross-hatched
area is shaded heavily and is the
region representing A' ∩ B. This
region represents the set of elements
that are **not** in A, but that **are** in B.

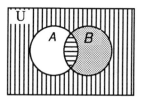

Another technique for representing sets with Venn Diagrams uses the fact that
two sets A and B separate the universal set into four distinct regions. These
four regions are shown in the figure below. The numbering is arbitrary.

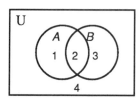

region 1   Elements of A, but not B: A ∩ B'

region 2   Elements of both A and B: A ∩ B

region 3   Elements of B, but not A: B ∩ A'

region 4   Elements of neither A nor B: A' ∩ B'

Set A can be considered equal to {1,2}.   (The elements in the sets
Set B can be considered equal to {2,3}.   represent the regions in the
Set U can be considered equal to {1,2,3,4}.   Venn diagram.)

With these considerations,
$$A' = \{3,4\}, B' = \{1,4\}, A \cap B = \{2\}, A \cup B = \{1,2,3\}$$

Problem 2   Draw a Venn diagram and determine the region represented
by $A' \cup B'$. Use the numbered regions technique.

Solution   **step 1**
In the figure at the right,
$$A = \{1,2\}$$
$$B = \{2,3\}$$
$$U = \{1,2,3,4\}$$

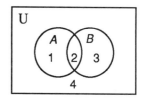

step 2
Since $A = \{1,2\}$ and $U = \{1,2,3,4\}$, $A' = \{3,4\}$.

step 3
Since $B = \{2,3\}$ and $U = \{1,2,3,4\}$, $B' = \{1,4\}$.

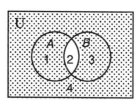

step 4
Since $A' \cup B' = \{3,4\} \cup \{1,4\} = \{1,3,4\}$,
region 1, region 3, and region 4 are shaded.
This shaded region represents $A' \cup B'$.

Notice that no regions need to be shaded using the numbered regions
technique until the final step. Many students find this technique preferable to
the shading technique. However, both will yield identical answers.

Problem 3   Draw a Venn diagram and determine the region represented
by $A' \cap B$. Use the numbered regions approach.

Solution   **step I**
Since $A = \{1,2\}$ and
$U = \{1,2,3,4\}$, $A' = \{3,4\}$.

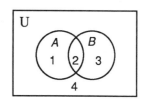

Chapter 1

step 2
Since A' = {3,4} and B = {2,3},
A' ∩ B = {3,4} ∩ {2,3} = {3}

step 3
The figure at the right shows region 3
which is A' ∩ B.

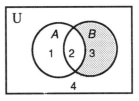

Notice that the shaded region obtained using this technique is exactly the same
as that obtained by solving the identical problem in Problem 1 using the
shading technique.

Problem Set 5.1

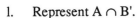

Use the figure at the right
for problems 1-4. Use the
shading technique or the
numbered regions
technique.

1. Represent A ∩ B'.
2. Represent A' ∪ B.
3. Represent A ∪ B'.
4. Represent A' ∩ B'.

Use the figure at the
right for problems 5-6.

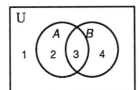

5. Which region(s) is/are
   in set B and not in
   set A?

6. Which region(s) is/are in set A
   but not in both set A and set B?

Review Competencies

7. If U = {1,4,5,10}, A = {1,4,10}, and
   B = {4,5,10}, find (A' ∪ B')'.

8. If U = {1,2,3,4,5}, A = {1,2,3}, and
   B = {2,3,5}, verify that (A ∪ B)' = A' ∩ B'
   (one of **De Morgan's Laws**).

9. If A and B are two sets, is the following true?
   A ⊆ (A ∪ B)

10. If A = {1,2,3,...,10}, B = {2,4,6,8},
    and C = {1,2,3,4,5}, verify that
    A ∩ (B ∪ C) = (A ∩ B) ∪ (A ∩ C).

11. Express 17.6% as a decimal.

The next problem illustrates how a Venn diagram can be used to represent an expression involving parentheses.

Problem 4   Draw a Venn diagram to determine the region represented by (A ∪ B)'. Use the shading technique.

Solution

step 1

Determine the region for the expression within the parentheses. The figure at the right shows the result for A ∪ B.

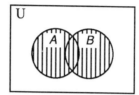

step 2

Find the complement of the figure found in step 1 by shading the areas left unmarked in step 1. The figure at the right shows the final result for (A ∪ B)'.

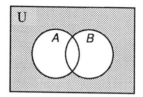

Recall that (A ∪ B)' = A' ∩ B' is one of **De Morgan's Laws**. If a Venn diagram had been drawn for A' ∩ B' instead, the region representing A' ∩ B' would have been identical to the region represented by (A ∪ B)'.

Problem Set 5.2

Use the figure at the right for problems 1-4. Use the shading technique or the numbered regions technique.

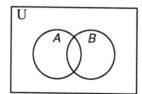

1. Represent (A ∩ B)'.

2. Represent (A' ∩ B)'.

3. Represent (A ∪ B')'.

4. Represent (A' ∪ B')'.

Review Competencies

5. If U = {1,3,6,9}, A = {1,3}, and B = {6,9}, verify that (A ∩ B)' = A' ∪ B' (one of **De Morgan's Laws**).

6. Using **De Morgan's Laws** and the facts that (A')' = A and (B')' = B, determine if (A' ∩ B)' = A ∪ B'.

7. If U = {1,2,3,4,5}, A = {1,2,3}, B = {3,4,5}, and C = {1,3,5}, verify that A ∪ (B ∩ C) = (A ∪ B) ∩ (A ∪ C).

8. Express 0.365 as a percent.

Chapter 1

Three sets separate the universal set U into the **eight** regions shown in the
figure below. The numbering is arbitrary.

region 1	Elements of A but not B or C
region 2	Elements of A and B but not C
region 3	Elements of B but not A or C
region 4	Elements of A and C but not B
region 5	Elements of A, B, and C
region 6	Elements of B and C but not A
region 7	Elements of C but not A or B
region 8	Elements of U that are not in A, B, or C

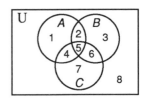

Set A can be considered to be equal to {1,2,4,5}.
Set B can be considered to be equal to {2,3,5,6}.   (The elements
Set C can be considered to be equal to {4,5,6,7}.   correspond to
Set U can be considered to be equal to {1,2,3,4,5,6,7,8}.   the regions in
Set A' can be considered to be equal to {3,6,7,8}.   the Venn
Set B' can be considered to be equal to {1,4,7,8}.   diagram.)
Set C' can be considered to be equal to {1,2,3,8}.

Problem 5   Draw a Venn diagram to determine the region represented by
(A ∩ B) ∩ C. Use the numbered regions technique.

Solution   **step 1**
The parentheses indicate that A ∩ B
must be determined first.
    A = {1,2,4,5} and B = {2,3,5,6}.
    Thus, A ∩ B = {2,5}.

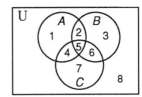

step 2
Since C = {4,5,6,7},
(A ∩ B) ∩ C = {2,5} ∩ {4,5,6,7} = {5}.
Thus, region 5 is shaded heavily. This is
the region represented by (A ∩ B) ∩ C.

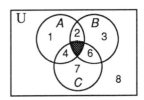

Problem Set 5.3

Use the Venn diagram shown at the right for problems 1-4. Use the shading technique or the numbered regions technique.

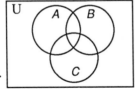

1. Represent A ∩ (B ∩ C).

2. Represent (A ∪ B) ∪ C.

3. Represent A ∪ (B ∩ C).

4. Represent (A ∪ B) ∩ C.

5. Using **De Morgan's Laws** and the facts that (A')' = A and (B')' = B, determine if (A' ∪ B')' = A ∩ B.

Use the Venn diagram at the right for problems 6 and 7.

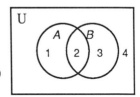

6. Which of the region(s) is/are not in set B?

7. Which of the region(s) is/are not in both set A and set B?

Let us now use a Venn diagram to solve a more complex problem.

Problem 6   Draw a Venn diagram to determine the region represented by (A ∩ B) ∪ (A ∩ C). Use the shading technique.

Solution

step 1
Determine the regions representing the first pair of parentheses. The intersection of sets A and B is shown at the right. Notice that A ∩ B is the area where the the two circles representing A and B overlap.

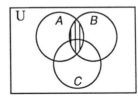

step 2
Determine the regions representing the second pair of parentheses. The intersection of sets A and C is shaded at the right using horizontal lines. A ∩ C is the area where the two circles representing A and C overlap.

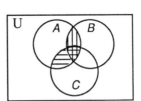

step 3
The union of the two expressions in parentheses includes all the regions shaded previously. This total area is heavily shaded and represents the final result for (A ∩ B) ∪ (A ∩ C).

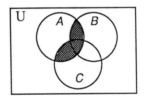

Problem Set 5.4

Use the figure below for problems 1 and 2. Use the shading technique or the numbered regions technique.

1. Represent
    $(A \cup B) \cap (A \cup C)$.

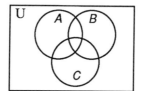

2. Represent
    $(A \cap C) \cup (B \cap C)$.

Use the figure below for problems 3 and 4.

3. Which of the region(s) is/are in set C, but not in Set A or Set B?

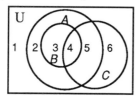

4. Which of the region(s) is/are not in set B or set C?

Use the figure below for problems 5 and 6.

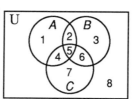

5. Which of the region(s) is/are in set B, but not in both set A and set C?

6. Which of the region(s) is/are in both set A and set C, but not in set B?

7. Which of the region(s) in the accompanying diagram is/are not in both set A and set B?

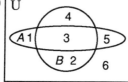

8. Which of the region(s) in the accompanying diagram is/are not in set C?

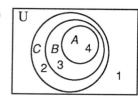

9. Which of the region(s) in the accompanying diagram is/are in set B but not in set A or set C?

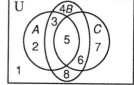

10. Which of the region(s) in the accompanying diagram is/are in both set A and set C but not in set B?

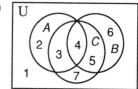

11. Which of the region(s) in the accompanying diagram is/are not in set B or set C?

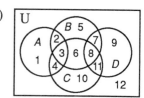

12. Which of the region(s) in the accompanying diagram is/are in set A and set D, but not set B?

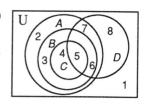

Review Competencies

13. State **De Morgan's Laws.**

14. Using **De Morgan's Laws** and the facts that $(A')' = A$ and $(B')' = B$, determine if $(A' \cap B')' = A \cup B$.

15. 2.5% of 35 = ?

16. What percent of 75 is 45?

Problem 7   Which statement listed below is true for the Venn diagram shown at the right? Assume that no region is empty.

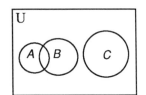

a.   No element is a member of both sets A and B.
b.   There are fewer than four elements that are shared by both set B and set C.
c.   Any element belonging to set U is also an element in set C.
d.   A ∩ U = U

Solution   Let us consider each statement in the list above.

Statement **a** is false. If an x is placed in the region where circles A and B overlap, x is an element of both sets A and B. The x indicates that some elements belong to all three sets A, B and U.

Statement **b** is true. Since B ∩ C = Ø, there are no elements that are shared by both set B and set C. Since they share zero elements, they certainly have fewer than four elements in common.

Statement **c** is false. An x may be placed anywhere in the rectangular region except inside circle C to show a counter-example for this statement. Although member-ship in set U is not a sufficient condition for member-ship in set C, observe that all elements belonging to set C are also members of set U.

Statement **d** is false. Set A intersected with set U is the set of elements in both A and U. Thus A ∩ U = A is true, but A ∩ U = U is false.

Problem Set 5.5

1.   Sets A, B, C, and U are related as shown in the diagram. Which one of the following is true assuming that no region is empty?

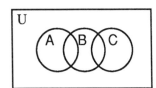

a.   Some elements are common to all four sets A, B, C, and U.
b.   Some elements belong to both sets B and C but not to set A.
c.   There are more than four elements that are shared by both set A and set C.
d.   All elements in set U are elements belonging to set C.

2. Sets A, B, C, D, and U are related as shown in the diagram. Which one of the following is true assuming that no region is empty?
   a. All elements in A ∩ B are also elements in set C.
   b. All elements of set U are elements of set D.
   c. Some elements are common to all three sets A, B, and D.
   d. Some elements belong to set A but not to set C.

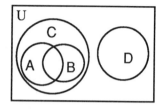

3. Sets A, B, C, D, and U are related as shown in the diagram. Which one of the following is true assuming that no region is empty?
   a. Some elements are common to all three sets B, C, and D.
   b. Some elements of set A are not elements of set B.
   c. All elements of A ∩ C are elements of set B.
   d. All elements of set U are elements of set D.

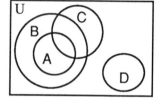

4. The diagram at the right shows sets A, B, C, and U. Each bounded region in the diagram is also given a lower case letter. None of these regions is empty. Which one statement listed below is true?
   a. Regions f and d make up all of set B.
   b. B ∪ C is represented by regions c and f.
   c. Region g represents the set of elements belonging to set C that do not belong to set A or set B.
   d. Region b represents the set of elements belonging to all three sets A, B, and C.

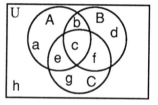

5. Sets A, B, C, and U are related as shown in the diagram. Which one of the following is true assuming that no region is empty?
   a. All elements of set B are elements of set A.
   b. Some elements are common to all three sets A, B, and C.
   c. A ∩ B = B
   d. Some elements are common to all three sets B, C, and U.

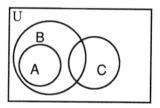

6. Sets A, B, C, and U are related as shown in the set diagram. Which one of the following is true assuming none of the regions is empty?
   a. All elements of set B are elements of set C.
   b. Some elements belong to all three sets A, B, and U.
   c. All elements of set U are elements of set B.
   d. A ∩ C is not the empty set.

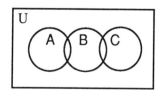

7. Sets A, B, C, and U are related as shown in the set diagram. Which one of the following is true assuming none of the regions is empty?

   a. All elements of set A are elements of set B.
   b. Some elements are common to all four sets A, B, C, and U.
   c. At least one element of set U is a member of both sets A and C.
   d. There are more than two elements that belong to $B \cap C$.

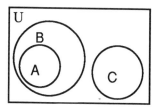

8. Sets A, B, C, and U are related as shown in the set diagram. Which one of the following is true assuming none of the regions is empty?

   a. No element is a member of all four sets A, B, C, and U.
   b. Any element that is a member of set B is also a member of set A.
   c. Some elements belong to set B but not to set C.
   d. $A \cap B = A$

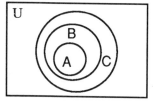

Review Competencies

9. Last year Tom Goetz earned $25,000 and Johanna Goetz earned $23,000. Their total family income, however, was $54,000 due to the fact that in addition to their combined earnings they also received income from an interest-bearing account. What fractional part of their family income came from the interest-bearing account?

10. In January, 40 people purchased a best-selling novel that was on sale for $30. In February the price of the novel was increased by 20% and 8 fewer customers purchased the book. Did the revenue from the book in February increase or decrease from January's revenue? By how much?

11. Find the sum of $-7\frac{5}{6}$ and $3\frac{8}{9}$, expressing the answer as a mixed number.

33

Chapter 1

We will now see that a Venn diagram can be used to solve certain
types of word problems.

Problem 8   Forty people were contacted in a movie survey. It showed that six people
liked musicals, westerns and comedies; thirteen liked musicals and
westerns; eleven liked musicals and comedies; ten liked westerns and
comedies; twenty-one liked musicals; twenty-three liked westerns; and
twenty-four liked comedies.
a. How many people surveyed liked only westerns?
b. How many people surveyed liked both musicals and
   westerns but not comedies?

Solution   **step 1**
Draw three intersecting circles to
represent the sets of people who like
musicals, westerns, and comedies.

step 2
Use the information that "six people
liked musicals,westerns, and
comedies" to place a 6 in the one region
where all three circles overlap.

step 3
Use the information that "thirteen liked
musicals and westerns." There are two
regions where the "musicals" and
"westerns" circles overlap. One of those
regions already has a 6 in it. Thus, the
other region must have a 7. This makes
the two overlapping circles have a total of 13.

step 4
Use the information that "eleven liked
musicals and comedies." Again, there
are two regions where the two circles,
"musicals" and "comedies," overlap.
One region already has a 6 in it. Thus,
the other region must have a 5.

34

step 5

The information that "ten liked westerns and comedies" is used as it was in Steps 3 and 4. There are two regions where the "westerns" and "comedies" circles overlap. One region already has a 6 in it. Thus, the other region must have a 4.

step 6

The information that "twenty-one liked musicals" means that there must be a total of 21 in the "musicals" circle. Since the circle has four regions and the numbers 6,7, and 5 are in three of them, the fourth region must have a 3 to give the desired total of 21.

step 7

The information that "twenty-three liked westerns" is used as it was in Step 6. The "westerns" circle has four regions and the numbers 6, 7, and 4 in three of those regions. The fourth region, therefore, must have a 6 to give the desired total of 23.

step 8

The information that "twenty-four liked comedies" is used as it was in Steps 6 and 7. Notice that there are already 15 people in the "comedies" circle (5 + 6 + 4). Thus, the remaining region of the circle must contain 9 people for a total of 24.

step 9

Now the answers to the original questions of this problem can be obtained.

a. How many surveyed people liked only westerns? The diagram shows the answer is 6.

b. How many surveyed people liked both musicals and westerns but not comedies? The diagram shows the answer is 7.

A helpful hint in solving this type of problem is to "work from the inside out." In other words, first find the number to be placed in the region representing the intersection of all three circles.

Notice that all 40 people are accounted for in the diagram. If the problem had stated that 45 people were contacted in a movie survey, then the extra 5 people would be placed in the region outside the three circles. It would be assumed that they did not like musicals, westerns, or comedies.

Problem Set 5.6

1. A group of teenagers were interviewed. It was found that 42 liked rock music; 34 liked folk music; 27 liked jazz; 7 liked rock music, folk music and jazz; 12 liked rock music and jazz; 14 liked rock music and folk music; and 10 liked folk music and jazz.
   a. How many liked rock music only?
   b. How many liked both folk music and jazz but not rock music?

2. A group of college students was interviewed, and it was found that 31 like sports cars, 19 like motorcycles, 12 like surfing, 16 like sports cars and motorcycles, 7 like motorcycles and surfing, 5 like sports cars and surfing, and 4 like sports cars, motorcycles and surfing.
   a. How many college students were interviewed?
   b. How many college students like both surfing and sports cars but not motorcycles?

3. In a high school there are 20 members of the swimming team, 14 members of the basketball team, and 16 members of the tennis team. Four students are on the swimming, basketball, and tennis teams; 8 are on the swimming and basketball teams; 10 are on the basketball and tennis teams; and 9 are on the swimming and tennis teams.
   a. How many are only on the swimming team?
   b. How many are on both the swimming and basketball teams but not on the tennis team?

Review Competencies

4. How many subsets and proper subsets can be formed from a set that contains three elements?

5. If $U = \{1,2,3,4,5,6,7\}$, $A = \{1,3,4,7\}$, $B = \{2,3,6,7\}$, and $C = \{2,3,4,5\}$, find:
   a. $(A' \cap B')'$ b. $A \cup (B \cap C)$

6. Using **De Morgan's Laws** and the facts that $(A')' = A$ and $(B')' = B$, determine if $(A' \cap B)' = A \cap B'$.

7. 22% of what number is 66?

8. A man bought a house for $50,000. He later sold it for $65,000. What was his percent of increase in value?

9. An **arithmetic progression** is a sequence of numbers each of which, after the first, is obtained from the preceding one by adding a constant number to it. For example, the following is an arithmetic progression with first number 5 and constant addend 3.
   5,8,11,14,17, etc.
   What is the next term in this arithmetic progression?

10. A **geometric progression** is a sequence of numbers each of which, after the first, is obtained from the preceding one by multiplying by a constant number. For example, the following is a geometric progression with first number 5 and constant multiplier 2.
    5, 10, 20, 40, 80, etc.
    What is the next term in this geometric progression?

Chapter 1 Summary

1. Definitions

a. A set is **well-defined** if it is possible to determine whether or not a given element belongs to it.

b. **Equal** (identical sets) have the exact same elements.

c. The **universal set**, U, is the set that contains all possible elements under consideration in a problem.

d. A is a **subset** of B (A $\subseteq$ B) if every element in set A is also in set B.

e. A is a **proper subset** of B (A $\subset$ B) if every element in set A is also in set B, and at least one element in set B is not in set A.

   A set with n elements has 2^n subsets and $2^n - 1$ proper subsets.

f. The **null set** (empty set: $\emptyset$) is a set with no elements.

g. **Equivalent** sets have the same number of distinct elements. When set A is equivalent to set B, the sets have the same **cardinal number**. $n(A) = n(B)$. Elements in equivalent sets can be placed in a one-to-one correspondence.

2. Set Operations

a. The **complement** of set A (A') is the set of elements that are not in set A, but still in set U.

b. A **intersection** B (A $\cap$ B) is a set consisting of elements that are in both set A **and** set B.

c. A **union** B (A $\cup$ B) is a set consisting of elements in set A **or** set B or in both set A and set B.

De Morgan's Laws

1. $(A \cup B)' = A' \cap B'$ 2. $(A \cap B)' = A' \cup B'$

3. Notations Used with Sets

a. $\in$: is an element of $\notin$: is not an element of
b. A $\subseteq$ B: A is a subset of B   c. A $\subset$ B: A is a proper subset of B.
d. U: universal set $\emptyset$: null set   e. A': complement of A
f. A $\cup$ B: A union B   g. A $\cap$ B: A intersection B

Sets can be defined by listing their elements, describing their elements in words, or using set-builder notation. Visual relationships among sets are shown using Venn diagrams.

Chapter 1 Self-Test

1. Which of the following sets is well-defined?
   a. The set of crazy drivers
   b. The set of mean mothers
   c. The set of beautiful actresses
   d. The set of Florida counties

2. Which of the following sets is written correctly?
   a. A = (2,4,6)   b. B = 2,...
   c. C = {1234}   d. D = {2,4,6}

3. Which of the following statements is true?
   a. 3 ∈ {2,4,6,8}   b. {2} ∈ {1,2,3,4}
   c. 3 ∈ {1,2,3,4}   d. 4 ⊆ {1,2,5,6}

4. Which of the following lists the elements specified in the set below?
   {x | x is an even counting number greater than 4}
   a. {4,6,8}   b. {6,8,10,...}
   c. {5,6,7,...}   d. {1,2,3}

5. Which of the following are equal sets?
   a. {1,2,3} and {4,5,6}
   b. {m,i,a,m,i} and {m,i,a}
   c. Ø and {Ø}   d. {2,4,6} and {2,4,6,Ø}

6. Supply a description for the set {4,5,6,...}
   a. The set of counting numbers that are greater than 4.
   b. The set of counting numbers that are less than 4.
   c. The set of odd counting numbers beginning with 4.
   d. The set of counting numbers that are greater than 3.

7. How many subsets can be formed from the set {p,q,r}?
   a. 3   b. 1
   c. 7   d. 8

8. If set A = {a,b,c} and set B = {a,b,c,d}, then
   a. A = B   b. B ⊆ A
   c. B ⊂ A   d. A ⊂ B and A ⊆ B

9. The number of proper subsets that can be formed from the set {1,2,Ø} is:
   a. 3   b. 7   c. 8   d. 4

10. If U = {a,b,c,d} and A = Ø, then A' =
    a. Ø   b. {a,b,c}
    c. {a,b,c,d}   d. {e,f,g,...}

11. Which of the following statements is not correct?
    a. The complement of the universal set is the null set.
    b. The complement of the null set is the universal set.
    c. The null set can be a proper subset of any set including itself.
    d. If a set contains two different elements, then three proper subsets can be formed.

12. If A = {2,4,6} and B = {2,4,6,6}, then which of the following is not true?
    a. Set A is not equal to set B.
    b. The cardinal number of set B is 3.
    c. n(A) = n(B)
    d. Set A is equivalent to set B.

13. If U is the universal set and Ø is the null set and A is any set whatever, what is A ∩ U ?
    a. U   b. Ø   c. A   d. A'

14. If A = {2,4,5,6,7} and B = {2,6,8,9,10}, then the number of proper subsets of A ∩ B is:
    a. 2   b. 3   c. 4   d. 7

15. If U = {0,1,2,...,9}, A = {1,3,4,8}, and B = {2,3,4,9}, then A' ∪ B is:
    a. set U   b. set B
    c. {2,9}   d. {0,2,3,4,5,6,7,9}

16. If U = {1,2,3,4,5}, A = {2,3,5}, and B = {1,4,5}, then (A ∩ B')' is:
    a.   {1,2,3,5}   b.   {1,2,3}   c.   {1,2,3,4,5}   d.   {1,4,5}

17. Which of the following Venn diagrams represents the set (A ∪ B)'?

a. 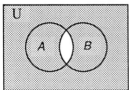   b. 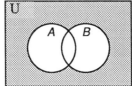   c.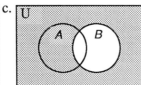

d. 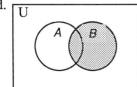   e.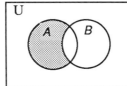

18. Which of the following Venn diagrams represents the set A ∪ (B ∩ C)?

a. 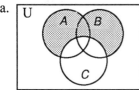   b. 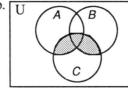   c.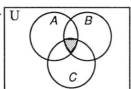

d. 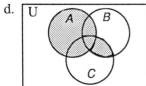   e.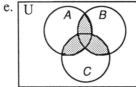

19. Which of the following Venn diagrams illustrates the case in which A
    and B are sets such that A ∩ B = A?

a. 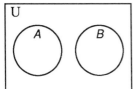   b. 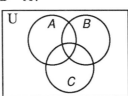   c.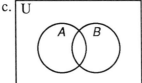

d. 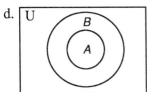   e.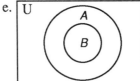

39

20. Sets A, B, C, and U are related as shown in the diagram. Which of the following statements is/are true assuming none of the regions is empty?
    i.   Any element of set U belongs to set C.
    ii. An element of set A is also an element of set B.
    iii. Some elements of set B belong to set C.
    iv. No element belongs to sets A, B, C, and U.

    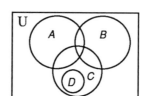

    a. i, ii and iii only   b. ii and iii only
    c. ii, iii and iv only   d. i and iv only

21. Sets A, B, C, D, and U are related as shown in the diagram. Which of the following statements is/are true assuming no regions are empty?
    i.   Some element belongs to all four sets A, B, C, and D.
    ii. Some element of set C is also an element of set D.
    iii. Some element belongs to set A and set B, but not to set C.

    a. i only   b. ii only
    c. ii and iii only   d. i and iii only

22. It was discovered in a group of 30 children that 4 like dogs, cats and goldfish as pets, 18 like dogs, 16 like cats, 18 like goldfish, 8 like dogs and cats, 10 like cats and goldfish, and 11 like dogs and goldfish. How many didn't like any of these as pets?

    a. 3   b. 5
    c. 6   d. none of these

two

Logic

"Elementary, my dear Watson!" Many have been thrilled by Sherlock Holmes' famous utterance. Every faithful reader knew he or she was to be treated to a marvelously logical explanation of some mysterious phenomena.

This chapter contains some of the basic principles of logic, namely, the language, the symbols, and some applications. Naturally this brief exposure cannot make you another Sherlock Holmes. Indeed, Lord Dunsany, an Irish poet, contended that logic, like whiskey, can lose its beneficial effect when taken in too large quantities. Still, this discussion might emphasize the need for clear thinking as well as increase your proficiency with the English language.

The study of mathematics can nurture the ability to reason. It can also focus attention on the importance of thinking logically. In all disciplines the ability to reason abstractly in solving problems is an important tool. One must be able to understand the written and spoken word of reasonable people, to analyze problems objectively, and to avoid arriving at conclusions hastily without considering all the evidence. Surely, intolerance and prejudice thrive in the absence of clear-thinking individuals.

Since the reasoning process is often expressed in terms of declarative English sentences, we begin our discussion by considering these sentences (called simple statements) and the ways in which they can be combined.

Section 1　Statements

> **Definition**
> A **simple statement** in logic is one which is either true or false. A simple statement contains no parts that are also simple statements.

Examples

1. The following is a simple statement: *Today is Sunday.*
   It certainly would be possible to determine whether this statement is true or false on any given day.

2. The following is a simple statement:
   There are ten billion grains of sand on the beaches of Daytona Beach, Florida.
   This statement is true or false, although it would not be easy to determine!

3. The following is not a simple statement: *Is logic illogical?*
   This is a question, and questions are not simple statements.

4. The following is not a simple statement: *Take a long walk on a short pier !*
   This exclamatory statement is in the form of a command and is not a declarative sentence.

5. The following is not a simple statement: *Logic is fascinating.*
   This declarative sentence is a matter of opinion and cannot clearly be labeled as true or false.

6. The following is not a simple statement:
   Carter was the second American President and Orlando is a city in Florida.
   This statement has two parts which are both simple statements.

Problem Set 1.1

Determine which of the following are simple statements.

1. Tallahassee is the capital of Florida.
2. Go jump in the lake!
3. The St. Johns River is the largest river in Florida and flows north.
4. What's a nice person like you doing in a place like this?
5. The state of Florida is a peninsula.
6. The first president of the United States was Abraham Lincoln.

7. $6 + 6 = 36$
8. $5 + 9 = 14$
9. $6 + 6 = 36$ and $5 + 9 = 14$
10. Playing baseball is more fun than practicing the piano.

Review Competencies

11. 5% of what number is 15?
12. What percent of 35 is 70?
13. A basketball team won 17 of its 25 games. What percent of its games did it win?

Now we will see that it is possible to combine simple statements.

> ## Definition
> A **compound statement** is one that is formed by combining two or more simple statements. A compound statement consisting of two simple statements joined by the connective "and" is called a **conjunction**. A compound statement consisting of two simple statements joined by the connective "or" is called a **disjunction.**

Examples

1. The following is a compound statement:
   Today is Sunday and this is the month of May.
   This statement is a conjunction.

2. The following is a compound statement:
   Today is Sunday or this is the month of May.
   This statement is a disjunction.

3. The following is a compound statement (conjunction):
   Thelma and Mary are both passing the course.
   What is implied is that "Thelma is passing the course and Mary is passing the course."

The connectives "and" and "or" are two connectives that we frequently use in English. From the standpoint of English, there are certain equivalent connectives that imply the idea of "and":

a. Today is Sunday, **but** this is the month of May.
b. Today is Sunday, **even though** this is the month of May.
c. Today is Sunday, **yet** this is the month of May.
d. Today is Sunday, **although** this is the month of May.

Of these equivalent connectives the word "but" will be of particular interest to us.

Problem Set 1.2

Identify the simple statements that form the following compound statements. Determine the connective in each.

1. It is windy and turning cold.
2. The Yankees and Mets won their games.
3. Roses are red or violets are blue.
4. Heathcliffe's friends have bad breath or need a bath.
5. It is a beautiful day today, but I am going to school.
6. I will take Mike or Carol to the basketball game.
7. Both Sandy and Larry are secretaries.
8. Tom reads rapidly and so does Karen.

9. Determine which of the following are simple statements.
   a. Give me a break!
   b. $7 + 5 = 13$
   c. It is more than 100 miles from Key West to Pensacola.

10. A television dealer sold 55 television sets one month. If $\frac{4}{5}$ of them were color sets, how many sets were sold that were not color?

11. A special sale is offering radios at $24 each plus 5% tax and televisions at $90 each plus 5% tax. If Debbie has $115, could she purchase a radio and a television?

Notations

Simple statements and connectives may be expressed using symbols. The letters p and q are often used to represent simple statements. The symbol ∧ is used to mean "and" and ∨ is used to mean "or."

Example

The letter p can be used to represent the simple statement
   Today is Monday.

The letter q can be used to represent the simple statement
   I am falling asleep.

The compound statement,
   Today is Monday and I am falling asleep.
is represented symbolically as p ∧ q.

Examples

Let p represent: *Today is Monday.*
Let q represent: *I am falling asleep.*
The compound statement,
    Today is Monday or I am falling asleep.
is written symbolically as p ∨ q.

Notation
The symbol ~ is translated as "not" and is used to
form the **negation** of a simple statement.

Examples

Let p be: *Today is Monday.*
Let q be: *I am falling asleep.*
Translate each of the following into an English statement.

1. ~p: *Today is not Monday.* This may equivalently be translated as:
    a. *It is **not true** that today is Monday.*
    b. *It is **not the case** that today is Monday.*
    c. *It is **false** that today is Monday.*

2. ~(~p): *It is false that today is not Monday.*
    This may be equivalently translated as: *Today is Monday.*

3. ~p ∧ q: *Today is not Monday and I am falling asleep.*

4. p ∨ ~q: *Today is Monday or I am not falling asleep.*

5. ~(p ∧ q): *It is not true that both today is Monday and I am falling asleep.*
    Notice that the compound statement is enclosed in parentheses and preceded
    by the negation symbol. The translation "*it is not true that **both**... *" is used
    to convey the idea of negating the entire expression (p ∧ q). The translation
    is a bit more precise than:
        It is not true that today is Monday and I am falling asleep
    which could be the translation for either ~(p∧ q) or ~p ∧ q. Observe that
    ~(p∧ q) and ~p ∧ q convey quite different meanings.

6. Let p be: *Today is not Monday.*
    The statement ~p is: *Today is Monday.*

Problem Set 1.3

1. Translate each of the following into English using:
   p to represent : *Mary has it.*
   q to represent: *Alice has had it.*
   a. p ∨ ~q   b. ~p ∨ q
   c. ~p ∧ ~q   d. ~(p ∨ q)
   e. ~(p ∧ ~q)   f. ~(~q)

2. Represent each of the following statements symbolically using:
   p to represent: *I study hard.*
   q to represent: *I pass the test.*
   a. *I do not study hard or I pass the test.*
   b. *It is not true that both I study hard and I do not pass the test.*

   c. *I do not study hard and I do not pass the test.*
   d. *Neither do I study hard nor do I pass the test. (careful!)*

Review Competencies

3. Identify the simple statements and the connective in each.
   a. *Both John and Sandy are efficient secretaries.*
   b. *The children are noisy or rude.*

4. $\frac{4}{6} = \frac{12}{?}$

Certain ideas in logic are consistent with concepts in set theory. Venn diagrams can describe a situation in which a statement (p) is true. We know all simple statements are either true or false. In the diagram at right, the region outside the circle (the complement) represents ~p (the falsity of a statement). Notice that the universal set of all simple statements contains all those which are true and those which are false.

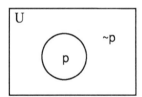

Section 2   Negations, Conjunctions, Disjunctions, Truth Tables, and De Morgan's Laws

In this section we consider some very basic concepts in our study of logic.

We know that it is possible to form the negation (~p) of a simple statement (p). Since both a simple statement and its negation have truth values, it should be obvious that a statement and its negation have **opposite** truth values.

Recall from Chapter 1 that the complement of the complement of a set is equal to the set itself. (A')' = A

Similarly, the negation of the negation of a statement is equivalent to the statement itself. That is, both ~(~p) and p have the same truth value.

> **~(~p) is equivalent to p**

Symbolically, we may write: ~(~p) ≡ p. The symbol "≡" means "is equivalent to." We will encounter this symbol again in later sections.

Examples

1. If statement p is a true statement, then ~p is a false statement. Consider each of the following statements:

   p: *Tallahassee is the capital of Florida.* (True)
   ~p: *Tallahassee is not the capital of Florida.* (False)

   Notice that ~(~p) is: It is false that Tallahassee is not the capital of Florida. This is a true statement and has the **same** truth value as statement p. Thus, ~(~p) is equivalent to p.

2. If statement p is a false statement, then ~p is a true statement. Consider each of the following statements:

   p: *Florida is located in Canada.* (False)
   ~p: *Florida is not located in Canada.* (True)

   Notice that ~(~p) is: *It is false that Florida is not located in Canada.* This is a false statement and has the same truth value as p. ~(~p) is equivalent to p.

The results of the two examples above may be summarized in the negation table shown at the right. The first row states that the negation of a true statement is a false statement. The second row tells us that the negation of a false statement is a true statement.

Negation Table

p	~p
T	F
F	T

Let us now consider the negations of other types of statements.

The following simple statement is true because "some" means "at least one."

> ***Some** professors are wealthy.* (True)

Notice that the English equivalent is:

> *There **is at least one** professor who is wealthy.*

The negation of this simple statement is:

No professors are wealthy. (False)

By negating the original statement, we are saying that the statement is not true for even one professor. The negation is also:

It is not true that some professors are wealthy.

This statement is also false.

Notice that the negation is not: *Some professors are not wealthy.* This statement is true and is not the negation because both the orginal statement and this particular statement are true. **A statement and its negation can never have the same truth value.**

Now consider the following simple statement:

All professors are wealthy. (False)

Notice that the English equivalent is:

*There are **no** professors who **are not** wealthy.*

The negation could be stated in any one of the four following ways.

*Some professors are **not** wealthy.* (True)

Not all professors are wealthy. (True)

At least one professor is not wealthy. (True)

It is not true that all professors are wealthy. (True)

Finally, consider the following simple statement:

No professors are wealthy. (False)

Notice that the English equivalent is: *All professors **are not** wealthy.*

The negation says: *Some professors are wealthy.* (True)

Notice that the English equivalent is: *At least one professor is wealthy.*

The following table summarizes the negations of statements involving such words as "all," "no," "some" and "some...not."

Original Statement	Negation
All...	Some...not...
Some... not...	All...
Some...	No...
No...	Some...

All...   No...

Negation

Some...   Some...not

In English we frequently encounter statements involving "all," "some," and "no." You should be able to express each of these statements in equivalent ways.

Statement	Equivalent Statement
All A are B.	There are no A that are not B.
Some A are not B.	Not all A are B.
Some A are B.	At least one A is a B.
No A are B.	All A are not B.

Problem Set 2.1

1. Consider the true statement:
   Tampa is in Florida.
   a. Write the negation of this statement.
   b. Is the negation a true statement or a false statement?

2. Consider the statement:
   You are not studying mathematics.
   a. What is the truth value of this statement as it relates to you?
   b. What is the truth value of the negation of this statement?

3. Write the negation of:
   Some people like baseball.

4. Write the negation of:
   All Floridians are nice people.

5. Write the negation of:
   No parakeets weigh fifty pounds.

6. Write the negation of: *All books have titles.*

7. Write the negation of:
   Some teachers are not interesting.

8. Write equivalent English statements for problems 3 through 7.

9. Let p be: *It is windy today.*
   Let q be: *I am going sailing.*
   Translate the following into words.
   a. $\sim p \vee \sim q$ b. $\sim(\sim p \wedge \sim q)$

10. Let p be: *Tallahassee is the capital of Florida.*
    Let q be: *Key West is the southernmost city in Florida.*
    Represent the following statements symbolically.
    a. Tallahassee is not the capital of Florida or Key West is the southernmost city in Florida.
    b. It is not true that both Tallahassee is the capital of Florida and Key West is not the southernmost city in Florida.

11. Add: $\frac{3}{7} + \frac{1}{4}$

12. Subtract: $6.1 - 0.005$

13. Express 48% as a decimal.

Chapter 2

If a compound statement is constructed using connectives and two simple statements, p and q, then **four** possibilities can occur:

		p	q
case 1	p could be true and q could be true.	T	T
case 2	p could be true and q could be false.	T	F
case 3	p could be false and q could be true.	F	T
case 4	p could be false and q could be false.	F	F

Recall that a compound statement in the form p ∧ q is called a conjunction. It is possible to determine the truth values of the conjunctions in terms of the four possibilities of p and q by using the following table:

Conjunction Table

	p	q	p ∧ q
(1)	T	T	T
(2)	T	F	F
(3)	F	T	F
(4)	F	F	F

Examples

1. The statement "2 + 2 = 4 (T) and 3 + 2 = 5 (T)" is true.
2. The statement "2 + 2 = 4 (T) and 3 + 2 = 6 (F)" is false.
3. The statement "2 + 2 = 5 (F) and 3 + 3 = 6 (T)" is false.
4. The statement "2 + 2 = 5 (F) and 3 + 3 = 7 (F)" is false.

Observe that a conjunction is true only when both of its simple statements are true (case (1) in the conjunction table).

The truth values of two statements p and q may also be illustrated by means of Venn diagrams. From the conjunction of p and q (p ∧ q) two statements p and q being true implies that the conjunction is also true. Observe the four possible regions in the figure below. The four regions represent the four possibilities in the conjunction table. Notice that the shaded region represents the only situation in which the compound statement p ∧ q is true.

p	q	p∧q	
T	T	T	(p ∧ q)
T	F	F	(p ∧ ~q)
F	T	F	(~p ∧ q)
F	F	F	(~p ∧ ~q)

Notice that p ∧ q is equivalent
to q ∧ p.

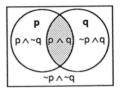

Observe the similarity between ∧ in logic and ∩, intersection, in set theory.

Recall that a compound statement in the form p ∨ q is a disjunction. It is possible to determine the truth values of the disjunction in terms of the four possibilities of p and q by means of the following table.

		Disjunction Table		
		p	q	p ∨ q
(1)		T	T	T
(2)		T	F	T
(3)		F	T	T
(4)		F	F	F

Examples

1. *Clearwater is in Florida or Jacksonville is in Florida.*
   This statement is true since both statements are true.
2. *Clearwater is in Florida or Jacksonville is in Texas.*
   This statement is true since one statement is true.
3. *Clearwater is in Georgia or Jacksonville is in Florida.*
   This statement is true since one statement is true.
4. *Clearwater is in Georgia or Jacksonville is in Texas.*
   This statement is false since both statements are false.

Observe that a disjunction is true when at least one of its simple statements is true (cases (1) through (3) in the disjunction table).

The word "or" in everyday English is commonly given either of two possible meanings. Consider the following statement:

Jose studies English or mathematics.

This statement is considered true if Jose studies English, if he studies mathematics, or if he studies both subjects. In this situation the word "or" is used in the **inclusive** sense and implies "one or the other or both." Consequently, we cannot conclude that:

If Jose studies Engish, then he does not study mathematics.
(He can be studying both disciplines.)

Similarly, the original statement does not tell us that:

If Jose studies mathematics, then he does not study English.

Notice, however, that when we are told that Jose studies English or mathematics, if we know with certainty that Jose does not study one of the subjects, we can conclude that he does study the other subject. Consequently, the original statement enables us to say:

If Jose does not study English, then he studies mathematics.
If Jose does not study mathematics, then he studies English.

Later in this chapter we will verify these informal observations with the principles of formal logic.

There is another, different, interpretation for the word "or." Consider the following statement:

The person in the distance is a man or a woman.

In this situation the truth of one of the possibilities excludes the truth of the other possibility. In the **exclusive** sense "or" implies "one or the other, but not both."

The ambiguity of "or" is resolved in English by adding the words "or both" or "but not both." Thus the inclusive sense is expressible in the unambiguous fashion:

Jose studies English or mathematics or both.

The exclusive sense is expressed as:

Jose studies English or mathematics, but not both.

In this chapter the word "or" is used in the inclusive sense. Thus, the **inclusive** disjunction, p ∨ q, is true if p alone is true or if q alone is true or if both p and q are true.

By means of the Venn diagram below, it can be seen that the meaning of ∨ in the disjunction is similar to the meaning of ∪, union, in set theory. In the disjunction, if either statement p or statement q is true or both are true, the statement, p ∨ q, is true.

p	q	p ∨ q	
T	T	T	(p ∨ q)
T	F	T	(p ∨ ~q)
F	T	T	(~p ∨ q)
F	F	F	(~p ∨ ~q)

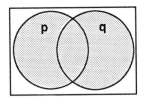

The shaded region on the diagram represents the situations in which the compound statement p ∨ q is true. Notice that p ∨ q is equivalent to q ∨ p.

Problem Set 2.2

Determine the truth values of:
1. $3 + 2 = 5$ and the fir tree gives us fur coats.
2. $3 + 2 = 5$ or the fir tree gives us fur coats.
3. $3 + 2 = 5$ and $2 + 5 = 7$.
4. $3 + 2 = 5$ or $2 + 5 = 7$.
5. Three is an even number or 4 is divisible by 2.
6. Three is an even number or 4 is not divisible by 2.
7. All birds have eyes and all dogs are mammals.
8. All birds have eyes and some dogs are not mammals.
9. Consider the statement: *Gill wrote the book or Blitzer wrote the book.*
   Which one of the following can we not conclude based upon using "or" in the inclusive sense?
   a. If Gill wrote the book, then Blitzer did not write the book.
   b. If Gill did not write the book, then Blitzer wrote the book.
   c. If Blitzer did not write the book, then Gill wrote the book.

Review Competencies

10. Write the negations of:
    a. *All books are well written.*
    b. *Some people are litterbugs.*
    c. *Some mathematics courses are not fun.*
    d. *No people are fifteen feet tall.*

11. Write a statement equivalent to: *Some cats are not cute.*

12. Express 0.014 as a percent.

13. Divide: $0.45 \div 1.5$

14. Subtract: $7.8 - 1.005$

The three basic truth tables (negation, conjunction, disjunction) can be used to construct truth tables for a variety of other statements. It is important for you to memorize these basic definitions. They can be remembered in table form or as compact rules.

1. Negation — The negation of a statement has the opposite truth value as the statement.

2. Conjunction — A conjunction is true only when all simple statements are true.

3. Disjunction — A disjunction is true when at least one simple statement is true.

55

Chapter 2

Problem 1 Construct a truth table for p ∨ ~q.

Solution

step 1

List the four possible truth values for p
and q. It is not necessary to list the
possibilities in this order, but it is standard
and makes for a consistent presentation.

p	q
T	T
T	F
F	T
F	F

step 2

Observe that the statement (p ∨ ~q) contains
~q. A column of truth values for ~q is
constructed next to the q column. The ~q
column is filled in by taking the opposite of
the truth values in the q column.

p	q	~q
T	T	F
T	F	T
F	T	F
F	F	T

step 3

The statement (p ∨ ~q) calls for the
disjunction of p and ~q. Another
column for p ∨ ~q is constructed
using the values in the p and ~q
columns. The p ∨ ~q column is
filled in by looking back at columns

p	q	~q	p ∨ ~q
T	T	F	T
T	F	T	T
F	T	F	F
F	F	T	T

1 (p) and 3 (~q) and remembering that a disjunction is true when at least
one simple statement is true. This occurs in all but the third row.

This completes the truth table for p ∨ ~q. The column at the far right shows
the values of p ∨ ~q for the four possibilities of the table. The final column
constructed is the "answer" column. Thus, the statement p ∨ ~q is false only
when p is false and q is true.

Problem Set 2.3

Construct truth tables for:

1. ~p ∨ q
2. p ∧ ~q
3. ~p ∨ ~q
4. ~p ∧ ~q

Review Competencies

5. Determine the truth values of:
    a. *All cows can fly or Disney World is
        in Florida.*
    b. *Two is an even number and 5 is an
        even number.*

6. Write the negations of:
    a. *No teachers are dull.*
    b. *All cars are lemons.*

7. Multiply: $\frac{5}{7} \cdot \frac{3}{5} \cdot \frac{14}{17}$

8. Express $\frac{3}{5}$ as a percent.

9. What is 15% of 30?

10. Express $3\frac{1}{2}\%$ as a decimal.

Problem 2   Construct a truth table for ~(p ∧ q).

Solution

step 1

As with all truth tables, first list
all possible truth values for the
letters representing statements. In
this case there are two letters
and four possibilities.

p	q
T	T
T	F
F	T
F	F

step 2

Make a column for the expression
within the parentheses using the
truth values from the conjunction
definition. A conjunction is true
only when all simple statements are
true (case 1).

p	q	p ∧ q
T	T	T
T	F	F
F	T	F
F	F	F

step 3

Construct one more column for ~(p ∧ q).
Fill in this last column by negating the values
in the p ∧ q column. Use the negation
definition, taking the opposite of the truth
values in column 3.
This completes the truth table for ~(p ∧ q).

p	q	p ∧ q	~(p ∧ q)
T	T	T	F
T	F	F	T
F	T	F	T
F	F	F	T

It is helpful to remember to first construct the column of truth values of the
compound statement that appears within the parentheses. After doing this,
negate each truth value for the expression in parentheses.

The final column in this table tells us that the statement ~(p ∧ q) is false only
when both p and q are true. For example, using p: *Miami is a city.* (true),
q: *Jacksonville is a city.* (true), we see that ~(p ∧ q) translates as: *It is not true
that both Miami and Jacksonville are cities.* This statement is false.

Chapter 2

Problem 3   Construct a truth table for (~p ∨ q) ∧ ~q.

Solution   The completed truth table is shown below. The first two columns on the left show the four possible truth values for p and q. The next three columns are successively completed for the expressions ~p, (~p ∨ q), and ~q. The final column for (~p ∨ q) ∧ ~q is completed by applying the conjunction to the columns labeled (~p ∨ q) and ~q.

p	q	~p	(~p ∨ q)	~q	(~p ∨ q) ∧ ~q
T	T	F	T	F	F
T	F	F	F	T	F
F	T	T	T	F	F
F	F	T	T	T	T

Problem Set 2.4

Construct truth tables for:

1. ~(p ∨ q)   2. ~(p ∨ ~q)   3. ~(p ∧ ~q)
4. ~(~p ∨ ~q)
5. Construct truth tables for:
    a. ~p ∨ (p ∧ ~q)   b. (~p ∨ ~q) ∧ (p ∧ q)   c. ~(p ∨ q) ∧ ~(~p ∧ q)
6. Construct a truth table for ~p ∧ (q ∨ ~r). This compound statement involves three simple statements: p, q and r. In this case there are **three** letters and **eight** combinations of truth values. A truth table showing the eight possibilities appears below. Complete each of the required columns.

p	q	r	~p	~r	(q ∨ ~r)	~p ∧ (q ∨ ~r)
T	T	T				
T	T	F				
T	F	T				
T	F	F				
F	T	T				
F	T	F				
F	F	T				
F	F	F				

A convenient way to remember the order for the possibilities is as follows. Let the p column contain four T's and four F's (TTTT,FFFF). Let the q column contain alternating groups of two T's and two F's (TT,FF,TT, etc.). Finally, let the r column contain alternating T's and F's (T,F,T,F, etc.).

7. Construct a truth table for the following: $(\sim p \lor \sim r) \land (\sim q \land r)$.
   The possible combinations of truth values for p, q, and r are listed.

p	q	r
T	T	T
T	T	F
T	F	T
T	F	F
F	T	T
F	T	F
F	F	T
F	F	F

8. Construct a truth table for $\sim(\sim p \land r) \lor (\sim q \lor \sim r)$.

Review Competencies

9. Write the negations of:
   a. *Some men are overweight.*   b. *Some adults are not arrogant.*

10. Write the statement equivalent to: *No triangles are squares.*

11. Express 35% as a fraction.   12. What percent of 40 is 8?

13. Express $2\frac{1}{4}\%$ as a decimal.

The results of a truth table can correspond with actual reality. A truth table should be something more than mere busy work, although that is what it may have seemed to you at times! Let us see how a truth table would apply to a real situation.

Problem Set 2.5

Use the following two simple statements:
   p: *You live in America.*   q: *You are studying math.*
Consider what happens to the statement p $\land \sim$q.

1. What does p $\land \sim$q say in words?

Chapter 2

2. Construct a truth table below to find the truth values of this compound statement in terms of the four possibilities of p and q.

	p	q	~q	p ∧ ~q
case 1	T	T		
case 2	T	F		
case 3	F	T		
case 4	F	F		

Consider each of the four cases listed above. See if the results in the final column of the truth table correspond to actual situations.

3. **case 1**
    Suppose you really live in America and you really are studying math. This would make both p and q true. Now suppose that someone says,
    You live in America and you are not studying math.
    a. Under these conditions, would this statement be true?
    b. Would this correspond to your result in the final column in the first row of the truth table?

4. **case 2**
    In this case suppose that you really live in America. Thus, p is true. Suppose also that you really are not studying math. Observe that this situation would make q false. Now, someone says,
    You live in America and you are not studying math.
    a. Under these conditions, would this be a true statement?
    b. Would this correspond with what you obtained in the final column in the second row of the truth table?

5. **case 3**
    In this case suppose you really do not live in America. Observe that this situation would make p false. Suppose that you really are studying math. This would make q true. Now, someone makes the statement,
    You live in America and you are not studying math.
    a. Under these conditions would this be a true statement?
    b. Would this correspond with your result in the final column of the third row of the truth table?

6. **case 4**

Finally, suppose you really do not live in America. This would make statement p false. Suppose also that you really are not studying math. Notice statement q is also false in this case. Now someone says, *You live in America and you are not studying math.*

   a. Under these conditions, would this really be a true statement?

   b. Would this correspond with what you obtained in the final column of the fourth row of the truth table?

Review Competencies

7. Write the negations of the following simple statements.

   a. *Some cars are not expensive.*

   b. *No person is above the law.*

8. Multiply: $\frac{4}{5} \cdot 3$ 10. 40% of what number is 28?

9. Divide: $5 \div 1\frac{1}{3}$ 11. Express $2\frac{1}{2}\%$ as a decimal.

A comparison of the final columns in the truth tables of $\sim(p \wedge q)$ and $\sim p \vee \sim q$ illustrates an important idea of logic. The tables are shown below.

p	q	$p \wedge q$	$\sim(p \wedge q)$
T	T	T	F
T	F	F	T
F	T	F	T
F	F	F	T

p	q	$\sim p$	$\sim q$	$\sim p \vee \sim q$
T	T	F	F	F
T	F	F	T	T
F	T	T	F	T
F	F	T	T	T

The final columns of the two tables are identical, and these statements are said to be **equivalent.** Equivalent statements have the same truth tables for corresponding truth values of the simple statements. The symbol used for equivalence is $\equiv$.

> **First De Morgan Law for Logic**
>
> $\sim(p \wedge q) \equiv \sim p \vee \sim q$

Observe that **the negation of a compound statement containing "and" is equivalent to negating each of the simple statements and changing "and" to "or."**

The meaning of $\wedge$ in the conjunction is similar to the meaning of $\cap$, intersection, in set theory, and $\vee$ in the disjunction is similar to $\cup$, union. Observe that this equivalence relationship in logic is a restatement of one of **De Morgan's Laws,** namely, $(A \cap B)' = A' \cup B'$.

This law is the basis for expressing the negation of certain compound statements. For example, consider the compound statement:

Canada is a country and Africa is a country.

If p represents *Canada is a country* and q represents *Africa is a country*, the compound statement is of the form p ∧ q. The negation of this statement, ~(p ∧ q), is equivalent to

~p ∨ ~q since ~(p ∧ q) ≡ ~p ∨ ~q.

Thus, the negation is:

*Canada is **not** a country or Africa is **not** a country.*

We can summarize this result in two ways:
1. The **negation** of *Canada is a country and Africa is a country* is: *Canada is not a country or Africa is not a country.*
2. *It is not true that both Canada and Africa are countries* is **logically equivalent** to *Canada is not a country or Africa is not a country.*

Observe that the second result tells us that if it is not true that both places are countries, then one is not a country **or** the other is not a country. The result does **not** assert that Canada is not a country and Africa is not a country. This can be reinforced by comparing the truth values for ~(p ∧ q) with those for ~p ∧ ~q.

p	q	p ∧ q	~(p ∧ q)		p	q	~p	~q	~p ∧ ~q
T	T	T	F		T	T	F	F	F
T	F	F	T		T	F	F	T	F
F	T	F	T		F	T	T	F	F
F	F	F	T		F	F	T	T	T

The final columns of the two tables are not identical, and these statements are not logically equivalent.

> **Equivalences and Nonequivalences for ~(p ∧ q)**
> 1. ~(p ∧ q) is logically equivalent to ~p ∨ ~q.
> 2. ~(p ∧ q) is not logically equivalent to ~p ∧ ~q.

A helpful procedure for transforming any given statement into an equivalent statement is:
1. Express the given statement symbolically.
2. Write the equivalent statement symbolically.
3. Express the equivalent symbolic statement in words.

Since ~(~p) ≡ p and ~(~q) ≡ q, we obtain the following **variation** of this **De Morgan Law** which can be verified using truth tables:

$$\sim(\sim p \wedge q) \equiv p \vee \sim q$$

If p is: *Ruth has black hair*, and q is : *Elizabeth wears glasses*, consider the statement:

Ruth does not have black hair and Elizabeth wears glasses. (~p ∧ q)

The negation is:

Ruth has black hair **or** *Elizabeth does* **not** *wear glasses.* (p ∨ ~q)

Other variations of this **De Morgan Law** can be verified using truth tables and appear as:

$$\sim(p \wedge \sim q) \equiv \sim p \vee q \quad \text{and} \quad \sim(\sim p \wedge \sim q) \equiv p \vee q$$

It is always necessary to first determine the simple statements in a given compound statement before considering the negation. Consider the statement:

Both Barbara and Roy are not cheerleaders.

This statement can be equivalently expressed as:

Barbara is **not** *a cheerleader and Roy is* **not** *a cheerleader.*

The compound statement is of the form ~p ∧ ~q. The negation of this statement, ~(~p ∧ ~q), is:

Barbara is a cheerleader **or** *Roy is a cheerleader.* (p ∨ q)

Finally, consider the statement:

Some cars are expensive and all people are lazy.

If p represents *Some cars are expensive,* and q represents *all people are lazy,* then the compound statement is of the form p ∧ q.

Recall that the negation of:

Some cars are expensive is: *No cars are expensive.*

The negation of: *All people are lazy* is: *Some people are not lazy.*

Thus, the negation (~p ∨ ~q) is:

No cars are expensive **or** *some people are not lazy.*

Note

Sometimes the symbol ↔ is used instead of ≡ to denote logical equivalence. Therefore, we can write either

~(p ∧ q) ≡ ~p ∨ ~q or ~(p ∧ q) ↔ ~p ∨ ~q.

Problem Set 2.6

Express the negations of:

1. *Jim is tall and John is fat.*
2. *Mary lives in Florida and Nancy lives in Georgia.*
3. *Estelle is both loud and abrasive.*
4. *Both the Dolphins and the Buccaneers are in the playoffs.*
5. *Clearwater is in Georgia and Atlanta is not in Florida.*
6. *The Strikers are not Jacksonville's team and the Bandits are Tampa's team.*
7. *Some movies are not exciting and no jokes are funny.*
8. *Vince is not tall and Mike is not fat.*
9. *All dogs are faithful and some cats are cute.*
10. Consider the statement:
    It is not true that both Frank and Judy appreciate rock music.

    Use De Morgan's Law to express the statement's logical equivalent.

11. Find the statement below which is not the logical equivalent of :
    It is not true that both Karen and Jim are students.
    a. Karen is not a student or Jim is not a student.
    b. Karen is not a student and Jim is not a student.

Review Competencies

12. Write negations of:
    a. *Some people are not dependable.*
    b. *No baby is self-sufficient.*

13. Determine the truth values of:.
    a. *All students are not geniuses and every pig has three legs.*
    b. *No individual is perfect or all cats have two tails.*

14. $6 = \frac{?}{5}$

15. Express $4\frac{2}{5}$ as an improper fraction.

16. Express $\frac{7}{25}$ as a percent.

17. Express 0.06% as a decimal.

We now compare the final columns in the truth tables of $\sim(p \vee q)$ and $\sim p \wedge \sim q$ which are shown below.

p	q	p ∨ q	~(p ∨ q)
T	T	T	F
T	F	T	F
F	T	T	F
F	F	F	T

p	q	~p	~q	~p ∧~q
T	T	F	F	F
T	F	F	T	F
F	T	T	F	F
F	F	T	T	T

Since the final columns are identical in each truth table, these statements are **equivalent**.

Second De Morgan Law for Logic

$$\sim(p \lor q) \equiv \sim p \land \sim q$$

In other words, **the negation of a compound statement containing "or" is equivalent to negating each of the simple statements and changing "or" to "and."** This is a restatement of the second of **De Morgan's Laws,** namely, $(A \cup B)' = A' \cap B'$. This law allows us to express the negation of other compound statements.

Consider the following compound statement:

She runs fast or she jumps high.

Since p represents *She runs fast* and q represents *She jumps high,* the compound statement is of the form $p \lor q$. The negation of this compound statement, $(p \lor q)$, is equivalent to $\sim p \land \sim q$ since $\sim(p \lor q) \equiv \sim p \land \sim q$.

Thus, the negation is:

*She does **not** run fast **and** she does **not** jump high..*

Three other variations of **De Morgan's Laws** are:

$$\sim(\sim p \lor q) \equiv p \land \sim q$$
$$\sim(p \lor \sim q) \equiv \sim p \land q$$
$$\sim(\sim p \lor \sim q) \equiv p \land q$$

Each of these variations can be verified using truth tables.

Consider the statement:

A majority of legislators must vote for a bill or the bill does not become law.

Observe that the two simple statements are:

A majority of legislators must vote for a bill (p). The bill does become law (q).

Thus, the statement is expressed as:

A majority of legislators must vote for a bill or the bill does not become law. $(p \lor \sim q)$

Since $\sim(p \lor \sim q) \equiv (\sim p \land q)$ the negation is:

*A majority of legislators do not vote for a bill **and** the bill does become law.* $(\sim p \land q)$

Imagine if a professor made both of the following statements to a class:

A majority of legislators must vote for a bill or the bill does not become law. A majority of legislators do not vote for a bill and the bill does become law.

Students would immediately be aware that the second statement **contradicts** the first statement. This, of course, is the case because the second statement is the negation of the first statement. Whenever we utter a negation immediately after a statement, we wind up contradicting ourselves (as well as confusing our listeners). This observation is reinforced in the following example.

Example

Consider the statement: $p \wedge {\sim}p$. Let us construct a truth table for this compound statement.

step 1
Since the statement is composed only of variations of simple statement p, there are only two cases for this table.

        p can be true or false.

p
T
F

step 2
Construct another column of values for $\sim$p. Complete the $\sim$p column by negating the values in the p column.

p	~p
T	F
F	T

step 3
Construct a column for $p \wedge {\sim}p$. Complete this column by applying the conjunction to the values in the p column with those in the $\sim$p column.

p	~p	p ∧ ~p
T	F	F
F	T	F

The final column shows that $p \wedge {\sim}p$ is false in all cases. A compound statement that is false in all cases is called a **contradiction** and the two simple statements p and ~p are said to be contradictory. In short, whether you form the negation of a statement using De Morgan's Laws or some other method, the negated statement will always contradict the original statement. If the original statement is true, the negated statement will be false and vice-versa.

Problem Set 2.7

Express the negations of:

1. *London is in England or Paris is in France.*
2. *The bill passes or it's not a law.*
3. *Dave visits San Francisco or London.*
4. *It is not hot or it is not humid.*
5. *All mathematics books have sample tests or they do not get published.*
6. *Antonio is not Prospero's brother or Romeo is Juliet's lover.*
7. *All people carry umbrellas or some people get wet.*

Review Competencies

8. Use DeMorgan's Law to express the logical equivalent of:
    a. *It is not true that both University of Florida and Sears are colleges.*
    b. *It is not true that both all musicians write lyrics and some Broadway musicals lose money.*

9. Find the statement below which is not the logical equivalent of:
    It is not true that both Carter and Reagan are Democrats.
    a. Carter is not a Democrat and Reagan is not a Democrat.
    b. Carter is not a Democrat or Reagan is not a Democrat.

10. Add: $2\frac{1}{3} + 3\frac{3}{4}$

11. Divide: $14 \div 0.002$

12. Express 2.5 as a percent.

13. What is 120% of 80?

Section 3   Conditional Statements

There are statements in logic that depend upon some given condition. Consider the following statement:

> *If the team wins the game, then I will be very happy.*

Observe that the condition is the team's winning the game, which implies that I will be very happy. If that condition is met, then I will be very happy. Observe also that this statement involves the use of the connective words "if" and "then."

This conditional statement can also be stated as follows:

1. *If the team wins the game, I will be very happy.*
2. *I will be very happy if the team wins the game.*
3. *When the team wins the game, I will be very happy.*
4. *I will be very happy when the team wins the game.*
5. *The team winning the game implies my being very happy.*

6. *The team wins the game only if I am very happy.*
7. *Only if I am very happy does the team win the game.*
8. *The team winning the game is a sufficient condition for my being very happy.*
9. *My being very happy is a necessary condition for the team winning the game.*
10. *The team does not win the game unless I am very happy.*

These statements can be expressed symbolically:

Let statement p be: *The team wins the game.*

Let statement q be: *I will be very happy.*

The statement could be written as: If p, then q   or   p implies q   or   p → q

---

Notation

The symbol → is translated as "if... then"

Definitions

Statements of the form p → q are called **conditional statements** or **implications**. In such a statement, p is called the **hypothesis** or **antecedent**, and q is called the **conclusion** or **consequent**.

Remember that the hypothesis always follows "if" or "when."

Examples

1. Consider the following statement:
   If you bring the teacher an apple, then you will make an A.
   The hypothesis (p) is the statement: *You bring the teacher an apple.*
   The conclusion (q) is the statement: *You will make an A.*

2. Consider the following simple statements:
   p: *You speak loudly.*   q: *I can hear you.*

   Express the following symbolic statements as sentences.
   a. ~p → ~q is: *If you do not speak loudly, then I cannot hear you.*
   b. q → p is: *If I can hear you, then you speak loudly.*
   c. ~q → ~p is: *If I cannot hear you, then you do not speak loudly.*

Conditional statements can appear in many forms. To express them in the familiar "if ... then" form it is often necessary to change the wording.

Examples

1. Consider the statement: *It pours when it rains.*
   This statement can be expressed as: *If it rains, then it pours.*

2. Consider the statement:
   Michelle experiences anxiety when she does not tell the truth.
   This statement can be expressed as:
   If Michelle does not tell the truth, then she experiences anxiety.

3. Consider the statement: *No dogs have wings.*
   This statement can be expressed as:
   If it is a dog, then it does not have wings.

4. Consider the statement: *Jane always eats dinner at a cafeteria.*
   This statement can be expressed as:
   If Jane eats dinner, then she eats at a cafeteria.

The ability to change the wording of a conditional statement to the familiar "if...then" form frequently depends upon understanding the English syntax, the way in which the words are put together to form phrases, clauses, and sentences. However, you may find the following table helpful.

> **Equivalent Ways Of Translating**
> $p \rightarrow q$ **Into English**
>
> a. If p, then q.
> b. If p, q.
> c. p implies q.
> d. p only if q.
> e. Only if q, p.
> f. p is a sufficient condition for q.
> g. q is a necessary condition for p.
> h. Not p unless q.

Example

Consider the statement: *If you live in New York, then you live in America.* In terms of sets, the set of New Yorkers is a subset of the set of Americans. Notice that living in New York is a **sufficient condition** (a guarantee) for living in America. This can be seen in the Venn Diagram at the right. However, living in America is certainly not a guarantee (not a sufficient condition) for living in New York.

Living in America is not a sufficient condition for living in New York, but it is a **necessary condition** because you do not live in New York unless you live in America. An x cannot be placed in the New Yorker circle unless it is also placed in the American circle. Living in America is a necessary prerequisite for living in New York. Obviously, if you don't live in America, then you can't possibly live in New York. You live in New York only if you live in America.

Let us use the previous table, with:
     p: *You live in New York.*
     q: *You live in America.*

We translate p → q into English as follows:

a. *If you live in New York, then you live in America.*

b. *If you live in New York, you live in America.*

c. *Living in New York implies living in America.*

d. *You live in New York only if you live in America.*

e. *Only if you live in America do you live in New York.*

f. *Living in New York is a sufficient condition for living in America.*

g. *Living in America is a necessary condition for living in New York.*

h. *You do not live in New York unless you live in America.*

Problem Set 3.1

1. Identify the hypothesis (antecedent) and conclusion (consequent) in each of the following conditional statements.
    a. If I work hard, I will pass the course.
    b. If a polygon is a quadrilateral, then the polygon has four sides.
    c. It is not raining if the sky is not overcast.

2. Consider the following simple statements:
    p: *You have long hair.*
    q: *You will get dandruff.*
    Express the following symbolic statements as sentences.
    a. $p \rightarrow q$ b. $q \rightarrow p$
    c. $\sim p \rightarrow \sim q$ d. $\sim q \rightarrow \sim p$

3. Express the following using the "if ... then" form.
    a. George always buys a Sony television.
    b. No soldier is afraid.
    c. The class is over when the bell rings.
    d. All students will pass this test.
    e. A Corvette won't enter this race.
    f. Criminals will be punished.
    g. None of your teachers is dull.
    h. Every tall person should play basketball.
    i. No one remains standing when the National Anthem is not being played.
    j. All members are required to attend the meeting unless excused.

4. Select the statement that is equivalent to:
    My throat is tense when I am not telling the truth.
    a. If my throat is tense, then I am not telling the truth.
    b. If I am telling the truth, then my throat is not tense.
    c. If I am not telling the truth, then my throat is tense.
    d. I am not telling the truth and my throat is not tense.

5. Express each of the following statements in "if - then" form.
    a. You avoid a ticket only if you observe the speed limit.
    b. Watering the flowers is a necessary condition for making them grow.
    c. Attending class regularly is a sufficient condition for passing the course.
    d. I can live in the North only if I can endure the winters.
    e. Turning on the ignition is a necessary condition for starting the car.
    f. Missing the bus is a sufficient condition for being late to class.

Review Competencies

6. Express the negations of:
    a. *Linda is not a good skater or Larry is a fast runner.*
    b. *All dogs are animals.*
    c. *Jim does not smoke and Laura does not drink.*
    d. *Some men are bad drivers.*

7. Add: $4\frac{1}{5} + 2\frac{1}{4}$

8. Subtract: $19 - 4.001$

9. Express 65% as a fraction.

10. Express $\frac{5}{8}$ as a decimal.

Chapter 2

We now wish to construct a truth table for the basic conditional statement.
Consider the following conditional statement:

    If it is a nice day, then I will go swimming.

Notice that the simple statements are:

    p: *It is a nice day.*   q: *I will go swimming.*

There are four possibilities that could occur with this conditional statement:

				p	q
case 1	*It is a nice day*	and	*I go swimming.*	T	T
case 2	*It is a nice day*	and	*I do not go swimming.*	T	F
case 3	*It is not a nice day*	and	*I go swimming.*	F	T
case 4	*It is not a nice day*	and	*I do not go swimming.*	F	F

Let us consider the truth values of the statement under each of these
four possible situations.

case 1

It is a nice day, and I go swimming. Notice that both
statements p and q are true in this case. Since I am doing
what I said I would do if it were a nice day, this conditional
statement is **true**.

p	q	p → q
T	T	T

case 2

It is a nice day, and I do not go swimming. In this case
p is true and q is false. Since I am not doing what I said
I would do if it were a nice day, the original conditional
statement is **false**.

p	q	p → q
T	F	F

case 3

It is not a nice day, and I go swimming. In this case,
p is false and q is true. Notice that this situation has
nothing whatsoever to do with the original statement.
The original statement is not affected at all. Nothing was said about what I
would do if it were not a nice day. Under these conditions the original statement is still **true** since it certainly has not been proven false.

p	q	p → q
F	T	T

case 4

It is not a nice day and I do not go swimming. In this
case p is false and q is also false. Here again, this
situation has nothing to do with the original statement.
Since it certainly has not been proven false, the original statement is still
considered **true**.

p	q	p → q
F	F	T

These results may be summarized in a basic table for the conditional statement. Since a conditional statement is also called an implication, the basic table is generally referred to as the implication table.

Implication Table

p	q	p → q
T	T	T
T	F	F
F	T	T
F	F	T

Observe that **an implication is false only when the hypothesis is true and the conclusion is false,** case (2) in the table.

Problem Set 3.2

Based upon the implication table, determine the truth or falsity of each of the following.

1. If $3 + 5 = 8$, then $6 \cdot 6 = 42$.
2. If $4 + 4 = 32$, then $5 + 5 = 10$.
3. If $3 + 5 = 8$, then $5 + 5 = 10$.
4. If $3 \cdot 9 = 28$, then $6 \cdot 5 = 35$.

5. Consider the statement:
If you see the concert, then you have a ticket.
Determine if the statement is true or false under each of the following conditions.
   a. You see the concert. You also have a ticket.
   b. You see the concert. However, you do not have a ticket.
   c. You do not see the concert even though you do have a ticket.
   d. You do not see the concert. You also do not have a ticket.

Review Competencies

6. Express the following using the "if... then" form.
   a. *All students will graduate.*
   b. *No one under seventeen is allowed in the theatre.*

7. Consider the following simple statements:
p: *You work hard.*   q: *You will succeed.*
Translate the following symbolic statements into English.
   a. $p \rightarrow q$
   b. $q \rightarrow p$
   c. $\sim p \rightarrow \sim q$
   d. $\sim q \rightarrow \sim p$

8. Express the negations of:
   a. *She is a good wife and she is a good mother.*
   b. *He likes steak or she does not like roast beef.*

9. Multiply: $2\frac{1}{3} \cdot \frac{1}{3}$
10. $\frac{3}{5} = \frac{15}{?}$

11. Express 125% as a decimal and a fraction.

12. What percent of 50 is 20?

13. Express $4\frac{3}{4}\%$ as a decimal.

14. Each day a small business owner sells 20 pizza slices at $1.50 per slice and 40 sandwiches at $2.50 each. If business expenses come to $60 per day, what is the owner's profit for a ten-day period?

Chapter 2

More complex implications may also be analyzed by means of truth tables.

Problem 1 Construct a truth table for the implication q → ~p

Solution

step 1

List the four truth possibilities
p and q in the table in the same
way as in previous truth tables.

p	q
T	T
T	F
F	T
F	F

step 2

Complete the table by constructing
columns for ~p and q → ~p. The ~p
column is completed by negating the
entries in the p column. The q → ~p
column is completed by applying the

p	q	~p	q →~p
T	T	F	F
T	F	F	T
F	T	T	T
F	F	T	T

implication definition to the q and ~p columns. Since an implication is
false only when the hypothesis (column 2) is true and the conclusion
(column 3) is false, the first row in the table is the only false entry.

Problem Set 3.3

Construct truth tables for each of the following.

1. ~p → q
2. q → p
3. ~p → ~q
4. ~q → ~p

5. Construct truth tables for p → q and ~q → ~p
    and compare the final columns.

6. Express ~q → ~p in words for each of the
    following in which p → q represents:
    a. *If I work hard, then I will pass the course.*
    b. *If Bob lives in Clearwater, then he lives
        in Florida.*
    c. *If x² is an even number, then x is an even
        number.*

7. Construct truth tables for q → p and ~p → ~q
    and compare the final columns.

Review Competencies

8. Determine the truth or falsity of each of the
    following statements.
    a. If $2 + 5 = 7$, then $2 + 5 = 8$.
    b. If $3 \cdot 7 = 20$, then $2 \cdot 4 = 7$.
    c. If $5 \cdot 6 = 32$, then $4 \cdot 7 = 28$.

9. Express the negations of:
    a. *He is not a good administrator and she
        is not a good teacher.*
    b. *Tom is not the father or he is the legal
        guardian.*
    c. *Some rivers are polluted.*
    d. *Some bridges are not safe.*

10. Subtract: $4\frac{1}{3} - 2\frac{2}{3}$

11. Express 25% as a fraction.

12. What is 9% of 30?

13. Evaluate: $(4 + 5)(5 - 2)$

The following problem illustrates that truth tables may be constructed for statements that require parentheses.

Problem 2   Construct a truth table for the statement $(p \vee q) \rightarrow q$.

Solution

step 1

As always, list the four truth possibilities for p and q in the customary order.

p	q
T	T
T	F
F	T
F	F

step 2

Construct a column for the statement $(p \vee q)$. Use the values in the p and q columns along with the definition of the disjunction to complete this column.

p	q	$(p \vee q)$
T	T	T
T	F	T
F	T	T
F	F	F

step 3

Construct a column for the entire implication statement, $(p \vee q) \rightarrow q$. Use the values in the $(p \vee q)$ column as the hypotheses and the values in the q column as conclusions to complete the final column.

p	q	$(p \vee q)$	$(p \vee q) \rightarrow q$
T	T	T	T
T	F	T	F
F	T	T	T
F	F	F	T

Problem Set 3.4

Construct truth tables for:

1. $(p \wedge q) \rightarrow p$ 2. $\sim p \rightarrow (p \vee q)$
3. $\sim(p \rightarrow q) \vee (q \rightarrow p)$ 4. $\sim q \rightarrow (p \wedge \sim q)$
5. $(\sim p \wedge q) \rightarrow (p \wedge q)$ 6. $[(p \rightarrow q) \wedge \sim p] \rightarrow \sim q$
7. $q \rightarrow (\sim r \vee p)$
8. $(p \rightarrow \sim r) \wedge (q \rightarrow \sim p)$

Review Competencies

9. Determine whether the following statements are true or false.
    a. *All goldfish can sing or a triangle has three sides.*
    b. *If 5 • 8 = 30, then 2 • 6 = 15.*

10. Express the negations of the following statements:
    a. *All television programs are educational.*
    b. *No car is designed to last forever.*
    c. *Bill is a good husband and he is not a good politician.*
    d. *Miami is not in Alabama or Florida is a fast growing state.*

11. Express $\sim q \rightarrow \sim p$ when $p \rightarrow q$ is:
    If the sun shines, then I will go to the beach.

12. $\frac{4}{7} = \frac{?}{21}$ 13. Divide: $2\frac{2}{3} \div 1\frac{3}{4}$

14. Evaluate: $12 \div 4 + 3 \cdot 5 - 6 \div 2$

Section 4 Forms of Conditional Statements

It is possible to classify conditional statements according to the form they assume.

> ### Definition
> If $p \rightarrow q$ is a conditional statement, then $q \rightarrow p$ is the **converse** of that conditional statement.

Recall that in $p \rightarrow q$, p is the hypothesis and q is the conclusion. **The converse of a conditional statement is obtained by reversing the hypothesis and the conclusion.**

Example

Consider the following conditional statement of the form $p \rightarrow q$:

If Tony is in Seattle, then he is in the West. (true)

Statement p is: *Tony is in Seattle.*

Statement q is: *Tony is in the West.*

The converse $(q \rightarrow p)$ of this statement is:

If Tony is in the West, then he is in Seattle.

Notice that the converse could be true, but could also be false. If Tony is in the West, he might or might not be in Seattle. This suggests that the converse of a conditional statement is not logically equivalent to the conditional statement.

> ### Definition
> If $p \rightarrow q$ is a conditional statement, then $\sim p \rightarrow \sim q$ is the **inverse** of the conditional statement.

The inverse of a conditional statement is obtained by negating both the hypothesis and the conclusion.

Example

Consider the following conditional statement of the form p → q:
>*If Tony is in Seattle, then he is in the West.* (true)
>>Statement p is: *Tony is in Seattle.*
>>Statement q is: *Tony is in the West.*

The inverse (~p → ~q) of this statement is:
>*If Tony is not in Seattle, then he is not in the West.*

Notice that the inverse could be true, but could also be false. If Tony is not in Seattle, he might or might not be in the West. Not only does this suggest that the inverse of a conditional statement is not logically equivalent to the conditional statement, but it also brings forth the idea that the inverse is not the negation of the conditional statement. If a statement is true, its negation must be false, and the inverse (If Tony is not in Seattle, then he is not in the West) is not necessarily false. The inverse does not contradict the original conditional statement.

> ### Definition
> If p → q is a conditional statement, then ~q → ~p is the **contrapositive** of that conditional statement.

The contrapositive of a conditional statement is obtained by reversing the hypothesis and the conclusion and negating both of them.

Example

Consider again the following conditional statement of the form p → q:
>*If Tony is in Seattle, then he is in the West.* (true)
>>Statement p is: *Tony is in Seattle.*
>>Statement q is: *Tony is in the West.*

The contrapositive (~q → ~p) of this statement is:
>*If Tony is not in the West, then he is not in Seattle.* (true)

Observe that the contrapositive must be true. If Tony is not in the Western United States, he cannot possibly be in Seattle. This suggests that the contrapositive of a conditional statement is logically equivalent to the conditional statement.

Problem 1   Write the contrapositive of the conditional statement:
If Mary and Ethel star, then the musical is a success.

Solution   Let p be: *Mary stars.*
Let q be: *Ethel stars.*
Let r be: *The musical is a success.*
The given conditional statement translates as: $(p \land q) \rightarrow r$
With $(p \land q)$ as the hypotheses and r as the conclusion, we form the contrapositive by reversing the hypothesis and the conclusion and negating both of them. The contrapositive is: $\sim r \rightarrow \sim(p \land q)$

Using one of De Morgan's Laws, we obtain:
$\sim r \rightarrow (\sim p \lor \sim q)$

This translates as:
If the musical is not a success, then Mary does not star or Ethel does not star.

Problem 2   Write the contrapositive of the conditional statement:
If it snows, then at least one road is open.

Solution   Let p be: *It snows.*
Let q be: *At least one road is open.* (Equivalently: *Some road is open.*)
The given statement $(p \rightarrow q)$ has as its contrapositive: $\sim q \rightarrow \sim p$
Since the negation of "some" is "no," the contrapositive is:
If no road is open, then it is not snowing.
Because "no road is open" means that all roads are closed, we can also express the contrapositve as:
If all roads are closed, then it is not snowing.

Problem Set 4.1

Write the converse, the inverse, and the contrapositive for each of the following conditional statements:

1. *If Nigel is in London, then he is in England.*

2. *If it is a banana, then it is yellow.*

3. *If Joan is at the movies, then she is having a good time.*

4. *If it's St. Patrick's Day, then Erin does not wear red clothing.*

5. *If one advocates censorship, then one does not advocate freedom.*

6. $p \rightarrow {\sim}q$

7. ${\sim}p \rightarrow {\sim}q$

8. *No animals are allowed in the restaurant.* (First write the statement in the "if…then" form.)

9. *All chemists are scientists.* (First write the statement in the "if…then" form.)

Write the contrapositive for each of the following conditional statements:

10 *If lines are parallel, then they do not share a common point.*

11. *If both Jose and Marsha play, then the team wins.*

12. *If one is in Atlanta or New Orleans, then one is in the South.*

13. ${\sim}p \rightarrow ({\sim}q \vee r)$

14. $(p \wedge {\sim}q) \rightarrow {\sim}r$

15. *If there is a hurricane, then all schools are closed.*

16. *If the review session is successful, then no students fail the test.*

17. Let p be: *I win the game.*
    Let q be: *I collect the prize.*
    Translate the following into English.
    a. ${\sim}p \wedge {\sim}q$ b. ${\sim}(p \vee {\sim}q)$

18. Express the negations of:
    a. *Some drivers are reckless.*
    b. *Irma is a fast typist and Theresa is a good cook.*

19. Express $\frac{25}{35}$ as an equivalent fraction in lowest terms.

20. Subtract: $4 - 1\frac{4}{5}$

21. Add: $12.45 + 1.007 + 513.6$

22. Express 175% as a decimal.

23. Evaluate: $6 - (2 \cdot 1 + 4 \div 2)$

24. An insurance company pays 100% emergency room treatment, 80% of the doctor bills after a $100 deductible, and 90% of all medication. If a patient submits bills of $125 for the emergency room treatment, $175 for doctor bills, and $95 for medication, what does the insurance company pay?

Below are shown truth tables for the conditional $(p \to q)$, its contrapositive $(\sim q \to \sim p)$, its converse $(q \to p)$ and its inverse $(\sim p \to \sim q)$.

Conditional Statement

p	q	$p \to q$
T	T	T
T	F	F
F	T	T
F	F	T

Contrapositive

p	q	$\sim q$	$\sim p$	$\sim q \to \sim p$
T	T	F	F	T
T	F	T	F	F
F	T	F	T	T
F	F	T	T	T

Converse

p	q	$q \to p$
T	T	T
T	F	T
F	T	F
F	F	T

Inverse

p	q	$\sim p$	$\sim q$	$\sim p \to \sim q$
T	T	F	F	T
T	F	F	T	T
F	T	T	F	F
F	F	T	T	T

Observe that the conditional $(p \to q)$ and the contrapositive $(\sim q \to \sim p)$ have the same truth values in the final column and thus are **logically equivalent**.

$$p \to q \; \equiv \sim q \to \sim p$$

This means that the contrapositive can replace any conditional without changing its meaning. Notice that the converse $(q \to p)$ and the inverse $(\sim p \to \sim q)$ also have the same truth values and are logically equivalent. However, a conditional statement is not equivalent to its converse or its inverse.

> **Equivalent and Non-Equivalences for a Conditional Statement**
> 1. $p \to q$ is logically equivalent to $\sim q \to \sim p$.
> 2. $p \to q$ is not logically equivalent to $q \to p$ (its converse) and $\sim p \to \sim q$ (its inverse).

Problem 3 Write two statements that are **not** logically equivalent to:
If one is a biologist, then one is a scientist. (true)

Solution Both the converse and the inverse are not logically equivalent to
the original conditional statement.
Let p be: *One is a biologist.*
Let q be: *One is a scientist.*
The non-equivalences are:
q → p (the converse): *If one is a scientist, then one is a
biologist.* (not necessarily true)
~p → ~q (the inverse): *If one is not a biologist, then
one is not a scientist.* (not necessarily true)

Problem 4 Write the logical equivalent of:
If all people take exams honestly, then no supervision is needed.

Solution The logical equivalent is the contrapositive.
Let p be: *All people take exams honestly.*
Let q be: *No supervision is needed.*
The logical equivalent is ~q → ~p. Since the negation
of "all" is "some...not" and the negation of "no" is
"some," the contrapositive is:
*If some supervision is needed, then some people do not
take exams honestly.*

Problem 5 Let p be: *One works hard.*
Let q be: *One falls behind.*
Write the rule of logical equivalence which directly (in
one step) transforms statement (1) into statement (2).
(1) If one works hard, then one does not fall behind.
(2) If one falls behind, then one does not work hard.

Solution Statement (1) translates into p → ~q.
Statement (2) translates into q → ~p.
Observe that statement (2) is the logical equivalent of
statement (1), obtained by taking the contrapositive of
statement (1). The equivalence rule is:
(p → ~q) ≡ (q → ~p)
This can be verified by comparing the final columns in the
truth tables of p → ~q and q → ~p.

Problem Set 4.2

1. Consider the following simple statements:
   p: *Sherry lives in Palm Beach.*
   q: *Sherry lives in Florida.*
   Express each of the following in English.
   Determine also whether each is always true
   or not necessarily true.
   a. $p \rightarrow q$ b. the converse of $p \rightarrow q$
   c. the inverse of $p \rightarrow q$
   d. the contrapositive of $p \rightarrow q$

Write two statements that are not logically
equivalent to each of the following.

2. *If I pass the test, then I am happy.*
3. *If it is St. Patrick's Day, then Erin wears green.*
4. *If one is a poet, then one is a writer.*
5. *If a number is divisible by 6, then it is divisible by 3.*
6. *If it's a banana, then it is not blue.*
7. *If Mary and Ethel star, then the musical is a success.*
8. *If one is in Atlanta or New Orleans, then one is in the South.*
9. *If the team wins, we celebrate and purchase new equipment.*

Write a statement that is logically equivalent to
each of the following.

10. *If one works hard, then one does not fail.*
11. *If the book does not have a test bank, it is not published.*
12. *Maggie has a good time when she attends the opera.*
13. *If all corporations place profit above human need, then some people suffer.*
14. Let p be: *One works.*   Let q be: *One is ill.*
    Write the rule of logical equivalence which
    directly (in one step) transforms statement (1)
    into statement (2).
    (1) *If one is not working, then one is ill.*
    (2) *If one is not ill, then one is working.*

15. Let p be: $|x| > 3$
    Let q be: $x > 3$
    Let r be: $x < -3$
    Write the rule of logical equivalence which
    directly (in one step) transforms statement (1)
    into statement (2).
    (1) If $|x| > 3$, then $x > 3$ or $x < -3$.
    (2) If $x \not> 3$ and $x \not< -3$, then $|x| \not> 3$.

16. Use the conditional statement:
    I will be happy if I pass the test.
    Express each of the following.
    a. its converse   b. its inverse
    c. its contrapositive

17. Express the converse, inverse, and
    contrapositive of $\sim p \rightarrow q$.

18. Express the negations of:
    a. *No elephants are ballet dancers.*
    b. *Jane is a good dancer or Martha is a good singer.*

19. Subtract: $8\frac{3}{4} - 4\frac{1}{3}$ 20. Express $\frac{12}{25}$ as a percent.

21. Divide: $5 \div 1\frac{3}{5}$ 22. Evaluate: $6(\frac{2}{3} \div \frac{4}{3} \cdot \frac{1}{4})$

23. A restaurant bill came to $60 before a 5% tax
    was added. How much money is saved by
    leaving a tip of 15% on the bill before the tax
    is added instead of 15% on the final bill inclu-
    sive of the tax?

We have seen that if a conditional statement is true, its converse is not necessarily true. However, there are certain English statements that are true in both directions.

Let us now consider a "two-way" conditional.

Let p be: *A set contains no elements.*
Let q be: *It is the null set.*

Notice that both of the following conditionals are true.

$p \rightarrow q$: *If a set contains no elements, then it is the null set.*
$q \rightarrow p$: *If a set is the null set, then it contains no elements.*

The conjunction of these two statements will therefore also be true and is the basis for defining the **biconditional** with symbol $\leftrightarrow$.

$$p \leftrightarrow q \text{ is equivalent to } (p \rightarrow q) \wedge (q \rightarrow p)$$

The choice of the two-way arrow ($\leftrightarrow$) is appropriate since, notationally speaking, this new connective indicates that the conditional can be considered in **both** directions. Furthermore, depending on the direction, both p and q are hypotheses and conclusions.

There are a number of ways of translating $p \leftrightarrow q$ into English.

Translations for $p \leftrightarrow q$

a. p is equivalent to q.
b. p implies q and q implies p.
c. p if and only if q. (abbreviated: p iff q)
d. q if and only if p. (abbreviated: q iff p)
e. p is both necessary and sufficient for q.
f. q is both necessary and sufficient for p.

Example

Let p be: *A set contains no elements.*
Let q be: *It is the empty set.*
p ↔ q can be translated as:

a   A set contains no elements is equivalent to it is the empty set.

b.   A set contains no elements implies that it is the empty set and a set is the empty set implies that it contains no elements.

c.   A set contains no elements if and only if it is the empty set.

d.   A set is the empty set if and only if it contains no elements.

e.   A set contains no elements is both necessary and sufficient for the set being the empty set.

f.   A set is the empty set is both necessary and sufficient for the set to contain no elements.

Problem 6

Consider the following simple statements:

p: You live in Florida.   q: You live in America.

Does the biconditional connective, ↔, relate these two statements?

Solution

The answer is no. The conditional does not work in both directions. If you live in Florida, then you do live in America. However, it is not necessarily true that if you live in America, then you live in Florida. Since the conditional does not work in both directions, it is incorrect to claim that living in Florida is both necessary and sufficient for living in America.

When a statement is formulated with "necessary and sufficient" or "if and only if" (iff), the statement is immediately the biconditional. This means that the first portion implies the second portion and, conversely, the second portion implies the first portion. The truth table for the biconditional is shown below. Notice that p ↔ q is true only when both statements have the same truth value.

Biconditional Table		
p	q	p ↔ q
T	T	T
T	F	F
F	T	F
F	F	T

Problem Set 4.3

1. Consider the following simple statements:
   p: *A number is a multiple of 2.*
   q: *A number is even.*
   Translate p ↔ q into English using the six
   ways indicated in this section.

2. Is the following statement true? *You are
   Albert Einstein if and only if you are a
   brilliant physicist.*
   Asked in another way, does the biconditional
   connective ↔ relate the two statements below?
   p: *You are Albert Einstein.*
   q: *You are a brilliant physicist.*

3. Is the following statement true? *x not being
   greater than 5 is both necessary and sufficient
   for x being less than or equal to 5.*

4. Is the following statement true?
   (a = b) iff (b = a)

5. Is the following statement true? *a is less than
   b iff b is greater than a.*

6. Is the following statement true? *x being
   greater than 5 is both necessary and
   sufficient for x being greater than 3.*

7. Is the following statement true? *x is greater
   than 0 iff x is positive.*

8. Is the following statement true? *x being divisible by 5 is both necessary and sufficient for
   x being divisible by 10.*

Review Competencies

9. Express each of the following in "if...then"
   form.
   a. You can go to the fair only if you clean
      up your room.
   b. Handing in the homework is a sufficient
      condition for receiving bonus points.
   c. Turning on the television set is a neces-
      sary condition for watching the soap
      opera.

10. Express the converse of: *If it rains, then I
    stay home.*

11. Write the contrapositive for: *If Martha is
    alone, then Martha is afraid.*

12. Express negations for:
    a. *Some students do not pay attention.*
    b. *All movies are exciting.*
    c. *Paul has the answer or Dan has the
       question.*
    d. *Myra is not cooperative and Lois is not
       dependable.*

13. Express 40% as a fraction.

14. What percent of 60 is 15?

15. Last year, Sarah Rosenberg earned $17,000
    and her son Josh earned $7,000. Their total
    yearly income was $28,000. Based on the as-
    sumption that their only other source of income
    was from interest on Sarah's savings account,
    what fractional part of their income was from
    interest on the account?

An important equivalence relating the disjunction and the conditional statement is shown below.

$$p \lor q \equiv {\sim}p \to q$$

In words, this says that **a disjunction may be expressed as a conditional statement by letting the negation of the first statement serve as the hypothesis and the second statement serve as the conclusion.**

To verify this equivalence we construct truth values for $p \lor q$ (the basic disjunction) and ${\sim}p \to q$:

p	q		$p \lor q$		p	q	$\sim p$		${\sim}p \to q$
T	T		T		T	T	F		T
T	F		T		T	F	F		T
F	T		T		F	T	T		T
F	F		F		F	F	T		F

Since the final columns are identical in each truth table, we have verified the equivalence: $p \lor q \equiv {\sim}p \to q$.

Example

Let p be: *Joe studies English.*   Let q be: *Joe studies mathematics.*

$p \lor q$ is: *Joe studies English or mathematics.*

This statement is logically equivalent to ${\sim}p \to q$ which says:

If Joe does not study English, then he studies mathematics.

This new statement is equivalent to its contrapositive, obtained by reversing and negating hypothesis and conclusion, which says:

If Joe does not study mathematics, then he studies English.

Both of these equivalences make sense, for if Joe studies one discipline or the other, if he does not study one, he must study the other. On the other hand, since we are dealing with the inclusive "or," the original statement ($p \lor q$: *Joe studies English or mathematics.*) is **not** logically equivalent to: *If Joe studies English, then Joe does not study mathematics.* (He might be studying both.)

To verify that $p \lor q$ is **not** equivalent to $p \to {\sim}q$, we once again construct truth tables.

p	q		$p \lor q$		p	q	$\sim q$		$p \to {\sim}q$
T	T		T		T	T	F		F
T	F		T		T	F	T		T
F	T		T		F	T	F		T
F	F		F		F	F	T		T

The final columns of the two tables are not identical, and these statements are not logically equivalent.

> ### Equivalences and Nonequivalences for $p \lor q$
> 1. $p \lor q$ is logically equivalent to both $\sim p \to q$ and $\sim q \to p$.
> 2. $p \lor q$ is not logically equivalent to either $p \to \sim q$ or $q \to \sim p$.

Other equivalences follow directly from the preceeding equivalence and could be verified with truth tables. They are:

$$p \lor \sim q \equiv \sim p \to \sim q \qquad \sim p \lor q \equiv p \to q \qquad \sim p \lor \sim q \equiv p \to \sim q$$

Notice again that a disjunction may be expressed as a conditional statement by letting the negation of the first statement serve as the hypothesis and the second statement serve as the conclusion.

Examples

Let p be: *My team wins the game.*   Let q be: *I will win the bet.*
$\sim p \lor q$ is: *My team does not win the game or I will win the bet.*
This statement is logically equivalent to $p \to q$ which says:
  If my team wins the game, then I will win the bet.
$\sim p \lor \sim q$ is: *My team does not win the game or I will not win the bet.*
This statement is logically equivalent to $p \to \sim q$ which says:
  If my team wins the game, then I will not win the bet.

Since it is possible to express a given disjunction as an equivalent conditional statement, it should also be possible to express a given conditional statement as an equivalent disjunction statement. It can be verified by truth tables that:

$$p \to q \equiv \sim p \lor q$$

Notice that the negation of the hypothesis serves as the first statement of the disjunction and the conclusion serves as the second statement of the disjunction. As we shall see, this equivalence might appear less evident when given certain translations into English.

Other variations are:

$$\sim p \to q \equiv p \lor q \qquad p \to \sim q \equiv \sim p \lor \sim q \qquad \sim p \to \sim q \equiv p \lor \sim q$$

Examples

Let p be: *The sun shines.*   Let q be: *I will go swimming.*

p → q is: *If the sun shines, then I will go swimming.*

This statement is logically equivalent to ~p ∨ q which says:

*The sun does **not** shine or I will go swimming. (~p ∨ q)*

~p → q is: *If the sun does not shine, then I will go swimming.*

This statement is logically equivalent to:

*The sun shines **or** I will go swimming. (p ∨ q)*

Note

Without careful thought one is likely to feel uncomfortable with the English translations in the examples above since they do not appear evident. This particular equivalence, p → q ≡ ~p ∨ q, is probably more clearly seen from the perspective of logic and truth tables than from the point of view of examples in English.

> Notice that we now have two statements that are logically equivalent to p → q:
>   1. p → q ≡ ~q → ~p (the contrapositive)
>   2. p → q ≡ ~p ∨ q

Problem 7

Express the negation of the following statement as a conditional statement.

Bob plans to both pay his bills and write a letter.

Solution

Let p be: *Bob plans to pay his bills.*

Let q be: *Bob plans to write a letter.*

p ∧ q is: *Bob plans to both pay his bills and write a letter.*

The negation of p ∧ q is ~(p ∧ q). Recall one of **De Morgan's Laws** which gives the equivalence ~(p ∧ q) ≡ ~p ∨ ~q. Since ~p ∨ ~q ≡ p → ~q, the negation of the conjunction expressed as a conditional statement is:

> ~(p ∧ q) ≡ p → ~q

Thus, the negation of the given statement expressed as a conditional statement is:

If Bob plans to pay his bills, then he does not plan to write a letter.

> Notice that we now have two statements that are logically equivalent to ~(p ∧ q):
> 1. ~(p ∧ q) ≡ p → ~q
> 2. ~(p ∧ q) ≡ ~p ∨ ~q   (A De Morgan Law)

Problem 8   Write two statements that are logically equivalent to:
It is not true that both Moby Dick and I Love Lucy are novels.

Solution   Let p be: *Moby Dick is a novel.*
Let q be: *I Love Lucy is a novel.*
The two equivalences for ~(p ∧ q) are:
1.   Moby Dick is not a novel or I Love Lucy is not a novel. (~p ∨ ~q)
2.   If Moby Dick is a novel, then I Love Lucy is not a novel. (p → ~q)

We can now summarize the equivalences and nonequivalences of various statements in logic.

Statement	Equivalences	Nonequivalences
p → q	~q → ~p ~p ∨ q	q → p (converse) ~p → ~q (inverse)
p ∨ q	~p → q ~q → p	p → ~q q → ~p
~(p ∧ q)	p → ~q ~p ∨ ~q	~p ∧ ~q

Problem Set 4.4

1.   Write equivalent conditional statements for the following disjunctions.
    a.   *Miami will win the division or Buffalo will win the division.*
    b.   *Kay or Joyce will go to the convention.*
    c.   *Professor Gill does not teach this course or I am withdrawing.*
    d.   *John is selected for the committee or I am not resigning.*
    e.   *Ron will not win the award or Doug will not win the award.*

2.   Write equivalent disjunction statements for:
    a.   *If Steve goes surfing, then he will miss class.*
    b.   *If I try to stand up, my back will hurt.*
    c.   *If she does not appear, we can conduct the meeting in her absence.*
    d.   *If Michelle discovers the flat tire, she will not be happy.*
    e.   *If Mark does not work, he cannot pay the rent.*

Chapter 2

3. Express the negations of each of the following statements as conditional statements.
    a. *Joan is both nervous and tense.*
    b. *Pete is going to the movies and the beach.*
    c. *Richard is rich and famous.*
    d. *Joe is our driver and our guide.*

4. Select the rule of logical equivalence which directly (in one step) transforms statement (1) into statement (2).
    (1) Louise is the teacher or Linda is the teacher.
    (2) If Louise is not the teacher, then Linda is the teacher.
    a. "Not (p or q)" is equivalent to "not p and not q."
    b. "p or q" is equivalent to "if not p, then q."
    c. "If p, then q," is equivalent to "not p or q."
    d. None of the above

5. Select the rule of logical equivalence which directly (in one step) transforms statement (1) into statement (2).
    (1) If you win, then I lose.
    (2) You do not win or I lose.
    a. "If p, then q" is equivalent to "if not q, then not p."
    b. "Not (not p)" is equivalent to "p."
    c. "If p, then q" is equivalent to "not p or q."
    d. None of the above

6. Select the statement that is logically equivalent to: *It is not true that both Johnson and Eisenhower were Democrats.*
    a. *If Johnson was a Democrat, then Eisenhower was not a Democrat.*
    b. *Johnson was not a Democrat and Eisenhower was not a Democrat.*
    c. *Johnson was a Democrat and Eisenhower was a Democrat.*
    d. *Johnson was a Democrat or Eisenhower was not a Democrat.*

7. Select the statement that is **not** logically equivalent to: *It is not true that both the Rolling Stones and De Morgan's Laws are rock groups.*
    a. *The Rolling Stones are not a rock group or De Morgan's Laws are not a rock group.*
    b. *If the Rolling Stones are a rock group, then De Morgan's Laws are not a rock group.*
    c. *If De Morgan's Laws are a rock group, then the Rolling Stones are not a rock group.*
    d. *The Rolling Stones are not a rock group and De Morgan's Laws are not a rock group.*

8. Select the statement that is logically equivalent to: *If the play is <u>Othello</u>, then Shakespeare is the author.*
    a. *If Shakespeare is the author, then the play is <u>Othello</u>.*
    b. *If the play is not <u>Othello</u>, then Shakespeare is not the author.*
    c. *If Shakespeare is not the author, then the play is not <u>Othello</u>.*
    d. *The play is <u>Othello</u> and Shakespeare is not the author.*

9. Select the statement that is **not** logically equivalent to:
   If the road is not repaired, then traffic problems will occur.
   a. *The road must be repaired or traffic problems will occur.*
   b. *Traffic problems will occur or the road is repaired.*
   c. *If traffic problems occur, the road was not repaired.*
   d. *If traffic problems do not occur, the road was repaired.*

10. Select the statement that is logically equivalent to: *A major dam on the upper Nile River must exist or the lower Nile will overflow its banks each year.*
    a. *If a major dam on the upper Nile River does exist, then the lower Nile will not overflow its banks each year.*
    b. *If no major dam on the upper Nile River exists, the lower Nile will overflow its banks each year.*
    c. *If the lower Nile overflows its banks each year, then a major dam on the upper Nile River does not exist.*
    d. *A major dam on the upper Nile River exists and the lower Nile does not overflow its banks each year.*

11. Select the statement that is not logically equivalent to: *Angela takes physics or chemistry.*
    a. *If Angela does not take physics, then she takes chemistry.*
    b. *If Angela does not take chemistry, then she takes physics.*
    c. *It is false that Angela does not take chemistry and does not take physics.*
    d. *If Angela takes chemistry, then she does not take physics.*

Review Competencies

12. Use the following conditional statement:
    If the sun is shining, then I will play tennis.
    Express each of the following.
    a. the converse
    b. the inverse
    c. the contrapositive

13. Express negations for:
    a. Joan is a good mathematician or Ruth is not a computer programmer.
    b. Sally is not obnoxious and Alice is not friendly.

14. Multiply: $4\frac{2}{3} \cdot 2\frac{3}{4}$

15. 20% of what number is 60?

16. Divide: $1 \div 0.05$

17. Evaluate: $10 \div 5 \cdot 3 \div 2$

18. If 160 is increased by 75% of itself, what is the result?

19. Each month Benjamin spends 25% of his income on food, 35% on housing, 10% on his car, and the rest on clothing. One particular month Benjamin spent $315 on clothing. What was his income that month?

It is now possible to find the negation of a simple statement, a conjunction, or a disjunction. One of **De Morgan's Laws** will now explain the negation of a conditional statement. Recall the following equivalence:

$$p \rightarrow q \equiv \sim p \vee q$$

By this equivalence, the negation of $p \rightarrow q$ will correspond to the negation of $\sim p \vee q$ which is $\sim(\sim p \vee q)$. By one of **De Morgan's Laws** we know that:

$$\sim(\sim p \vee q) \equiv p \wedge \sim q$$

Thus, we obtain the following **negation** of $p \rightarrow q$:

$$\sim(p \rightarrow q) \equiv p \wedge \sim q$$

Example

Let p be: *I work hard.*   Let q be: *I will succeed.*

$p \rightarrow q$ is: *If I work hard, then I will succeed.*

The negation $(p \wedge \sim q)$ is: *I work hard and I will **not** succeed.*

Problem 9

The city council of a northern city asserted that: *If it snows, then at least one major road will remain open.*

The council was wrong. What can we conclude?

Solution

Since the council was wrong, we can conclude the negation of their statement.

Let p be: *It snows.*

Let q be: *At least one major road will remain open.*

Since the negation of "at least one" (which means "some") is "no," the negation $(p \wedge \sim q)$ is:

It snows and no major road remains open.

Equivalently, we can conclude that:

It snows and all major roads are closed.

We can now summarize the negation of various statements in logic.

Statement	Negation
p	$\sim p$
$\sim p$	p
$p \wedge q$	$\sim p \vee \sim q$ or $p \rightarrow \sim q$
$p \vee q$	$\sim p \wedge \sim q$
$p \rightarrow q$	$p \wedge \sim q$

The final negation in this list forms the basis of saying that the negation of "all A are B" is "some A are not B." All A are B means that:

> *If x is in A, then x is in B.*
> Let p be: x is in A.
> Let q be: x is in B.

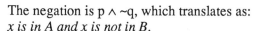

The negation is p ∧ ~q, which translates as:
x is in A and x is not in B.
This means that something is in A and not in B. Thus, the negation of "all A are B" is "some A are not B."

Problem Set 4.5

1. Express the negations of:
   a. *If Barbara wins the election, then I will be sorry.*
   b. *If you speak loudly, then I can hear you.*
   c. *If it is a nice day, then I will go to the beach.*
   d. *If John is a criminal, then he will be punished.*
   e. *If Bill is 18, then he is eligible to vote.*

2. Suppose that each statement listed below turns out to be false. What can we conclude in each case?
   a. *If a person studies Latin, then that person has an excellent vocabulary.*
   b. *If it's hot, then at least one of the city's pools will be open.*
   c. *If both Mary and Ethel star, the musical is a success.*
   d. *If there is a lottery, then all schools will receive increased funding.*

Review Competencies

3. Write equivalent disjunction statements for:
   a. *If Joe does not arrive in time, then he will miss the show.*
   b. *If Scott does not get married, then he will be very lonely.*

4. Write equivalent conditional statements for each of the following disjunctions.
   a. *Liz is the nominee or Millie will resign from the committee.*
   b. *Jack is not going on vacation this year or Jim will travel to Europe.*

5. Write the converse of: *If she speaks, everyone will listen.*

6. Express negations for:
   a. *All mathematics books are exciting.*
   b. *Don is a stable person and Ron is not even-tempered.*

7. What is 40% of 60? 8. Express $4\frac{1}{8}$ as an improper fraction.

9. Evaluate: $12 - 2 \cdot 3 + 4 \div 2$

10. Express 5% as a decimal.

11. The formula for finding the value (V) of an investment (P) after one year at a given rate (r) is $V = P(1 + r)$. If $5,500 is invested at 12%, determine the value after one year.

12. A store received 200 containers of juice to be sold by April 1. Each container cost the store $0.75 and sold for $1.25. The store signed a contract with the manufacturer in which the manufacturer agreed to a $0.50 refund for every container not sold by April 1. If 150 containers were sold by April 1, how much profit did the store make?

Section 5 The Structure of Certain Valid and Invalid Arguments

> **Definition**
> A statement that is true in all cases is a **tautology**.

Example

Consider the statement: $p \lor \sim p$. Let us construct a truth table for this compound statement.

step 1

Since the statement is composed only of variations of simple statement p, there are only two cases for this table.

        p can be true or false.

p
T
F

step 2

Construct another column of values for $\sim p$. Complete the $\sim p$ column by negating the values in the p column.

p	$\sim p$
T	F
F	T

step 3

Construct a column for $p \lor \sim p$. Complete this column by applying the disjunction to the values in the p column with those in the $\sim p$ column.

p	$\sim p$	$p \lor \sim p$
T	F	T
F	T	T

The truth table verifies that $p \lor \sim p$ is a tautology because the statement is true in all cases.

Let p represent: *You live in Florida.*
A tautology is:

    You live in Florida or you do not live in Florida. ($p \lor \sim p$)
This statement is true in all cases.

Many of the equivalences and negations that we previously discussed can be thought of as tautologies. For example, consider:

$$\sim(p \land q) \equiv \sim p \lor \sim q$$

Both $\sim(p \wedge q) \to (\sim p \vee \sim q)$ and $\sim(p \wedge q) \leftrightarrow (\sim p \vee \sim q)$ are tautologies. If we show the truth table for $\sim(p \wedge q) \to (\sim p \vee \sim q)$, we have:

p	q	$p \wedge q$	$\sim(p \wedge q)$	$\sim p$	$\sim q$	$\sim p \vee \sim q$	$\sim(p \wedge q) \to (\sim p \vee \sim q)$
T	T	T	F	F	F	F	T
T	F	F	T	F	T	T	T
F	T	F	T	T	F	T	T
F	F	F	T	T	T	T	T

Since the statement in the last column is true in all cases, it is a tautology.

Problem Set 5.1

1. Construct truth tables for each of the following and determine whether they are tautologies.
   a. $p \wedge \sim p$ b. $q \vee \sim q$
   c. $\sim(p \vee q) \to (\sim p \wedge \sim q)$
   d. $(p \to q) \to (\sim q \to \sim p)$
   e. $\sim(p \to q) \to (p \wedge \sim q)$
   f. $(p \to q) \to (\sim p \vee q)$

Review Competencies

2. Select the statement that is **not** logically equivalent : *If Pilar is in Soweto, then she is in South Africa.*
   a. *If Pilar is not in South Africa, then she is not in Soweto.*
   b. *Pilar is not in Soweto unless she is in South Africa.*
   c. *If Pilar is in South Africa, then she is in Soweto.*
   d. *Unless Pilar is in South Africa, she is not in Soweto.*

3. Select the statement that is logically equivalent to: *Janine plays tennis or basketball.*
   a. *If Janine plays tennis, then she does not play basketball.*
   b. *If Janine does not play tennis, then she plays basketball.*
   c. *If Janine plays basketball, then she does not play tennis.*
   d. *It is not true that Janine does not play tennis or does not play basketball.*

4. Express negations for:
   a. *No foods are harmful to the body.*
   b. *The sun is not shining and I am going to the beach.*
   c. *If she is feeling well, then she will go to the party.*

5. At a masquerade party the judges eliminate $\frac{1}{4}$ of the eligible contestants after each half hour. If 256 contestants were present at the start of the party, how many would still be eligible for a prize after two hours?

6. A plant nursery offers free delivery on orders of $50 or more to one address. The nursery charges $5 for deliveries under $50. A delivery truck left with three palms for $100 each, four oak trees for $60 each, and six ficus trees for $40 each. Each tree was delivered to a different address. If the delivery person can keep 12% of all money collected, how much was kept after all trees were delivered?

No study of logic can be complete without a discussion of the principles involved in correct reasoning. In this section we shall examine several forms of reasoning.

Definitions

Forms of reasoning are known as **arguments**. Arguments contain statements assumed to be true which are called **premises**. Arguments also contain statements called **conclusions** which are derived from the premises. An argument is said to be **valid** if the conclusion follows directly from the premises. More formally, an argument is valid when the conjunction of the premises (all given statements connected by "and") implying the conclusion forms a tautology. An argument that is invalid is called a **fallacy** or **fallacious argument**.

Most arguments that we shall consider consist of two given premises and a conclusion. These statements can be arranged as shown below.

given premises	Statement 1 (is true)
	Statement 2 (is true)
conclusion	Statement 3

It is important to remember that valid does not mean "true," and invalid does not mean "false." The conclusion of an argument generally follows such words as "so," "therefore," "thus," and "hence." Mathematicians use the symbol $\therefore$ to represent these words that have the effect of emphasizing the conclusion.

Example

Consider the following argument:

If Joan is ill, then she will miss school. Joan is certainly ill.
So, *Joan will miss school.*

This argument can be arranged formally as:

premise	If Joan is ill, then she will miss school.
premise	<u>Joan is ill.</u>
conclusion	Joan will miss school.

This argument contains two premises which imply its conclusion. The argument is **valid** because the conclusion follows directly from the premises. If the two premises are assumed true, then this particular conclusion must be true also.

It is possible to express the argument of the last example in symbols.

Let p be: *Joan is ill.*

Let q be: *She will miss school.*

The first premise may be expressed as: $p \rightarrow q$

The second premise may be expressed as: p

The conclusion may be expressed as: q

Valid Form 1

An argument is **valid** (non-fallacious) if it appears in the form:

premise	$p \rightarrow q$
premise	p
conclusion	$\therefore \ q$

This is the case because $[(p \rightarrow q) \wedge p] \rightarrow q$ is a tautology.

This form for a valid argument is called **modus ponens** or the **Law of Detachment**. Modus ponens is derived from the Latin and means "a manner of affirming." What is being affirmed is the hypothesis of the conditional statement in the first premise.

You will be asked to verify the tautology of Valid Form 1 in the next problem set by demonstrating, through tables, that the conditional statement is true in all cases.

Example

Consider the following argument:

If you eat too much you will get fat. But you are not getting fat.
So, you did not eat too much.

This argument can be arranged as:

premise	If you eat too much, you will get fat.
premise	You are not getting fat.
conclusion	You did not eat too much.

The argument contains two premises which imply the conclusion. The argument is **valid** because the conclusion follows directly from the premises. If you assume the first premise is true and also assume the second premise is true, then you must accept the conclusion as being true.

Let p be: *You eat too much.*
Let q be: *You will get fat.*

The first premise may be expressed as: $p \rightarrow q$
The second premise may be expressed as: $\sim q$
The conclusion may be expressed as: $\sim p$

Valid Form 2

An argument is **valid** if it appears in the form:

premise	$p \rightarrow q$
premise	$\sim q$
conclusion	$\therefore \sim p$

This is the case because $[(p \rightarrow q) \wedge \sim q] \rightarrow \sim p$ is a tautology.

This form for a valid argument is called **modus tollens** or the **Law of Contraposition**. Modus tollens, also derived from the Latin, means "a manner of denying." What is being denied is the conclusion of the conditional statement in the first premise. Also, notice the similarity of this form to the contrapositive of a conditional statement. Recall that the contrapositive of a conditional statement is equivalent to the original statement.

You will be asked to verify the tautology of Valid Form 2 in the next problem set.

Example

Consider the following argument:

If a polygon is a triangle, then it has three sides.
If a polygon has three sides then it has three angles.
Thus, if a polygon is a triangle, then it has three angles.

This argument is **valid** because the conclusion follows directly from the premises. The argument can be arranged as:

premise	If a polygon is a triangle, then it has three sides.
premise	If a polygon has three sides, then it has three angles.
conclusion	If a polygon is a triangle, then it has three angles.

Let p be: *A polygon is a triangle.*
Let q be: *A polygon has three sides.*
Let r be: *A polygon has three angles.*

The first premise may be expressed as: $p \rightarrow q$
The second premise may be expressed as: $q \rightarrow r$
The conclusion may be expressed as: $p \rightarrow r$

Valid Form 3

An argument is **valid** if it appears in the form:
premise	$p \rightarrow q$
premise	$q \rightarrow r$
conclusion	$\therefore \ p \rightarrow r$

This is the case because $[(p \rightarrow q) \wedge (q \rightarrow r)] \rightarrow (p \rightarrow r)$ is a tautology.

This tautology of Valid Form 3 is called the **transitive tautology**. This form of valid argument is called a **syllogism**. You will be asked to verify the tautology of Valid Form 3 in the next problem set.

Example

Consider the following argument:

If you know logic, you will appreciate Sherlock Holmes. It's obvious you do not know logic. So, *you will not appreciate Sherlock Holmes.*

This argument can be arranged as:

premise	If you know logic, you will appreciate Sherlock Holmes.
premise	<u>You do not know logic.</u>
conclusion	You will not appreciate Sherlock Holmes.

This argument is **invalid** because this conclusion does not necessarily follow from the two premises. It is possible to assume the truth of the two premises and find the conclusion is false. If you do not know logic, it does not necessarily follow that you will not appreciate Sherlock Holmes. The first premise merely states what will happen if you know logic. It says nothing about what will happen if you do not know logic. You certainly could still appreciate Sherlock Holmes.

Let p be: *You know logic.*
Let q be: *You will appreciate Sherlock Holmes.*
The first premise may be expressed as: $p \rightarrow q$
The second premise may be expressed as: $\sim p$
The conclusion may be expressed as: $\sim q$

Invalid Form 1

An argument is **invalid** (fallacious) if it appears in the form:

premise	$p \rightarrow q$
premise	$\underline{\sim p}$
conclusion	$\therefore \ \sim q$

This is the case because $[(p \rightarrow q) \wedge \sim p] \rightarrow \sim q$ is not a tautology.

This invalid argument is known as the **fallacy of the inverse**. Can you see the inverse of the conditional statement ($\sim p \rightarrow \sim q$) suggested by the argument? Recall that the inverse of a conditional statement is not equivalent to the conditional statement.

You will be asked to verify that the statement of Invalid Form 1 is not a tautology in the next problem set.

Example

Consider this argument:
If you like math, then you are a nut, and you certainly are a nut.
Therefore, you like math.

This argument can be arranged as:

premise	If you like math, then you are a nut.
premise	<u>You are a nut.</u>
conclusion	You like math.

Let p be: *You like math.*
Let q be: *You are a nut.*
The first premise may be expressed as: $p \rightarrow q$
The second premise may be expressed as: q
The conclusion may be expressed as: p

This argument is **invalid**. The conclusion does not necessarily follow from the premises. If you are indeed a nut, one is not forced to accept the particular conclusion that you like math. There could certainly be other possible conclusions.

Invalid Form 2

An argument is **invalid** (fallacious) if it appears in the form:

premise	$p \rightarrow q$
premise	q
conclusion	$\therefore \ p$

This is the case because $[(p \rightarrow q) \wedge q] \rightarrow p$ is not a tautology.

This invalid argument is known as the **fallacy of the converse.** Can you see the converse of a conditional statement $(q \rightarrow p)$ suggested by this argument? Recall that the converse of a conditional statement is not equivalent to the conditional statement.

You will be asked to verify that the statement of Invalid Form 2 is not a tautology in the next problem set.

Example

Let us now consider an argument that does not contain any conditional statements in the premises or the conclusion.

> *The radiator is clogged. I know this because the radiator is leaking or it is clogged, and the radiator definitely is not leaking.*

Notice that the conclusion appears **first** in the argument. The two premises follow the words "I know this because." In any argument presented in this chapter you should be able to distinguish between the premises and the conclusion since arguments will often be stated in this format.

This argument can be aranged as:

premise	The radiator is leaking or it is clogged.
premise	The radiator is not leaking.
conclusion	The radiator is clogged.

In this argument the first premise limits the possibilities to two. If it is asserted that one of the possibilities is false, then the other must be true. If one discovers that the problem with the car was neither a leaking nor a clogged radiator, the error lies with the first premise. The argument, however, is **valid**. To be valid, the conclusion must assert that the possibility not eliminated by the second premise is true. Again, the validity of the argument can best be seen by expressing the argument symbolically.

Let p be: The radiator is leaking.
Let q be: The radiator is clogged.

The first premise may be expressed as: $p \vee q$
The second premise may be expressed as: $\sim p$
The conclusion may be expressed as: q

Valid Form 4

An argument is **valid** if it appears in the form:

premise	$p \vee q$
premise	$\sim p$
conclusion	$\therefore \quad q$

This is the case because $[(p \vee q) \wedge \sim p] \rightarrow q$ is a tautology.

This argument should suggest an equivalence that has been discussed in a previous section: $p \vee q \equiv \sim p \rightarrow q$.

You will be asked to verify the tautology of Valid Form 4 in the next problem set.

Example

An argument similar to the last example is:

This tree is a royal palm. I know this because this tree is a royal palm or a coconut palm, and it is not a coconut palm.

This argument may be arranged as:

premise	This tree is a royal palm or a coconut palm.
premise	<u>This tree is not a coconut palm.</u>
conclusion	This tree is a royal palm.

This argument is **valid** for the same reasons as the preceding argument.

Let p be: *This tree is a royal palm.*
Let q be: *This tree is a coconut palm.*

The first premise may be expressed as: $p \vee q$
The second premise may be expressed as: $\sim q$
The conclusion may be expressed as: p

> ### Valid Form 5
>
> An argument is **valid** if it appears in the form:
>
> | | |
> |---|---|
> | **premise** | $p \vee q$ |
> | **premise** | $\underline{\sim q}$ |
> | **conclusion** | $\therefore\ p$ |
>
> This is the case because $[(p \vee q) \wedge \sim q] \rightarrow p$ is a tautology.

You will be asked to verify the tautology of Valid Form 5 in the next problem set.

Example

Consider the following argument :

The stranger is a liar or a thief. The stranger certainly is a liar. Therefore, the stranger is not a thief.

This argument can be arranged as:

premise	The stranger is a liar or a thief.
premise	<u>The stranger is a liar.</u>
conclusion	The stranger is not a thief.

Notice that the information in the first premise does not prohibit the stranger from being **both** a liar and a thief. It simply asserts that the description of the stranger is limited to the two possibilities. Both alternatives could occur together. Thus, this argument is **invalid** since finding one possibility true does not eliminate the other.

Let p be: *The stranger is a liar.*
Let q be: *The stranger is a thief.*
The first premise may be expressed as: p ∨ q
The second premise may be expressed as: p
The conclusion may be expressed as: ~q

> ### Invalid Form 3
>
> An argument is **invalid** if it appears in the form:
>
> | | |
> |---|---|
> | **premise** | p ∨ q |
> | **premise** | p |
> | **conclusion** | ∴ ~q |
>
> This is the case because [(p ∨ q) ∧ p] → ~q is not a tautology.

You will be asked to verify that this is not a tautology in the next problem set.

Example

An argument similar to the last example is:

> *John is a student or a soccer player. John definitely is a soccer player. So, John is not a student.*

This argument may be arranged as:

premise	John is a student or a soccer player.
premise	John is a soccer player.
conclusion	John is not a student.

This argument is **invalid** by the same reasoning used in the previous argument.

Let p be: *John is a student.*
Let q be: *John is a soccer player.*
The first premise may be expressed as: p ∨ q
The second premise may be expressed as: q
The conclusion may be expressed as: ~p

> ### Invalid Form 4
>
> An argument is **invalid** if it appears in the form:
>
> | | |
> |---|---|
> | **premise** | p ∨ q |
> | **premise** | q |
> | **conclusion** | ∴ ~p |
>
> This is the case because [(p ∨ q) ∧ q] → ~p is not a tautology.

You will be asked to verify that this is not a tautology in the next problem set.

Ideally, you should know all nine forms of the arguments previously discussed. However, at the very least, you should know the five valid forms. For our purposes, any argument that does not conform to one of the valid forms will be invalid. It should be emphasized that merely knowing the forms is not enough. You also need to be able to symbolize any given argument correctly.

To Summarize:

An argument is **valid (non-fallacious)** if it appears in any of the following forms:

1.	2.	3.	4.	5.
$p \to q$	$p \to q$	$p \to q$	$p \lor q$	$p \lor q$
p	$\sim q$	$q \to r$	$\sim p$	$\sim q$
$\therefore q$	$\therefore \sim p$	$\therefore p \to r$	$\therefore q$	$\therefore p$

An argument is **invalid (fallacious)** if it appears in any of the following forms:

6.	7.	8.	9.
$p \to q$	$p \to q$	$p \lor q$	$p \lor q$
$\sim p$	q	p	q
$\therefore \sim q$	$\therefore p$	$\therefore \sim q$	$\therefore \sim p$

It must be emphasized that the **validity of an argument depends only on its form** and not whether the premises and conclusions are true or false statements in the real world. An argument may not be valid even though its conclusion is true. The following **fallacious** argument illustrates this:

premise	If a number is divisible by 10, then it is divisible by 5.	$p \to q$
premise	20 is divisible by 5.	q
conclusion	20 is divisible by 10.	$\therefore p$

It is certainly true that 20 is divisible by 10. The problem is that an invalid argument with true premises can result in a conclusion that might be true, but might be false. Change the number in question from 20 to 15 and it becomes more evident that the argument is fallacious.

premise	If a number is divisible by 10, then it is divisible by 5.	$p \to q$
premise	15 is divisible by 5.	q
conclusion	15 is divisible by 10.	$\therefore p$

Problem Set 5.2

1. A valid argument is one that is based upon a tautology. Construct truth tables for each of the following statements and determine whether or not they are tautologies.

   a. $[(p \rightarrow q) \wedge p] \rightarrow q$ b. $[(p \rightarrow q) \wedge q] \rightarrow p$

   c. $[(p \rightarrow q) \wedge (q \rightarrow r)] \rightarrow (p \rightarrow r)$

   Hint for c. For the truth table involving three statements (p, q, and r), complete each of the following columns.

p	q	r	$(p \rightarrow q)$	$(q \rightarrow r)$	$(p \rightarrow r)$
T	T	T			
T	T	F			
T	F	T			
T	F	F			
F	T	T			
F	T	F			
F	F	T			
F	F	F			

   d. $[(p \rightarrow q) \wedge \sim q] \rightarrow \sim p$ e. $[(p \rightarrow q) \wedge \sim p] \rightarrow \sim q$

2. One should not infer that the fallacies that we have discussed are the only possible forms that yield arguments that are invalid. Construct truth tables for each of the following statements and determine if they form invalid arguments.

   a. $[(p \rightarrow q) \wedge (q \rightarrow r)] \rightarrow (r \rightarrow p)$

   b. $[(p \rightarrow q) \wedge p] \rightarrow \sim q$ c. $[(p \rightarrow q) \wedge q] \rightarrow \sim p$

   d. $[(p \rightarrow q) \wedge \sim p] \rightarrow q$ e. $[(p \rightarrow q) \wedge \sim q] \rightarrow p$

3. Determine which of the following arguments are valid.

   a. If Engelbert is an adolescent, then Engelbert has pimples. Engelbert is an adolescent. Therefore, Engelbert has pimples.

   b. If the Jaybirds beat their opponent, they will win their division. The Jaybirds will not win their division. So, the Jaybirds did not beat their opponent.

   c. If Miss Fink watches too much television, then Miss Fink will go crazy. Miss Fink has definitely gone crazy. Thus, Miss Fink has watched too much television.

   d. 17 is not divisible by 10. I know this because if a number is divisible by 10, then it is divisible by 5, and 17 is not divisible by 5.

   e. You will be cool if you like logic. But you do not like logic. So, you will not be cool.

f. Mr. Clutz is a good teacher. I know this because if a man is a good teacher, he will be the talk of the school, and Mr. Clutz is certainly the talk of the school.

g. If you make an A, then you have earned it. If you have earned it, then the teacher will give you a gold star. Therefore, if you make an A, then the teacher will give you a gold star.

h. If Brenda is intelligent, then she earns A's in her courses. If she earns A's in her courses, then she feels satisfied. So, if Brenda feels satisfied, then she is intelligent.

i. If today is Tuesday, two days from today will be Thursday. But today is not Tuesday. So, two days from today will not be Thursday.

j. If logic makes you sick, then you are weird, and you are weird. As a result, logic must make you sick.

k. If you are tall, then you will have more fun and if you have more fun, then you will get in trouble. So, if you do not get in trouble, then you are not tall.

l. There is less noise when we close the door. There is less noise. Thus, we have closed the door.

4. Premises and conclusions can appear in a variety of ways in a given argument. Validity can still be determined by rearranging the statements and comparing the argument to one of the basic forms for validity. Determine the validity for each of the following.

a. The defendant is not guilty because if the defendant is guilty the shoe found at the scene of the crime would fit him. This shoe is much too large.

b. Chickens can talk if ducks can sing. Thus, ducks can't sing if chickens can't talk.

5. A premise is missing from each of the following arguments. Determine the premise which will make the argument valid.

a. _
   Joe is an outstanding mathematician.
   Joe should excel in engineering.

b. _ _ _ _ _ _ _ _ _ _ _ _ _ _ _ _ _ _
   I do not find this course enjoyable.
   This course is not meaningful.

c. _
   If one is mature, one is responsible.
   If one is socially intelligent, then one is responsible.

6. The following arguments are expressed in paragraph form. The conclusion is omitted. If the premises imply a valid conclusion, state the conclusion. Indicate also if the premises do not yield a valid conclusion.

   a. The administration can claim in the next election to be the party of fiscal responsibility if it succeeds in balancing the budget. It now appears certain that the administration will not balance the budget.

   b. If the enemy were planning to invade the island, large numbers of troops would be gathering on the coast. An informant reported overhearing a conversation among some enemy soldiers that large numbers of troops had recently been gathering on the coast.

7. Verify that each of the following is a tautology.

   a. $[(p \vee q) \wedge \sim p] \to q$ b. $[(p \vee q) \wedge \sim q] \to p$

8. Determine whether the following arguments are valid.

   a. John must attend class or he will be dropped from the course. John is not attending class. Therefore, John will be dropped from the course.

   b. Ron has received numerous speeding tickets or Ron is a slow driver. Ron is not a slow driver. So, Ron has not received any speeding tickets.

   c. I will go to college. I know this because I will go to college or get a job, but there is no way that I will get a job.

   d. I have tail feathers or I am a turkey. I certainly don't have tail feathers. So, I am not a turkey.

   e. Rick did not commit the crime or he was at the scene of the crime. Rick was not at the scene of the crime. So, Rick did not commit the crime.

9. Supply a conclusion that will make the following arguments valid.

   a. I will work or I will play.
      I will not play.
      Therefore, . . .

   b. The Dolphins finished in first place in their division or the Dolphins did not enter the playoffs. The Dolphins entered the playoffs. Therefore, . . .

   c. $a = 0$ or $b = 0$
      $a = 3$
      Therefore, . . .

10. Verify that each of the following is not a tautology.
    a. $[(p \lor q) \land p] \to \sim q$ b. $[(p \lor q) \land q] \to \sim p$

11. Determine whether the following arguments are valid.
    a. I have a head cold or an upset stomach. I have a head cold. So, I don't have an upset stomach.
    b. The car battery is not dead. I know this because the car battery is dead or the gas tank is empty. There is no doubt that the gas tank is empty.

12. Two valid arguments are given below. Select the symbolic form of the reasoning pattern used in both arguments.
    a. Mitch is a student or a soccer player. He certainly is not a student. Thus, Mitch must be a soccer player.
    b. Lory's car is a Trans Am. I know this because Lory's car is a Firebird or a Trans Am, and Lory's car definitely is not a Firebird.

i.	ii.	iii.	iv.	v.
$p \lor q$	$p \lor q$	$p \lor q$	$p \lor q$	$p \lor q$
p	$\sim p$	q	$\sim p$	p
$\therefore \sim q$	$\therefore q$	$\therefore \sim p$	$\therefore \sim q$	$\therefore q$

13. Select the symbolic form for the following argument.
    If Debbie takes vitamins and exercises, then she will be healthy. Debbie is not taking vitamins. Therefore, if Debbie exercises, then she will be healthy.

a.	b.	c.	d.
$(q \lor r) \to p$	$(q \land r) \to p$	$(q \land r) \to p$	$(q \lor r) \to p$
$\sim r$	$\sim r$	$\sim q$	$\sim q$
$\therefore q \to p$	$\therefore q \to p$	$\therefore r \to p$	$\therefore r \to p$

14. Select the symbolic form for the following argument.
    If I study for two hours, I get tired or irritable. I am studying for two hours and not getting irritable. Therefore, I am getting tired.

a.	b.	c.	d.
$r \to (p \lor q)$	$r \to (p \lor q)$	$r \to (p \land q)$	$r \to (p \land q)$
$r \land \sim p$	$r \land \sim q$	$r \lor \sim p$	$r \lor \sim q$
$\therefore \sim q$	$\therefore p$	$\therefore q$	$\therefore p$

15. Many arguments that initially appear to be long when expressed in English are quite compact when written in symbolic notation. Symbolize each of the following arguments and then select the argument/s that is/are not valid.

    a. Don is not afflicted with the psychopathology of schizophrenia. I know this because Don is an artistic genius or afflicted with the psychopathology of schizophrenia, and he certainly is an artistic genius.

    b. If a subway system is not in operation in Mexico City, then automobile traffic causes hazardous air pollution. On May 11, 1984, the subway system was not in operation due to a strike by the city's transit workers. Consequently, on May 11, 1984, automobile traffic caused hazardous air pollution in Mexico City.

    c. If a subway system is not in operation in Mexico City, automobile traffic causes hazardous air pollution. On March 1, 1967, automobile traffic caused hazardous air pollution in Mexico City. Consequently, on March 1, 1967, a subway system was not in operation in Mexico City.

16. Given that: If the figure is a square, then the figure is a quadrilateral.
    If the figure is a quadrilateral, then the figure is a polygon.
    Determine which of the following conclusions can be validly deduced from these premises.

    a. If the figure is a polygon, then the figure is a square.
    b. If the figure is not a square, then the figure is not a polygon.
    c. If the figure is not a polygon, then the figure is not a square.
    d. If the figure is a quadrilateral, then the figure is a square.

17. Given that: If the animal is a dog, it barks.
    If the animal barks, it is a good guardian.
    Morris is not a good guardian.
    Determine which of the following conclusions can be validly deduced from these three premises.

    a. Morris is a dog.   b. Morris barks.
    c. Morris is not a dog.   d. None of these conclusions can be validly deduced.

Review Competencies

18. Select the statement that is logically equivalent to:
    If the temperature drops below 20°, then my orange tree will die.

    a. If my orange tree dies, then the temperature dropped below 20°.
    b. If the temperature did not drop below 20°, then my orange tree did not die.
    c. If my orange tree did not die, then the temperature did not drop below 20°.
    d. The temperature dropped below 20° and my orange tree did not die.

19. Which option in the preceding problem represents the negation of the original statement?

20. Select the statement that is **not** logically equivalent to:
    Janine plays tennis or basketball.
    a.   If Janine does not play tennis, then she plays basketball.
    b.   If Janine does not play basketball, then she plays tennis.
    c.   If Janine plays tennis, then she does not play basketball.
    d.   It is not true that Janine does not play tennis and does not play basketball.

21. After several tryouts, 20% of a group was eliminated. At that point, 32 people remained. How many people were present at first?

22. One-tenth of a person's salary is spent for clothing, $\frac{1}{3}$ for food, and $\frac{1}{5}$ for rent. What fraction of the salary is left?

In the introduction to this chapter we stated that one must be able to avoid arriving at conclusions hastily without considering all the evidence. One means of drawing conclusions that are warranted is through the use of the valid forms of arguments, as illustrated in the problems that follow.

Problem 1   Draw a conclusion that will make the following argument valid.
            If all people take exams honestly, then no supervision is needed.
            Some supervision is needed.

Solution   Let p be: *All people take exams honestly.*
            Let q be: *No supervision is needed.*
            The form of the argument is:
            $p \rightarrow q$ If all people take exams honestly, then no supervision is needed.
            $\underline{\sim q}$ Some supervision is needed. (Recall that the negation of "no" is "some.")
            $\therefore \sim p$

            The conclusion ~p makes the argument valid. Since the negation of "all" is "some…not," we can conclude that:
            Some people do not take exams honestly.

Problem 2

If one studies mathematics, then one will be logical. If one studies English, then one will have an excellent vocabulary. Judy studies mathematics and not English. Thus,

a. Judy will be logical but not have an excellent vocabulary.
b. Judy will be logical.
c. Judy will not have an excellent vocabulary.
d. None of the above is warranted.

Solution

Use letters to represent the simple statements.
Let p represent: *One studies mathematics.*
Let q represent: *One will be logical.*
Let r represent: *One studies English.*
Let s represent: *One will have an excellent vocabulary.*

We now arrange the premises to form two separate arguments.

$p \rightarrow q$ (If one studies mathematics, one will be logical.) $r \rightarrow s$ (If one studies English, one will have an excellent vocabulary.)

p (Judy studies mathematics.)   ~r (Judy does not study English.)

In the first argument a valid conclusion is q since

$$p \rightarrow q$$
$$\underline{p}$$
$$\therefore q$$

is a valid form. Thus, Judy will be logical.

In the second argument ~s is not a valid conclusion since

$$r \rightarrow s$$
$$\underline{\sim r}$$
$$\therefore \sim s$$

is not a valid form. We cannot validly conclude that Judy will not have an excellent vocabulary. The valid conclusion, then, corresponds to choice b. (Judy will be logical.)

Problem 3   We will go to the beach or we will go to the movies. We will not go to the movies. If it snows, we will not go to the beach. What can we logically (validly) deduce from these premises?

Solution   Use letters to represent the simple statements.
Let p represent: *We will go to the beach.*
Let q represent: *We will go to the movies.*
Let r represent: *It snows.*
By the first two premises, we have:

$p \lor q$ (We will go to the beach or the movies.)
$\underline{{\sim}q}$ (We will not go to the movies.)

A valid conclusion is p, we will go to the beach.

By the last premise, we have $r \rightarrow {\sim}p$.   (If it snows, we will not go to the beach.)
We now combine this last premise with the valid conclusion p.
We now have:

$r \rightarrow {\sim}p$ (If it snows, we will not go to the beach.)
$\underline{p}$ (We will go to the beach.)

A valid conclusion is ~r. We can deduce: It is not snowing.

Since we have two true conclusions, we can conclude:
We will go to the beach *and* it is not snowing.

Problem 4 Draw a conclusion that will make the following argument valid.
If Sarah and Tim play, the team wins. The team lost on Tuesday and Tim played.

Solution You may be able to immediately conclude that Sarah did not play. After all, had she played, the team would have won and we are told that the team lost. The conclusion (Sarah did not play) can also be shown to be valid using valid forms.

Let p be: Sarah plays.
Let q be: Tim plays.
Let r be: The team wins.
We now translate from English into symbolic language.

$(p \wedge q) \to r$ (If Sarah and Tim play, the team wins.)

$\underline{\sim r \qquad\qquad}$ (The team lost on Tuesday.)

$\therefore \sim(p \wedge q)$ (It is not true that both Sarah and Tim played.)

Notice that the valid form that we used could be thought of as:

Our conclusion, $\sim(p \wedge q)$, can be rewritten by one of **De Morgan's Laws**. $\sim(p \wedge q)$ is equivalent to $\sim p \vee \sim q$ (Sarah did not play or Tim did not play.) We still have not incorporated the fact that Tim did play (q). Writing this underneath $\sim p \vee \sim q$, we obtain:

$$\sim p \vee \sim q$$
$$\underline{\qquad q \qquad}$$

Although it might not be obvious that this will lead to a valid conclusion, it will become more evident if we think of $\sim p$ as □ and $\sim q$ as △ . The form of our argument is:

The conclusion that we have boxed, namely $\sim p$, forms a valid argument. The valid conclusion ($\sim p$) translates as: *Sarah did not play.*

Problem Set 5.3

Draw a conclusion that will make each of the following arguments valid.

1. If all people obey rules, then no prisons are needed. Some prisons are needed.
2. If no people smoke cigarettes, then all people save money on health insurance. Some people do not save money on health insurance.
3. If one studies mathematics, then one will be logical. If one studies English, then one will have an excellent vocabulary. Joel is not logical. Joel does not study English.
4. If a person has a knowledge of music, then that person enjoys Mozart. If a person has a knowledge of theater, then that person enjoys Sondheim. Kathy enjoys Mozart and Kathy does not enjoy Sondheim.
5. If a person studies mathematics, then that person is logical. If a person studies English, then that person appreciates language. Angie studies English but not mathematics.
6. If it snows, we read or watch television. It snowed on Friday. We did not read on Friday.
7. It is not true that both Carmen and Marco study biology. Carmen does study biology.
8. If Susan studies and attends lecture, then she does well. Susan is not doing well and she is attending lecture.
9. If a student attends all lectures and does some homework problems, then that student passes the course. Fran failed the course and attended all lectures.
10. If it is Tuesday, then Javier has his philosophy course. When Javier has philosophy, he contemplates the meaning of existence.
11. If the medication is effective, then I will not miss class. If I do not miss class, then I will not fall behind.
12. Living in Florida is a necessary condition for living in Tampa. One lives in North America if one lives in Florida.

Draw two conclusions that will make each of the following arguments valid.

13. If a political system does not embody justice, then that system has no peace and that system must spend a great deal of money building jails. South Africa is a political system that does not embody justice.
14. If a person has good eye-hand coordination, then that person can hit and throw a ball. An outstanding baseball player has good eye-hand coordination.
15. If it rains or snows, then Pierre reads. Pierre did not read on Friday.
16. If a person studies mathematics, then that person is logical. If a person studies English, then that person appreciates language. Samir is not logical, but he did study English.
17. We will study or watch television. We will not study. If we pass the exam, we will not watch television.

Review Competencies

Write the negation of each of the following.
18. *All winter days are cold.*
19. *It is cold and it is not humid.*
20. *Tim will do the homework or will not pass the course.*
21. *If it rains, we will not use the pool.*
22. Which whole number is divisible by 7 and also is a factor of 42?
    a. 84   b. 35   c. 21   d. 9
23. $7\frac{5}{6} - 3\frac{8}{9} =$
24. $\frac{37}{1000} =$
    a. 0.037% b. 0.37% c. 3.7%   d. 37%
25. If 25 is reduced to 20, what is the percent decrease? (Hint: The fraction for percent decrease is the amount of decrease divided by the original amount.)

Section 6   Euler Circles

Let us now consider logical arguments that involve words such as "all," "some," and "no." We will approach these arguments in a different way from those in the previous section.

It is possible to determine the validity of such arguments by using a visual approach known as **Euler circles**. These are attributed to the Swiss mathematician Leonard Euler who used diagrams to study the validity of arguments. In these diagrams, circles are used to indicate relationships of premises to conclusions just as circles are used in Venn Diagrams to visualize set relationships.

Problem 1   Use Euler circles to determine the validity of the following argument.
> *All dogs are faithful animals. Snoopy is a dog.*
> *Therefore, Snoopy is a faithful animal.*

This argument may be stated in the following form:

premise	All dogs are faithful animals.
premise	<u>Snoopy is a dog.</u>
conclusion	Snoopy is a faithful animal.

Solution   **step 1**

The first premise *All dogs are faithful animals* involves the word "all." This statement claims that the set of dogs is included in the set of faithful animals. This situation is diagrammed using Euler circles in the accompanying figure. The "dog" circle is **inside** the "faithful animal" circle to show that all dogs are also faithful animals. The interior of the small circle represents the set of all dogs, and the interior of the large circle represents the set of all faithful animals. Notice that the statement *All dogs are faithful animals* is equivalent to the statement *There are no dogs that are not faithful animals.*

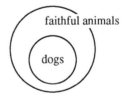

A helpful hint is that whatever word immediately follows the word "all" will represent the set of elements contained in the smaller circle. Sometimes the word "all" is not stated in the premise but only implied. Thus, this first premise *All dogs are faithful animals* could also have been stated merely as *Dogs are faithful animals.*

step 2

By the second premise *Snoopy is a dog*, Snoopy would appear as a point within the circle representing the set of dogs. Notice that there is no other way to draw the circles that would support the given premises.

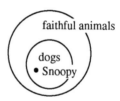

step 3

Finally, the conclusion *Therefore, Snoopy is a faithful animal* can be clearly seen in the diagram. The point representing Snoopy lies within the set of faithful animals. Since the diagram also supports the conclusion, the argument is **valid**. The important consideration is that **if there is only one way to diagram the given premises and the diagram agrees with the conclusion of the argument, then the argument is valid.**

The argument above begins with a general statement *All dogs are faithful animals* and proceeds to a specific instance *Snoopy is a faithful animal.* This argument is an example of **deductive reasoning.** Deductive reasoning proceeds from general statements to specific instances.

Problem 2

Use Euler circles to determine the validity of the following argument.
Dogs are faithful animals, and Fido is a faithful animal. Therefore, Fido is a dog.

This argument may be stated in the following form:

premise	All dogs are faithful animals.
premise	<u>Fido is a faithful animal.</u>
conclusion	Fido is a dog.

Solution

step 1

As in the previous example, circles are used to represent the sets of dogs and faithful animals. The "dogs" circle is inside the "faithful animals" circle because of the first premise *All dogs are faithful animals.*

117

step 2

Using the second premise *Fido is a faithful animal*, a point needs to be placed in the diagram to represent Fido. Two possibilities exist:

i. It could be that Fido should be placed in the circle representing the set of dogs. This possibility would agree with both premises. The diagram for this possibility appears at the right.

ii. However, a second possibility could be that Fido is not a dog, but is still a faithful animal. This possibility also agrees with both premises. The diagram for this possibility appears at the right.

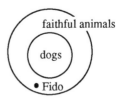

step 3

The conclusion *Therefore, Fido is a dog* agrees with the first possibility, but not the second. **Every possibility agreeing with the premises must agree with the conclusion for an argument to be valid.** Therefore, this argument is **fallacious** (invalid). In short, if there is even one possible diagram that contradicts the conclusion, then the argument is invalid.

Problem Set 6.1

Determine the validity of each argument by using Euler circles.

1. All slobs are revolting and Cecil is a slob. So, Cecil is revolting.

2. Mules are stubborn and Horatio is stubborn. Therefore, Horatio is a mule.

3. Babies are cute and Samantha is cute. So, Samantha is a baby.

4. Clem is an animal. I know this because Clem is an athlete and all athletes are animals.

5. All freshmen are college students. I know this because all college students are doing it and all freshmen are doing it.

6. All clocks keep time accurately. All time-measuring devices keep time accurately. Therefore, all clocks are time-measuring devices.

7. Plants that are native to the Florida Keys can withstand prolonged droughts. The gumbo limbo is native to the Florida Keys. Therefore, the gumbo limbo can withstand prolonged droughts.

Review Competencies

8. Which one of the following is not valid?
   a. If one is a biologist, then one is a scientist. Janet is not a scientist. Consequently, Janet is not a biologist.

b. If a number is divisible by 12, then it is divisible by 6. Sixty is divisible by 12. Thus, 60 is divisible by 6.

c. Jane is taking French or Spanish. Jane isn't taking Spanish. Accordingly, Jane is taking French.

d. If Ian is in Liverpool, he is in England. Ian is in England. Thus, Ian is in Liverpool.

9. Select the conclusion that will make the following argument valid.
   If all people work hard, then no people fail.
   Some people fail.
   a. Some people work hard.
   b. If no people fail, then all people work hard.
   c. All people work hard.
   d. Some people do not work hard.

10. If a person wins on <u>Jeopardy</u>, then that person knows many things. If a person studied Latin, then that person has an excellent vocabulary. Ramon knows many things. Ramon did not study Latin. What can we logically (validly) conclude in this situation?
    i. *Ramon won on <u>Jeopardy</u>.*
    ii. *Ramon does not have an excellent vocabulary.*
    a. i only   b. ii only
    c. Both i and ii   d. Neither i nor ii

11. Select the rule of logical equivalence which directly (in one step) transforms statement (1) into statement (2).
    (1) It is not true that both Atlanta and Haiti are cities.
    (2) If Atlanta is a city, then Haiti is not a city.
    a. $\sim(p \wedge q) \equiv (\sim p \vee \sim q)$
    b. $(p \wedge q) \equiv (p \rightarrow \sim q)$
    c. $\sim(p \vee q) \equiv (p \rightarrow \sim q)$
    d. The correct equivalence rule is not given.

12. Calculators costing $20 each are reduced by 30%. At the reduced rate, what is the cost of 25 calculators?

13. If the numerator of a fraction lies between 12 and 20 inclusively, and the denominator lies between $\frac{1}{4}$ and $\frac{1}{2}$ inclusively, what is the greatest number possible?

14. $6(4 + 3 \div 2 - 8) =$

Problem 3   Use Euler circles to determine the validity of the following argument.
    You are a turkey. I know this because all turkeys are birds, and all birds have legs. You have legs.

This argument may be stated in the following form:

premise	All turkeys are birds.
premise	All birds have legs.
premise	<u>You have legs.</u>
conclusion	You are a turkey.

Solution

step 1

Since the word "all" appears in the first premise *All turkeys are birds* the circle representing turkeys is placed completely **inside** the circle representing birds. The diagram at the right shows this relationship.

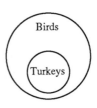

Notice again that the premise *All turkeys are birds* could have been stated simply as *Turkeys are birds*. This premise is also equivalent to *There are no turkeys that are not birds*.

step 2

The second premise *All birds have legs* contains the word "all" and another circle representing "creatures with legs" must completely contain the "birds" circle. The diagram at the right shows these relationships.

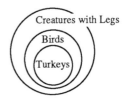

step 3

The third premise *You have legs* means that "you" must be a point in the "creatures with legs" circle. However, this point could have three different positions relative to the circles.

    i.   "You" could be placed in the "creatures with legs" circle outside the other two smaller circles. This situation is pictured at the right.

    ii.   "You" could also be placed in the "birds" circle but outside the "turkeys" circle. This situation is pictured at the right.

    iii.   "You" could be placed in the circle of turkeys. This places the point inside the other two circles. This situation is pictured at the right.

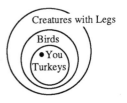

Only this third possibility agrees with the conclusion that *you are a turkey*. The first two possibilities agree with the premises, but contradict the conclusion. The argument is **invalid.** Every possibility supported by the premises must agree with the conclusion for a valid argument.

When **every** possibility justifies the conclusion, a Euler circle demonstration makes the argument valid. When **any one** possibility contradicts the conclusion, a Euler circle demonstration makes the argument fallacious.

Problem 4

Use Euler circles to determine the validity of the following argument.
Jill is either a poet or a journalist. I know this because all poets are writers, all journalists are writers, all writers appreciate language, and Jill is a writer.

This argument may be stated in the following form:

premise	All poets are writers.
premise	All journalists are writers.
premise	All writers appreciate language.
premise	Jill is a writer.
conclusion	Jill is either a poet or a journalist.

Solution

step 1

Since the word "all" appears in the first premise (*All poets are writers*), the circle representing poets is placed completely inside the circle representing writers. The diagram at the right shows this relationship.

step 2

The second premise (*All journalists are writers*) contains the word "all" and another circle representing journalists must be placed completely inside the circle representing writers. As shown in the accompanying diagrams, numerous possibilities exist.

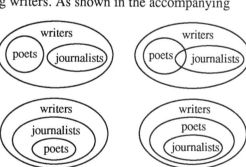

step 3

The third premise (*All writers appreciate language*) again contains the word "all" and the circle representing writers must be placed within another circle representing the set of people who appreciate language. Each of the four possible diagrams in step 2 must show the "writers" circle inside the "appreciate language" circle. This is shown in the accompanying diagrams.

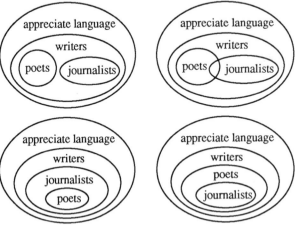

step 4

The last premise (*Jill is a writer*) means that Jill will be shown as a point in the writer circle. We will take the four possible diagrams shown so far and in each diagram we will place Jill in as many different positions as possible within the writer circle.

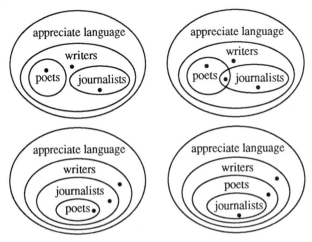

If just one possible diagram does not illustrate the conclusion (*Jill is either a poet or a journalist*), this means that the conclusion might be false relative to the premises, and therefore the argument is invalid. If you focus on the one possible diagram shown at the right, we see that in this case Jill is a writer but is neither a poet nor a journalist. This diagram contradicts the conclusion and the argument is **invalid**.

On the other hand, if you study the many diagrams shown in step 4, observe that in every case both the circles representing poets and journalists fall within the "appreciate language" circle. Furthermore, Jill always is contained within this outtermost circle. Consequently, we can logically (validly) conclude that:

        All poets appreciate language.
        All journalists appreciate language.
        Jill appreciates language.

Problem 5   Use Euler circles to determine the validity of the following argument.
        All insects have six legs and horses have four legs. Accordingly, horses are not insects.

This argument may be stated in the following form:

premise	All insects have six legs.
premise	<u>Horses have four legs.</u>
conclusion	Horses are not insects.

Solution   **step 1**
Since *all insects have six legs*, the circle representing insects is placed completely inside the circle representing six-legged creatures, as shown at the right.

step 2
The second premise (*Horses have four legs*) implies that no horses have six legs. This means that the circle representing horses must be drawn outside the circle representing six-legged creatures, as shown at the right.

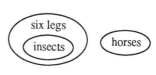

We have only one possible diagram, showing the "horses" as a circle drawn separately from the "insect" circle. This one possible diagram shows that *horses are not insects*. (Equivalently, no horses are insects.) This Euler circle demonstration makes the argument **valid**.

Chapter 2

Problem 6   Use Euler circles to determine the validity of the following argument.
            *No parrots have four legs. This can be concluded because dogs
            have four legs and parrots are not dogs.*

            The argument may be stated in the following form:
            premise   All dogs have four legs.
            premise   Parrots are not dogs.
            conclusion   No parrots have four legs.

Solution   **step 1**
            The first premise (*All dogs have four legs*) is
            diagrammed at the right.

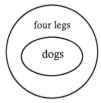

            step 2
            We now construct Euler circles for the second premise (*Parrots are not
            dogs*). Since no parrots are dogs, the "parrot" circle must be drawn outside
            the "dog" circle. There are at least three possible ways to do this, as shown
            in the accompanying diagram.

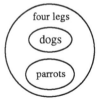 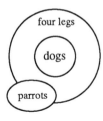 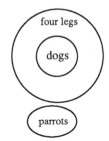

            Only the last set diagram justifies the conclusion "*No parrots have four
            legs.*" Since the conclusion does not necessarily follow from the premises,
            as seen in the first two set diagrams, the argument is **invalid** (a fallacy).

This example once again illustrates that an argument may not be valid even
though its conclusion is true in real life. Although it is true that no parrots have
four legs, this statement may or may not be true when deduced from the given
premises.

Problem Set 6.2

Determine the validity of each of the following arguments by using Euler circles.

1. Stan is a dancer. I know this because all dancers are athletes, and all athletes are in good shape. Stan is in good shape.

2. All dogs have hair and all dalmatians are dogs. Therefore, all dalmatians have hair.

3. Buzzards have wings. I know this because birds have wings and buzzards are birds.

4. All Londoners wear trenchcoats and all spies wear trenchcoats. Therefore, all Londoners are spies.

5. All birds have feathers and dogs do not have feathers. Thus, no dogs are birds.

6. All birds have feathers and no dogs are birds. Thus, no dogs have feathers.

7. No good students talk during lecture. Zane, based upon his 4.0 grade point average, is certainly a good student. We can conclude that Zane does not talk during lecture.

8. All moccasins hurt when they strike. I know this because all moccasins are poisonous snakes and all poisonous snakes hurt when they strike.

9. Citrus is rich in vitamin C and grapefruit is rich in Vitamin C. Thus, grapefruit is citrus.

10. All horses have legs and snakes are not horses. Thus, no snakes have legs.

11. All multiples of 6 are multiples of 3. Seventeen is not a multiple of 3. Consequently, 17 is not a multiple of 6.

For each of the following examples, use Euler circles to select a logical conclusion that will result in a valid argument.

12. *All physicists are scientists. All biologists are scientists. All scientists are college graduates. Florence is a scientist. Therefore,*
    a. Florence is a physicist or a biologist.
    b. All biologists are college graduates.
    c. Florence is not a college graduate.
    d. None of the above is warranted.

13. *All spies have trenchcoats. Every Londoner has a trenchcoat. Sean is a spy. Therefore,*
    a. No Londoners are spies.
    b. Sean is a Londoner.
    c. Sean has a trenchcoat.
    d. None of the above is warranted.

14. *No polite people talk at live theater. No theater lovers talk at live theater. Jerry loves theater. Therefore,*
    a. Jerry is polite and does not talk at live theater.
    b. Jerry is polite and talks at live theater.
    c. Jerry does not talk at live theater.
    d. None of the above is warranted.

15. *All logicians are strange and all strange people are unexciting. Phil is exciting. Therefore,*
    a. Phil is strange but is not a logician.
    b. Phil is a strange logician.
    c. Phil is neither strange nor a logician.
    d. None of the above is warranted.

16. *No creatures that eat meat are vegetarians. No cat is a vegetarian. Avram is a cat. Therefore,*
    a. Avram eats meat.
    b. All creatures that do not eat meat are vegetarians.
    c. Avram is not a vegetarian.
    d. None of the above is warranted.

17. Select the rule of logical equivalence which directly (in one step) transforms statement (1) into statement (2).

(1) If one is not a hard worker, then one falls behind.

(2) If one does not fall behind, then one is a hard worker.

a. $(\sim p \to q) \equiv (\sim q \to p)$
b. $(\sim p \to q) \equiv (p \to \sim q)$
c. $\sim(\sim p \to q) \equiv (p \wedge \sim q)$
d. $(p \to \sim q) \equiv (\sim q \to p)$

18. Select the statement below which is logically equivalent to:

Gene is an actor or a musician.

a. If Gene is an actor, then he is not a musician.
b. If Gene is not an actor, then he is a musician.
c. It is false that Gene is not an actor or not a musician.
d. If Gene is an actor, then he is a musician.

19. $(-0.03) \div (-0.75) =$

20. Imelda travels at a constant rate, traveling 110 miles at 11:45 A.M., 125 miles at 12:15 P.M. and 140 miles at 12:45 P.M. At what time will Imelda have traveled 200 miles?

We have seen that when the premises of an argument contain "all" or "no," the Euler circles shown below are useful in determining whether or not an argument is valid.

All A are B.   No A are B.

Another fundamental idea is considered in the following problem in which the word "some" appears in a premise. As we shall see, some A are B means at least one member of set A is a member of set B. We will diagram statements involving the word "some" by using intersecting circles, as shown in the accompanying Euler diagram.

Some A are B.

Problem 7   Use Euler circles to determine the validity of the following argument.
Some teachers are clowns. I know this because all teachers are funny people, and some clowns are funny people.

This argument may be stated in the following form:

premise   All teachers are funny people.
premise   Some clowns are funny people.
conclusion   Some teachers are clowns.

Solution

step 1

Diagram the first premise *All teachers are funny people* by a circle representing the set of teachers completely contained in a circle representing the set of funny people. The diagram would appear as shown at the right.

Notice that the premise *All teachers are funny people* is equivalent to: *There are no teachers who are not funny people.* It is also equivalent to: *Teachers are funny people.*

step 2

The word "some" in the premise *Some clowns are funny people* claims that there is **at least one** clown who is a funny person. Furthermore, the premise *Some clowns are funny people* states no relationship between clowns and teachers. Two possibilities exist for diagramming the information of this premise.

i. Diagram statements involving the word "some" by using **intersecting** circles. The premise *Some clowns are funny people* diagrams as two intersecting circles for clowns and funny people. In the possibility shown at the right, the clown and teacher circles do not intersect, which translates as *No clowns are teachers*.

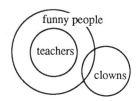

ii. Again, the clown and funny people circles intersect because *Some clowns are funny people*. However, in the possibility shown at the right, the clown and teacher circles also intersect, which translates as *At least one clown is a teacher*.

It is important to note that whenever the word "some" occurs in the second premise of an argument, more than one diagram must be drawn that will support the premise.

step 3

Finally, consider the conclusion: *Therefore, some teachers are clowns.* Notice that one of the two possibilities for drawing the diagrams does not agree with this conclusion, namely case (i). Since it is not correct to select only the diagram that does agree, this argument is **fallacious**.

A common misimpression with Euler diagrams is that if there is more than one possible way to diagram the premises, an argument is invalid. Our next problem shows that this is not the case.

Problem 8 Use Euler circles to determine the validity of the following argument.
All dogs have fleas and some dogs have rabies. Consequently, all dogs with rabies have fleas.

Solution **step 1**
Diagram the first premise (*All dogs have fleas*) as shown at the right.

step 2
Diagram the second premise (*Some dogs have rabies*) by showing the set of dogs intersecting the set of creatures that have rabies. Two possibilities exist, as shown in the accompanying Euler diagrams.

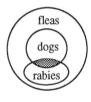 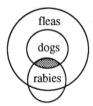

In both Euler diagrams, the shaded region represents dogs with rabies. In both diagrams this region falls inside the circle representing animals with fleas. Both diagrams illustrate that *all dogs with rabies have fleas* and the argument is **valid**.

Problem Set 6.3

Use Euler circles to determine the validity of the following arguments.

1. All logic problems make sense and some jokes make sense. Therefore, some logic problems are jokes.

2. Some immoral acts are justifiable. I know this because all thefts are immoral acts and some thefts are justifiable.

3. Flight attendants are pleasant. Some people with beards are pleasant. Therefore, some people with beards are flight attendants.

4. Some people enjoy reading and some people enjoy television. Therefore, some people who enjoy reading enjoy television.

5. All mathematicians are human beings and all physics teachers are human beings. Therefore, some physics teachers are mathematicians.

6. All philosophers are wise and some Greeks are philosophers. Therefore, some Greeks are wise.

7. No philosophers are wicked. Some Greeks are philosophers. Therefore, some Greeks are not wicked.

8. Some philosophers are wicked. Some Greeks are philosophers. Therefore, some Greek philosophers are wicked.

9. All parrots have feathers and some parrots talk. Therefore, all parrots that talk have feathers.

For each of the following examples, use Euler circles to select a logical conclusion that will result in a valid argument.

10. All poets appreciate language and all writers appreciate language. Therefore,
    a. All poets are writers.
    b. Some poets are writers.
    c. No poets are writers.
    d. None of the above is warranted.

11. All politicians make promises, but some politicians are liars. Therefore,
    a. Some people who make promises are politicians.
    b. All liars make promises.
    c. All people who make promises are liars or politicians.
    d. None of the above is warranted.

12. All timid creatures are bunnies, some timid creatures are dumb, and some students are timid creatures. Select the conclusion(s) that makes the argument valid.
    a. Some bunnies are dumb
    b. Some students are bunnies.
    c. Some students are dumb-bunnies.
    d. None of the above is warranted.

Review Competencies

Write the negation of the following statements.
13. *All politicians are dishonest.*
14. *It is cold and it is not snowing.*
15. *If it is cloudy, we will not go to the beach.*
16. *If all people follow rules, then no prisons are necessary.*
17. A dinner came to $80 before 5% tax. How much would be saved by leaving a 15% tip on the dinner bill before adding the tax rather than on the dinner bill that includes the tax?
18. If 32 is decreased by 25% of itself, what is the result?

Section 7   Logic in Everyday Life

We have seen that valid forms of arguments and Euler circles are useful in arriving at conclusions that are warranted because the conclusions result in valid arguments. However, it is also possible to use more informal methods of reasoning to draw logical conclusions. Some of these methods are presented in the following problems.

Problem 1

   i. *Any student who gets an F on a final exam does not pass the course.*
  ii. *Receiving a grade of A, B, C, or D on a final exam is no guarantee of passing the course.*
 iii. *On a final exam, 3 students scored A, 12 scored B, 20 scored C, 5 scored D, and 2 scored F.*

What can we conclude in this situation?
a. Two or more students did not pass the course.
b. Exactly two students did not pass the course.
c. At most two students did not pass the course.
d. None of the above can logically be concluded.

Solution

Since two students scored F, we know that by the first premise they did not pass the course. However, by the second premise, we are told that the other 40 students who scored A, B, C, or D on the final might not have passed the course. We can conclude that two or more students did not pass the course, option a.

Problem 2

Read the requirements and each applicant's qualifications for obtaining a $50,000 loan. Then identify which of the applicants would qualify for the loan.

To qualify for a loan of $50,000, an applicant must have a good credit rating and have an income of at least $35,000 if single or a combined income of more than $60,000 if married.

Evita Sanchez is single, earns $90,000, and has a bad credit rating.

Rudy Valentino is single, has a good credit rating, and works at two jobs. He makes $20,000 on one job and 50% of $20,000 on the second job.

Tom Mix is married, has a good credit rating, and earns $40,000. His wife earns $20,000.

a. Evita   b. Rudy   c. Tom   d. None of these

Solution   Evita Sanchez is not eligible for the loan because of her bad credit rating.

Rudy Valentino has good credit and earns $20,000 + (0.50)($20,000) = $20,000 + $10,000 = $30,000$. Since he is single and does not have an income of at least $35,000, he does not qualify for the loan.

The Mix family brings in $60,000 a year. They do not qualify because they must earn more than $60,000.

None of these people is qualified for the loan, making the correct answer option d.

Problem 3   Marilyn, Bob, and Tony each have two different occupations which are also different from one another's. Their occupations are writer, architect, teacher, doctor, lawyer, and artist. Each character in the following premises is a distinct person.

1. The teacher and writer went skiing with Marilyn.
2. The doctor hired the artist to paint a landscape.
3. The doctor met with the teacher.
4. The artist is related to the architect.
5. Tony beat both Bob and the artist at tennis.
6. Bob lives next door to the writer.

Use the given statements to find each person's occupations.

Solution   Premise 5 tells us that Marilyn must be an artist.

Using premises 1, 4, and 2, we can conclude that Marilyn is not a teacher, writer, architect, or doctor. Thus, Marilyn is a lawyer.

Premise 6 tells us that Bob is not a writer. Therefore, he can be a teacher, architect, or doctor. From premise 3 we conclude that the doctor and the teacher are not the same person. This means that Bob must be an architect.

At this point, Bob is also either a teacher or a doctor, but not both. If Bob were a doctor, then Tony would be a teacher and a writer, which premise 1 indicates is not possible, since the teacher and writer are different people. Thus, Bob's other job is a teacher.

Marilyn - artist and lawyer
Bob - architect and teacher
Tony - (the two remaining occupations) - doctor and writer

Problem Set 7.1

1. i.   *Any student who gets lower than C on a final exam does not pass the course.*
    ii. *Receiving a grade of A, B, or C on a final exam is no guarantee of passing the course.*
    iii. *On a final exam, 5 students scored A, 5 scored B, 5 scored C, 5 scored D, and 5 scored F.*

    What can we logically conclude in this situation?
    a. Exactly 10 students did not pass the course.
    b. At most 10 students did not pass the course.
    c. 10 or more students did not pass the course.
    d. None of the above can logically be concluded.

2. In order to rent an apartment at Kendall Estates, one must pay the rent for the first and last month in advance, in addition to making a deposit for possible damages equal to 50% of the monthly rent. Kendall Estates charges $400 per month for a studio apartment and $550 per month for a one-bedroom apartment. Lena and Saul saved $1,350 to move into Kendall Estates. Which statement following can be correctly concluded?
    a. They did not save enough money to rent either apartment.
    b. They saved enough money to rent the studio and the one bedroom apartment.
    c. They saved enough money to rent the studio apartment and still have one additional month's rent.
    d. They saved enough money to rent the studio apartment and be precisely $50 short of paying an additional month's rent.

3. In order to get a loan, one must have a good credit rating and be single making at least $18,000 per year or be married with a combined salary of at least $35,000 per year.

    Tim is single and earns $90,000 per year. He has a bad credit rating.

    Ethel and Lee Fisher, a married couple, have a good credit rating. Ethel earns $20,000 per year and Lee earns $14,900 per year.

    Bob and Jane Hope, brother and sister (both unmarried), have good credit ratings. Jane earns $20,000 per year and Bob makes $17,000 per year.

    a. Tim is eligible for the loan.
    b. The Fishers are eligible for the loan.
    c. Bob is eligible for the loan.
    d. Jane is eligible for the loan.

4. (This logic problem dates back to eighth century writing.) A farmer needs to take his goat, wolf, and cabbage across a stream. His boat can hold him and one other passenger (the goat, wolf, or cabbage). If he takes the wolf with him, the goat will eat the cabbage. If he takes the cabbage, the wolf will eat the goat. Only when the farmer is present are the cabbage and goat safe from their respective predators. How does the farmer get everything across the stream?

5. As shown in the accompanying diagram, men (1), (2), and (3) are placed in a line, blindfolded, facing a wall.

(1)　　(2)　　(3)

Three hats are then taken from a container of three tan hats and two black hats. The men are given that information, and the blindfolds are then removed. Each man, still in the same position, is asked to determine what color hat he is wearing. Man (3) responds, "I do not know what color hat I am wearing." Man (2), hearing the response and seeing the man and hat ahead of him, responds, "I do not know what color hat I am wearing." Man (1), seeing only the wall and hearing the two replies, says, "I know what color hat I am wearing." Which color is he wearing and how did he logically determine the color?

6. Observe the following series of letters. Then determine the logical way in which the missing letters have been removed.

   A B D E F I J K L P Q R S T Y Z A B C D J K L M N O P

7. All but one of these three-digit numbers share a common feature. Which number does not logically belong in the list?

   495   583   374   276   891
   165   363   396

8. John, Jack, Joan, and Jane have occupations that consist of art critic, architect, acrobat, and aviator, but not necessarily in that order. Each person has one occupation and all people have different occupations. Joan has never heard of "perspective," and has no idea of who Pablo Picasso is. The aviator, who is happily married, earns more than his sister, the art critic. John is not married. Who can you conclude is the architect?

Review Competencies

9. All of the following arguments have true conclusions, but one of the arguments is not valid. Select the argument that is **not** valid.

a. All biologists are scientists. All scientists attended college. Thus, people who did not attend college are not biologists.
b. Poets appreciate language and writers appreciate language. Therefore, poets are writers.
c. No democracies are an outgrowth of a military coup. In October, 1991, Haiti's president was overthrown by a military coup. Therefore, the government that took over Haiti in October, 1991, was not a democracy.
d. All babies are illogical. No illogical being can manage a crocodile. Therefore, babies cannot manage a crocodile.

10. Which one of the following is not valid?
a. If a person earns an A in calculus, then that person does not miss more than two classroom lectures. Jacques, a calculus student, missed five classroom lectures. Thus, Jacques did not earn an A in calculus.
b. Maggie does not have a good time when she listens to sexist humor. At this moment Maggie is listening to a comedian whose material is sexist and racist throughout. Consequently, at this moment Maggie is not having a good time.
c. A majority of teachers must vote in favor of a textbook or the book will not be adopted. Over 60% of the teachers did not vote in favor of Tropic of Calculus. Therefore, Tropic of Calculus was not adopted.
d. If one is to answer the question correctly, then the first step is to find the length of the hypotenuse. Harry did not answer the question correctly. Therefore, Harry did not find the length of the hypotenuse.

11. A home has an assessed valuation of $78,500. The tax rate is $3.40 per $100 of assessed valuation. A 3% discount is allowed if the tax is paid within the first quarter of the year. How much will taxes come to if the homeowner takes advantage of the discount?

Section 8   Fallacies in Everyday Life

Learning to recognize fallacies is important. Some people use fallacies to intentionally deceive. Others use fallacies innocently; they are not even aware they are using them. The reasoning employed to reach a conclusion may be erroneous. Such faulty reasoning can lead to poor decision making.

This section presents some fallacies that can occur everyday.

The **Fallacy of Emotion** consists of appealing to emotion, rather than reason, in an argument.

Example

The following argument illustrates the Fallacy of Emotion.
Professor Smith will pass me in psychology. I know this because, despite my 30% exam average, Professor Smith is a humane person. He certainly realizes that if he fails me, I might not be able to handle the emotional stress. Humane people certainly don't cause others to have breakdowns.

The **Fallacy of False Authority** consists of assuming that a person who is an authority in one field can speak as an authority on an issue totally unrelated to his/her field of competence.

Example

The following argument illustrates the Fallacy of False Authority.
The Cadillac must be the best car on the road. I'm led to this conclusion because Dr. Jones is a famous psychiatrist and she told me that the Cadillac is the best car made.

The **Fallacy of Hasty Generalization** consists of drawing a conclusion about a group of events when not all the events have been observed.

Example

The following illustrates the Fallacy of Hasty Generalization.
This entire logic exam must be difficult. I know this because I've worked the first two problems and they are difficult.

The **Fallacy of False Cause** consists of assuming that simply because two events occur together, that one is the cause and the other is the effect.

Example

The following argument illustrates the Fallacy of False Cause.
Heavy rains cause war. I know this because it rained heavily on the same day that war was declared.

The **Fallacy of Ambiguity** consists of using a word in more than one sense in an argument.

Example

The following argument illustrates the Fallacy of Ambiguity.
Paula has contributed to the development of America. I know this because the people who have contributed most to the development of America have been progressive and Paula is running for election on the Progressive Party ticket.

The **Fallacy of Composition** consists of assuming that if each of the members of a group has a certain characteristic, then the group also has that characteristic.

Example

The following argument illustrates the Fallacy of Composition.
The President's cabinet will always make the intelligent decision. I know this because each member of the President's cabinet is intelligent.

The **Fallacy of Division** consists of assuming that if an entire group has a certain characteristic, then each member of the group also has that characteristic.

Example

The following argument illustrates the Fallacy of Division.
Question 4 on this exam must be difficult. I know this because this exam was prepared by Professor Flunkem and his exams are always difficult.

The **Fallacy of the Complex Question** directs that an individual answer a self-incriminating question.

When this fallacy is committed an individual is required to admit something he/she may not wish to admit simply by answering a question with "yes" or "no."

Example

The following argument illustrates the Fallacy of the Complex Question. *Amy once cheated on tests. I know this because I once asked her if she had stopped cheating on tests and she answered, "yes."*

Observe that the fallacies in everyday life are fairly easy to recognize when presented in exaggerated examples. However, these fallacies often appear in more subtle ways, creating a very fine line between acceptable and fallacious reasoning. Consider the following examples.

Examples

1. Rat poison causes death. I know this because Ben drank rat poison and died.

2. This particular book is boring. I know this because John Brown wrote it and most of his books are boring.

We now have another procedure to determine the validity or invalidity of numerous arguments. Let us summarize these procedures.
1. If the premises contain the connectives "if. . . then" or "or," use the valid and invalid forms.

2. If the premises contain such quantifying words as "all," "some," or "no," use Euler circles.

3. Recognize the fallacies in everyday life which often appear in exaggerated examples.

Finally, there are many arguments that are evidently valid which do not need to be approached through any of the methods we have discussed (valid - invalid forms, Euler circles, or fallacies in everyday life). Two examples of these valid arguments appear below.

Examples

1. The May mathematics convention had the highest attendance to date. I know this because 400 people attended the May convention and before the May convention no more than 350 people ever attended a mathematics convention.

2. Rita is not in Professor Taylor's class. I know this because Rita makes an A on every exam she takes, never misses a class or an exam, and nobody made an A on Professor Taylor's last exam.

Problem Set 8.1

1. Indicate the fallacy that is illustrated in each of the following. More than one answer might be possible.

    a. Georgians certainly have bad manners. I know this because I just observed a driver in a car bearing Georgia license plates pushing rudely through the crowd leaving Disney World.

    b. Lucy is to be greatly admired. I know this because a person who is truly democratic is to be greatly admired. Lucy has always voted for every Democratic candidate.

    c. The quartet will be great. I know this because each individual in the quartet is a great musician.

    d. Socialized medicine must be wrong. I know this because Ed is a great actor and disapproves of socialized medicine.

    e. Bob will not consider going fishing without wearing his old hat. I know this because on four successive fishing trips Bob wore his old hat and had good luck. On the fifth trip he forgot to wear his hat and returned with no fish.

    f. My automobile engine will never give me a problem. I know this because I once read that my car is the most trouble-free car in America.

    g. I deserve a raise. I know this because my car is falling apart, my wife is sick, and my children are starving.

    h. Maria used to smoke cigarettes. I know this because Maria once answered "yes" when asked if she had stopped smoking cigarettes.

In each of the following problems, use valid-invalid forms, Euler circles, everyday fallacies, or self-evident valid arguments to select the one argument that is fallacious.

2. a. No snakes are horses. I know this because all horses have legs and snakes do not have legs.

    b. This semester resulted in the highest average SAT score to date. I know this because this semester the average SAT score was 620 and, before this, the highest average SAT score was only 550.

c. Black cats bring bad luck. I know this because a black cat walked in front of my path and I tripped.

d. If a number is divisible by 9, then it is divisible by 3. Sixteen is not divisible by 3. So, 16 is not divisible by 9.

3. a. All dogs have hair. I know this because all mammals have hair and all dogs are mammals.

b. Today's broiled fish at this restaurant is terrible. I know this because the broiled fish is today's special and all their specials are terrible.

c. Forty is divisible by 10. I know this because if a number is divisible by 20, then it is divisible by 10. Furthermore, 40 is divisible by 20.

d. At one time, Allan kicked his dog. I know this because when I asked him, "Have you stopped kicking your dog?" he said, "yes."

4. a. Don did not bake this cake. I know this because all of Don's cakes are chocolate and this cake is not chocolate.

b. Champ is a good puppy. I know this because Champ comes from the best kennel in the country.

c. Some fleas are on cats with long fur. I know this because all cats have fleas and some cats have long fur.

d. If this job gets finished, then Connie will be unemployed. If Connie is unemployed, then she will attend college full time. Thus, I can conclude that if this job gets finished, then Connie will attend college full time.

Review Competencies

5. Write the negation of:
   A bill must receive majority approval or it does not become law.

6. What can we logically conclude from the following premises?
   No polite people chat during lecture. No excellent student chats during lecture. Bud is an excellent student.
   a. Bud is a polite person.
   b. All people who are not polite chat during lecture.
   c. Bud chats during lecture.
   d. Bud does not chat during lecture.

7. Select the rule of logical equivalence which directly (in one step) transforms statement i into statement ii.
   i. *Ethel studies French or Spanish.*
   ii. *If Ethel does not study Spanish, then she studies French.*
   a. $(p \lor q) \equiv (\sim q \to p)$
   b. $(p \lor q) \equiv (q \to \sim p)$
   c. $(p \land q) \equiv (\sim q \to p)$
   d. $(p \land q) \equiv (q \to \sim p)$

8. If a person travels extensively, then that person is not ethnocentric. If a person studies philosophy, then that person is open to new ideas. Diana is ethnocentric. Diana is open to new ideas. What can we logically (validly) conclude about Diana?
   i. *Diana does not travel extensively.*
   ii. *Diana studies philosophy*
   a. i only   b. ii only
   c. Both i and ii   d. Neither i nor ii

9. Evaluate: $7 \cdot 0 + 0 \div 3$

10. Evaluate: $4 + 3 \cdot 6 + 5 \div 5 + 6 \div 2$.

11. Determine the common relationship between the numbers in each pair. Then identify the missing term.
    $(4,2)$ $(10,5)$ $(0.6,0.3)$ $(\frac{1}{10}, \frac{1}{20})$ $(\frac{1}{4},\underline{\quad})$

12. Income tax for taxable income between $8,000 and $12,000 is $4,200 plus 38% of the excess over $8,000. How much tax must be paid on a taxable income of $9,500?

13. Find the smallest positive multiple of 4 which leaves a remainder of 1 when divided by 5 and a remainder of 4 when divided by 6.

Chapter 2 Summary

1. Basic Definitions of Logic

Negation	p	~p
	T	F
	F	T

		Conjunction	Disjunction	Implication	Biconditional
p	q	p ∧ q	p ∨ q	p → q	p ↔ q
T	T	T	T	T	T
T	F	F	T	F	F
F	T	F	T	T	F
F	F	F	F	T	T

Conjunction	Disjunction	Implication	Biconditional
True only when both components are true.	True when at least one component is true.	False only with a true hypothesis and a false conclusion.	True when both components have the same truth value.

2. Equivalences and Nonequivalences

Equivalences for p → q
a.   (p → q) ≡ (~q → ~p), ~q → ~p is the contrapositive
b.   (p → q) ≡ (~p ∨ q)
Nonequivalences for p → q
c.   (p → q) is not equivalent to (q → p), the converse.
d.   (p → q) is not equivalent to (~p → ~q), the inverse.
Equivalences for p ∨ q
e.   (p ∨ q) ≡ (~p → q)
f.   (p ∨ q) ≡ (~q → p)
Nonequivalences for p ∨ q
g.   (p ∨ q) is not equivalent to (p → ~q).
h.   (p ∨ q) is not equivalent to (q → ~p).

Equivalences for ~(p ∧ q)

i.   ~(p ∧ q) ≡ (~p ∨ ~q)

j.   ~(p ∧ q) ≡ (p → ~q)

Nonequivalences for ~(p ∧ q)

k.   ~(p ∧ q) is not equivalent to (~p ∧ ~q).

l.   ~(p ∧ q) is not equivalent to (p ∧ q).

Equivalences for Writing Negations

m.   ~(p ∧ q) ≡ ~p ∨ ~q

n.   ~(p ∨ q) ≡ ~p ∧ ~q

o.   ~(p → q) ≡ p ∧ ~q

p.   ~(p ∧ q) ≡ p → ~q

3. Negations For "All," "Some," and "No"

statement	negation
All. . .	Some. . . not
Some. . . not	All. . .
Some. . .	No. . .
No. . .	Some. . .

4. English Equivalences

statement	equivalent(s)
All A are B.	There are no A that are not B.
Some A are not B.	Not all A are B.
Some A are B.	At least one A is a B.
No A are B.	All A are not B.
If p, then q	p implies q; p only if q; only if q, p; p is sufficient for q; q is necessary for p; not p unless q.
p if and only if q	p implies q and q implies p; p is equivalent to q; p is necessary and sufficient for q; q is necessary and sufficient for p.

5. Determining Whether or Not an Argument is Valid

a. If premises contain "if. . . then" or "or," use:

Valid Forms:

$$p \rightarrow q \qquad p \rightarrow q \qquad p \rightarrow q \qquad p \lor q \qquad p \lor q$$
$$\underline{p} \qquad\qquad \underline{\sim q} \qquad\qquad \underline{q \rightarrow r} \qquad \underline{\sim p} \qquad\qquad \underline{\sim q}$$
$$\therefore q \qquad\quad \therefore \sim p \qquad\quad \therefore p \rightarrow r \qquad \therefore q \qquad\qquad \therefore p$$
$$\qquad\qquad\qquad\qquad\qquad\qquad \therefore \sim r \rightarrow \sim p$$

Invalid Forms:

$$p \rightarrow q \qquad p \rightarrow q \qquad p \lor q \qquad p \lor q$$
$$\underline{q} \qquad\qquad \underline{\sim p} \qquad\qquad \underline{p} \qquad\qquad \underline{q}$$
$$\therefore p \qquad\quad \therefore \sim q \qquad\quad \therefore \sim q \qquad \therefore \sim p$$

b. If premises contain "all," "some," or "no," use Euler circles:

    All A are B   Some A are B   No A are B

 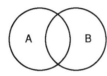 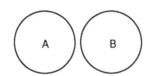

c. Be able to recognize common fallacies that arise in everyday life.

Chapter 2 Self-Test

1. Select the negation of: *The lawn must be fertilized monthly or it will not be green.*
   a. The lawn is not fertilized monthly or it is green.
   b. The lawn is not fertilized monthly and it is not green.
   c. The lawn is not fertilized monthly and it is green.
   d. If the lawn is not fertilized monthly, then it is green.

2. Select the negation of: *It is hot and it is not humid.*
   a. It is not hot and it is humid.
   b. It is not hot or it is humid.
   c. It is not hot and it is not humid.
   d. It is not hot or it is not humid.

3. Select the statement which is logically equivalent to: *It is not true that both the Rolling Stones and Pat Boone are rock groups.*
   a. The Rolling Stones are not a rock group and Pat Boone is not a rock group.
   b. The Rolling Stones are a rock group or Pat Boone is a rock group.
   c. If the Rolling Stones are not a rock group, then Pat Boone is not a rock group.
   d. If the Rolling Stones are a rock group, then Pat Boone is not a rock group.

4. Select the negation of: *Some people do not have good manners.*
   a. Some people have good manners.
   b. No people have good manners.
   c. All people have good manners.
   d. All people do not have good manners.

5. Select the negation of: *If one works hard, then one will be successful.*
   a. One does not work hard and one will not be successful.
   b. One works hard or one will not be successful.
   c. One works hard and one will not be successful.
   d. One works hard and one will be successful.

6. Select the statement that negates: *Terry plans to both do her homework and listen to the radio.*
   a. If Terry does not plan to do her homework, then she plans to listen to the radio.
   b. If Terry plans to do her homework, then she does not plan to listen to the radio.
   c. If Terry does not plan to do her homework, then she does not plan to listen to the radio.
   d. If Terry plans to do her homework, then she plans to listen to the radio.

7. Select the statement which is not logically equivalent to: *It is not true that both <u>Married with Children</u> and <u>Moby Dick</u> are television shows.*
   a. <u>Married with Children</u> is not a television show or <u>Moby Dick</u> is not a television show.
   b. If <u>Married with Children</u> is a television show, then <u>Moby Dick</u> is not a television show.
   c. It is not true that both <u>Moby Dick</u> and <u>Married with Children</u> are television shows.
   d. <u>Married with Children</u> is a television show and <u>Moby Dick</u> is a television show.

8. Select the statement which is logically equivalent to: *All politicians are corrupt.*
   a. There are no corrupt people who are politicians.
   b. There are no politicians who are not corrupt.

c. There are some politicians who are not corrupt.

d. There are no corrupt people who are not politicians.

9. Select the statement which is logically equivalent to: *If the play is Macbeth, Shakespeare is the author.*
   a. If Shakespeare is the author, the play is Macbeth.
   b. If Shakespeare is not the author, the play is not Macbeth.
   c. If the play is not Macbeth, Shakespeare is not the author.
   d. The play is Macbeth and Shakespeare is not the author.

10. Select the statement which is not logically equivalent to: *If Jones is in Boston, then he is in the East.*
    a. If Jones is not in the East, then he is not in Boston.
    b. Jones is not in Boston unless he is in the East.
    c. Jones is in the East when he is in Boston.
    d. If Jones is in the East, then he is in Boston.

11. Select the statement that is not logically equivalent to: *Anita is a chemist or owns a business.*
    a. If Anita is not a chemist, she owns a business.
    b. If Anita does not own a business, she is a chemist.
    c. If Anita is a chemist, she does not own a business.
    d. It is false that Anita is not a chemist and does not own a business.

12. Given that:
    i. *All biologists are scientists.*
    ii. *All chemists are scientists.*
    iii. *All scientists have college degrees.*
    iv. *Ray has a college degree.*
    Determine which of the following conclusions can be logically deduced.
    a. Some biologists are chemists.
    b. No biologists are chemists.
    c. Ray is either a biologist or a chemist.
    d. Some biologists do not have college degrees.
    e. People without college degrees are neither biologists nor chemists.

13. All of the following arguments have true conclusions, but one of the arguments is not valid. Select the argument that is not valid.
    a. Prolonged periods of cold will damage trees that are native to the Florida Keys. The gumbo limbo is a native Florida Keys tree. Therefore, prolonged periods of cold will damage the gumbo limbo.
    b. All insects have wings and no mules have wings. Therefore, mules are not insects.
    c. All people who do well in solid geometry must sketch three-dimensional figures on two-dimensional surfaces. The ability to sketch three-dimensional figures on a two-dimensional surface necessitates good spatial relationships. Therefore, doing well in solid geometry necessitates good spatial relatonships.
    d. Trees are plants. Palms are plants. Therefore, palms are trees.

14. Which one of the following is not valid?
    a. If I purchase season tickets to the football games, then I will not attend all lectures. If I do not attend all lectures, then I will not do well in school. Therefore, if I purchase season tickets to the football games, then I will not do well in school.

b. If both the microphone and the air conditioning fail, the workshop will not be held. This Saturday the workshop was held although the air conditioning did fail. Therefore, this Saturday the microphone did not fail.

c. No democracies are totalitarian. All dictatorships are totalitarian. Stalin's Russia was a dictatorship. Thus, Stalin's Russia was not a democracy.

d. Miguel is blushing or he is sunburned. Miguel is definitely sunburned. Therefore, Miguel is not blushing.

15. Which one of the following is not valid?
   a. If one is a good baseball player, then one must have good eye-hand coordination. Todd does not have good eye-hand coordination. Therefore, Todd is not a good baseball player.
   b. If a parrot talks, it is intelligent. This parrot is intelligent. Therefore, this parrot talks.
   c. If one applies oneself, then one does not fall behind. Ana applies herself. Therefore, Ana does not fall behind.
   d. A majority of legislators must vote for a bill or that bill will not become law. A majority of legislators did not vote for legislation X. Therefore, legislation X did not become law.

16. Select the conclusion that will make the following argument valid. *If all people follow rules, then no criminal judges are needed. Some criminal judges are needed.*
   a. Some people follow rules.
   b. If no criminal judges are needed, then all people follow rules.
   c. No people follow rules.
   d. Some people do not follow rules.

17. Which one argument listed below is fallacious?
   a. Brooke isn't coming to the party. I know this because if Brooke is coming, then Don is coming and Don, due to his persistent cough from a lingering cold, isn't coming to the party.
   b. A train must connect Miami to Key West or traffic delays will occur during hurricane evacuations. A train did not connect Miami to Key West in 1966. Consquently, traffic delays did occur during the hurricane evacuation in 1966.
   c. Mike's new carpet won't last for five years. I know this because Mike just purchased a carpet that cost $300, and I know for a fact that no carpet that costs less than $500 ever lasts for more than four years.
   d. Ed was once rude to his mathematics professor. I know this because when I asked Ed, "Have you stopped being rude to your mathematics professor?" he replied, "Yes."

18. Select the statement that is the negation of: If it rains, we do not go to the beach.
   a. If it does not rain, then we go to the beach.
   b. It is raining and we do not go to the beach.
   c. If we go to the beach, then it is not raining.
   d. It rains and we go to the beach.

19. Two valid arguments are given below. Determine the symbolic form of the reasoning pattern used in both arguments.
   i. *If you are playing the stereo softly, then I can hear you. I cannot hear you. Therefore, you are not playing the stereo softly.*
   ii. *Maggie has a good time if she plays a video game. Maggie is not having a good time. So, she is not playing a video game.*

   a. $p \rightarrow q$ b. $p \rightarrow q$ c. $p \rightarrow q$ d. $p \rightarrow q$
   $\underline{\sim p}$ $\underline{q}$ $\underline{\sim q}$ $\underline{q \rightarrow r}$
   $\therefore \sim q$ $\therefore p$ $\therefore \sim p$ $\therefore p \rightarrow r$

20. Select the symbolic form for each of the following arguments.

    i. *If the Dolphins win or lose, then I will remain calm. The Dolphins win. Thus, I will remain calm.*

    ii. *If you read or watch television, you will fall asleep. You are reading. So, you will fall asleep.*

    a. $(p \wedge q) \to r$
       $\underline{q}$
       $\therefore$ r

    b. $(p \vee q) \to r$
       $\underline{p}$
       $\therefore$ r

    c. $p \to (q \vee r)$
       $\underline{p}$
       $\therefore$ r

    d. $(p \wedge q) \to r$
       $\underline{p}$
       $\therefore$ r

21. Select the symbolic form for each of the following arguments.

    i. *If the sun shines, I go to the beach or go shopping. The sun is shining and I am not going to the beach. So, I am going shopping.*

    ii. *If I study, I get tired or frustrated. I am studying and not getting tired. Therefore, I am getting frustrated.*

    a. $(r \vee p) \to q$
       $\underline{r \wedge {\sim}p}$
       $\therefore$ q

    b. $r \to (p \vee q)$
       $\underline{r \wedge {\sim}p}$
       $\therefore$ q

    c. $r \to (p \vee q)$
       $\underline{{\sim}r \wedge {\sim}p}$
       $\therefore$ q

    d. $r \to (p \vee q)$
       $\underline{r \wedge p}$
       $\therefore$ q

22. Study the premises below. Select a logical conclusion if one is warranted. *No people who eat junk food as a steady diet are healthy. No substance abusers are healthy. Siham is healthy.*

    a. Some substance abusers eat junk food as a steady diet.

    b. All people who do not eat junk food as a steady diet are healthy.

    c. Siham is neither a substance abuser nor a person who eats junk food as a steady diet.

    d. None of the above is warranted.

23. Select the conclusion which will make the following argument valid. *If the Dolphins win the final game, then they are the best team in the country. If they are the best team in the country, then their supporters will be proud.*

    a. If their supporters are proud, then the Dolphins are the best team in the country.

    b. If their supporters are proud, then the Dolphins won the final game.

    c. If the Dolphins do not win the final game, then their supporters will not be proud.

    d. If their supporters are not proud, then the Dolphins did not win the final game.

24. Study the premises below. Select a conclusion that can be logically deduced if one is warranted. *All physicists are scientists. All biologists are scientists.*

    a. All physicists are biologists.

    b. Some physicists are biologists.

    c. No physicists are biologists.

    d. None of the above conclusions can be logically deduced.

25. Study the premises below. Select a conclusion if one is warranted. *All professional athletes are in good condition. Some teachers are professional athletes. Emile is a teacher.*

    a. All teachers who are professional athletes are in good condition and some teachers are in good condition.

    b. No person is a teacher, a professional athlete, and in good condition.

    c. Emile is in good condition.

    d. None of the above conclusions can be logically deduced.

26. Study the premises below. Select a logical conclusion if one is warranted.
    If both Mary and Thelma play, the team always wins. The team lost Saturday, but Thelma played Saturday.
    a. Mary played on Saturday.
    b. Mary did not play on Saturday.
    c. If Mary does not play or Thelma does not play, the team always loses.
    d. None of the above conclusions is warranted.

27. *I told Marilyn that if the river overflows, at least one of the three roads will remain open. I was wrong.*
    Select a logical conclusion if one is warranted.
    a. The river did not overflow and all three roads remained open.
    b. The river overflowed and all three roads remained open.
    c. The river overflowed and all three roads were closed.
    d. None of the above conclusions is warranted.

28. *If a person studies Latin, then that person has an excellent vocabulary. If a person studies mathematics, then that person thinks logically. Mark has an excellent vocabulary. Mark, however, did not study mathematics.*
    What can be validly deduced about Mark?
    i. *Mark studied Latin.*
    ii. *Mark does not think logically.*
    a. i only   b. ii only
    c. i and ii only
    d. None of the above statements can be validly deduced.

29. Select the rule of logical equivalence which directly (in one step) transforms statement (i) into statement (ii).
    i. *If roses are red, then violets are blue.*
    ii. *If violets are not blue, then roses are not red.*
    a. "If p, then q" is equivalent to "not p or q."
    b. "If p, then q" is equivalent to "if not q, then not p."
    c. "Not (p and q)" is equivalent to "not p or not q."
    d. "Not (not p)" is equivalent to "p."

30. Select the rule of logical equivalence which directly (in one step) transforms statement (i) into statement (ii).
    i. *If it is cold and snowing, then Bob is energetic.*
    ii. *If Bob is not energetic, then it is not cold or it is not snowing.*
    a. "If p or q, then r" is equivalent to "If not r, then not p and not q."
    b. "If p or q, then r" is equivalent to "If not r, then not p or not q."
    c. "If p and q, then r" is equivalent to "If not r, then not p and not q."
    d. "If p and q, then r" is equivalent to "If not r, then not p or not q."

31. Select the rule of logical equivalence illustrated by the following pair of equivalent statements:
    i. *If $|x| > 4$, then $x > 4$ or $x < -4$.*
    ii. *If $x \not> 4$ and $x \not< -4$, then $|x| \not> 4$.*
    a. "If p, then q or r" is equivalent to "if not q and not r, then not p."
    b. "If p, then q or r" is equivalent to "not p or q or r."
    c. The negation of "if p, then q or r" is equivalent to "p and not q and not r."
    d. None of the above

32. Any student who receives a grade of D or F on a final examination does not pass the course. On the other hand, a higher grade on the final examination is no guarantee of passing the course. This semester, final examination grades consisted of 2 As, 4 Bs, 8 Cs, 5 Ds, and 1 F. What can we logically conclude?
    a. Precisely 30% of the students did not pass the course.
    b. At most 70% of the students passed the course.
    c. No more than 30% of the students failed the course.
    d. None of the above conclusions is warranted.

33. To be eligible for admissions to a selective college, one must have a GPA of at least 3.85 and SAT scores of at least 650 or a GPA of at least 3.5 and SAT scores of at least 725. Isabel has a 3.9 GPA and an SAT score of 645. Eddie has a 3.52 GPA and an SAT score of 730. Mike has a 3.47 GPA and a perfect SAT score of 800. Who is eligible for admission?
    a. Isabel b. Eddie
    c. Mike d. None of these

34. Construct a truth table for $(p \vee {\sim}q) \to {\sim}p$ if the statements p and q occur in the following order:

p	q
T	T
T	F
F	T
F	F

Which of the following represents the final column of the truth values in the table?

a.	b.	c.	d.
F	T	F	T
F	F	F	T
T	T	F	F
T	T	T	F

35. The table at the right shows the numbers of cases for a truth table containing 1, 2, or 3 simple statements.

1 statement	2 statements	3 statements
p	p q	p q r
T	T T	T T T
F	T F	T T F
	F T	T F T
	F F	T F F
		F T T
		F T F
		F F T
		F F F

If a compound statement consists of 5 simple statements, how many cases would there be in the truth table?
a. 15 b. 16 c. 32 d. None of these

36. Select the rule of logical equivalence which directly (in one step) transforms statement (i) into statement (ii).
    i. *It is not true that both George and Ira wrote music.*
    ii. *If George wrote music, then Ira did not write music.*

    a. The negation of "p and q" is equivalent to "if p, then not q."
    b. It is not true that both "p and q" is equivalent to "if not p, then q."
    c. It is not true that both "p and q" is equivalent to "p and not q."
    d. The correct equivalence rule is not given.

37. Select the statement which is logically equivalent to: *Joe grows mangos or oranges.*
    a. If Joe grows mangos, he does not grow oranges.
    b. If Joe grows oranges, he does not grow mangos.
    c. If Joe does not grow mangos, he grows oranges.
    d. Joe grows both mangos and oranges.

38. Why is the following argument valid?

    $$\sim p \to r$$
    $$\underline{\sim r}$$
    $$\therefore\ p$$

    a. Because $[(\sim p \to r) \vee \sim r] \to p$ is a tautology.
    b. Because $[(\sim p \to r) \wedge \sim r] \to p$ is a tautology.
    c. Because $[(\sim p \to r) \to \sim r] \wedge p$ is a tautology.
    d. Because the form of every argument must be valid.

39. To move into an apartment at Euclid Estates, one must sign a one year lease, pay the first and last months' rent in advance, and pay a deposit according to the following schedule:

Apartment	Rent	Deposit
2-bedroom	$625 monthly	$500
1-bedroom	$500 monthly	$400

    Harry and Leona have $1,760. What can we conclude in this situation?
    a. They can rent the 1-bedroom, but not the 2-bedroom apartment.
    b. They can rent the 2-bedroom apartment and have $20 left.
    c. They can rent either apartment they choose.
    d. They can rent the 1-bedroom apartment and have enough money left to purchase a $375 dishwasher.

40. If a person is a bigot, then that person suffers from a form of mental illness. If a person cheats on exams, then that person lacks basic ethics. Seth does not suffer from any form of mental illness. However, Seth does cheat on exams. What can be validly deduced about Seth?
    i. *Seth is not a bigot.*
    ii. *Seth lacks basic ethics.*

    a. i only   b. ii only
    c. Both i and ii   d. Neither i nor ii

three

Algebra
and
Number Theory

Students in most academic programs are expected to possess some skills in algebra. These skills might include the ability to compute with signed numbers, to solve basic equations and inequalities, to manipulate various formulas, and to graph relationships. The language of algebra is important.

Algebra may not seem to be directly useful for everyone in everyday life. Its importance for a student, however, lies in the future since it lays the groundwork for more advanced study in higher mathematics, statistics, and computer technology.

A knowledge of basic algebra will help a student survive the seemingly endless succession of tests necessary for graduation from high school and college, for admission to special schools and universities, and for hiring and promotion. Thus, a lack of understanding of basic algebra can limit the attainment of a variety of goals.

Section 1   The Basic Operations and
the Closure Property

In this section, we will consider some sets of numbers, some operations on these numbers, and some properties (laws) that direct the operations. A **property** of a set of numbers is a feature that characterizes or describes it. Properties determine what happens to the numbers. An **operation** is a process that can be performed on the numbers. The most familiar operations are addition, subtraction, multiplication, and division.

Chapter 3

Let us begin by discussing the two most important operations—addition and multiplication.

Definitions

If a and b are any two elements of a set, then the operation **addition** associates a unique element, called the **sum**, denoted by a + b, with these two elements (called **addends**).

If a and b are any two elements of a set, then the operation **multiplication** associates a unique element, called the **product**, denoted by a • b, with these two elements (called **factors**). The product a • b may also be expressed as ab.

The word "unique" in the above definitions means "one and only one." Since we are associating a pair of elements (numbers) with a unique element, we say that addition and multiplication are **binary operations**. For example, in addition, the unique number associated with the pair 3 and 5 is 8. In multiplication, the unique number associated with this pair is 15.

Subtraction (a − b) and division (a ÷ b) are other operations on numbers, but are of less importance than addition and multiplication. Division, for example, is limited by the fact that no number can be divided by zero. Although addition, multiplication, subtraction, and division are called the four basic operations, addition and multiplication will be of particular interest here.

Are there relationships between operations? Yes. One example is that multiplication is repeated addition. Thus, 3 • 5 means 5 + 5 + 5 or 3 + 3 + 3 + 3 + 3. In both cases, the product is 15.

Let us consider other relationships between operations. Can you think of instances in which you do something and then undo it? For example, you go out of a room, and you come back into it. You go to sleep, and you wake up. You turn the car ignition on, and you turn it off. Each of the pairs of activities is said to be the **inverse** of the other. One "undoes" the other. Mathematical operations have inverses too. The subtraction operation is the inverse of the addition operation. The inverse of adding 2 to a number is subtracting 2. For example, if 2 is added to 4, the sum is 6. If 2 is then subtracted from 6, the result is again 4. The division operation is the inverse of the multiplication operation. The inverse of multiplying a number by 4 is dividing by 4. For example, if 3 is multiplied by 4, the product is 12. If 12 is then divided by 4, the result is again 3.

Let us now recall a set of numbers that was first mentioned in Chapter 1.

The numbers used by a person to count are called **counting numbers**. They are also called **natural numbers**.

Definition
The set of **counting numbers** is the set $\{1,2,3,\ldots\}$.

Numbers such as 0, $\frac{2}{3}$, 0.45 and -3 are not counting numbers.

Now that we have described the set of counting numbers, let us establish some properties which we expect these counting numbers to obey under the operations of addition and multiplication. To begin, if we add two counting numbers (like 17 and 76), is the number representing the unique sum (93) still in the set of counting numbers? Yes. It would not be necessary to "move outside" the set to find the sum. In the same way, if we multiply two counting numbers (like 14 and 92), the number representing the product (1,288) is still a unique counting number. The result is found within the set of counting numbers. Thus, the sum and product of two counting numbers are always unique counting numbers. This discussion leads us to state our first property.

Property 1
THE CLOSURE PROPERTY
A set is said to be **closed** for some operation if, when any two elements in the set are combined under the operation, there is only *one* result that is found in the set.

From our discussion, we can state that the set of counting numbers is closed for addition, and the set of counting numbers is closed for multiplication.

In our examples, we will consider various sets of numbers and confine ourselves to the four basic operations.

Examples

1. The set of even counting numbers {2,4,6,...} is closed for addition because the sum of any two even counting numbers is an even counting number. Consider the sum of 4 and 8 (4 + 8). The unique sum (12) is found within the set of even counting numbers. It is not necessary to look outside the set to find the sum. The set of even counting numbers is also closed for multiplication. The product of two even counting numbers is an even counting number. Consider the product of 2 and 4 (2 • 4). The unique product (8) is found in the set of even counting numbers.

2. The set of odd counting numbers {1,3,5,...} is not closed for addition. The sum of two odd numbers (1 + 3, for example) is not an odd counting number. Observe that this set is closed for multiplication. Consider the product of 3 and 5 (3 • 5). The unique product (15) is an odd number.

3. The set of counting numbers {1,2,3,...} is not closed for subtraction. When a counting number is subtracted from a counting number, the result may or may not be a counting number. Consider 3 − 3. The result (0) is not a counting number. The set of counting numbers is not closed for division. When a counting number is divided by a counting number, the result may or may not be a counting number. Consider 2 ÷ 3. The result ($\frac{2}{3}$) is not a counting number.

4. The set {1,2} is not closed for addition. Consider the sum of 1 and 2 (1 + 2). The sum (3) is not found in the set. This set is not closed for multiplication. Consider the product of 2 and 2 (2 • 2). This product (4) is not found in the set.

Problem Set 1.1

1. With what number does the set of counting numbers begin?

2. Is the set of counting numbers an infinite set?

3. Is there always at least one counting number between any two counting numbers?

4. Is there always a counting number larger and smaller than any given counting number?

5. Is the set of counting numbers closed for division?

6. Is the set {0,1} closed for subtraction? Division?

7. Is the set {1} closed for addition? Multiplication? Division?

8. Is the set {0} closed for addition? Multiplication? Subtraction?

9. Is the set of one-digit counting numbers closed for multiplication?

10. Add: $\frac{7}{6} + 2\frac{1}{4}$

11. Subtract: $1\frac{7}{10} - \frac{2}{3}$

12. Multiply: $2\frac{1}{3} \cdot 3\frac{1}{4}$

13. Divide: $1\frac{1}{4} \div 2\frac{2}{3}$

14. Add: $2.637 + 3 + 0.04$

15. Subtract: $7 - 0.475$

16. Divide: $0.15 \div 0.003$

17. Express 0.035 as a percent.

18. A football team won 7 games and lost 4 games. What fraction of games did the team win?

Section 2 Important Properties of Addition and Multiplication

In this section, we will consider some very important properties of addition and multiplication.

Let us mention some physical situations where it is important to observe the order in which activities are carried out. Our task will be to determine whether the end result is the same no matter the order in which we perform the individual operations. For example, let us consider the sequence of activities of first putting sugar in your coffee and then cream in your coffee. Would there be the same end result if we had changed the order of the sequence and had first put cream in your coffee and then sugar? The answer is yes. We can say that this pair of activities is **commutative.** (A commuter is a person who travels to and from some location.) These activities, putting sugar and cream in the coffee, can be commuted in the order in which they are performed. The order can be reversed.

What about the physical situation in which you start the engine of your car and then begin driving it? Is the end result the same if we reverse the order of the activities and first begin driving the car and then start the engine? The answer is, of course, no. This pair of activities is not commutative.

The expression x + y tells us to add y to x, and the expression y + x tells us to add x to y. Physically, these are different situations. The end result, however, is the same. Consider the child when asked to add 4 and 3. The child may start with 4 and add 3 to find 4 + 3, or the child may begin with 3 and add 4 to find 3 + 4. There is no question that 4 + 3 is different from 3 + 4, but the result in both cases is 7.

155

Chapter 3

From this discussion let us state an important property.

Examples

1. It is true that $2 + 3 = 3 + 2$.

2. It is true that $x + (y + z) = (y + z) + x$. The commutative property of addition is illustrated because the order in which x and (y + z) are added is reversed.

3. It is true that $(x + y) + z = (y + x) + z$. The commutative property of addition is illustrated because the order in which x and y appear in the parentheses is reversed.

4. It is not true that $4 - 2 = 2 - 4$. Subtraction is not a commutative operation even though subtraction is the inverse operation of addition.

Problem Set 2.1

1. Complete the following using the commutative property of addition:
   a. $(p + q) + \underline{Y} = r + (p + q)$
   b. $(x + y) + z = (y + \underline{x}) + \underline{z}$
   c. $(a + \underline{b}) + (\underline{c} + d) = (c + d) + (a + b)$

2. Is loading a gun and then firing it a commutative operation? why

3. Is putting on a shirt and then putting on a tie a commutative operation? no

4. Is the following an illustration of the commutative property of addition?
   $$[(a + b) + c] + d = d + [(a + b) + c]$$

Review Competencies

5. Is the set {1} closed for addition?

6. Is the set of even counting numbers closed for multiplication?

7. Is the set of odd counting numbers closed for addition?

8. Divide: $8 \div 0.02$

9. Express 125% as a decimal.

10. Express $\frac{7}{10}$ as a decimal.

11. A family spends $\frac{1}{3}$ of its income for food and $\frac{1}{4}$ for rent. What fractional part of the income has been spent?

156

Let us now state a related property for the multiplication operation.

> **Property 3**
> THE COMMUTATIVE PROPERTY OF MULTIPLICATION
> If x and y are two counting numbers, then $x \cdot y = y \cdot x$.

Examples

1. It is true that $3 \cdot 4 = 4 \cdot 3$.

2. It is true that $x \cdot (y \cdot z) = (y \cdot z) \cdot x$. The commutative property of multiplication is illustrated because the order in which x and (y $\cdot$ z) are multiplied is reversed.

3. It is true that $(5 \cdot 3) + 2 = (3 \cdot 5) + 2$. Even though two operations are involved (multiplication and addition), the commutative property of multiplication is illustrated because the order in which 5 and 3 are multiplied is reversed within the parentheses.

4. It is not true that $4 \div 2 = 2 \div 4$. Division is not a commutative operation even though division is the inverse operation of multiplication.

Even though subtraction and division are not commutative operations, there are other operations that are commutative. Recall, for example, that set union and set intersection are commutative operations. Certainly $A \cup B = B \cup A$ and $A \cap B = B \cap A$ for any sets A and B. Similarly, conjunction and disjunction are commutative because $p \wedge q$ is equivalent to $q \wedge p$ and $p \vee q$ is equivalent to $q \vee p$.

Problem Set 2.2

1. Complete the following, illustrating the commutative property of multiplication.
   a. $(\underline{x} \cdot y) \cdot z = \underline{z} \cdot (x \cdot y)$
   b. $(a \cdot b) + (a \cdot c) = (\underline{b} \cdot a) + (c \cdot \underline{a})$
   c. $(b + c) \cdot a = \underline{a} \cdot (b + c)$

2. Is the following an illustration of the commutative property of multiplication?
   $[2 + (1 + 3)] \cdot 4 = 4 \cdot [2 + (1 + 3)]$ ✓

3. Is the connective "if… then" (in logic) commutative? In other words, is $p \rightarrow q$ equivalent to $q \rightarrow p$? _yes_

Review Competencies

4. If a and b are counting numbers, state the commutative property of addition.

5. Is putting on your socks and then your shoes a commutative operation?

6. Is the set of counting numbers closed for addition?

7. Express $\frac{7}{8}$ as a percent.

8. Express $5\frac{1}{4}\%$ as a decimal.

9. Subtract: $3\frac{3}{5} - 1\frac{1}{4}$

10. Express $\frac{29}{8}$ as a mixed number.

157

Let us now consider the physical situation in which a boy is asked to put on his socks, his shoes, and his shirt to get ready for school. He first puts on his socks and then his shoes, and, after doing this, puts on his shirt. Would the end result have been the same if he had grouped his activities differently and, say, put on his shoes first and then his shirt, and then finally put on his socks? The answer is no. If we use parentheses to indicate the pair of activities to be completed first, then:

pair to be completed first

(putting on socks and putting on shoes) and putting on shirt

is not the same as

pair to be completed first

putting on socks and (putting on shoes and putting on shirt).

If one is asked to add $4 + 6 + 3$, one could proceed in two possible ways. Using parentheses to show which addition is completed first, the addition of $4 + 6 + 3$ could be completed as:

$$(4 + 6) + 3 \quad \text{or as} \quad 4 + (6 + 3).$$

Observe that the end result is the same no matter which method of **grouping** or **associating** one uses. The two computations are shown to have the same end result as follows:

$$(4 + 6) + 3 = 4 + (6 + 3)$$
$$10 \quad + 3 = 4 + \quad 9 \quad \text{(adding within parentheses first)}$$
$$13 = 13$$

This idea of grouping leads us directly to another important property.

> Property 4
> THE ASSOCIATIVE PROPERTY OF ADDITION
> If x, y and z are counting numbers, then:
> $(x + y) + z = x + (y + z)$.

NO
SUBSTRACTION

Examples

1. It is true that $(2 + 3) + 7 = 2 + (3 + 7)$

   because
   $$
   \begin{aligned}
   (2 + 3) + 7 &= 2 + (3 + 7) \\
   5 + 7 &= 2 + 10 \\
   12 &= 12
   \end{aligned}
   $$

2. It is not true that $(7 - 4) - 2 = 7 - (4 - 2)$

   because $(7 - 4) - 2 = 3 - 2 = 1$
   and $7 - (4 - 2) = 7 - 2 = 5$

The associative property does not hold for subtraction. The associative property provides a guide for adding three (or more) quantities together. The operation within the parentheses is to be performed **first.** What is important is the fact that the order of the addends is identical in both the left member and the right member of the associative property. The elements do not change their order in the sequence when reading from left to right.

The associative property can be extended to hold for more than three addends. In such cases brackets are used along with parentheses to indicate the manner of grouping. For example:
$$
1 + [2 + (3 + 4)] = 1 + [(2 + 3) + 4]
$$

Problem Set 2.3

1. Complete the following to illustrate the associative property of addition.
   $$(p + q) + r = p + (q + \square)$$

2. Is the following an illustration of the associative property of addition and no other property?
   $$(x + y) + z = x + (z + y) \quad \checkmark$$

3. Determine whether the associative property for addition is illustrated in each of the following:
   a. $5 + [(1 + 2) + 3] = 5 + [1 + (2 + 3)]$ ✗
   b. $[2 + (3 + 4)] + 5 = 2 + [(3 + 4) + 5]$ ✓
   c. $(1 + 2) + (3 + 4) = 1 + (2 + 3) + 4$ ✓

4. Verify that $(10 - 6) - 4$ is not equal to $10 - (6 - 4)$.

Review Competencies

5. If a and b are counting numbers, state the commutative properties of addition and multiplication.

6. Is the set $\{0,1\}$ closed for addition? For multiplication?

7. What is 40% of 72?

8. Express 285% as a decimal.

9. Two years ago a college had an enrollment of 10,000 students. This year the enrollment has increased by 20%. What is the enrollment for this year?

Chapter 3

Let us now state a multiplication property directly related to the associative property of addition.

Property 5
THE ASSOCIATIVE PROPERTY OF MULTIPLICATION
If x, y, and z are counting numbers, then $(x \bullet y) \bullet z = x \bullet (y \bullet z)$.

Examples

1. It is true that

$$(4 \bullet 5) \bullet 6 = 4 \bullet (5 \bullet 6).$$
$$(4 \bullet 5) \bullet 6 = 4 \bullet (5 \bullet 6)$$
$$20 \bullet 6 = 4 \bullet \quad 30$$
$$120 = 120$$

2. It is not true that $(100 \div 10) \div 5 = 100 \div (10 \div 5)$.
$$(100 \div 10) \div 5 = 10 \div 5 = 2 \qquad 100 \div (10 \div 5) = 100 \div 2 = 50$$
The associative property does not hold for division.

Addition and multiplication are not the only operations that are associative. Set union and set intersection are other examples of associative operations. Recall that:

$$(A \cup B) \cup C = A \cup (B \cup C) \quad \text{and} \quad (A \cap B) \cap C = A \cap (B \cap C).$$

Similarly, conjunction and disjunction are associative. Thus,

$$[(p \wedge q) \wedge r] \equiv [p \wedge (q \wedge r)] \quad \text{and} \quad [(p \vee q) \vee r] \equiv [p \vee (q \vee r)].$$

Problem Set 2.4

1. Complete the following, illustrating the associative property of multiplication.
$$2 \bullet (3 \bullet 4) = (2 \bullet 3) \bullet 4$$

2. Is the following an illustration of the associative property of multiplication and no other property?
$$(a \bullet b) \bullet c = a \bullet (c \bullet b)$$

3. Is the following an illustration of the associative property of multiplication?
$$[a \bullet (b \bullet c)] \bullet d = a \bullet [(b \bullet c) \bullet d]$$

4. Verify that $(12 \div 6) \div 2$ is not equal to $12 \div (6 \div 2)$. True

Review Competencies

5. If a, b and c are counting numbers, state the associative property of addition.

6. Complete the following illustration of the associative property of addition.
$$(\_ + \_) + z = x + (y + z)$$

7. Is the set {0,1} closed for division?

8. Express 0.53% as a decimal.

9. George weighed 180 pounds. After dieting, his weight went down to 162 pounds. What was the percent of decrease in his weight?

$1 \neq 4$

At times it is necessary to group quantities together as well as rearrange them to achieve the desired result. Thus, the commutative and associative properties are often used together. For example, observe below the following chain of reasoning.

Problem 1   Verify that $(a + b) + c = b + (a + c)$.

Solution   The equality to be verified is not a direct application of either the commutative or associative properties. We must use both of them to make our verification.

step 1
$(a + b) + c = (b + a) + c$ (commutative property of addition)

step 2
$(b + a) + c = b + (a + c)$ (associative property of addition)

∧ conjuction
∨ disjunction

step 3
Therefore, $(a + b) + c = b + (a + c)$ because each expression was shown to be equal to $(b + a) + c$ in Steps 1 and 2.

Notice the similarity between this chain of reasoning and the syllogism used in logic for proving valid arguments.

$$[(p \rightarrow q) \wedge (q \rightarrow r)] \rightarrow (p \rightarrow r)$$

Problem Set 2.5

Provide reasons for each step in problems 1 and 2.   Review Competencies

1. Verify that $(x + y) + z = (z + y) + x$
   step 1 $(x + y) + z = x + (y + z)$ *Commutative*
   step 2 $x + (y + z) = x + (z + y)$ *Associative*
   step 3 $x + (z + y) = (z + y) + x$ *Both*

2. Verify that $p \cdot (q \cdot r) = (r \cdot p) \cdot q$
   step 1 $p \cdot (q \cdot r) = p \cdot (r \cdot q)$ *Commutative*
   step 2 $p \cdot (r \cdot q) = (p \cdot r) \cdot q$ *Associative*
   step 3 $(p \cdot r) \cdot q = (r \cdot p) \cdot q$ *Both.*

3. If the product $5 \cdot 7 \cdot 20 \cdot 6$ is to be quickly calculated, which pair of numbers should be multiplied first?
   (6·5) (7·20)
   30·140 =

4. If a, b, and c are counting numbers, state the associative property of multiplication.

5. If a and b are counting numbers, state the commutative properties of addition and multiplication.

6. Does the commutative property hold for subtraction of counting numbers?

7. Is the following true?
   $(20 \div 10) \div 2 = 20 \div (10 \div 2)$

8. Is the set $\{0,1\}$ closed for subtraction?

9. 30% of what number is 60?

Chapter 3

In multiplication, the number one (1) plays a unique role. The product of any counting number and 1 is that identical counting number. Thus, we now state another important property for the counting numbers.

> ## Property 6
> THE IDENTITY PROPERTY FOR MULTIPLICATION
> If x is a counting number, then $x \cdot 1 = 1 \cdot x = x$.
>
> 1 is called the **identity element** for multiplication or the **multiplicative identity**.

Examples

1. It is true that $4 \cdot 1 = 4$.

2. It is true that $1 \cdot 0 = 0$.

3. It is true that $(2 + 3) \cdot 1 = (2 + 3)$

The role of the multiplicative identity is similar to the role of the universal set in the set operation intersection. Recall that $A \cap U = A$ for any set A.

Problem Set 2.6

1. Is it possible for a set to contain more than one multiplicative identity? *NO*

2. Does the set {0,1} contain the multiplicative identity? *Yes*

Review Competencies

3. If a, b, and c are counting numbers, state the associative properties of addition and multiplication.

4. Complete the following illustration of the associative property of multiplication:
$$x \cdot (y \cdot \_\_) = (x \cdot \_\_) \cdot z$$

5. Is the set of counting numbers closed for subtraction? For division?

6. 20% of what number is 80?

7. Express 0.031 as a percent.

8. A woman bought a house for $50,000. She later sold it for $75,000. What was her percent of increase?

Does the set of counting (natural) numbers have an identity element for addition? If x is any counting number, is there a counting number such that x + [ ] = x? The answer to both these questions is no. The set of counting numbers has an identity element for multiplication but none for addition. Thus, a weakness has been identified for the set of counting numbers, and we must correct it.

The German mathematician Leopold Kronecker stated, "God created the natural numbers; all else is the work of man." Perhaps this means that humans possess the ability to understand the idea of the basic counting numbers. Any other numbers conceived have originated out of necessity or as a result of human creativity. Let us therefore be creative and enlarge the set of counting numbers to include the number zero (0). This brings us to another definition.

Definition
The set of **whole numbers** is the set $\{0,1,2,3,...\}$.

Numbers such as $\frac{2}{3}$, 0.45 and, -4 are not included in the set of whole numbers, but every counting number is also a whole number. Recall that the set of whole numbers was first mentioned in Chapter 1.

The set of whole numbers is closed for both addition and multiplication. If any two whole numbers are added or multiplied, the result is a whole number. The set of whole numbers is not closed for subtraction. Certainly $2 - 3$ does not represent a whole number. The set of whole numbers is not closed for division because $2 \div 3$ does not represent a whole number.

The commutative properties for both addition and multiplication hold for the set of whole numbers. For example, it is true that $3 + 0 = 0 + 3$ and $5 \cdot 0 = 0 \cdot 5$.

The associative properties for both addition and multiplication hold for the set of whole numbers. For example, it is true that $(2 + 0) + 4 = 2 + (0 + 4)$ and $(3 \cdot 4) \cdot 0 = 3 \cdot (4 \cdot 0)$.

Finally, the set of whole numbers contains the multiplicative identity (1). If x is any whole number, then $x \cdot 1 = 1 \cdot x = x$.

Problem Set 2.7

1. True or false?
   a. Every counting number is also a whole number, but not every whole number is a counting number.
   b. Since the set of whole numbers contains an element 0 that is not found in the set of counting numbers, there is not a one-to-one correspondence between the two sets.

2. Do subtraction and division of whole numbers obey either the commutative or the associative properties?

Review Competencies

3. Which number is the multiplicative identity for the set of counting numbers?

4. Does the set {0,1,2} contain the multiplicative identity?

5. Is the set {0} closed for subtraction?

6. 60 is 20% of what number?

7. Subtract: $10 - 0.567$

8. Express $\frac{19}{9}$ as a mixed number.

The number zero plays a very important role in addition. The sum of any whole number and zero is that identical whole number. We can now state a property.

> ## Property 7
> THE IDENTITY PROPERTY FOR ADDITION
> If x is any whole number, then $x + 0 = 0 + x = x$.
> 0 is called the **identity element** for addition or the **additive identity**.

Examples

It is true that $3 + 0 = 3$ and $0 + 6 = 6$. The number zero plays a similar role in addition as the number one plays in multiplication. The addition of any counting number and zero is that identical counting number. The role of the additive identity is similar to the role of the null set in the set operation union. Recall that $A \cup \emptyset = A$ for any set A.

Algebra and Number Theory

Problem Set 2.8

1. Is there a whole number x such that $x + 0 = 0$? If so, what is it?

2. Is the additive identity for the whole numbers a unique element? That is, is there one and only one additive identity?

Review Competencies

3. Is the following true or false? Every counting number is also a whole number, but not every whole number is a counting number.

4. What is the multiplicative identity for the set of whole numbers?

5. Does the commutative property hold for division of whole numbers?

6. Does the associative property hold for subtraction of whole numbers?

7. Is the set of one-digit counting numbers closed for addition?

8. 25% of what number is 75?

9. Multiply: $2\frac{1}{4} \cdot 2\frac{1}{5} \cdot \frac{4}{9}$

10. A basketball team won 19 of its 25 games. What percent of its games did it lose?

So far each of the properties we have discussed in this chapter have involved either addition or multiplication but not both operations together. Let us now consider an extremely important property for whole numbers that involves both operations.

Property 8
THE DISTRIBUTIVE PROPERTY OF
MULTIPLICATION OVER ADDITION
If x, y, and z are whole numbers, then
$$x(y + z) = (x \cdot y) + (x \cdot z).$$
The property may also be expressed as $x(y + z) = xy + xz$.

Examples

1. It is true that $2 \cdot (3 + 4) = (2 \cdot 3) + (2 \cdot 4)$.
   This equality can be verified as follows:
   $$
   \begin{aligned}
   2 \cdot (3 + 4) &= (2 \cdot 3) + (2 \cdot 4) \\
   2 \cdot 7 &= 6 + 8 \\
   14 &= 14
   \end{aligned}
   $$

2. It is true that $2 \cdot (a + b) = 2(a + b) = 2a + 2b$.

3. It is true that $3 \cdot (a + 2) = 3(a + 2) = 3a + 3 \cdot 2 = 3a + 6$.

Chapter 3

Since two operations are involved in this property, we say that **multiplication is distributed over addition.** This property may also be stated as:

$$(b + c) \cdot a = (b \cdot a) + (c \cdot a) \quad \text{or} \quad (b + c)a = ba + ca$$

This form is often called the right distributive property as opposed to the left distributive property stated in Property 8.

The distributive property may also be extended to include more than two terms in the parentheses. For example:

$$a(b + c + d) = (a \cdot b) + (a \cdot c) + (a \cdot d)$$

or

$$a(b + c + d) = ab + ac + ad$$

Forms of the distributive property were encountered in set theory. Recall that:
$A \cup (B \cap C) = (A \cup B) \cap (A \cup C)$ and $A \cap (B \cup C) = (A \cap B) \cup (A \cap C)$.

Problem Set 2.9

1. Find a replacement for "a" to illustrate the distributive property:
   a. $3 \cdot (2 + 4) = (3 \cdot 2) + (3 \cdot a)$
   b. $a \cdot (4 + 5) = (2 \cdot 4) + (2 \cdot 5)$
   c. $(3 + a) \cdot 4 = (3 \cdot 4) + (2 \cdot 4)$
   d. $2 \cdot (a + 3) = (2 \cdot 4) + (2 \cdot 3)$
   e. $(3 \cdot 5) + (3 \cdot a) = 3(5 + 4)$

2. Is addition distributive over multiplication? That is, for any three whole numbers a, b, and c, is it always true that
   $a + (b \cdot c) = (a + b) \cdot (a + c)$?

3. Is addition distributive over addition? For any three whole numbers a, b, and c, is it always true that $a + (b + c) = (a + b) + (a + c)$?

4. Is multiplication distributive over multiplication? For any three whole numbers a, b, and c, is it always true that $a \cdot (b \cdot c) = (a \cdot b) \cdot (a \cdot c)$?

5. Verify whether $4 \cdot (3 - 2) = (4 \cdot 3) - (4 \cdot 2)$ and $5 \cdot (4 - 1) = (5 \cdot 4) - (5 \cdot 1)$. What can be concluded from these examples?

6. Verify whether
   $12 \div (4 + 2) = (12 \div 4) + (12 \div 2)$. Does the left distributive property appear to hold for division over addition?

7. Verify whether
   $(12 + 4) \div 2 = (12 \div 2) + (4 \div 2)$. Does the right distributive property appear to hold for division over addition?

Review Competencies

8. What number is the additive identity?

9. Is the set of whole numbers closed for multiplication?

10. Complete the following using the commutative property of addition:
    $$(x + \underline{\quad}) + (\underline{\quad} + z) = (y + z) + (x + y)$$

11. Is the following an illustration of the associative property for addition and only that property? $(a + b) + c = a + (c + b)$

12. Which number is the multiplicative identity?

13. A television dealer sold 40 television sets one month. If $\frac{3}{8}$ of the sets were color sets, how many color sets did she sell?

14. 24% of 50 is ____.

15. Subtract: $5\frac{2}{3} - 2\frac{3}{4}$

Section 3   Operations with Signed Numbers

Since subtraction is the inverse operation of addition, it is possible to state the **difference** of two numbers in terms of a **sum** of numbers.

> ## Definition
> If a and b are any two elements of a set, then
> the operation **subtraction**, denoted by a – b,
> associates a unique element c, called the
> **difference**, with these two elements such that
> a = b + c.

In the expression a – b (read "a minus b"), a is called the **minuend** and b the **subtrahend** (the quantity to be subtracted).

In the definition of subtraction, what number c can be added to b that produces a as a result? In subtraction, we are asked to find an unknown addend which is called the difference. Thus, the expression 5 – 2 names the whole number 3 because 3 is the number which can be added to 2 to produce a sum of 5.

Do you see any problem here? If we define the difference a – b to be that whole number c such that a = b + c, then expressions such as 3 – 4 have no meaning. Is there a whole number c such that 3 = 4 + c? No. Recall that subtraction of whole numbers is not closed, and 3 – 4 has no whole number result. It will be necessary to invent new numbers that will provide answers for all problems of the form a - b where a and b are whole numbers.

If you earn $5 and then promptly spend $5, the final amount you have is $0. Let us represent the money spent as (-5). We could describe this situation as 5 – 5 = 0 or 5 + (-5) = 0. If it is 4 degrees above zero on a wintery night and the temperature drops 4 degrees, we know that the temperature is 0 degrees. We are saying that 4 – 4 = 0 or 4 + (-4) = 0. These ideas are generalized in the following important property:

> ## Property 9
> THE INVERSE PROPERTY FOR ADDITION
> If x is a whole number, then (-x) is called
> its **additive inverse** and is a unique
> number such that x + (-x) = 0.

Examples

1. Since 2 is a whole number, then (-2) is its unique inverse such that
   $2 + (-2) = 0$. Parentheses are placed around -2. Parentheses are used
   whenever one sign is followed by another sign.

2. Since 0 is a whole number, then (-0) is its unique inverse such that
   $0 + (-0) = 0$.

Problem Set 3.1

1. If $6 - 2 = 4$, then $6 - 4 = 2$. Is it true that any subtraction fact implies another one by simply interchanging the subtrahend and the difference?

2. Is it true that $9 - 4 - 2 = 9 - 2 - 4$? Is it true that the order of the subtrahends may be interchanged in a subtraction expression without affecting the result?

3. What is the additive inverse of 6?

Review Competencies

4. If a, b, and c are whole numbers, state the distributive property of multiplication over addition.

5. Find a replacement for x to illustrate the distributive property.
   $(2 + x) \cdot 4 = (2 \cdot 4) + (3 \cdot 4)$

6. If a and b are whole numbers, state the commutative properties of addition and multiplication.

7. If a, b, and c are whole numbers, state the associative properties of addition and multiplication.

8. Which number is the additive identity? The multiplicative identity?

9. Is the set of whole numbers closed for subtraction?

10. $\frac{3}{5} = \frac{?}{20}$

11. A ten foot length of rope was cut into 4 equal pieces. How long was each piece?

12. Divide: $1.24 \div 0.4$

13. Express 85% as a fraction.

When x is a counting number, its additive inverse (-x) is called a **negative number.** The definition below defines the set of additive inverses of the counting numbers.

> ### Definition
> The set of **negatives of the counting numbers** is the
> set $\{-1,-2,-3,\ldots\}$.

Examples

1. The negative of the counting number 4 is -4 (read "negative 4").

2. The negative of the counting number 5 is -5.

The counting numbers can be thought of as **positive** numbers. 3 is the same as +3, but the positive sign is generally omitted. The negative counting numbers cannot have the negative sign omitted. The whole number zero is neither positive nor negative. Positive and negative numbers are referred to as **signed numbers.**

The dash used to indicate a negative number should not be read as "minus." -3 is "negative 3" and should not be read as "minus 3." The word "minus" is a term that implies the subtraction operation.

There is a difference between a "negative number" and the "negative of a number." The negative of a number is simply the number with the opposite sign. The negative of 4 is -4, but the negative of -4 is 4.

Problem Set 3.2

1. Is there a one-to-one correspondence between the set of counting numbers and the set of the negatives of the counting numbers?

2. Is the negative of a number always a negative number?

3. What is the intersection of the set of counting numbers and the set of the negatives of the counting numbers?

Review Competencies

4. What is the additive inverse of 5?

5. Find a replacement for x to illustrate the distributive property.
$$2 \cdot (3 + 4) = (2 \cdot 3) + (2 \cdot x)$$

6. Is the additive identity for the whole numbers unique?

7. Is the following an illustration of the associative property of multiplication?
$$[a \cdot (b \cdot c)] \cdot x = a \cdot [(b \cdot c) \cdot x]$$

8. Complete the following using the commutative property of addition.
$$(a + b) + c = (b + \underline{\ \ }) + \underline{\ \ }$$

9. Is the set {1} closed for division?

10. $\frac{8}{3} = \frac{24}{?}$

11. Express $\frac{27}{10}$ as a mixed number.

12. Miss Jones weighed 139 pounds before going on a diet. She has now lost $21\frac{2}{3}$ pounds. What is her present weight?

Chapter 3

We have observed that subtraction for whole numbers is neither commutative nor associative. There is no identity element for subtraction because zero only satisfies the requirement when it is the subtrahend. $2 - 0 = 2$, but $0 - 2$ is not 2.

We also know that the set of whole numbers is not closed for subtraction. If we subtract any two whole numbers, the difference is not always a whole number.

To overcome these weaknesses, let us form a new set of numbers which consists of the counting numbers, 0, and the negatives of the counting numbers. We simply enlarge our present set of whole numbers and include the negatives of the counting numbers.

Definitions
The set of **integers** is $\{\ldots,-3,-2,-1,0,1,2,3,\ldots\}$.
The set $\{1,2,3,\ldots\}$ is called the set of **positive integers**.
The set $\{-1,-2,-3,\ldots\}$ is called the set of **negative integers**.
The set $\{0,1,2,3,\ldots\}$ is sometimes called the set of **nonnegative integers.**

Problem Set 3.3

1. Is there any difference between the set of positive integers and the set of counting numbers?

2. Is the union of the set of positive integers and the set of negative integers identical to the entire set of integers? Why?

3. Does the set of integers contain the additive identity and the multiplicative identity?

4. Does every integer have an additive inverse?

Review Competencies

5. What is the additive inverse of -10?

6. True or false? $0 \cdot a = 0$

7. What is the negative of -2?

8. Find a replacement for x to illustrate the distributive property.
$$x \cdot (2 + 3) = (x \cdot 2) + (x \cdot 3)$$

9. True or false? $(5 - 3) - 2 = 5 - (3 - 2)$

10. Multiply: $\frac{3}{4} \cdot 12$

11. Divide: $\frac{5}{6} \div 5$

12. Express $\frac{7}{25}$ as a percent.

A **number line** is shown below this paragraph. Its **origin** is the point marked zero (0). Points to the right of zero are marked with positive integers and those to the left with negative integers. Corresponding to each integer there is precisely one point on the number line.

Positive integers on the number line extend to the right of 0 indefinitely. Negative integers extend to the left of 0 indefinitely. On the number line, the integers are ordered from left to right. An integer is greater than another if it appears to the right of that integer on the number line.

We represent an integer on the number line by drawing an arrow from zero (0) and continuing until the end of the arrow corresponds to the point representing the integer. The figures below show the representations of the integers 3 and -2:

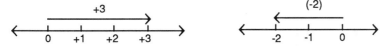

Notice that arrows representing positive integers proceed from left to right, and arrows representing negative integers proceed from right to left. Positive integers and negative integers are represented by **directed distances.** Arrows(directed distances) on the number line are useful in illustrating addition of integers.

Problem 1   Use the number line to add 1 + 2.

Solution   The problem asks for the sum of +1 and +2. Adding positive integers on the number line will imply moving to the right on the line. Thus, 1 + 2 appears as the following on the number line.

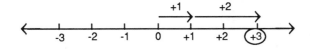

Since the combination of arrows ends at +3, 1 + 2 = 3.

Chapter 3

Problem 2   Use the number line to add -1 + 2.

Solution   The problem asks for the sum of -1 and +2. In the figure below, notice that the arrow representing -1 moves to the left. The arrow representing +2 begins where the arrow representing -1 ends and moves to the right.

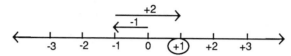

Since the combination of the arrows ends at +1, -1 + 2 = 1.

Problem 3   Use the number line to add -3 + 2.

Solution   The arrows representing -3 and +2 appear below:

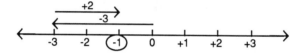

The combination of arrows ends at -1. Thus, -3 + 2 = -1.

Problem 4   Use the number line to add -1 + (-2).

Solution   Adding negative integers on the number line will imply moving to the left on the line. The arrows representing -1 and -2 appear below:

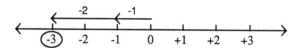

Since the combination of the arrows ends at -3, -1 + (-2) = -3.

The term 2x represents 2 • x. Likewise, the term 3x represents 3 • x. Also, the term x represents 1 • x and 2(3x) is 2 • 3x or 6x. The term ab implies a • b. The terms 2x and 3x are called **similar** or **like** terms because the letter component of each term is identical. It is only possible to add or subtract terms when they are similar. It is not possible to add 2x and 3y because they are not similar terms.

Problem 5   Add 2x + 3x.

Solution   The number line can be used to add 2x and 3x. After the integers +2 and +3 are added, the result is combined with x. Therefore, 2x + 3x = 5x.

$-1 + (-3) = -4$

Problem 6 Add -2x + (-4x)

Solution The sum of -2 and -4 is -6. This result is then combined with x because -2x and -4x are similar terms. Therefore, -2x + (-4x) = -6x.

Problem 7 Add the following terms: -2y + 7x + 5y

Solution Only the terms -2y and 5y are similar terms and can be added. Since -2y + 5y = 3y, the sum for -2y + 7x + 5y is 3y + 7x or 7x + 3y.

Problem 8 Add the following terms: a + 4y + 2b + 3a + y

Solution First combine the similar terms: a + 3a = 4a and 4y + y = 5y.
a + 4y + 2b + 3a + y = 4a + 5y + 2b.

Problem Set 3.4

1. Use the number line to find sums for the following:
   a. -3 + 5 = 2   b. -2 + 2 0   c. 2 + (-4) -2
   d. -2 + 7 5   e. -1 + (-3) +4

2. Add each of the following without using the number line:
   a. -3 + 6 +3   b. 4 + (-8) -4 c. 2 + (-6) -4
   d. -7 + 4 -3   e. -4 + (-5) -9

3. Add the following:
   a. 3y + 4y 7y   b. -5x + 3x -2x
   c. -3a + 6a +3a   d. -3xy + 3xy
   e. -4b + (-3b) -7b f. 6π + 3π 9π
   g. -7ab + 2ab + 3ab h. -3x + 7x + 3a
   i. a + 4y + b + 2y a+b+6y
   j. $x^2 + x + 3x^2 + (-2x)$ $4x^2 + x - 2x$ $4x^2 - x$

Review Competencies

4. Is the set of integers closed for addition?

5. Does the set of integers satisfy the commutative property of addition?

6. Does every integer have an additive inverse?

7. Is the following true or false?
   a + (b • c) = (a + b) • (a + c)

8. What number is the multiplicative identity? The additive identity?

9. Does the following illustrate the associative property of addition?
   [1 + (2 + 3)] + 4 = 1 + [(2 + 3) + 4]

10. Is the set of integers closed for multiplication?

11. Express 1.7 as a percent.

12. What is 120% of 70?

13. 5% of what number is 50?

14. Evaluate: $5 - \frac{3}{4}$

15. Express 65% as a fraction.

16. Express $\frac{7}{4}$ as a decimal.

173

We will now see that it is possible to subtract integers using the number line. To subtract $2 - 3$, recall that $2 - 3$ is the number c such that $2 = 3 + c$. If we start at 3 on the number line, we must move one unit to the left to obtain 2. We have moved one unit in the negative direction.

Thus, $2 = 3 + (-1)$ which means that $c = -1$. Therefore, $2 - 3 = -1$. Notice that $2 - 3$ is the directed distance from 3 to 2. Thus, **a – b is the directed distance from b to a.**

To add $2 + (-3)$ on the number line, the following figure would appear:

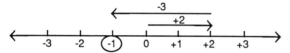

Since we are at -1 on the number line, we say $2 + (-3) = -1$. We have demonstrated that $2 - 3 = -1$ and $2 + (-3) = -1$ which implies that:
$$2 - 3 = 2 + (-3)$$
Thus, subtracting 3 from 2 produces the same result as adding the additive inverse of 3 to 2.

The existence of an additive inverse for each whole number allows us to consider subtraction in terms of addition. **Subtracting b from a** is the same as **adding the additive inverse of b to a.** That is:

$$a - b = a + (-b)$$

Thus, to subtract algebraically, simply change the sign of the subtrahend and add. We can rewrite any statement of subtraction as a statement of addition.

Examples

1. $4 - 1 = 4 + (-1) = 3$

2. $1 - 3 = 1 + (-3) = -2$

3. $-2 - 4 = -2 + (-4) = -6$

4. $3 - (-2) = 3 + (+2) = 5$

5. $-2 - (-1) = -2 + (+1) = -1$

Problem 9   Use the number line to subtract -1 – 2.

Solution   Since a – b is the directed distance from b to a, the expression -1 – 2 is the directed distance from 2 to -1. If we start at +2 on the number line, we must move 3 units to the **left** (-3) to stop at -1. The figure below shows that the directed distance from 2 to -1 is -3. Thus -1 – 2 = -3.

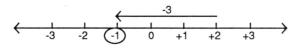

Algebraically, -1 – 2 = -1 + (-2). The arrows representing -1 and -2 appear below and also produce -3 as the result.

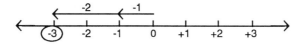

Problem 10   Subtract: -2x – 3x

Solution   Since -2x and 3x are similar terms, we can subtract them. -2 – 3 is the same as -2 + (-3) or -5. Thus, -2x – 3x = -5x.

Problem Set 3.5

1. Use the number line to subtract.
   a. 4 – 1 3
   b. 2 – 5 -3
   c. 0 – 3 -3
   d. -1 – 4 -5

2. Subtract :
   a. 3 – 8 -5
   b. -1 – 6 -7
   c. -4 – 9 -13
   d. 2 – (-4) $+6$
   e. -3 – (-1) -2
   f. 4x – 2x $-2x$
   g. 3y – y $2y$
   h. 2a – 3a $-a$
   i. -2b – 3b $-5b$
   j. 5y – 7y $-2y$
   k. 7x – 2(3x) – 2x $-x$
   l. -3a – (2a)4 – 2a $-13a$
   m. -5x – a – 2x $-7x - a$
   n. bc – 3y – 2bc – y $bc - 4y$
   o. 2xy – 3x – y – x – y $-4x + 2xy - 2y$

3. Is the set of integers closed for subtraction?

4. Does the set of integers satisfy the commutative property of subtraction? no

5. Add:
   a. -6 + 2 -4
   b. 3 + (-5) -2
   c. -4 + 7 $+3$

6. Add:
   a. 4x + 3x $7x$
   b. 2y – 3y $-y$
   c. -3ab + 3ab 0
   d. -7c + 4c $-3c$
   e. -4π + 5π $1π$
   f. 3xy – 6xy + 5xy + y $2xy + y$

7. True or false? a + (b + c) = (a + b) + (a + c) no

8. Evaluate: 7 – 0.003 6.997

9. Express 3% as a decimal. 0.03 ~~0.30~~

10. Express $\frac{1}{8}$ as a decimal. 0.125

Review Competencies

We will now consider the multiplication of two integers. Recall that multiplication is repeated addition. If we wish to multiply two positive integers, +3 and +6, we know that $3 \cdot 6 = 6 + 6 + 6 = 18$. This suggests the following rule:

> ### Rule 1
> **The product of two positive numbers is positive.** $(+x)(+y) = +(xy)$

Using the fact that multiplication is repeated addition, we can multiply a positive integer by a negative integer. Therefore,
$$3 \cdot (-6) = (-6) + (-6) + (-6) = -18.$$
To multiply $(-6) \cdot 3$, we recall the commutative property for multiplication which allows us to restate the problem as $3 \cdot (-6)$. Consequently,
$$(-6) \cdot 3 = 3 \cdot (-6) = -18.$$
This discussion suggests another rule:

> ### Rule 2
> **The product of two numbers having unlike signs is negative.** $(+x)(-y) = -(xy); (-x)(+y) = -(xy)$

Examples

1. $3 \cdot (-5) = 3(-5) = -15.$
2. $(-4) \cdot 6 = (-4)6 = -4 \cdot 6 = -24.$
3. $-2(3) = -2 \cdot 3 = (-2) \cdot 3 = (-2)(3) = -6.$

To consider the product of two negative integers, we recall the distributive property of multiplication over addition: $x(y + z) = (x \cdot y) + (x \cdot z)$. To multiply -3 by -6, consider the following in which $x = -3$, $y = 6$ and $z = -6$:

$x(y + z)$	$= (x \cdot y) + (x \cdot z)$		(Distributive Property)
$-3[6 + (-6)]$	$= (-3)(6) + (-3)(-6)$		(Substitution)
$-3 \cdot 0$	$= -18$	$+ (-3)(-6)$	(Additive Inverse Property)
			$x + (-x) = 0$ and the product of two numbers with unlike signs is negative.
0	$= -18$	$+ (-3)(-6)$	The product of a number and 0 is 0.

This last statement implies that $(-3)(-6)$ must be $+18$ to make the statement true. Thus, $(-3)(-6) = +18$. This discussion suggests the following rule.

> ### Rule 3
> **The product of two negative numbers is positive**
> $(-x)(-y) = +(xy)$

Note
Rules 1 and 3 are often combined to state: **The product of two numbers having like signs is positive.**

Examples

1. (-4) • (-5) = (-4)(-5) = -4 • -5 = 20
2. -2(-4) = (-2)(-4) = 8
3. -(-4) = -1(-4) = 4
4. (-a)(-b) = ab
5. -(1 – 3) = -(-2) = -1(-2) = 2 (First evaluate what is within the parentheses.)
6. -3(-2x) = 6x
7. -5(-2xy) = 10xy

When a product involves more than two signed numbers, count the number of negatives. If there are an **even** number of negative numbers, then the product is **positive.** If there are an **odd** number of negatives, then the product is **negative**.

Examples

1. (-8)(-2)(-5) = -80. There are 3 (odd number) negative numbers, so the product is negative.
2. (-2)(-1)(-3)(-6) = +36. There are 4 (even number) negative numbers, so the product is positive.

Recall again the distributive property of multiplication over addition.
$$x(y + z) = (xy) + (xz)$$

Examples

The distributive property, properly applied, justifies each of the following.
1.   3(a + b) = 3a + 3b
2.   -(x + y) = -1(x + y) = (-1 • x) + (-1 • y) = -x + (-y) = -x – y
3.   q(r – s) = qr – qs
4.   2(w – k) = 2w – 2k
5.   -(a – b) = -1(a – b) = (-1a) – (-1b) = -a – (-b) = -a + b
6.   a(b + c + d) = ab + ac + ad
7.   a(b – c – d) = ab – ac – ad
8.   -3(a – b) = -3a + 3b
9.   -(x + 2y) = -x – 2y
10. -2(-2x + 3y) = 4x – 6y
11. -2(2x – y – 3z) = -4x + 2y + 6z
12. -(x + 2y – 4z) = -x – 2y + 4z

Chapter 3

Evaluate:

1. $3 \cdot (-5)$ -15
2. $5 \cdot (-4)$ -20
3. $0 \cdot (-2)$ 0
4. $(-3) \cdot 6$ -18
5. $(-5) \cdot 6$ -30
6. $(-6)(-6)(-6)$ -216

7. $(-2)(-5)$ +10
8. $(-4)(-7)(-1)(-2)$ 56
9. $-(-4)$ +4
10. $-(-6)$ +6
11. $-(2-3)$ +1
12. $-(1-4)$ +3
13. $-(-6+3)$ +3 (-3)

14. $-(-5-2)$ +7
15. $-4-(1-5)$ 0
16. $-6-(2-3)$ -5
17. $-4-(5-2)$ -7
18. $-7-(2-6)-4$ -7
19. $3(-5x)$ -15x

20. $-2(3yz)$ -6yz
21. $-4(-5t)$ +20t
22. $-4(2x-3y)$ -8x-12y
23. $-2(-3x+4y)$ +6x-8y
24. $-(2x+y-3z)$ -2x-y+3z
25. $-3(-x-2y+3z)$ +3x+6y-9z

26. Subtract:
    a. $5-7$
    b. $0-5$
    c. $-3-2$
    d. $4y-y$
    e. $4x-6x$
    f. $-3a-4a$

27. Add:
    a. $-1+5$
    b. $3+(-4)$
    c. $-8+3$
    d. $-4a+7a$
    e. $5\pi - 8\pi$
    f. $-10x+7x$

28. What is the additive inverse of 8?

29. What is the negative of -6?

30. $\frac{3}{7} = \frac{12}{?}$

31. Add: $2\frac{3}{5} + 1\frac{1}{6}$

32. Divide: $1\frac{1}{3} \div 2\frac{1}{4}$

33. A football team won 5 games and lost 6 games. What fraction of the games did the team lose?

34. Multiply: $2.3 \cdot 0.002$

35. Express 35% as a fraction.

36. Express $4\frac{1}{10}$ as an improper fraction.

37. Express $8\frac{1}{2}\%$ as a decimal.

We now consider the fourth basic operation — division. We observed that subtraction can be defined in terms of addition, which is its inverse operation. Thus, we are motivated to define division in terms of its inverse operation, which is multiplication.

> ## Definition
>
> If a and b are two elements of a set (b not zero), then the operation **division**, denoted by $a \div b$, associates a unique number c, called the **quotient**, with these two elements such that $a = b \cdot c$.

The expression a ÷ b may also be represented as $\frac{a}{b}$. The non-zero number b is said to **divide** a. The quantity a is called the **dividend** and the quantity b is called the **divisor**.

Certainly 3 divides 6 (6 ÷ 3 or $\frac{6}{3}$) since there is a number, 2, such that $6 = 3 \cdot 2$. If $\frac{6}{3} = 2$, then $6 = 3 \cdot 2$.

It is also true that 4 divides 0 (0 ÷ 4 or $\frac{0}{4}$) since there is a number, 0, such that $0 = 4 \cdot 0$. It is also true that 3 divides 0, 5 divides 0 and, in general, any non-zero integer divides 0 and provides 0 as the quotient.

Can it also be said that 0 divides 4 (4 ÷ 0 or $\frac{4}{0}$)? Is there a number c such that $4 = 0 \cdot c$?

There can be no such number since the product of zero and any other number is 0. Thus, an expression like $\frac{4}{0}$ has no meaning. What about $\frac{0}{0}$?

$$\text{If } \frac{0}{0} = c \text{ then } 0 = 0 \cdot c.$$

In this case, c can be **any** number. Thus $\frac{0}{0}$ is said to be **indeterminate**. We say that **division by 0 is undefined** and not permitted. This is why our definition of division indicated that if the number b divides a $(\frac{a}{b})$, then b cannot be zero.

The quotient $\frac{a}{b}$ may be restated as $a \cdot \frac{1}{b}$. Thus, $\frac{2}{3} = 2 \cdot \frac{1}{3}$ and $\frac{x}{3} = x \cdot \frac{1}{3}$ or $\frac{1}{3} \cdot x$.

The quotient $\frac{a}{b}$ can represent an integer. Certainly $3 = \frac{3}{1}$ and $-4 = \frac{-4}{1}$. Perhaps you have already observed that the quotient a ÷ b does not always represent an integer. Certainly 2 ÷ 3 $\left(\frac{2}{3}\right)$ is not an integer.

We know numbers such as $\frac{2}{3}, \frac{4}{5}, \frac{2}{1}$, and $\frac{-3}{2}$ are **fractions** which we use to make comparisons of two quantities.

We agree that fractions are numbers, yet many of them are not contained in the set of integers. Have we discovered a weakness in the set of integers? Yes. It appears that the set of integers is not closed for division. We must form a

Chapter 3

new set of numbers that contains these additional numbers as well as the ones we have already been using. We will then determine whether this new set behaves like the set of integers.

The existence of an additive inverse motivates us to explore the possibility of a multiplicative inverse. We know that addition of a number and its additive inverse produces the additive identity. Might there be some special number that can be multiplied by another number to produce the multiplicative identity (1)?

For example, is there a number x such that $4 \cdot x = 1$? Yes. The number is $\frac{1}{4}$. Is there a number x such that $\frac{2}{3} \cdot x = 1$? Yes. The number is $\frac{3}{2}$. This discussion leads us to another property.

Property 10

THE INVERSE PROPERTY FOR MULTIPLICATION

If $\frac{x}{y}$ is a number, $x \neq 0$ and $y \neq 0$, then $\frac{y}{x}$ is its

multiplicative inverse such that $\frac{x}{y} \cdot \frac{y}{x} = 1$.

$\frac{y}{x}$ is also called the **reciprocal** of $\frac{x}{y}$.

Examples

1. The inverse of 3 $\left(\frac{3}{1}\right)$ is $\frac{1}{3}$ because $3 \cdot \frac{1}{3} = 1$.

2. The inverse of $\frac{1}{2}$ is 2 $\left(\frac{2}{1}\right)$ because $\frac{1}{2} \cdot 2 = 1$.

3. The inverse of $\frac{3}{4}$ is $\frac{4}{3}$ because $\frac{3}{4} \cdot \frac{4}{3} = 1$. Observe that $\frac{4}{3}$ is neither an integer nor an inverse of an integer.

You may find it helpful to remember that inverses restore the identity . The additive inverse brings us back to the identity of addition, zero. For example:
$$3 + (-3) = 0 \quad \text{The identity, (0), is restored.}$$

The multiplicative inverse brings us back to the identity of multiplication, one. For example:
$$3 \cdot \frac{1}{3} = 1 \quad \text{The identity, (1), is again restored.}$$

Problem Set 3.7

1. Why is there no multiplicative inverse for 0?

2. Express the multiplicative inverse for:

    a.   5   b. $\frac{1}{4}$ c. $\frac{5}{4}$

3. The set of integers contains the additive inverses of the counting numbers. Does the set of integers contain the multiplicative inverses of the counting numbers?

4. Does $\frac{a}{b}$ equal $a \cdot \frac{1}{b}$?

5. Perform the following operations.
   a. $2 \cdot (-5)$ b. $(-4)(6)(-1)$ c. $(-3)(-9)$
   d. $-(-8)$ e. $-(-4 - 3)$ f. $-5 - (1 - 6)$
   g. $4 - (5 - 3)$ h. $5 + (-9)$ i. $-6 + 3$
   j. $-2 + 1$ k. $-7x - 2x$ l. $-10x + 4x$
   m. $-5 - 5$

6. Express $\frac{11}{25}$ as a percent.

7. Find 130% of 90.

8. 15% of what number is 45?

9. Express $\frac{31}{8}$ as a mixed number.

10. Express 240% as a decimal.

We are now prepared to form a new set of all the numbers that can be expressed in fractional form using integers.

> ## Definition
> If x and y are two integers, y not zero, then a **rational number** is one that is expressed in the form $\frac{x}{y}$.

Rational numbers are numbers that appear in fractional form or as a quotient of two integers, such as $\frac{3}{4}, \frac{5}{1}, \frac{0}{3}, \frac{-3}{4}, \frac{2}{-3}$, and $\frac{-3}{-4}$. Notice the word **ratio** in rational.

A ratio of two numbers x and y is a comparison of x to y by division. A ratio is often expressed as a fraction. Ratios will be considered more fully in a later chapter.

When expressed in decimal form, a rational number either comes to an end or has a repeating pattern. For example, $\frac{1}{2} = 0.5$ (terminates or comes to an end) and $\frac{1}{3} = 0.333\ldots$ (repeats the digit 3 forever which is expressed as $0.\overline{3}$).

Notice that the rational number $\frac{3}{11}$ can be changed to its decimal equivalent by dividing $3.000\ldots$ by 11. $\frac{3}{11} = 0.272727\ldots = 0.\overline{27}$

Problem Set 3.8

1. The set of rational numbers is an extension of the set of integers. Can every integer also be expressed as a rational number? Can every rational number be expressed as an integer?

2. Is the set of integers closed for division (except by 0)?

3. Express the multiplicative inverses for:
   a. 10   b. $\frac{1}{5}$ c. $\frac{2}{3}$

4. Perform the following operations.
   a. $(-2)(8)$ b. $(-4)(-9)(-1)$ c. $-(-10)$
   d. $-(-2 - 5)$ e. $-7 - (2 - 3)$ f. $4 + (-8)$
   g. $-4a - 2a$ h. $-7a + 5a$ i. $-2\pi + 5\pi$
   j. $-3 - 5$

5. Express 5 as a percent.

6. What percent of 70 is 28?

7. Express 17.5% as a decimal.

8. Divide: $0.366 \div 0.6$

9. Divide: $6 \div 2\frac{2}{3}$

10. Multiply: $\frac{5}{6} \cdot \frac{3}{4} \cdot \frac{3}{10}$

Just as it is possible to express the negative of a positive integer, it is possible to express the negative of a rational number. The negative of $\frac{2}{3}$ is $\frac{-2}{3}$ or $-\frac{2}{3}$ or $\frac{2}{-3}$. The negative of $\frac{4}{2}$ is $\frac{-4}{2}$ or $-\frac{4}{2}$ or $\frac{4}{-2}$. Since $\frac{4}{2} = 2$, the simplest way to express its negative is -2.

Recall that the definition of division states that the expression $a \div b$ is that number c such that $a = b \cdot c$. That definition is crucial in understanding the following rules which govern the signs for dividing rational numbers.

> **Rule 1**
>
> **The quotient of two positive numbers is positive.**
>
> $\frac{+x}{+y} = +\frac{x}{y}$

Example

$10 \div 5 = 2$ because $10 = 5 \cdot 2$.

> **Rule 2**
>
> **The quotient of two numbers having unlike signs is negative.**
>
> $\frac{+x}{-y} = -\frac{x}{y}$ and $\frac{-x}{+y} = -\frac{x}{y}$

Examples

1. $6 \div (-2) = -3$ because $6 = (-2) \cdot (-3)$.
2. $-6 \div 2 = -3$ because $-6 = 2 \cdot (-3)$.
3. $-2 \div 6 = \frac{-1}{3}$ because $-2 = 6 \cdot \left(\frac{-1}{3}\right)$.
4. $\frac{6}{-2} = -3$
5. $\frac{-8}{4} = -2$
6. $\frac{-10x}{5} = -2x$
7. $\frac{6ab}{-3} = -2ab$
8. $-5 \cdot \frac{x}{5} = -5 \cdot \frac{1}{5} \cdot x = -1 \cdot x = -x$
9. $-8 \cdot \frac{x}{2} = -8 \cdot \frac{1}{2} x = -4x$

> ### Rule 3
> **The quotient of two negative numbers is positive.**
> $$\frac{-x}{-y} = +\frac{x}{y}$$

Examples

1. $(-12) \div (-4) = 3$ because $-12 = (-4) \cdot 3$.

2. $(-4) \div (-8) = \frac{1}{2}$ because $-4 = (-8) \cdot \frac{1}{2}$.

3. $\frac{-8}{-2} = 4$

4. $\frac{-10a}{-2} = 5a$

5. $\frac{-7abc}{-7} = abc$

Note

As in the case of multiplication, Rules 1 and 3 can be combined as follows: **The quotient of two numbers having like signs is positive.**

Problem Set 3.9

1. Perform each of the following divisions.
    a. $8 \div (-2)$　　b. $10 \div (-5)$ -2　c. $-4 \div 2$ -2
    d. $-9 \div (-3)$ 3　e. $-8 \div 8$ -1　　f. $-2 \div (-4)$ 0.5
    g. $-3 \div 9$ 0.333　h. $4 \div (-12)$ 0.333

2. Divide each of the following.
    a. $-18 \div \frac{3}{4}$ 0.25　b. $\frac{4}{5} \div \frac{-3}{7}$ 0.428　c. $\frac{-3}{8} \div \frac{-4}{7}$ 0.65625

3. Find the result for each of the following.
    a. $\frac{4x}{-2}$ $-2x$　　b. $\frac{-6x}{3}$ $-2x$　　c. $\frac{-4x}{-4}$ $1x$
    d. $4 \cdot \frac{x}{4}$ $4x$ $1x$　e. $6 \cdot \frac{x}{3}$ $2x$　f. $-4 \cdot \frac{x}{2}$ $-2x$
    g. $\frac{8xy}{-2}$ $-4xy$　h. $\frac{-6ab}{-3}$ $2ab$　i. $\frac{-10abc}{5}$ $-2abc$

4. Is the set of rational numbers closed for addition? For multiplication?

5. Do the commutative properties of addition and multiplication hold for the set of rational numbers?

6. Do the associative properties of addition and multiplication hold for the set of rational numbers?

7. Does the set of rational numbers contain the additive identity and the multiplicative identity?

Review Competencies

8. State the multiplicative inverse for:
    a. 4　　　b. $\frac{1}{7}$　　　c. $\frac{5}{6}$

9. Perform the following operations.
    a. $5 \cdot (-4)$　　b. $(-5)(-8)$　　c. $-(-12)$
    d. $-(2 - 10)$　　e. $5 - (4 - 1)$
    f. $-7 - (-2 - 6)$　g. $3 + (-3)$　h. $-7 + 1$
    i. $-6 - 7$　　j. $10a - 11a$　k. $-9x - 3x$

10. Express 55% as a fraction.

11. What percent of 80 is 8?

12. Express $\frac{12}{50}$ as a percent.

13. Subtract: $7 - 0.004$

14. Divide: $3\frac{2}{3} \div 1\frac{2}{5}$

183

Section 4   Square Roots and their Operations

In an expression such as y^2, the 2 is called an **exponent**. (Exponents will be discussed more fully later.) For our purposes, $y^2 = y \cdot y$. Thus, $3^2 = 3 \cdot 3$ and $(-4)^2 = (-4)(-4)$. This discussion leads to another type of number.

> ## Definition
> A **square root** of a non-negative number x is one of two equal numbers whose product is x.

Every positive rational number x has two square roots, y and -y, such that $y^2 = x$ and $(-y)^2 = x$. The positive square root of x is the **principal square root** and is indicated by the use of a radical symbol $\sqrt{\ }$. $\sqrt{x}$ is called a **radical expression.**

Examples

1. $2^2 = 4$ and $(-2)^2 = 4$. Both 2 and -2 are square roots of 4, but 2 is the principal square root and $\sqrt{4} = 2$. If both square roots are desired, we write $\pm\sqrt{4} = \pm 2$. If only the negative square root is desired, we write $-\sqrt{4} = -2$.

2. $\sqrt{\frac{1}{9}} = \frac{1}{3}$. Both fractions, $\frac{1}{3}$ and $\frac{-1}{3}$, are square roots of $\frac{1}{9}$, but the radical sign not prefixed by $\pm$ indicates the principal square root, which is $\frac{1}{3}$.

3. $\sqrt{5}$ cannot be exactly expressed as a rational number. However, 5 has two square roots, one positive and one negative, and its principal square root is approximately 2.236. $\sqrt{5}$ is not a rational number.

4. $\sqrt{-4}$ is a radical expression that names no number on the number line because there is no rational number that can be used as a factor two times to produce -4 .

Problem Set 4.1

1. Determine each of the following exactly, if possible.
   a. $\sqrt{25}$ b. $\sqrt{20}$ c. $\sqrt{36}$
   d. $\sqrt{\frac{4}{25}}$ e. $\sqrt{6}$ f. $\sqrt{0}$

2. Which of the following are rational numbers?
   a. $\sqrt{49}$ b. $\sqrt{17}$
   c. $\sqrt{9}$ d. $\sqrt{2}$

3. If x is a rational number, is $\sqrt{x}$ necessarily a rational number?

5. Express the negative of $\frac{3}{4}$ in three ways.

6. Subtract: $\frac{7}{8} - \frac{2}{3}$

7. What is 15% of 40?

8. Divide: $5 \div 0.001$

9. Express $3\frac{8}{9}$ as an improper fraction.

10. Express $4\frac{7}{10}$ % as a decimal.

Review Competencies

4. Perform the following operations.
   a. $6 \div (-3)$ b. $(-10) \div 10$ c. $(-5) \div (-5)$
   d. $(-4) \div (-12)$ e. $(-2) \div (-8)$ f. $(-20) \div \frac{4}{5}$
   g. $\frac{2}{3} \div \left(\frac{-8}{9}\right)$ h. $\left(\frac{-4}{3}\right) \div \left(\frac{-1}{12}\right)$ i. $\frac{6x}{-3}$
   j. $\frac{-9x}{-3}$ k. $-8 \cdot \frac{x}{2}$ l. $(-6)(-7)$
   m. $-3 - 8$ n. $-9 - (4 - 3)$

Let us now consider the various operations with square roots (radical expressions). We will see that radical expressions can be simplified, added, subtracted, multiplied, and divided.

Certain radical expressions can be immediately simplified. For example, $\sqrt{25} = 5$ and $\sqrt{36} = 6$. Other radical expressions, such as $\sqrt{3}$ or $\sqrt{5}$, cannot be simplified other than to write a decimal approximation for the square root. For instance, $\sqrt{3}$ is not a rational number, but is approximately equal to 1.73 correct to the nearest hundredth.

Sometimes it is necessary to break a radical expression into two factors such that one of the factors can be simplified. The basic justification for this simplification is the following:

$$\sqrt{ab} = \sqrt{a} \cdot \sqrt{b} \text{ where both a and b are non-negative.}$$

185

Chapter 3

Problem 1   Simplify $\sqrt{75}$

Solution $\sqrt{75} = \sqrt{25 \cdot 3} = \sqrt{25} \cdot \sqrt{3} = 5\sqrt{3}$.

Problem 2   Simplify $\sqrt{125}$

Solution $\sqrt{125} = \sqrt{25 \cdot 5} = \sqrt{25} \cdot \sqrt{5} = 5\sqrt{5}$

Problem 3   Simplify $\sqrt{21}$

Solution $\sqrt{21} = \sqrt{7 \cdot 3} = \sqrt{7} \cdot \sqrt{3}$. Since neither $\sqrt{7}$ nor $\sqrt{3}$ can be further simplified, it is better to leave $\sqrt{21}$ in its original form.

Problem Set 4.2

1. Where possible, simplify each of the following radical expressions.

    a. $\sqrt{12}$ b. $\sqrt{150}$ c. $\sqrt{18}$
    d. $\sqrt{20}$ e. $\sqrt{63}$ f. $\sqrt{70}$
    g. $\sqrt{72}$

Review Competencies

2. Determine the following.

    a. $\sqrt{64}$ b. $\sqrt{\dfrac{16}{25}}$ c. $\sqrt{0}$

3. Perform the following operations.

    a. $-28 \div 7$ b. $\dfrac{-7}{8} \div \dfrac{-3}{4}$

    c. $\dfrac{-10 \cdot x}{x}$ d. $-2 - 10$

4. Express 250% as a decimal and a fraction.

5. Express 0.058 as a percent.

Recall that it is only possible to add or subtract similar terms. Radical expressions may be added or subtracted when they are similar. For example, $3\sqrt{2}$ and $4\sqrt{2}$ are similar radical expressions and may be added or subtracted. However, $2\sqrt{2}$ and $4\sqrt{3}$ are not similar radical expressions and may not be added or subtracted.

Problem 4   Simplify $7\sqrt{3} + 4\sqrt{3}$.

Solution   There are two terms in the expression $7\sqrt{3} + 4\sqrt{3}$, and the terms are separated by the addition sign. These are similar terms because they have the same radical expression, $\sqrt{3}$. Since $7x + 4x = 11x$, the same idea is used in simplifying $7\sqrt{3} + 4\sqrt{3}$.

$$\text{Thus, } 7\sqrt{3} + 4\sqrt{3} = 11\sqrt{3}$$

Problem 5   Simplify $\sqrt{5} - 3\sqrt{5}$.

Solution $\sqrt{5}$ and $3\sqrt{5}$ are similar terms. $\sqrt{5}$ means $1\sqrt{5}$. Since $x - 3x$ simplifies to $-2x$, we apply the same process.

$$\text{Thus, } \sqrt{5} - 3\sqrt{5} = -2\sqrt{5}$$

Problem 6   Simplify $7\sqrt{3} + 3\sqrt{7}$.

Solution $\sqrt{3}$ and $\sqrt{7}$ are different radical expressions. Therefore, $7\sqrt{3}$ and $3\sqrt{7}$ are not similar terms, and $7\sqrt{3} + 3\sqrt{7}$ cannot be simplified.

In some instances it is necessary to simplify individual terms before adding or subtracting. That process is shown in the following problem.

Problem 7   Simplify $\sqrt{80} + 7\sqrt{5}$.

Solution   It initially appears that $\sqrt{80}$ and $7\sqrt{5}$ are not similar terms, but $\sqrt{80}$ needs to be simplified before making that decision.

$$\begin{aligned}
\sqrt{80} + 7\sqrt{5} &= \sqrt{16 \cdot 5} + 7\sqrt{5} \\
&= \sqrt{16} \cdot \sqrt{5} + 7\sqrt{5} \\
&= 4\sqrt{5} + 7\sqrt{5} \\
&= 11\sqrt{5}
\end{aligned}$$

Let us now summarize the procedure for adding or subtracting radical expressions.

> **To add or subtract radical expressions:**
> 1. If possible, simplify each radical expression.
> 2. Add or subtract the similar terms.

Chapter 3

Problem 8   Simplify $3\sqrt{54} + \sqrt{24} - 10\sqrt{6}$.

Solution   The first two terms must be simplified before any addition and subtraction can be performed.

$$\begin{aligned}
3\sqrt{54} + \sqrt{24} - 10\sqrt{6} &= 3\sqrt{9 \cdot 6} + \sqrt{4 \cdot 6} - 10\sqrt{6} \\
&= 3\sqrt{9} \cdot \sqrt{6} + \sqrt{4} \cdot \sqrt{6} - 10\sqrt{6} \\
&= 3 \cdot 3\sqrt{6} + 2\sqrt{6} - 10\sqrt{6} \\
&= 9\sqrt{6} + 2\sqrt{6} - 10\sqrt{6} \\
&= \sqrt{6}
\end{aligned}$$

Problem 9   Simplify $7\sqrt{5} - \sqrt{125} + \sqrt{18}$.

Solution

$$\begin{aligned}
7\sqrt{5} - \sqrt{125} + \sqrt{18} &= 7\sqrt{5} - \sqrt{25 \cdot 5} + \sqrt{9 \cdot 2} \\
&= 7\sqrt{5} - \sqrt{25} \cdot \sqrt{5} + \sqrt{9} \cdot \sqrt{2} \\
&= 7\sqrt{5} - 5\sqrt{5} + 3\sqrt{2} \\
&= 2\sqrt{5} + 3\sqrt{2}
\end{aligned}$$

The result can be simplified no further because the radical expressions are not similar.

Problem Set 4.3

1. If possible, simplify:

    a. $8\sqrt{5} + 11\sqrt{5}$ b. $9\sqrt{6} + 5\sqrt{6} - 15\sqrt{6}$

    c. $7\sqrt{11} + 7\sqrt{5}$ d. $\sqrt{48} + \sqrt{12}$

    e. $\sqrt{50} - \sqrt{18} - 8\sqrt{2}$ f. $11\sqrt{7} + \sqrt{28} - 3\sqrt{63}$

    g. $\sqrt{48} - \sqrt{75} + \sqrt{18} - 2\sqrt{8}$

Review Competencies

2. Simplify:
    a. $\sqrt{32}$ b. $\sqrt{72}$

3. Which of the following numbers are rational?
    a. $\sqrt{5}$ b. $\sqrt{36}$ c. $\sqrt{0}$ d. $\sqrt{15}$

4. What is the multiplicative inverse of $\frac{4}{3}$?

5. What is the additive inverse of -6?

The following problems deal with the multiplication and division of radical expressions. You will notice that it is not necessary to have similar terms for multiplication and division.

Let us summarize the procedures for multiplying and dividing radical expressions on the next page.

> To **multiply radicals**, use the following:
> $\sqrt{a} \cdot \sqrt{b} = \sqrt{ab}$ when a and b are non-negative.
>
> To **divide radicals**, use the following:
> $\dfrac{\sqrt{a}}{\sqrt{b}} = \sqrt{\dfrac{a}{b}}$, $b \neq 0$
>
> Always simplify, if possible, the radical expression that results.

Problem 10   Multiply: $\sqrt{5} \cdot \sqrt{6}$

Solution $\sqrt{5} \cdot \sqrt{6} = \sqrt{5 \cdot 6} = \sqrt{30}.$ $\sqrt{30}$ cannot be simplified.

Problem 11   Divide: $\dfrac{\sqrt{6}}{\sqrt{2}}$

Solution $\dfrac{\sqrt{6}}{\sqrt{2}} = \sqrt{\dfrac{6}{2}} = \sqrt{3}$

Problem 12   Multiply: $3\sqrt{5} \cdot 2\sqrt{5}$

Solution
$$
\begin{aligned}
3\sqrt{5} \cdot 2\sqrt{5} &= 3 \cdot 2 \cdot \sqrt{5} \cdot \sqrt{5} \\
&= 6 \cdot \sqrt{25} \\
&= 6 \cdot 5 \\
&= 30
\end{aligned}
$$

Problem 13   Multiply: $(7\sqrt{3})(-2\sqrt{6})$

Solution
$$
\begin{aligned}
(7\sqrt{3})(-2\sqrt{6}) &= 7(-2) \cdot \sqrt{3} \cdot \sqrt{6} \\
&= -14 \cdot \sqrt{18} \\
&= -14 \cdot \sqrt{9 \cdot 2} \\
&= -14 \cdot \sqrt{9} \cdot \sqrt{2} \\
&= -14 \cdot 3 \cdot \sqrt{2} \\
&= -42\sqrt{2}
\end{aligned}
$$

Chapter 3

Problem 14 Divide: $\dfrac{8\sqrt{45}}{-2\sqrt{5}}$

Solution $\qquad \dfrac{8\sqrt{45}}{-2\sqrt{5}} = \dfrac{8}{-2} \cdot \sqrt{\dfrac{45}{5}} = -4\sqrt{9} = -4 \cdot 3 = -12$

Problem Set 4.4

1. Perform the indicated operations.

   a. $\sqrt{7} \cdot \sqrt{3}$ b. $\dfrac{\sqrt{10}}{\sqrt{2}}$

   c. $2\sqrt{6} \cdot 4\sqrt{6}$ d. $(9\sqrt{2})(-4\sqrt{12})$

   e. $(-3\sqrt{5})(-2\sqrt{10})$ f. $\dfrac{16\sqrt{12}}{-2\sqrt{3}}$

   g. $\dfrac{-25\sqrt{15}}{-5\sqrt{3}}$

Review Competencies

2. Simplify if possible:
   a. $3\sqrt{8} - 5\sqrt{8} + 7\sqrt{8}$ b. $\sqrt{24} - \sqrt{54}$

3. Perform the following operations.

   a. $(-4)(-2)$ b. $-(-3)$ c. $-(-3 - 4)$
   d. $-6 - (1 - 5)$ e. $5 + (-6)$ f. $-3 + (-7)$
   g. $-2b - 6b$ h. $-1 - 7$

4. Express $\dfrac{37}{8}$ as a mixed number.

5. Express $\dfrac{3}{5}$ as a percent.

6. How many ribbons $2\frac{1}{4}$ feet long can be cut from a ribbon 12 feet long?

One type of simplification of fractions containing radical expressions is called **rationalizing the denominator.** The process of rationalizing the denominator refers to the elimination of a radical expression in the denominator of a fraction without changing the value of the fraction. With square roots, this can be accomplished by multiplying the numerator and denominator of the fraction by whatever radical expression appears in the denominator.

Problem 15 Rationalize the denominator of $\dfrac{1}{\sqrt{2}}$.

Solution

To remove the radical expression from the denominator, both numerator and denominator are multiplied by $\sqrt{2}$.

$$\frac{1}{\sqrt{2}} = \frac{1}{\sqrt{2}} \cdot \frac{\sqrt{2}}{\sqrt{2}} = \frac{\sqrt{2}}{\sqrt{4}} = \frac{\sqrt{2}}{2}$$

Observe that by multiplying by $\dfrac{\sqrt{2}}{\sqrt{2}}$, we are really multiplying by 1, the multiplicative identity. Any fraction can be multiplied by the same expression in the numerator and denominator without changing the value of the fraction.

In some situations it is easier to first simplify the expression in the denominator and then rationalize the denominator. The general procedure for rationalizing a denominator is stated below.

> **To rationalize the denominator of a fraction:**
> 1. If possible, simplify the radical expression in the denominator.
> 2. Multiply the numerator and denominator by the radical expression in the denominator.

Problem 16 Rationalize the denominator of $\dfrac{1}{\sqrt{75}}$.

Solution Simplify $\sqrt{75}$. $\qquad \sqrt{75} = \sqrt{25 \cdot 3} = \sqrt{25} \cdot \sqrt{3} = 5\sqrt{3}$

Rationalize the denominator by multiplying by $\dfrac{\sqrt{3}}{\sqrt{3}}$ which is equal to 1. Thus,

$$\frac{1}{\sqrt{75}} = \frac{1}{5\sqrt{3}} = \frac{1}{5\sqrt{3}} \cdot \frac{\sqrt{3}}{\sqrt{3}} = \frac{\sqrt{3}}{5\sqrt{9}} = \frac{\sqrt{3}}{5 \cdot 3} = \frac{\sqrt{3}}{15}$$

Since all the problems we are considering involve square roots, we always obtain the following pattern occurring in the denominator.

$$\frac{1}{\sqrt{a}} = \frac{1}{\sqrt{a}} \cdot \frac{\sqrt{a}}{\sqrt{a}} = \frac{\sqrt{a}}{\sqrt{a^2}} = \frac{\sqrt{a}}{a}$$

Problem 17 Rationalize the denominator of $\dfrac{3}{\sqrt{60}}$.

Solution Simplify the denominator.

$$\frac{3}{\sqrt{60}} = \frac{3}{\sqrt{4 \cdot 15}} = \frac{\sqrt{3}}{\sqrt{4} \cdot \sqrt{15}} = \frac{3}{2\sqrt{15}}$$

Rationalize the denominator. Thus,

$$\frac{3}{\sqrt{60}} = \frac{3}{2\sqrt{15}} = \frac{3}{2\sqrt{15}} \cdot \frac{\sqrt{15}}{\sqrt{15}} = \frac{3\sqrt{15}}{2\sqrt{225}} = \frac{3\sqrt{15}}{2 \cdot 15} = \frac{3\sqrt{15}}{30} = \frac{\sqrt{15}}{10}$$

Problem Set 4.5

1. Rationalize the denominators of:

a. $\dfrac{1}{\sqrt{5}}$ b. $\dfrac{1}{\sqrt{48}}$ c. $\dfrac{3}{\sqrt{7}}$

2. Perform the indicated operations.

a. $\sqrt{6} \cdot \sqrt{8}$ b. $(-2\sqrt{4})(3\sqrt{9})$

d. $\dfrac{4}{\sqrt{24}}$ e. $\dfrac{-7}{\sqrt{6}}$ f. $\dfrac{-8}{\sqrt{60}}$

c. $\dfrac{20\sqrt{18}}{-4\sqrt{2}}$

g. $\dfrac{15}{\sqrt{27}}$

3. Simplify, if possible: $\sqrt{12} - \sqrt{27}$

Recall that every rational number, when expressed in decimal form, either terminates or has a repeating pattern. This is not true of numbers such as $\sqrt{2}$. If we attempt to express $\sqrt{2}$ in decimal form, the decimal will never precisely equal 2 when squared. For example, $\sqrt{2}$ may approximately equal 1.4142, but

$$(1.4142)(1.4142) = 1.9999616 (\text{not quite } 2).$$

A better approximation for $\sqrt{2}$ is 1.4142135, but even this number, when used as a factor two times, will not produce 2. The decimal for $\sqrt{2}$ appears to be longer with no repeating pattern.

> **Definition**
> An **irrational number** is a number that, when expressed in decimal form, neither terminates nor repeats.

Examples

Numbers such as 6, $2\sqrt{3}$, $3 + 2\sqrt{8}$, and π, (read **pi** and only approximately equal to 3.14) are irrational numbers.

Since a number cannot be both rational and irrational, irrational numbers cannot be expressed as the ratio of two integers.

Problem Set 4.6

1. Which of the following numbers are rational and which are irrational?

    a. $\sqrt{7}$ b. $\sqrt{1}$ c. $\sqrt{24}$
    d. $\sqrt{25}$ e. $\sqrt{0}$

2. What is the intersection of the set of rational numbers and the set of irrational numbers?

3. If x is a rational number, is $\sqrt{x}$ always an irrational number?

Review Competencies

4. Rationalize the denominators of:

    a. $\dfrac{3}{\sqrt{8}}$ b. $\dfrac{-2}{\sqrt{32}}$

Let us now combine the set of rational numbers with the set of irrational numbers to form a new set of numbers.

> ### Definition
> The set of **real numbers** is the union of the set of rational numbers and the set of irrational numbers.

The fact that the set of real numbers contains both the set of rational numbers and the set of irrational numbers is reinforced by the following diagram.

Real Numbers

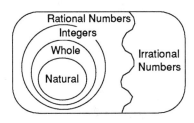

The number line discussed earlier in this chapter can be called the **real number line**. All the rules for multiplying and dividing signed numbers hold for the set of real numbers as do all the properties (commutative, associative, distributive, identities, and inverses).

Problem Set 4.7

1. Is it true that every rational number is a real number, but not every real number is a rational number?

2. Given the set of real numbers as the universal set, what is the complement of the set of irrational numbers?

3. Is the set of real numbers closed for addition and multiplication?

4. Do the commutative properties of addition and multiplication hold for the set of real numbers?

5. Do the associative properties of addition and multiplication hold for the set of real numbers?

6. Does the distributive property hold for the set of real numbers?

7. Does the set of real numbers contain the additive and multiplicative identities?

8. Which one of the following is not true for all real numbers?
   a. $7x + 7y = 7(x + y)$
   b. $5xy(3x + y) = 5xy(3y + x)$
   c. $(x - y)(2x + y) = (2x + y)(x - y)$
   d. $4(xy) = (4x)y$

9. Which one of the following is not true for all real numbers?
   a. $8 + (xy) = (8 + x)(8 + y)$
   b. $6x + 5y = y5 + 6x$
   c. $10y + 2x = 2(5y + x)$
   d. $(2x + 5y)(2x - 5y) = (2x - 5y)(5y + 2x)$

10. Which one of the following is not true for all real numbers?
    a. $7 - 3(x + 2y) = 7 - 3x - 6y$
    b. $8y + 2x = 2(x + 4y)$
    c. $7x + (-7x) = 0$
    d. $7(x + y) = 7x + y$

11. Which one of the following is not true for all real numbers?
    a. $(x - 3y)(5x + y) = (y + 5x)(x - 3y)$
    b. $6(ab) = (6a)(6b)$
    c. $7 - 4(a - 3b) = 7 - 4a + 12b$
    d. $15xy + 0 = (xy)15$

Review Competencies

12. Label each of the following as rational or irrational.
    a. $\sqrt{10}$ b. $\sqrt{49}$ c. $\sqrt{63}$

13. Rationalize the denominator of $\dfrac{5}{\sqrt{20}}$

14. Simplify, if possible: $2\sqrt{3} + \sqrt{75} - \sqrt{48}$

15. $(-6) \cdot 3\frac{1}{3} =$

16. $(-\frac{1}{5}) \div (-\frac{3}{4}) =$

17. A motor home rents for $350 per week plus $.50 per mile. What is the rental cost for a three-week trip of 400 miles?

18. Two classes each have an enrollment of 32 students. On a particular day, $\frac{3}{4}$ of one class and $\frac{5}{8}$ of the other are present. Determine the total number of students absent from the two classes.

Section 5   Additional Topics in Number Theory

In this section, we consider some important topics related to the integers and counting numbers.

> ## Definition
> Let a and b represent integers. If there exists an integer c such that $a = b \cdot c$, then b is a **factor** of a, b **divides** a, a is **divisible** by b, and a is a **multiple** of b.

Examples

1. 9 is a factor of 27, since $27 = 9 \cdot 3$.

2. -7 divides 35, since $35 = (-7)(-5)$.

3. 35 is divisible by 5, since $35 = 7 \cdot 5$.

4. 24 is a multiple of 4, since $24 = 4 \cdot 6$.

Problem 1   List all the positive factors of 18.

Solution   1,2,3,6,9,18

Problem 2   List the first four positive multiples of 5.

Solution   5,10,15,20

> ## Definitions
> A **prime number** is a natural number greater than 1 whose only positive factors are 1 and itself. A counting number greater than 1 which has factors in addition to 1 and itself is a **composite number**.

195

Examples

1. 17 is a prime number since the only positive factors of 17 are 17 and 1.

2. 14 is a composite number, since $14 = 7 \cdot 2$.

3. 1 is neither a prime number nor a composite number.

Problem 3 How many positive factors of 20 are even and divisible by 5?

Solution Positive factors of 20 are 1,2,4,5,10, and 20. Even factors are 2,4,10, and 20. Only 10 and 20 are divisible by 5, so the answer is two.

Problem 4 Find the smallest positive multiple of 4 which yields a remainder of 1 when divided by 5.

Solution Positive multiples of 4 are 4,8,12,16,20,24,... . Observe that 16 is the smallest number from this collection that yields a remainder of 1 when divided by 5.

Problem 5 Which number has a divisor of 5 and is also a factor of 60?
   a. 7   b. 30   c. 35   d. 120

Solution We can immediately eliminate 7 because it does not have a divisor of 5. Observe that 5 does divide into 30, 35, and 120, so the other options all have divisors of 5. In addition, since $60 = 30 \cdot 2$, we see that 30 is a factor of 60. Neither 35 nor 120 is a factor of 60 because these numbers cannot be multiplied by another integer to obtain the product 60. Thus, 30 has a divisor of 5 and is a factor of 60, making option b the correct answer.

Problem 6 How many prime numbers less than 25 and greater than 10 will yield a remainder of 3 when divided by 5?

Solution Prime numbers between 10 and 25 include 11, 13, 17, 19 and 23. Each of these five numbers should be divided by 5 to see which give remainders of 3. Both 13 and 23 yield a remainder of 3 when divided by 5, so the answer is two.

Problem Set 5.1

1. List all the positive factors of 24.

2. List the first five positive multiples of 7.

3. Which one of the following numbers is a prime number?
   a. 63   b. 57   c. 51   d. 47

4. How many positive factors of 42 are even, less than 42, and also divisible by 3?

5. Find the smallest positive multiple of 7 which yields a remainder of 1 when divided by 3.

6. How many prime factors of 24 are also prime factors of 30?

7. Which number has a divisor of 7 and is also a factor of 42?
   a. 8   b. 21   c. 35   d. 84

8. Which number has both a divisor of 7 and a factor of 42?
   a. 6   b. 21   c. 35   d. 84

9. What prime numbers less than 25 and greater than 10 will yield a remainder of 1 when divided by 4?

10. Rationalize the denominator: $\frac{12}{\sqrt{2}}$

11. Simplify: $\sqrt{5} \cdot \sqrt{25}$

12. In November, Fran had 35 customers who each purchased $30 worth of merchandise. In December, 6 fewer customers made purchases but the remaining customers each increased their purchase by 20%. What was the change in Fran's sales from November to December?

13. $(-0.03) \div (-0.75) =$

14. $\frac{27}{1000} =$
   a. 0.027%   b. 0.27%
   c. 2.7%   d. 27%

15. 203.9% =
   a. $20\frac{39}{100}$ b. $20\frac{39}{1000}$
   c. $2\frac{39}{100}$ d. $2\frac{39}{1000}$

Although there are many ways to factor a given composite number, there is only one way to factor it into prime numbers.

Definition
The factorization of a composite number into prime numbers is called its **prime factorization**.

Chapter 3

Problem 7   Find the prime factorization of 180.

Solution

By inspection,
$$180 = 2 \cdot 90 = 2 \cdot 2 \cdot 45 = 2 \cdot 2 \cdot 3 \cdot 15 = 2 \cdot 2 \cdot 3 \cdot 3 \cdot 5.$$
Notice that the last factorization of 180 contains nothing but prime factors. Thus, the prime factorization of 180 is $2 \cdot 2 \cdot 3 \cdot 3 \cdot 5$ which can also be written $2^2 \cdot 3^2 \cdot 5$.

Problem 8   How many prime factors of 280 are also prime factors of 120?

Solution

By inspection,
$$280 = 2 \cdot 140 = 2 \cdot 2 \cdot 70 = 2 \cdot 2 \cdot 2 \cdot 35 = 2 \cdot 2 \cdot 2 \cdot 5 \cdot 7 = 2^3 \cdot 5 \cdot 7$$
Two prime factors of 280 (2 and 5) are also prime factors of 120.

Problem 9   Find the greatest common divisor of 98 and 84.

Solution

Find the prime factorization of 98 and 84.
$98 = 2 \cdot 49 = 2 \cdot 7 \cdot 7$
$84 = 2 \cdot 42 = 2 \cdot 7 \cdot 6 = 2 \cdot 7 \cdot 3 \cdot 2$
Two prime factors of 84, namely 2 and 7, are also prime factors of 98. We use these shared prime factors to find the largest natural number that divides both 98 and 84. The greatest common divisor is $2 \cdot 7$ or 14.

Problem Set 5.2

1. Find the prime factorization of 630.

2. Begin with the fact that $792 = 24 \cdot 33$ to find the prime factorization of 792.

3. How many prime factors of 84 are also prime factors of 35?

4. How many prime factors of 378 are also prime factors of 162?

5. How many prime factors of 546 are also prime factors of 173?

6. What is the greatest common divisor of 385 and 105?

Review Competencies

7. Simplify: $\sqrt{36} + \sqrt{5}$

8. If x represents an integer, is $\frac{1}{x^2}$ greater than zero for every integer?

9. If the numerator of a number lies between 12 and 20 inclusively, and the denominator lies between $\frac{1}{4}$ and $\frac{1}{2}$ inclusively, what is the greatest number possible?

10. $(-0.73) + (0.296) =$

11. $-7\frac{5}{6} + 3\frac{8}{9} =$

12. If Jill drives between 60 and 70 miles per hour for 3 to 4 hours, which one of the following is the best estimate of the distance traveled?
    a.   90 miles   b.   190 miles
    c.   225 miles   d.   275 miles

Section 6   Order of Operations

Adding, subtracting, multiplying, and dividing real numbers are important skills, and methods for performing each of these operations have been discussed in this chapter. In this section, the problems involve more than one operation. For example:

$$9 \div 3 + 6 = ?$$

In this problem there are three numbers and two operations. The first task is to determine whether the division or addition should be performed first. If the problem is done from left to right by completing the division first, the answer is 9. However, if the problem is completed by first adding $3 + 6$, the final answer is 1. To avoid the chaos that would result from arriving at two correct answers to a single problem, it is desirable that everyone agree to a standard method for approaching such evaluations. Mathematicians have agreed upon an **order of operations** which consists of the following steps:

Order Of Operations

1. If the problem contains parentheses, perform any operations inside the parentheses first.

2. Begin at the left of the problem and perform any multiplication and/or division as they appear from left to right.

3. Finally, perform any addition and subtraction from left to right until a final result is obtained.

Problem 1 Evaluate $3 + 2 \cdot 6 - 8 \div 4$.

Note: Evaluate means "find the value of."

Solution

step 1

There are no parentheses to consider. Therefore the entire problem is read from left to right, and the first multiplication (underlined) or division is performed.

$$3 + (2 \cdot 6) - 8 \div 4 = 3 + 12 - 8 \div 4$$

step 2

Again, reading from left to right, the next multiplication or division is performed.

$$3 + 12 - (8 \div 4) = 3 + 12 - 2$$

step 3

Since the multiplication and division have now been completed, the problem is read again from left to right and the first addition or subtraction is performed.

$$3 + 12 - 2 = 15 - 2$$

step 4

There is now only one operation to complete.

$$15 - 2 = 13$$

The rules governing the order of operations apply also to fractions and decimals.

Problem 2 Evaluate $10 - 8 \div \frac{1}{2} \cdot \frac{1}{4}$.

Solution

Each step in the sequence is indicated by the use of parentheses in the problem at the right.

$$\begin{aligned}
10 - 8 \div \tfrac{1}{2} \cdot \tfrac{1}{4} &= 10 - (8 \div \tfrac{1}{2}) \cdot \tfrac{1}{4} \\
&= 10 - (16 \cdot \tfrac{1}{4}) \\
&= (10 - 4) \\
&= \quad 6
\end{aligned}$$

Problem Set 6.1   *(handwritten: 18)*

Evaluate:

1. $4 \cdot 8 - 6 \cdot 3 = 14$
2. $4 + 7 \cdot 3 - 8 \div 4 = 23$
3. $2 \cdot 5 + 3 \cdot 2 - 4 \cdot 3 + 5 \cdot 0 = 4$
4. $3 + 2 \cdot 6 + 4 \div 4 + 9 \div 3 = 19$
5. $\frac{1}{4} \cdot 8 + \frac{3}{9} \div \frac{1}{3} + \frac{7}{5} \cdot \frac{5}{8}$
6. $10 \div 5 + 2 \cdot 7 - 8 \div 4 \cdot 2$
7. $8 - 3 \cdot 2 + 7 \div 7 \cdot 0$
8. $10 \div 5 \cdot 3 \div 2 \cdot 6 - 3 \cdot 2$
9. $4 \cdot 3 \cdot 2 - 8 \div 4 \div 2 + 0 \cdot 3 \div 5$
10. $8 \div \frac{3}{2} + 5 \cdot \frac{1}{3} - \frac{1}{3} \div \frac{1}{5}$

11. 40% of what number is 28?

12. What percent of 75 is 15?

13. What is 150% of 80?

14. A basketball team won 22 of its 25 games. What percent of its games did it lose?

15. A theatre has 750 seats. At one performance $\frac{2}{3}$ of the seats were occupied. How many seats were not occupied?

(handwritten work: $\frac{3}{9} \times \frac{3}{1} = \frac{2}{?}$)

Parentheses may be placed around a given expression in a problem. This grouping symbol indicates that the operations **inside** the parentheses must be completed before evaluating the rest of the problem. The parentheses are dropped once the operations inside are completed.

Examples

1. $5 + (2 \cdot 4) = 5 + 8 = 13$
2. $7 - (6 \div 2) = 7 - 3 = 4$
3. $4 + (2 \cdot 4 + 3) = 7 + (8 + 3) = 7 + 11 = 18$

If a number appears in front of the parentheses, the number should be **multiplied** by the quantity obtained inside the parentheses.

Examples

1. $2(4 \cdot 3) = 2 \cdot 12 = 24$

2. $5(4 - 2 + 3) = 5(5) = 25$

3. $4(3 + 4) - 2(1 + 3) = 4(7) - 2(4) = 28 - 8 = 20$

4. $\frac{2}{5}(8 + 17) - \frac{1}{3}(6 + 9) = \frac{2}{5}(25) - \frac{1}{3}(15) = 10 - 5 = 5$

5. $6(4 + 3 \div 2 - 8) = 6(4 + 1.5 - 8) = 6(5.5 - 8) = 6(-2.5) = -15$

Chapter 3

Evaluate:

1. $10 - (3 \cdot 1 + 9 \div 3)$ 4
2. $3(2 + 5) + 2(6 - 1)$ 31
3. $10 - (2 \cdot 7 - 4)$ 0
4. $\frac{1}{2}(10 - 2 \cdot 3)$ 2
5. $3(\frac{1}{3} + \frac{2}{3}) - \frac{1}{8}(12 - 4)$ 2
6. $(2 + 7)(7 - 3)$ 36
7. $12 \div (5 - 2) - 3(2 \cdot 4 \div 8)$ 1
8. $(6 \cdot 0) + (0 \div 3) - 0(5 - 1) + 0$ 0
9. $6(\frac{2}{3} \div \frac{4}{3} \cdot \frac{1}{4} \cdot \frac{1}{3}) + \frac{1}{4}(7 - 3 - 2)$
10. $(4 + 3 - 1 + 2 - 5)(7 - 2 - 1 + 4 - 3)$ 15
11. $8(6 + 5 \div 4 - 10)$ -22

Review Competencies

12. Evaluate:

    a. $\frac{1}{3} \cdot 9 + \frac{4}{5} \div \frac{1}{5} + \frac{2}{3} \cdot \frac{3}{2}$

    b. $5 + 6 \div 2 - 9 \div 3 + 2 \cdot 4$

13. Add: $\frac{3}{4} + 1\frac{1}{5}$

14. Subtract: $13.4 - 6.602$

15. Express 325% as a decimal.

16. Express $4\frac{1}{4}\%$ as a decimal.

17. A team played 70 games and won 60% of the games. If there are 40 games left, how many do they need to win to ultimately win 70% of all games played?

Many situations require us to evaluate formulas and algebraic expressions. Once we substitute the given value(s) for the letter(s) that appear in the formula, we then use the agreed-upon order of operations.

Problem 3

The formula for finding the dealer's cost for a car (C) having list price (L) and shipping charge (S) is $C = 0.8(L - S) + S$. Find the dealer's cost for a car having a list price of $14,950 and a shipping charge of $600.

Solution

$$\begin{aligned} C &= 0.8(L - S) + S \\ &= 0.8(14,950 - 600) + 600 \\ &= 0.8(14,350) + 600 \\ &= 11,480 + 600 = 12,080 \end{aligned}$$

Thus, the dealer's cost is $12,080.

Problem 4   Evaluate $2(x-5)^2 + 2x(y-z)$ if $x = -2$, $y = 3$ and $z = -1$.

Solution

$$
\begin{aligned}
2(x-5)^2 + 2x(y-z) &= 2(-2-5)^2 + 2(-2)[3-(-1)] \\
&= 2(-7)^2 + 2(-2)[3+1] \\
&= 2(-7)^2 + 2(-2)(4) \\
&= 2(49) + 2(-2)(4) \\
&= 98 + (-16) \\
&= 82
\end{aligned}
$$

The equality $f(x) = 3x^2 - 5x + 2$ is read "f at x equals $3x^2 - 5x + 2$" and gives the value of the **function** f for every specified value of x. Substitute -4 for x to find f(-4).

Problem 5   Given the function defined by $f(x) = 3x^2 - 5x + 2$, find f(-4).

Solution

$$
\begin{aligned}
f(x) &= 3x^2 - 5x + 2 \\
f(-4) &= 3(-4)^2 - 5(-4) + 2 \\
&= 3(16) - 5(-4) + 2 \\
&= 48 + 20 + 2 \\
&= 70
\end{aligned}
$$

We say that f at -4 is 70. When the value of the independent variable (or domain) is -4, the value of the dependent variable (or range) is 70. Observe that a function is a relationship between two variables such that for each value of the independent variable there corresponds exactly one value of the dependent variable. The set of all values of the independent variable is called the domain of the function. The set of all values taken on by the dependent variable is called the range of the function.

Problem Set 6.3

1. The formula for finding what an investment (P) at a fixed interest rate (R) is worth after one year is given by the formula $W = P(R + 1)$, where W represents the worth of the investment. What will an investment of $12,500 at 14% be worth after one year?

2. The formula for changing a Fahrenheit temperature (F) to a Celsius temperature (C) is $C = \frac{5}{9}(F - 32)$. What is the Celsius temperature when the Fahrenheit temperature is 50°?

3. The formula for the sales price (P) of an article with dealer's cost (D) and markup (M) is $P = D + MD$. What is the sales price of an article costing a dealer $12 with a markup of 40%?

4. The formula used to find rate of speed (R) in miles per hour for the distance (D) in miles and the time (T) in hours is $R = \frac{D}{T}$. Find the rate of speed of an object that travels 164 miles in $1\frac{1}{3}$ hours.

5. Evaluate $2x^2 + 5xy$ if $x = -3$ and $y = 2$.

6. Evaluate $3(x - 1)^2 + 2x(y - z)$ if $x = -3$, $y = 2$ and $z = -4$.

7. If $f(x) = x^2 + 2x - 1$, find $f(-6)$.

8. Given the function defined by $f(x) = x^3 - 2x^2 - x$, find $f(-3)$.

Review Competencies

9. Evaluate:
   a. $36 - 24 \div 4 \cdot 2 - 1$
   b. $\frac{3}{4} - 8(\frac{1}{2} - \frac{1}{4})$
   c. $\frac{1}{3} \cdot \frac{1}{2} - \frac{3}{4} + \frac{1}{2}$

10. A telephone service charges $20.70 per month for the first 66 calls and 15 cents for each additional call. A 12% tax is also added to the final bill. In one month 78 calls were made. What was the bill?

The same order of operations agreement is observed when expressions involve letters.

Problem 6 Simplify $4x^2 \cdot 3 - 5x^2 \div 5 + x - 3x$.

Solution

step 1
Working from left to right, the multiplication ($4x^2 \cdot 3$) and the division ($5x^2 \div 5$) are completed in that order.
$$4x^2 \cdot 3 - 5x^2 \div 5 + x - 3x = 12x^2 - x^2 + x - 3x$$

step 2
Working from left to right, the addition and subtraction of similar terms are now completed. $12x^2 - x^2 + x - 3x = 11x^2 - 2x$

Problem Set 6.4

Simplify:

1. $x - 4x \cdot 3 + 6x \div 2 + 4x$

2. $15a \div 5 - 2 \cdot 5a + 7a + 1$

3. $4b^2 \cdot 2 \cdot 3 \div 12 + 6b^2 \div 3 - b^2$

4. $8x - 3x \cdot 2 + 14x^2 \div 7 \cdot 2$

5. $\frac{1}{4} \cdot 8c + 3c \div \frac{1}{3} - 4c \div 4 + c^2$

6. $2(3a \cdot 2) - (6a \div 2 \cdot 4)$

7. $8x - (4x - 2x \div 2)$

8. $12b \div (8b - 16b \div 2 - 6)$

9. $-5(10p \div 5 + 2p) + 3(2 \cdot 3p \div 6 + 3p - p)$

10. $4(2x + 8x \div 2 \div 2 \div 2) - 2(2 \cdot 2x - 10x \div 2)$

Review Competencies

11. Express $\frac{5}{4}$ as a decimal.

12. Express 0.01% as a decimal.

13. Divide: $4 \div 0.002$

14. Evaluate: $15 - (2 \cdot 3 + 6 \div 2)$

We have seen that parentheses are used to group or include numbers and variables together. Other symbols of grouping or inclusion are brackets and braces.

Brackets [ ]
Braces   { }

To simplify an expression containing more than one of the grouping symbols, begin with the **innermost** grouping symbol and simplify the expression within it.

Problem 7   Simplify $4 - 2[3 + (5 - 2)]$.

Solution

$$\begin{aligned}4 - 2[3 + (5 - 2)] &= 4 - 2[3 + 3] &&\text{(simplifying in parentheses first)}\\ &= 4 - 2(6) &&\text{(simplifying in brackets)}\\ &= 4 - 12\\ &= -8\end{aligned}$$

Problem 8   Simplify $2 - \{4 - 3[2 - (4 - 5)]\}$.

Solution

$$\begin{aligned}2 - \{4 - 3[2 - (4 - 5)]\} &= 2 - \{4 - 3[2 - (-1)]\} &&\text{(simplify parentheses)}\\ &= 2 - \{4 - 3[3]\} &&\text{(simplify brackets)}\\ &= 2 - (-5) &&\text{(simplify braces)}\\ &= 7\end{aligned}$$

Problem 9   Simplify $\{-2x - [6x - 2(3x - x)]\}$.

Solution

$$\begin{aligned}\{-2x - [6x - 2(3x - x)]\} &= \{-2x - [6x - 2(2x)]\} &&\text{(simplify parentheses)}\\ &= \{-2x - [2x]\} &&\text{(simplify brackets)}\\ &= -4x &&\text{(simplify braces)}\end{aligned}$$

Chapter 3

Problem 10

Simplify $2x[x - 3(2x - 1)] - 4x^2$.

Solution

$$
\begin{aligned}
2x[x - 3(2x - 1)] - 4x^2 &= 2x[x - 6x + 3] - 4x^2 \\
&= 2x[-5x + 3] - 4x^2 \\
&= -10x^2 + 6x - 4x^2 \\
&= -14x^2 + 6x
\end{aligned}
$$

(This can also be written $6x - 14x^2$.)

Problem Set 6.5

Simplify:

1. $6 - 2[3 - (1 - 4)]$

2. $-4 - \{2 + 3[1 - (2 - 3)]\}$

3. $-5a - \{3a + 3[4a - (6a - 4a)]\}$

4. $-2\{x + [3x - (x - 2x)]\} + 7x$

5. $-3\{x - 3[x - 3(x - 3x)]\}$

6. $-2(3 - 2x) - (6 - 5x)$

7. $6x - [3x - 2(x - 5)]$

8. $(x^2y + 9xy - 4xy^2) - (2xy^2 - x^2y + 9xy)$

9. $4x[2x^2 - 2x(x + 3) - 5]$

10. $x^2z - x[xy - x(y - z)]$

Review Competencies

11. Express 60% as a fraction.

12. Express $4\frac{7}{10}\%$ as a decimal.

13. Determine the common relationship between the numbers in each pair. Then identify the missing term.

$(3,1)$ $(0.9, 0.3)$ $(-6, -2)$ $(\frac{1}{10}, \frac{1}{30})$

$(\frac{-1}{3}, \underline{\quad})$

14. Find the missing number in the geometric progression. A definition of a geometric progression appears on page 36, problem #10.

$9, \ -3, \ 1, \ -\frac{1}{3}, \ \underline{\quad\quad}, \ -\frac{1}{27}$

15. Ten people purchased calculators. Half bought a calculator that cost $6.95. The other half spent between $4.95 and $12.95 for a calculator. What is a reasonable estimate of the total amount of money spent by the ten people?
    a. $55   b. $75   c. $105   d. $160

Chapter 3 Summary

1. The Real Numbers

a. Natural (Counting) Numbers
$\{1,2,3,\ldots\}$

b. Whole Numbers
$\{0,1,2,\ldots\}$

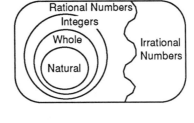

c. Integers
$\{\ldots,-3,-2,-1,0,1,2,3,\ldots\}$

d. Rational Numbers
Numbers in the form $\frac{x}{y}$, where x and y are integers, $y \neq 0$.

e. Irrational Numbers
Numbers that, when expressed as decimals, neither terminate nor repeat.

2. Properties of the Real Numbers

In each of the following properties, x, y, and z represent real numbers.

a. Closure of Addition $x + y$ is a unique real number.

b. Closure of Multiplication xy is a unique real number.

c. Commutative of Addition $x + y = y + x$

d. Commutative of Multiplication $xy = yx$

e. Associative of Addition $(x + y) + z = x + (y + z)$

f. Associative of Multiplication $(xy)z = x(yz)$

g. Identity of Multiplication $x \cdot 1 = 1 \cdot x = x$

h. Identity of Addition $x + 0 = 0 + x = x$

i. Inverses of Addition $x + (-x) = 0$

j. Inverses of Multiplication $\frac{x}{y} \cdot \frac{y}{x} = 1, x \neq 0, y \neq 0$

k. Distributive $x(y + z) = xy + xz$

Chapter 3

3. Multiplying and Dividing Signed Numbers

The product or quotient of two numbers having like signs is positive. The product or quotient of two numbers having unlike signs is negative.

4. Operations With Square Roots

a. Addition-Subtraction
    If possible, simplify each radical expression. Then combine like terms.
b. Multiplication
    Use the fact that $\sqrt{a} \cdot \sqrt{b} = \sqrt{ab}$ (a and b are non-negative).
c. Division
    Use the fact that $\dfrac{\sqrt{a}}{\sqrt{b}} = \sqrt{\dfrac{a}{b}}$ (a and b are non-negative and b $\neq$ 0.)
d. Rationalization of the Denominator
    If possible, simplify the radical expression in the denominator. Multiply numerator and denominator by the radical expression in the denominator.

5. Basic Vocabulary of Number Theory

a. Let a and b represent integers. If there exists an integer c
    such that a = b $\cdot$ c, b is a **factor** of a, b **divides** a, a is **divisible** by b,
    and a is a **multiple** of b.
b. Prime Number:
    A natural number greater than 1 whose only positive factors
    are 1 and itself.
c. Composite Number:
    A natural number greater than 1 which is not prime.

6. Order of Operations

a. Work all operations within grouping symbols, working within innermost
    grouping symbols first.
b. Work all multiplication and division from left to right.
c. Work all addition and subtraction from left to right.

Chapter 3 Self-Test

1. Which statement illustrates the commutative property of multiplication?
   a. $c + (a \cdot b) = c + (b \cdot a)$
   b. $(a + b) \cdot 5 = a(5) + b(5)$
   c. $a + b = b + a$
   d. $a(8) = a(3 + 5)$
   e. None of these

2. Which statement illustrates the associative property of addition?
   a. $[a + (b + c)] + d = a + [(b + c) + d]$
   b. $(a \cdot b) \cdot c = a \cdot (b \cdot c)$
   c. $(a + b) + c = a + (c + b)$
   d. $5(a + b) = 5a + 5b$
   e. None of these

3. Select the property or properties that could be used to simplify the following numerical expression in the least number of computational steps.
   $$.37(26.5) + .37(65.2) = ?$$
   a. Associative property of addition and commutative property of addition
   b. Distributive property and associative property of addition
   c. Distributive property only
   d. Distributive property and commutative property of addition

4. Select the property or properties illustrating the following relationship:
   $$3(4 + 5) = 3 \cdot 5 + 3 \cdot 4$$
   a. Associative property of multiplication only
   b. Commutative property of addition
   c. Distributive property and associative property of multiplication
   d. Distributive property only
   e. Distributive property and commutative property of addition

5. Identify the property of operations illustrated by $2(x) + 2(y) = 2(x + y)$
   a. Commutative property of addition
   b. Associative property of addition
   c. Commutative property of multiplication
   d. Associative property of multiplication
   e. Distributive property of multiplication over addition

6. Choose the equivalent expression for $3(5) + 3(a)$.
   a. $18a$
   b. $3(5 \cdot a)$
   c. $(3)(3) + (5)a$
   d. $3(5 + a)$
   e. $(3 + 3)(5 + a)$

7. Select the multiplicative inverse for the real number $\frac{2}{3}$.
   a. $\frac{-2}{3}$ b. $\frac{3}{2}$ c. 0 d. 1
   e. None of these

8. Select the correct answer for $36 - 12 \div 4 \cdot 3 - 1$.
   a. 34 b. 28 c. 17 d. 1
   e. None of these

9. Select the correct answer for $3[\frac{2}{3} - 3(\frac{1}{3} + 1)]$.
   a. $\frac{14}{3}$ b. $\frac{-10}{3}$ c. $\frac{2}{3}$ d. -10
   e. None of these

10. Select the correct answer for $4b^2 \cdot 2 \cdot 3 \div 12 + 12b^2 \div 3 - b^2$.
    a. $6b^2$ b. $4b^2$ c. $35b^2$ d. $5b^2$
    e. None of these

11. Choose the correct answer for $5\pi + 8\pi - 3$.
   a. $13\pi^2 - 3$ b. 10π
   c. $13\pi - 3$ d. $4\pi^2 - 3$
   e. None of these

12. Choose the correct answer for $-2\sqrt{5} - \sqrt{3} + \sqrt{5}$.
   a. $-3\sqrt{5} - \sqrt{3}$ b. $-2 - \sqrt{3}$
   c. $-\sqrt{5} - \sqrt{3}$ d. $-\sqrt{5} + \sqrt{3}$
   e. None of these

13. Choose the correct answer for $\sqrt{3} + \sqrt{12}$.
   a. $3\sqrt{3}$ b. $5\sqrt{3}$ c. 6   d. $\sqrt{15}$
   e. None of these

14. Choose the correct answer for $\sqrt{3} \cdot \sqrt{8}$.
   a. 24   b. $4\sqrt{6}$ c. 73   d. $2\sqrt{6}$
   e. None of these

15. Choose the correct answer for $\dfrac{9}{\sqrt{3}}$.
   a. $\dfrac{\sqrt{3}}{3}$ b. $3\sqrt{3}$ c. $9\sqrt{3}$ d. $\sqrt{3}$
   e. None of these

16. The difference of two whole numbers is an odd number. Which of the following statements is true about these two numbers?
   a. Both of the numbers may be odd.
   b. Both of the numbers may be even.
   c. Only one of the numbers is even.
   d. The sum of the numbers may be even.

17. If $x \# y * z = \dfrac{yz}{x}$, $p \# q * r = \dfrac{qr}{p}$, and $a \# b * c = \dfrac{bc}{a}$ then $k \# m * n = $ _____
   a. $\dfrac{km}{n}$ b. $\dfrac{kn}{m}$ c. $\dfrac{mn}{k}$ d. kmn
   e. None of these

18. Which whole number is divisible by 5 and also a factor of 20?
   a. 10   b. 15   c. 55   d. 60
   e. None of these

19. Find the smallest positive multiple of 8 which yields a remainder of 2 when divided by 5.
   a. 16   b. 24   c. 32   d. 40
   e. None of these

20. How many prime factors of 126 are also factors of 15?
   a. one   b. two   c. three   d. four
   e. five

21. Choose the expression equivalent to $7(x + y)$.
   a. $7xy$ b. $7x + y$
   c. $7x + 7y$ d. $7 + (x + y)$

22. Choose the expression equivalent to $(2a + 6b)(2a - 6b)$.
   a. $2(a + 3b)(a - 3b)$
   b. $(2a - 6b)(6b + 2a)$
   c. $8ab(2a - 6b)$
   d. $(2a + 6b)(6b - 2a)$

23. Choose the expression equivalent to $12y + 4x$.
   a. $16xy$ b. $y(12 + 4x)$
   c. $3y + x$ d. $4(x + 3y)$

24. Choose the expression equivalent to $6x + 4y$.
   a. $4x + 6y$ b. $4y - 6x$
   c. $10(x + y)$ d. $4y + 6x$

25. Choose the equation that is not true for all real numbers.
    a. $8x + 8y = 8(x + y)$
    b. $4xy(3x + y) = 4xy(3y + x)$
    c. $(x - y)(x + y) = (y + x)(x - y)$
    d. $(5x)y = 5(xy)$

26. Choose the equation that is not true for all real numbers.
    a. $7ab(x - 4y) = 7abx - 28aby$
    b. $(5 + x) + y = 5 + (x + y)$
    c. $3(xy) = (3x)(3y)$
    d. $5a - 10b = -5(2b - a)$

27. How many prime numbers less than 30 and greater than 10 will yield a remainder of 2 when divided by 3?
    a. five   b. four
    c. three   d. two

28. $36 - 12 \div 4 \cdot 3 - 2 =$
    a. 33   b. 25   c. 16   d. 0

29. If $f(x) = 200 + x + 0.4(200 + x)$, find $f(50)$.
    a. 1250   b. 500
    c. 350   d. 260

30. How many factors of 12 are divisible by 3?
    a. two   b. three
    c. four   d. five

31. Find $f(-3)$ given that $f(x) = 4x^2 - 5x - 6$.
    a. 45   b. 33   c. 15   d. -27

32. $7 - 5(2a - b) =$
    a. $7 - 10a + 5b$ b. $7 - 10a - 5b$
    c. $4a - 2b$ d. $4a - b$

33. $\sqrt{36} + \sqrt{3} =$
    a. $\sqrt{39}$ b. $6\sqrt{3}$ c. $36\sqrt{3}$ d. $\sqrt{3} + 6$

34. Which whole number is divisible by 3 and is also a factor of 45?
    a. 7   b. 9   c. 18   d. 90

35. If $y = (2x + 3)^2$, find y when $x = 5$.
    a. 169   b. 121   c. 109   d. 26

Problem 1 Select the equivalent expression for $2^2 + 7^2$.

    a. $(2 + 7)^2$ b. $(2 + 7)^4$

    c. $2 \cdot 2 + 7 \cdot 2$ d. $2 \cdot 2 + 7 \cdot 7$

Solution By definition, $2^2 + 7^2 = 2 \cdot 2 + 7 \cdot 7$ which is d. Notice that
$(2 + 7)^2 = 9^2 = 81$ and $2^2 + 7^2 = 4 + 49 = 53$.

In general $(x + y)^a$ is **not** equal to $x^a + y^a$.

Problem Set 1.1

Evaluate:

1. 2^3 2. 4^3

3. 2^5 4. 10^1

5. $(-6)^3$ 6. $(-10)^2$

7. 10^3 8. $(0.5)^2$

9. $2 \cdot 3^2$ 10. $4^2 \cdot 4$

11. $2 \cdot 10^1$ 12. $3 \cdot 10^2$

13. $4 \cdot 10^3$ 14. $2^2 \cdot 3^2$

15. $3(5^2)$ 16. $(2 \cdot 3)^2$

17. $2(4)^2$ 18. $2^2 + 3^2$

19. $(2^3)^2$ 20. $(2^2)(4^2)$

21. $3^3 - 2^4$ 22. $\dfrac{2^5}{2^3}$

23. Is the square of a whole number always a whole number?

24. Can the square of a whole number ever be less than that whole number?

25. Select a choice equivalent to $(5^3)(3^4)$.

    a. $15 \cdot 12$

    b. $(5 + 5 + 5) \cdot (3 + 3 + 3 + 3)$

    c. $(5 + 3)^7$ d. None of these

26. Select a choice equivalent to $x^3 + y^3$.

    a. $(x + y)^3$ b. $x^3 \cdot y^3$

    c. $(xy)^6$ d. $(x)(x)(x) + (y)(y)(y)$

27. Select a choice equivalent to $(x + y)^2$.

    a. $x^2 + y^2$ b. $(x + y)(x + y)$

    c. $(xy)^2$ d. $(xy) + (xy)$

Review Competencies

28. $6 = \dfrac{?}{4}$ 29. Multiply: $3\frac{1}{4} \cdot 2\frac{1}{2}$

30. Add: $2\frac{1}{3} + 3\frac{1}{4}$

31. Express 45% as a fraction.

32. Express $\frac{13}{25}$ as a percent.

33. Express 0.375 as a percent.

34. 25% of what number is 75?

35. Express $2\frac{1}{2}$ % as a decimal.

It is true that $x^1 = x$. Usually, the exponent 1 is not written. It is also possible to use 0 as an exponent as is stated in the following definition:

> ### Definition
> Any numeral, other than zero, that is raised to a zero exponent is equal to 1. That is, if x is any non-zero numeral, then $x^0 = 1$.

The motivation for this definition will be discussed later.

Examples

1. $10^0 = 1$
2. $8^0 = 1$
3. $(100)^0 = 1$
4. $5 \cdot 7^0 = 5 \cdot 1 = 5$

Problem Set 1.2

Evaluate:

1. 4^0
2. $(1,000)^0$
3. $(\frac{3}{4})^0$
4. $5 \cdot 10^0$
5. $8 \cdot 10^0$
6. $4^0 \cdot 8$
7. $6 \cdot 0 \cdot 6^0$

Review Competencies

8. Evaluate:
   a. 3^3
   b. $2(5^3)$
   c. $3 \cdot 3^2$
   d. $2 \cdot 10^2$
   e. $2^3 \cdot 3^2$
   f. $(2 \cdot 4)^2$
   g. $2^4 + 3^3$
   h. $2^5 - 3^2$

9. Subtract: $\frac{7}{9} - \frac{1}{2}$

10. Subtract: $6 - 0.003$

11. What is 20% of 85?

12. Evaluate: $3 + 2 \cdot 6 + 4 \div 4$

We can now use the idea of exponents to express a numeral in expanded form more conveniently. This preferred method is how we shall agree to express a numeral in expanded form from now on.

Problem 2   Express the numeral 333 in expanded form using exponents.

Solution   Previously, the numeral was expressed as:
$$333 = (3 \cdot 100) + (3 \cdot 10) + (3 \cdot 1)$$

Using exponents, the numeral can now be written as:
$$333 = (3 \cdot 10^2) + (3 \cdot 10^1) + (3 \cdot 10^0)$$

We can now say that our decimal system of numeration is based on groups of ten or powers of ten. The **place value** of the digit 5 in the numeral 435 is **ones** or 10^0. The place value of the digit 3 in 435 is **tens** or 10^1.

The place value of the digit 4 in 435 is **hundreds** or 10^2. The place value of the digit 6 in 6,153 is **thousands** or 10^3.

Problem Set 1.3

1. Express each of the following in expanded form using exponents.
   a. 69   b. 587   c. 3,256   d. 409

2. What is the place value of each of the under-lined digits?
   a. 8<u>1</u>   b. 6<u>0</u>3   c. 4,<u>0</u>06   d. <u>2</u>,679

3. Evaluate:
   a. 10^0
   b. $(\frac{2}{3})^0$
   c. $6 \cdot 1^0$
   d. $7 \cdot 0 \cdot 70^0$
   e. 4^2
   f. $3(5^2)$
   g. $3 \cdot 10^2$
   h. $(3 \cdot 3)^2$

4. Divide: $0.21 \div 0.07$

5. Express 135% as a decimal.

6. What percent of 150 is 30?

7. Evaluate: $10 \div (6 - 4) - (2 \cdot 4 \div 8)$

We have seen that positive integers as well as zero can be exponents. By the following definition, we see that negative integers can also be exponents.

> **Definition**
> If x is any non-zero number, and a is any positive integer,
> then $x^{-a} = \dfrac{1}{x^a}$.

The motivation for this definition will be discussed later.

Examples

1. $3^{-2} = \frac{1}{3^2} = \frac{1}{9} = 0.11\ldots = 0.\overline{1}$

2. $10^{-3} = \frac{1}{10^3} = \frac{1}{1000} = 0.001$

3. $5^{-1} = \frac{1}{5^1} = \frac{1}{5} = 0.2$

4. $2 \cdot 10^{-2} = 2 \cdot \frac{1}{10^2} = 2 \cdot \frac{1}{100} = 0.02$

5. $4^{-2} + 4^{-2} = \frac{1}{4^2} + \frac{1}{4^2} = \frac{1}{16} + \frac{1}{16} = \frac{2}{16} = \frac{1}{8} = 0.125$

6. $(-2)^{-3} = \frac{1}{(-2)^3} = \frac{1}{(-2)(-2)(-2)} = -\frac{1}{8}$

Problem Set 1.4

Express each of the following as both a fraction and its decimal equivalent.

1. 4^{-2} 2. 2^{-3} 3. 2^{-4}

4. 10^{-1} 5. 10^{-4} 6. $3 \cdot 4^{-1}$

7. $5 \cdot 10^{-2}$ 8. $7 \cdot 10^{-3}$ 9. $3 \cdot 10^{-1}$

10. $2^{-1} + 2^{-1}$

Review Competencies

11. Express 4,357 in expanded form.

12. Evaluate.

   a. $(\frac{3}{5})^0$ b. $4^0 \cdot 6$

   c. $3(2)^2$ d. $2^3 \cdot 3^3$

13. Divide: $\frac{3}{5} \div \frac{2}{7}$

14. Add: $4\frac{1}{7} + 2\frac{1}{2}$

15. Multiply: $\frac{3}{5} \cdot 6$

16. Express 9% as a decimal.

17. 7% of what number is 49?

18. Evaluate: $(5 + 3 - 2 + 1)(10 - 3 - 2 + 1 - 2)$

19. What is the place value of the underlined digit in 2,534?

We use negative exponents to express certain numerals in expanded form.

Problem 3   Express the numeral 24.3 in expanded form.

Solution   Notice that 24.3 means $24\frac{3}{10}$. In expanded form, this is:
$$24.3 = 24\frac{3}{10} = (2 \cdot 10^1) + (4 \cdot 10^0) + (3 \cdot 10^{-1})$$
Observe that the expression $(3 \cdot 10^{-1})$ means $3 \cdot \frac{1}{10}$ or $\frac{3}{10}$ which is 3 tenths (0.3). Thus, the **place value** of the digit 3 is expressed as 10^{-1} or $\frac{1}{10}$ or tenths.

Problem 4 Express the numeral 643.25 in expanded form.

Solution $643.25 = (6 \cdot 10^2)+(4 \cdot 10^1)+(3 \cdot 10^0)+(2 \cdot 10^{-1})+(5 \cdot 10^{-2})$

Thus the place value of the digit 5 is 10^{-2} or $\frac{1}{100}$ or hundredths. The

place value of the digit 2 is 10^{-1} or $\frac{1}{10}$ or tenths.

Problem Set 1.5

1. Express each of the following in expanded form.
   a. 4.39
   b. 23.296
   c. 0.009
   d. 6,000.08

2. What is the place value of the underlined digits?
   a. 2.6_7_8
   b. 0.00_9_

Review Competencies

3. Evaluate.
   a. 3^{-2}
   b. 10^{-1}
   c. 10^{-3}
   d. $2 \cdot 3^{-2}$
   e. $7 \cdot 10^{-2}$
   f. 3^0
   g. $3 \cdot 10^0$
   h. $4 \cdot 10^1$
   i. $2(3)^3$

4. Express 806 in expanded form.

5. $\frac{4}{9} = \frac{?}{27}$

6. Subtract: $2\frac{1}{9} - \frac{1}{6}$

7. Divide: $\frac{5}{6} \div 5$

8. Evaluate: $3(\frac{1}{3} + \frac{2}{3}) - \frac{1}{4}(8-4)$

9. What is the place value of the underlined digit in 4,00_1_?

10. Select a choice equivalent to $x^{-3} + y^0$.
    a. $x \cdot x \cdot x + y$
    b. $x \cdot x \cdot x + 1$
    c. $\frac{1}{x \cdot x \cdot x} + 1$
    d. $\frac{1}{x \cdot x \cdot x + 1}$

Let us now turn to the task of determining the place value numeral of a number expressed in expanded form.

Problem 5 What place value numeral is represented by the following expression?
$$(3 \cdot 10^1) + (4 \cdot 10^0) + (5 \cdot 10^{-1})$$

Solution Evaluate the expression in each grouping.

$(3 \cdot 10^1) = 3 \cdot 10 = 30$ and $(4 \cdot 10^0) = 4 \cdot 1 = 4$ and $(5 \cdot 10^{-1}) = 5 \cdot \frac{1}{10} = 0.5$

Thus, $(3 \cdot 10^1) + (4 \cdot 10^0) + (5 \cdot 10^{-1}) = 30 + 4 + 0.5 = 34.5$.

Problem Set 1.6

Determine the place value numeral represented by each of the following.

1. $(5 \cdot 10^2) + (3 \cdot 10^1) + (2 \cdot 10^0) + (3 \cdot 10^{-1}) + (6 \cdot 10^{-2})$

2. $(8 \cdot 10^2) + (4 \cdot 10^0) + (3 \cdot 10^{-2})$

3. $(3 \cdot 10^{-1}) + (6 \cdot 10^{-2}) + (7 \cdot 10^{-3})$

4. $(3 \cdot 10^{-2}) + (5 \cdot 10^{-3}) + (4 \cdot 10^{-4})$

5. Select the expanded notation for 3006.002

   a. $(3 \cdot 10^4) + (6 \cdot 10) + (2 \cdot \frac{1}{10^3})$

   b. $(3 \cdot 10^3) + (6 \cdot 10^0) + (2 \cdot \frac{1}{10^2})$

   c. $(3 \cdot 10^4) + (6 \cdot 10^7) + (2 \cdot 10^9)$

   d. $(3 \cdot 10^3) + (6 \cdot 10^0) + (2 \cdot \frac{1}{10^3})$

6. Select the expanded notation for 500.03

   a. $(5 \cdot 10^2) + (3 \cdot \frac{1}{10^2})$

   b. $(5 \cdot 10^3) + (3 \cdot \frac{1}{10})$

   c. $(5 \cdot 10^3) + (3 \cdot \frac{1}{10^2})$

   d. $(5 \cdot 10^2) + (3 \cdot \frac{1}{10})$

Review Competencies

7. Express 12.031 in expanded form.

8. Evaluate:
   a. 2^{-4} b. 10^{-3}
   c. $4 \cdot 0 \cdot 5^0$ d. 2^5
   e. $3 \cdot 10^2$ f. $2(3)^2$

9. Evaluate: $(4 \cdot 0) + (0 \div 2) - 0(5 - 7)$

10. Express $\frac{3}{10}$ as a percent.

11. Express 35% as a fraction.

12. Divide: $0.123 \div 0.3$

13. What is the place value of the underlined digit in 1.05$\underline{7}$?

14. A truck that was purchased for $1,500 was worth $265 at the end of ten years. Assuming that the value depreciated steadily each year, what was it worth at the end of seven years after its purchase?

15. A nursery will get a price 5% off the regular price if they order more than 100 small trees which normally sell for $4.00 per tree. The nursery will get an additional 2% discount on the discounted order if they pay their bill by the fifth of the month. If they purchase 500 small trees and pay their bill by May 2, how much will they pay for the order?

16. Calculators that normally cost $20 each will be reduced by 40%. At the reduced rate, how much will 25 calculators cost?

Section 2   Laws of Exponents

Consider the following problem: $x^2 \cdot x^3$. Since x^2 can be expressed as $x \cdot x$ and x^3 can be expressed as $x \cdot x \cdot x$, the product $x^2 \cdot x^3$ can be expressed as $(x \cdot x) \cdot (x \cdot x \cdot x)$ or x^5. Notice that the exponent in the result (x^5) could have been obtained by **adding** the exponents of x^2 and x^3.

$$x^2 \cdot x^3 = x^{2+3} = x^5$$

The addition of exponents was possible only because the base of x^2 was the same as the base of x^3. This will always be the case and gives rise to the first law of exponents.

$$x^a \cdot x^b = x^{a+b}$$

Now consider the division problem $\frac{x^5}{x^2}$. The expression x^5 can be written as $x \cdot x \cdot x \cdot x \cdot x$ and x^2 can be written as $x \cdot x$. Thus,

$$\frac{x^5}{x^2} = \frac{x \cdot x \cdot x \cdot \cancel{x} \cdot \cancel{x}}{\cancel{x} \cdot \cancel{x}} = x^3$$

Notice that the exponent in the result (x^3) could have been obtained by **subtracting** the exponent of x^2 from the exponent of x^5.

$$\frac{x^5}{x^2} = x^{5-2} = x^3$$

Notice that the base of x^5 is the same as the base of x^2. The process for dividing when the exponents have the same base is generalized in the second law of exponents.

$$\frac{x^a}{x^b} = x^{a-b} \quad \text{when } x \neq 0$$

This second law of exponents is used to justify that $x^0 = 1$.

We know both: $\quad \frac{x^2}{x^2} = \frac{\cancel{x} \cdot \cancel{x}}{\cancel{x} \cdot \cancel{x}} = 1 \quad$ and $\quad \frac{x^2}{x^2} = x^{2-2} = x^0$

Therefore, $1 = x^0$ or $x^0 = 1$.

This second law can also be used to justify that $x^{-2} = \dfrac{1}{x^2}$.

We know both:

$$\frac{x^2}{x^4} = \frac{\cancel{x} \cdot \cancel{x}}{x \cdot x \cdot \cancel{x} \cdot \cancel{x}} = \frac{1}{x \cdot x} = \frac{1}{x^2} \quad \text{and} \quad \frac{x^2}{x^4} = x^{2-4} = x^{-2}$$

Therefore, $\dfrac{1}{x^2} = x^{-2}$ or $x^{-2} = \dfrac{1}{x^2}$.

These laws of exponents can be used when the exponents are positive or negative integers.

Examples

1. $x^{-2} \cdot x^5 = x^{-2+5} = x^3$

2. $(2a^4)(3a^{-2}) = (2 \cdot 3)(a^{4+(-2)}) = 6a^2$

3. $\dfrac{x^3}{x^{-5}} = x^{3-(-5)} = x^{3+(+5)} = x^8$

4. $\dfrac{10y^{-2}}{-5y} = -2y^{-2-1} = -2y^{-3} = -2 \cdot \dfrac{1}{y^3} = \dfrac{-2}{y^3}$

(The result is expressed with a positive exponent.)

Let us now discuss some additional laws of exponents.

To simplify an expression such as $(2^3)^2$, we apply the definition of an exponent in which (2^3) is the base. Thus,

$$(2^3)^2 = (2^3)(2^3) = (2 \cdot 2 \cdot 2)(2 \cdot 2 \cdot 2) = 2^6$$

Since $(2^3)^2 = 2^6$, we see that we can simply **multiply** the exponents of $(2^3)^2$ to obtain $3 \cdot 2 = 6$.

This discussion leads us to the third law of exponents.

$$(x^a)^b = x^{ab}$$

Examples

1. $(2^2)^4 = 2^{2 \cdot 4} = 2^8$
2. $(x^4)^5 = x^{4 \cdot 5} = x^{20}$
3. $(y^{-2})^{-3} = y^{(-2)(-3)} = y^6$

To simplify an expression such as $(2 \cdot 3)^3$, we again apply the definition of an exponent in which $(2 \cdot 3)$ is the base. Thus,

$$(2 \cdot 3)^3 = (2 \cdot 3)\,(2 \cdot 3)\,(2 \cdot 3)$$

$$= 2 \cdot 2 \cdot 2 \cdot 3 \cdot 3 \cdot 3 \quad \text{(Using the commutative and associative}$$
$$\text{properties of multiplication)}$$

$$= 2^3 \cdot 3^3$$

Since $(2 \cdot 3)^3 = 2^3 \cdot 3^3$, we can generalize and state the fourth law of exponents.

$$(xy)^a = x^a y^a$$

Examples

1. $(xy)^4 = x^4 y^4$
2. $(3a)^2 = 3^2 \cdot a^2 = 9a^2$

Finally, consider an expression such as $(\frac{2}{3})^3$. We again apply the definition of an exponent in which $\frac{2}{3}$ is the base. Thus,

$$\left(\frac{2}{3}\right)^3 = \frac{2}{3} \cdot \frac{2}{3} \cdot \frac{2}{3} = \frac{2^3}{3^3}$$

Since $\left(\frac{2}{3}\right)^3 = \frac{2^3}{3^3}$, we can generalize and state the fifth law of exponents.

$$\left(\frac{x}{y}\right)^a = \frac{x^a}{y^a}, \, y \neq 0$$

Examples

1. $\left(\frac{2}{3}\right)^2 = \frac{2^2}{3^2} = \frac{4}{9}$

2. $\left(\frac{x}{y}\right)^5 = \frac{x^5}{y^5}$

Some exponential simplifications require us to use more than one law of exponents.

Examples

1. $(x^2 y^3)^4 = (x^2)^4 (y^3)^4 = x^8 y^{12}$

2. $\left(\frac{x^2}{y^5}\right)^3 = \frac{(x^2)^3}{(y^5)^3} = \frac{x^6}{y^{15}}$

In certain cases we use laws of exponents after applying the distributive property.

Example

1. $3x^4(2x^2 + y) = (3x^4)(2x^2) + (3x^4)(y) = 6x^6 + 3x^4 y$

Our next problem incorporates exponential properties with the agreed-upon order of operations.

Problem 1 Simplify $2 - (2y)(3y^2) + \frac{(4y^2)^2}{2y}$

Solution **step 1**
$(2y)(3y^2) = (2)(3)(y^{1+2}) = 6y^3$

step 2
$\frac{(4y^2)^2}{2y} = \frac{4^2(y^2)^2}{2y} = \frac{16y^4}{2y} = 8y^{4-1} = 8y^3$

step 3
$2 - (2y)(3y^2) + \frac{(4y^2)^2}{2y}$
$= 2 - 6y^3 + 8y^3$
$= 2 + 2y^3$

Problem Set 2.1

1. Use the laws of exponents to simplify the following and express the answers with positive exponents.

   a. $a^4 \cdot a^7$ b. $x^5 \cdot x$ c. $2x^4x^5$

   d. $(3x^3)(2x^4)$ e. $y^4 \cdot y^{-7}$ f. $x^{-2} \cdot x^{-5}$

   g. $\dfrac{x^7}{x^4}$ h. $\dfrac{x^5}{x^5}$ i. $\dfrac{x^{-3}}{x^4}$

   j. $\dfrac{-12x^6}{4x^2}$ k. $(x^{-3})^6$ l. $(2 \cdot a)^4$

   m. $\left(\dfrac{x}{y}\right)^3$

2. Is it true that $a^x \cdot a^y = a^{xy}$?

3. Is it true that $2^2 \cdot 2^3 = 4^5$?

4. Simplify each of the following.

   a. $(x^5y^3)^4$ b. $\left(\dfrac{x^3}{y^7}\right)^2$

   c. $(4y^2)^3$ d. $(3y^3)^2 - (2y^2)^3$

   e. $\dfrac{(10y^4)^3}{25y}$ f. $\dfrac{2(3b^2)^3}{2b^4}$

5. Choose the expression that is equivalent to $4a^3(a^3b^4)$.

   a. $5a^3b^4$ b. $4a^3(b^3a^4)$

   c. $4a^9b^4$ d. $(2a^3b^2)^2$

6. Choose the expression that is equivalent to $5x^2y(3x + y^2)$.

   a. $15x^2y + 5x^2y^2$

   b. $8x^3y + 5x^2y^3$

   c. $3x(5x^2y + y^2)$

   d. $5x^2y(y^2 + 3x)$

7. Choose the expression that is equivalent to $2a^3(a^2b)^4$.

   a. $2a^3(a^6b^4)$ b. $2a^3(b^2a)^4$

   c. $2a^{11}b^4$ d. $(2a^5 + 2a^3b)^4$

8. Choose the expression that is not true for all real numbers.

   a. $4x^3y^2(5x - 2y^7) = 20x^4y^2 - 8x^3y^9$

   b. $(3x^3)(2y^4) = 6(xy)^7$

   c. $5x^3y^2(4x + y^7) = (y^7 + 4x)5x^3y^2$

   d. $4x^3(x^2y^7) = (4x^3x^2)y^7$

Review Competencies

9. Determine the place value numeral represented by $(2 \cdot 10^0) + (3 \cdot 10^{-2})$.

10. Express 201.004 in expanded form.

11. Evaluate:

    a. 2^{-2} b. 10^{-1} c. $(50)^0$

    d. 3^4 e. $4 \cdot 10^3$ f. $3 \cdot 2^{-2}$

    g. $(2^2)^3$ h. $2^4 - 3^2$

12. Divide: $21 \div \dfrac{7}{8}$

13. Express $\dfrac{17}{11}$ as a mixed number.

14. $\dfrac{?}{4} = \dfrac{5}{2}$

15. Evaluate: $8 \div \dfrac{3}{2} - \dfrac{1}{3} \div \dfrac{1}{5}$

16. What is the place value of the underlined digit in 23.2_04_?

17. $-\dfrac{3}{4} - (-2) =$

18. $\dfrac{17}{25} =$

    a. 0.68   b. 0.068   c. 6.8%   d. 0.68%

19. A tutoring center pays tutors $4.50 per hour up through 25 hours a week, paying an additional $2.25 per hour for each hour over 25 hours per week. How much will a tutor earn in a two-week period if the hours worked come to 30 for the first week and 35 for the second week?

Section 3 Scientific Notation

In the modern world, very large and very small numbers are quite common. For example, the population of the United States is approximately 220,000,000. In calculations, such numbers can often become very awkward and cumbersome. Fortunately, they can be easily expressed by writing them in what is known as **scientific notation**.

A number that is written in **scientific notation** means that it is written as a **product** of a decimal number greater than or equal to 1, but less than 10, and a power of ten.

A number expressed in scientific notation is:

$$\left(\begin{array}{l} \text{a decimal number greater} \\ \text{than or equal to 1, but} \\ \text{less than 10} \end{array} \right) \quad \text{x} \quad 10^{\text{(some power)}}$$

Let us review below the powers of ten frequently encountered in this chapter.

$10^5 = 10 \cdot 10 \cdot 10 \cdot 10 \cdot 10 = 100,000$ $10^{-5} = \frac{1}{100,000} = 0.00001$

$10^4 = 10 \cdot 10 \cdot 10 \cdot 10 \quad = 10,000$ $10^{-4} = \frac{1}{10,000} = 0.0001$

$10^3 = 10 \cdot 10 \cdot 10 \quad\quad = 1,000$ $10^{-3} = \frac{1}{1,000} = 0.001$

$10^2 = 10 \cdot 10 \quad\quad\quad = 100$ $10^{-2} = \frac{1}{100} = 0.01$

$10^1 = 10 \quad\quad\quad\quad = 10$ $10^{-1} = \frac{1}{10} = 0.1$

$10^0 = 1 \quad\quad\quad\quad = 1$

Examples

1. The number 2,510 expressed in scientific notation is:
    $$2.51 \cdot 10^3 \quad (2,510 = 2.51 \cdot 1000)$$

    Notice that the decimal point is placed so that the first factor is a number greater than or equal to 1, but less than 10 (2.51). This will always result in having **one** non-zero digit to the **left** of the decimal point. Since the decimal point has been moved 3 places to the left, the other factor is 10 with 3 as its exponent. Thus, **the power of ten will be equal to the number of places the decimal point is moved.** Observe that 2,510 is **not** expressed in scientific notation as:
    $$0.251 \cdot 10^4 \text{ or } 25.1 \cdot 10^2.$$

2. The number 0.027 expressed in scientific notation is $2.7 \cdot 10^{-2}$. We again place the decimal point so that the first factor is greater than or equal to 1, but less than 10, giving 2.7. Since we began with 0.027 and now have 2.7, the decimal point has been moved two places to the **right**. To compensate for this change, the other factor is 10 with -2 as its exponent. Observe that the power of ten is **positive** when the decimal point is moved to the **left** and **negative** when the decimal point is moved to the **right**. Again, in both cases, the power of ten is equal to the number of places the decimal point is moved.

These two examples illustrate the two general steps used to write a number in scientific notation:

Steps in writing a decimal numeral in scientific notation

A. Move the decimal point so that the number written has only one non-zero digit to the left of the decimal point. This is equivalent to writing the first factor as a number greater than or equal to 1, but less than 10.

B. Look at the number of places the decimal point has been moved in considering the change from the original decimal numeral to the new numeral obtained in Step 1.
   i. If the decimal point has been moved to the left, the number of places it has been moved is the positive exponent for 10.
   ii. If the decimal point has been moved to the right, the number of places it has been moved is the negative exponent for 10.

---

3. The decimal numeral represented by $3.5 \cdot 10^{3}$ is 3,500. Since the exponent is 3, it is necessary to move the decimal point **three** places to the **right** to determine the decimal numeral.

4. The decimal numeral represented by $8.3 \cdot 10^{-4}$ is 0.00083. Since the exponent is -4, it is necessary to move the decimal point four places to the **left** to determine the decimal numeral.

Problem Set 3.1

1. Express the following in scientific notation.
   a. 325,000   b. 485,000,000
   c. 71.2   d. 3.86
   e. 0.0006   f. 0.175
   g. 0.000000796   h. 0.1

2. Write the decimal numeral represented by each of the following.

   a. $2.9 \cdot 10^4$ b. $5.3 \cdot 10^{-3}$ c. $6.9 \cdot 10^0$

Review Competencies

3. Evaluate the following and write the results using only positive exponents.

   a. $x^2 \cdot x^5$ b. $(2x^3)(4x^6)$ c. $a^{-3} \cdot a^{-4}$

   d. $y^7 \cdot y^3$ e. $\frac{a^{10}}{a^7}$ f. $\frac{20x^5}{4x^2}$

   g. $(x^{-3})^{-3}$ h. $(3y)^3$ i. $(\frac{3}{4})^2$

4. Determine the numeral represented by
   $(6 \cdot 10^3) + (2 \cdot 10^1) + (3 \cdot 10^{-2})$

5. Express 0.007 in expanded form.

6. Evaluate:

   a. 4^{-3} b. $6 \cdot 10^3$ c. $10^0 \cdot 4$

   d $4(2)^3$ e. $5 \cdot 5^{-1}$

7. What is 12% of 40?

8. Select a choice equivalent to $(7^4)^3$.

   i. $7^4 \cdot 7^4 \cdot 7^4$ ii. 7^{21} iii. 7^7

   iv. 7^{12} v. $7^4 + 7^4 + 7^4$

   a. i and ii only   b. iii only
   c. iv only   d. iv and v only
   e. i and iv only

Expressing extremely large numbers or extremely small numbers in scientific notation can simplify tedious calculations involving multiplication and division.

Examples

$$
\begin{aligned}
1. \ (12{,}000{,}000)(0.000003) &= (1.2 \cdot 10^7)(3 \cdot 10^{-6}) \\
&= (1.2 \cdot 3)(10^7 \cdot 10^{-6}) \qquad \text{rearranging factors} \\
&= 3.6 \cdot (10^7 \cdot 10^{-6}) \qquad\qquad 1.2 \cdot 3 = 3.6 \\
&= 3.6 \cdot 10 \qquad\qquad 10^7 \cdot 10^{-6} = 10^{7+(-6)} = 10^1 = 10 \\
&= 36
\end{aligned}
$$

2. $\dfrac{7,500,000,000}{150,000}$ = $\dfrac{7.5 \cdot 10^9}{1.5 \cdot 10^5}$

                   = $\dfrac{5 \cdot 10^9}{10^5}$ $\dfrac{7.5}{1.5} = 5$

                   = $5 \cdot 10^4$ $\dfrac{10^9}{10^5} = 10^4$

                   = $50,000$

3. $\dfrac{0.00025}{500,000}$ = $\dfrac{2.5 \cdot 10^{-4}}{5 \cdot 10^5}$

                   = $\dfrac{0.5 \cdot 10^{-4}}{10^5}$ $\dfrac{2.5}{5} = 0.5$

                   = $0.5 \cdot 10^{-9}$ $\dfrac{10^{-4}}{10^5} = 10^{-4-5} = 10^{-9}$

                   = $5 \cdot 10^{-1} \cdot 10^{-9}$ $0.5 = 5 \cdot 10^{-1}$

                   = $5 \cdot 10^{-10}$ $10^{-1} \cdot 10^{-9} = 10^{-1+(-9)} = 10^{-10}$

                   = 0.0000000005

4. $\dfrac{(800,000)(60,000)}{6,000,000}$ = $\dfrac{(8 \cdot 10^5)(6 \cdot 10^4)}{6 \cdot 10^6}$

                   = $\dfrac{(8 \cdot 6)(10^5 \cdot 10^4)}{6 \cdot 10^6}$

                   = $\dfrac{48 \cdot 10^9}{6 \cdot 10^6}$

                   = $8 \cdot 10^3$

                   = $8,000$

Problem Set 3.2

1. Use scientific notation to perform each of the following calculations.
   a. $(3,000,000)(0.00002)$
   b. $(4,000,000)(100,000)(0.0002)$
   c. $\dfrac{80,000,000}{40,000}$
   d. $\dfrac{(2,000,000)(0.004)(100,000)}{80,000}$
   e. $\dfrac{(150,000)(60,000)}{(9,000)(0.01)}$
   f. $\dfrac{0.00064}{3,200,000}$
   g. $\dfrac{0.00036}{40,000}$

2. Find the decimal numeral for each of the following.
   a. $(2.6 \cdot 10^{-3})(2 \cdot 10^2)$
   b. $(1.5 \cdot 10^4)(2.3 \cdot 10^{-6})$
   c. $\dfrac{3.5 \cdot 10^3}{5 \cdot 10^5}$
   d. $\dfrac{2.8 \cdot 10^{-2}}{1.4 \cdot 10^2}$

Review Competencies

3. Express the following in scientific notation.
   a. $925,000,000$ b. 0.000062

4. Write the decimal numeral for $7.5 \cdot 10^4$.

5. Determine the place value numeral for $(2 \cdot 10^0) + (3 \cdot 10^{-1})$.

6. Simplify and write with positive exponents:
   a. $a^4 \cdot a$
   b. $(-5a^4)(3a^3)$
   c. $\dfrac{25x^7}{5x^2}$
   d. $2 \cdot 10^{-2}$
   e. $3(5)^2$

7. Express 0.071 as a percent.

8. Evaluate: $15 \div 5 \cdot 3 - 4$

9. What percent of 50 is 19?

10. What is the place value for the underlined digit in 23.00$\underline{6}$?

11. Select the item equivalent to $(x + y)^3$.
    a. $x^3 + y^3$
    b. $(x + y) + (x + y) + (x + y)$
    c. $(x)(x)(x)(y)(y)(y)$
    d. None of these

Scientific notation numerals can simplify the addition and subtraction of some very large or very small numbers.

Problem 1 Write the sum of $(2.75 \cdot 10^4) + (3.5 \cdot 10^5)$ using both a scientific notation numeral and a decimal numeral.

Solution First, write the problem so that **both** expressions in parentheses contain the same power of 10, namely 10^5. In general, the **higher** exponent that appears in the problem will be used.

Since $2.75 = 2.75 \cdot 10^0$, we can write:

$$2.75 = 0.275 \cdot 10^1.$$

Thus, $(2.75 \cdot 10^4) = (0.275 \cdot 10^1 \cdot 10^4) = (0.275 \cdot 10^5)$
The problem now is: $(0.275 \cdot 10^5) + (3.5 \cdot 10^5)$
Using the distributive property, this is equivalent to:
$$(0.275 + 3.5) \cdot 10^5$$
Thus, the final result is $3.775 \cdot 10^5$ or 377,500.

Problem 2 Write the difference of $(3.65 \cdot 10^6) - (1.05 \cdot 10^5)$ using both a scientific notation numeral and a decimal numeral.

Solution Both expressions must contain the **same** power of 10 -- in this case, 10^6.
$1.05 \cdot 10^5$ is written as $0.105 \cdot 10^6$.
The problem now is completed as:
$$(3.65 \cdot 10^6) - (0.105 \cdot 10^6)$$

$$= (3.65 - 0.105) \cdot 10^6$$

$$= 3.545 \cdot 10^6$$

Thus, the final result is $3.545 \cdot 10^6$ or 3,545,000.

Problem Set 3.3

1. Find the sum or difference of:
    a. $(1.67 \cdot 10^6) + (2.52 \cdot 10^5)$
    b. $(3.15 \cdot 10^4) + (1.5 \cdot 10^6)$
    c. $(2.65 \cdot 10^5) - (1.35 \cdot 10^4)$
    d. $(4.52 \cdot 10^6) - (2.35 \cdot 10^4)$

Review Competencies

2. Use scientific notation to compute
    $(8,000,000) \cdot (1,000,000) \cdot (0.00004)$

3. Write the decimal numeral represented by $4.1 \cdot 10^{-3}$.

4. Simplify:
   a. $x^{-5} \cdot x^8$ b. $\frac{x^{-2}}{x^5}$ c. $\frac{-30x^7}{6x}$

5. Determine the place value numeral for
   $(1 \cdot 10^{-2}) + (3 \cdot 10^{-3}) + (5 \cdot 10^{-4})$.

6. Express 5297.31 in expanded form.

7. Evaluate:
   a. 5^{-2} b. $6 \cdot 10^{-3}$ c. $6^0 \cdot 6$
   d. $(3 \cdot 4)^2$ e. $3(4)^2$

8. Express $7\frac{5}{6}$ as an improper fraction.

9. Subtract: $4\frac{1}{9} - 1\frac{5}{6}$

10. Divide: $\frac{11}{12} \div \frac{5}{6}$

11. Subtract: $12 - 0.025$

12. Express 160% as a fraction and a decimal.

13. 150% of 100 is ___.

Section 4   The Base of a System of Numeration

In the decimal system of numeration, all the groupings are by tens or powers of ten. The numeral 24 implies 2 groups of ten (10^1) and 4 ones (10^0). Since we are grouping by tens, the decimal numeral has a **base** of ten.

> ## Definition
> The **base** of a system of numeration refers to the number of individual digit symbols that can be used in that system as well as to the way in which grouping is to occur.

There are ten digit symbols in the base ten system. They are:
$$0,1,2,3,4,5,6,7,8 \text{ and } 9.$$

In the following problems we will see that we can express numerals in base systems other than ten.

Problem 1   Consider the following collection of x's: x x x x x x x
                                                                 x x x x x x x

Write a numeral representing this collection of x's in base 9.

Solution   A base 9 numeral requires **grouping by nines**. Therefore the collection of x's might be separated as:

x x x x | x x x
x x x x x | x x

Notice that there is 1 group of 9 and 5 individual x's left or, stated another way, 1 group of 9 and 5 ones. Thus, the base 9 numeral is written as 15_9. The small 9 to the lower right of 15 is a subscript and indicates that 15_9 is a base 9 numeral. This can also appear as 15 (base nine).

Problem 2   Use the collection of 14 x's to write a base 7 numeral.

Solution   The collection of x's is first separated into groups of 7.

x x x x x x x
x x x x x x x

Notice that there are two groups of 7 and no individual x's left or 2 groups of 7 and 0 ones. The base 7 numeral is 20_7 or 20 (base seven).

Problem 3   Use the collection of 14 x's to write a base 5 numeral.

Solution   The collection of x's is first separated into groups of 5.

x x x | x x | x x
x x | x x x | x x

Notice that there are two groups of 5 and 4 individual x's left or 2 groups of 5 and 4 ones. Therefore, the base 5 numeral is 24_5 or 24 (base five).

There were fourteen x's in each of the preceding three problems. "Fourteen" is 14 (rather than 14_{10}) in the base ten system, but 15_9, 20_7, and 24_5 are three other base system numerals for fourteen.

$$14 = 15_9 = 20_7 = 24_5$$

There are many numerals for 14. A numeral is simply the name of a number.

Problem Set 4.1

Use the collection of x's shown at the right to write numerals in the following bases:

```
x x x x x x x x x
x x x x x x x x x
```

1. base 7   2. base 9
3. base 8   4. base 6

Review Competencies

5. Find the sum of $(2.45 \cdot 10^7) + (1.85 \cdot 10^6)$ in scientific notation.

6. Use scientific notation to compute:
$$\frac{(6,000,000)(0.004)}{8,000}$$

7. Write the place value numeral for $7.5 \cdot 10^0$.

8. Simplify the following and write the results with positive exponents.

    a. $a^{-2} \cdot a^{-3}$ b. $\frac{x^{10}}{x^6}$ c. $\frac{-40a^5}{-4a^3}$

    d. 10^{-4} e. 3^{-3} f. $(1 \cdot 8)^2$

    g. $(2^2)^{-2}$ h. $(2b)^3$ i. $\left(\frac{4}{5}\right)^2$

9. $10 = \frac{?}{4}$

10. Express $\frac{8}{3}$ as a mixed number.

11. Subtract: $10 - 3\frac{1}{5}$

12. What percent of 17 is 34?

13. Evaluate: $(3 + 5)(3 - 5)$

14. Which of the following equals $(4^3)(3^2)$?
    a. $(4 + 4 + 4)(3 + 3)$
    b. $(4 \cdot 4 \cdot 4)(3 \cdot 3)$
    c. $(4 \cdot 3)(3 \cdot 2)$
    d. $(4 \cdot 3)^6$

It is possible to convert two-digit numerals expressed in other bases to base ten numerals.

Problem 4   Convert the numeral 34_5 to a numeral in base ten.

Solution 34_5 means there are 3 groups of 5 and 4 ones, which can be shown as:
$$34_5 = 3(5) + 4(1) = 15 + 4 = 19$$

Problem 5   Convert the numeral 23_6 to a base ten numeral.

Solution   The numeral 23_6 implies 2 groups of six and 3 ones. The conversion is accomplished in the following steps.
$$23_6 = 2(6) + 3(1) = 12 + 3 = 15$$

Problem Set 4.2

1. Write each of the following as a base ten numeral.
   a. 43_5 b. 34_7 c. 32_4 d. 42_9
   e. 21_3 f. 52_6 g. 43_8 h. 11_2
   i. 61_7 j. 30_6 k. 4_5 l. 2_7
   m. 6_8

2. Is there any difference between 20_4 and 2_4?

3. Is $2_3 = 2_4 = 2_5$?

Review Competencies

4. Use the collection of x's shown at the right to find numerals in:

   x x x x x x x
   x x x x x x x

   a. base 9   b. base 5   c. base 4

5. Express the following in scientific notation.
   a. 128,000   b. 0.0026

6. Use scientific notation to perform the computation at the right. $\dfrac{(30,000)(200,000)}{60,000}$

7. Simplify and write the results with only positive exponents.
   a. $x \cdot x^3$ b. $(-5a^3)(-4a^5)$ c. $\dfrac{x^{-2}}{x^4}$
   d. 2^{-5} e. 10^0 f. 10^{-3}
   g. $4 \cdot 10^{-2}$ h. $4(2)^2$ i. $(1 \cdot 4)^2$

8. Express 23.125 in expanded form.

9. Determine the place value numeral for $(1 \cdot 10^{-1}) + (3 \cdot 10^{-3})$.

10. Subtract: $\dfrac{4}{5} - \dfrac{1}{4}$

11. Divide: $\dfrac{7}{8} \div 4$

12. Express $\dfrac{13}{25}$ as a percent.

13. What is 30% of 180?

14. Evaluate: $\dfrac{1}{4} \cdot 8 + \dfrac{1}{3} \div \dfrac{1}{3}$

15. What is the place value of the underlined digit in 0.056?

16. Select the item equivalent to $x^{-2} + y^0 + z^3$.

   a. $\dfrac{1}{x \cdot x + 1 + z \cdot z \cdot z}$
   b. $-2 \cdot x \cdot x + 1 + z \cdot z \cdot z$
   c. $-\dfrac{1}{x \cdot x} + 0 + z \cdot z \cdot z$
   d. $\dfrac{1}{x \cdot x} + 1 + z \cdot z \cdot z$
   e. None of the above

Section 5   Bases other than Ten

Recall that in the base ten system of numeration the only symbols permitted for digits are 0, 1, 2, 3, 4, 5, 6, 7, 8, and 9. There are **ten** digit symbols for base **ten**.

In a base six system of numeration the only digit symbols permitted are 0, 1, 2, 3, 4 and 5. There are six digit symbols for a base **six** numeration system.

Chapter 4

Notice that it is impossible to write a base six symbol such as 67_6 because 6 and 7 are not among the digit symbols permitted in base 6.

In a base four system of numeration the only digit symbols permitted are 0, 1, 2 and 3. Again, it is not possible to write a base four numeral such as 43_4 because 4 is not among the digit symbols permitted in base four.

Problem 1   Express 436_7 in expanded form.

Solution   Recall that the **base ten** numeral 436 in expanded form is:
$$(4 \bullet 10^2) + (3 \bullet 10^1) + (6 \bullet 10^0)$$
Notice that the exponents are applied to the base within the parentheses. Similarly, 436_7 is written as:
$$(4 \bullet 7^2) + (3 \bullet 7^1) + (6 \bullet 7^0)$$
Notice that the place value of the digit 6 in 436_7 is ones because $7^0 = 1$. The place value of the digit 3 is 7^1 or 7 and the place value of the digit 4 is 7^2 or 49.

The table below indicates the digit symbols and place values for the base ten and some other selected base numeration systems.

Base	Digits		Place Values					
10	0,1,2,3,4,5,6,7,8,9	10^3	10^2	10^1	10^0 .	10^{-1}	10^{-2}	10^{-3}
2	0,1	2^3	2^2	2^1	2^0 .	2^{-1}	2^{-2}	2^{-3}
3	0,1,2	3^3	3^2	3^1	3^0 .	3^{-1}	3^{-2}	3^{-3}
5	0,1,2,3,4	5^3	5^2	5^1	5^0 .	5^{-1}	5^{-2}	5^{-3}
8	0,1,2,3,4,5,6,7	8^3	8^2	8^1	8^0 .	8^{-1}	8^{-2}	8^{-3}

Examples

1. The place value of the underlined digit in $3\underline{2}04_5$ is 5^2 or 25.

2. The place value of the underlined digit in $764\underline{5}_8$ is 8^0 or 1.

3. The place value of the underlined digit in $211.0\underline{2}1_3$ is 3^{-2} or $\frac{1}{9}$.

Problem Set 5.1

1. Write each of the following numerals in expanded form.
   a. 354_6 b. 2754_8

2. If a base system involves only three digit symbols, what is the base?

3. Which of the following symbols is (are) expressed incorrectly?
   a. 423_5 b. 678_9 c. 678_7 d. 12345_6

4. Express the place value of the underlined digit in each of the following.
   a. $1\underline{1}01_2$ b. $\underline{2}102_3$ c. $3240\underline{3}_5$
   d. $\underline{6}7542_8$ e. $3.1\underline{2}_4$

Review Competencies

5. Express each of the following as a base ten numeral.
   a. 44_8 b. 40_5 c. 5_6

6. Use the collection of x's at the right to express numerals in: x x x x x x x x
   a. base 7   b. base 5   c. base 9   x x x x x x x x

7. Express 0.0000065 in scientific notation.

8. Use scientific notation to compute $(500,000)(0.003)$.

9. Simplify and write the results with positive exponents.
   a. $a^{-3} \cdot a^5$ b. $x^{-2} \cdot x^{-4}$ c. $\frac{x^{12}}{x^7}$
   d. $\frac{15x^5}{-5x^2}$ e. $(1000)^0$ f. $4 \cdot 10^0$
   g. $2(5)^2$ h. $4 \cdot 2^{-2}$ i. $(a^3)^7$
   j. $(ab)^6$ k. $(\frac{x}{y})^7$

10. Express 212.015 in expanded form.

11. What percent of 30 is 12?

12. Divide: $0.25 \div 0.0005$

13. Evaluate: $15 - (4 \cdot 1 + 6 \div 3)$

14. Express $4\frac{3}{4}\%$ as a decimal.

15. Evaluate: $6(\frac{2}{3} \div \frac{4}{3} \cdot \frac{1}{4} \cdot \frac{1}{3})$

Let us now turn our attention to converting numerals in other bases to those in base ten.

Problem 2   Convert the numeral 32_6 to a numeral in base ten.

Solution   We know from the previous section that 32_6 implies 3 groups of six and 2 ones or $3(6) + 2(1) = 20$. The numeral 32_6 can also be written in expanded form and evaluated as follows:
$$32_6 = (3 \cdot 6^1) + (2 \cdot 6^0)$$
$$= (3 \cdot 6) + (2 \cdot 1)$$
$$= 18 + 2$$
$$= 20$$
Therefore, $32_6 = 20$.

Chapter 4

This problem leads to the following generalization:

> To convert a numeral in any other base to base ten, first
> write it in expanded form. Then evaluate the expanded
> form expression.

Problem 3   Convert 234_5 to a base ten numeral.

Solution   First write 234_5 in expanded form.
$$234_5 = (2 \cdot 5^2) + (3 \cdot 5^1) + (4 \cdot 5^0)$$
Evaluate the expanded form expression:
$$
\begin{aligned}
(2 \cdot 5^2) + (3 \cdot 5^1) + (4 \cdot 5^0) &= (2 \cdot 25) + (3 \cdot 5) + (4 \cdot 1) \\
&= \quad 50 \quad + \quad 15 \quad + \ 4 \\
&= 69
\end{aligned}
$$
Therefore, $234_5 = 69$.

Problem 4   Convert 111_2 to a base ten numeral.

Solution

$$
\begin{aligned}
111_2 &= (1 \cdot 2^2) + (1 \cdot 2^1) + (1 \cdot 2^0) \\
&= (1 \cdot 4) + (1 \cdot 2) + (1 \cdot 1) \\
&= \ 4 \ + \ 2 \ + \ 1 \\
&= \ 7
\end{aligned}
$$

Therefore, $111_2 = 7$.

Problem 5   Convert 1203_4 to a base ten numeral.

Solution

$$
\begin{aligned}
1203_4 &= (1 \cdot 4^3) + (2 \cdot 4^2) + (0 \cdot 4^1) + (3 \cdot 4^0) \\
&= (1 \cdot 64) + (1 \cdot 16) + (0 \cdot 4) + (3 \cdot 1) \\
&= \ 64 \ + \ 32 \ + \ 0 \ + \ 3 \\
&= 99
\end{aligned}
$$

Therefore, $1203_4 = 99$.

Problem Set 5.2

1. Express each of the following as a base ten numeral.
   a. 423_5 b. 3143_5 c. 3000_5
   d. 1011_2 e. 226_8 f. 3312_4

Review Competencies

2. Use the collection of x's at the   x x x x x x
   right to express numerals in: x x x x x x
   a. base 8   b. base 7   c. base 4

3. Use scientific notation to
   evaluate the fraction $\dfrac{(20,000)(3,000)}{(1,000)(0.01)}$
   at the right.

4. Simplify and write each of the results only with positive exponents.
   a. $a \cdot a$ b. $(-6x^2)(5x^4)$ c. $y^{-1} \cdot y^{-2}$
   d. $\dfrac{x^4}{x^4}$ e. $\dfrac{-12x^3}{-3x^3}$ f. 15^0
   g. $(2 \cdot 5)^2$ h. $3(3)^2$ i. $3 \cdot 4^{-1}$

5. Express 0.0001 in expanded form.

6. Multiply: $\dfrac{4}{7} \cdot 21$

7. Express 7% as a decimal.

8. Express 85% as a fraction.

9. Evaluate: $\frac{1}{2}(10 - 2 \cdot 3)$

10. What is the place value of the underlined digit in $\underline{6}575_8$?

11. Determine the common relationship between the numbers in each pair. Then identify the missing term.
    $(0.8, 0.64)$ $(0,0)$ $(-\frac{1}{3}, \frac{1}{9})$ $(\frac{1}{5}, \underline{\quad})$

12. What is the next term in the following arithmetic progression? (The definition of an arithmetic progression is on page 36, #9.)
    4, -1, -6, -11, _____

13. What is the next term in the following geometric progression? (The definition of a geometric progression is on page 36, #10.)
    2, -6, 18, _____

14. Find the smallest positive multiple of 7 which leaves a remainder of 1 when divided by 4 and a remainder of 1 when divided by 5.

Chapter 4 Summary

1. Definitions involving Exponents

a. x^n means that x is used as a factor n times, where n is a counting number greater than or equal to 1.

b. $x^1 = x$ c. $x^0 = 1, x \neq 0$ d. $x^{-a} = \frac{1}{x^a}$

2. Laws of Exponents

a. $x^a \cdot x^b = x^{a+b}$ b. $\frac{x^a}{x^b} = x^{a-b}, x \neq 0$ c. $(x^a)^b = x^{ab}$

d. $(xy)^a = x^a y^a$ e. $\left(\frac{x}{y}\right)^a = \frac{x^a}{y^a}, y \neq 0$

3. Scientific Notation

a. The original number is written as a product of a decimal number greater than or equal to 1, but less than 10, and a power of ten.

b. The first two laws of exponents are used to perform multiplication and division problems in scientific notation. The base is always 10.

4. Systems of Numeration

a. In base B, the only symbols permitted for the digits are 0,1,2,3,...,B-1. (For example, numerals in base 5 use the symbols 0,1,2,3, and 4.)

b. Place values in base B are: $B^3 B^2 B^1 B^0. \ B^{-1} B^{-2} B^{-3}$
    (For example, place values in base 5 are: $5^3 5^2 5^1 5^0. \ 5^{-1} 5^{-2} 5^{-3}$.)

The base determines both the symbols permitted for the digits and the place value of each digit.

To convert a number in any base to base ten, write it in expanded form (expressing the sum of each digit multiplied by its place value). Then evaluate the expanded form expression.

Chapter 4 Self-Test

1. Select the equivalent expression for $(2^3)(5^2)$.
   a. 10^6 b. $(2 + 2 + 2)(5 + 5)$
   c. $(2 \cdot 2 \cdot 2)(5 \cdot 5)$
   d. $(2 \cdot 5)^6$ e. None of these

2. Select the equivalent expression for $3^2 + 4^2$.
   a. $(3 + 4)^2$ b. $(3 + 4)^4$
   c. $(3)(2) + (4)(2)$ d. $(3)(3) + (4)(4)$
   e. None of these

3. Select the equivalent expression for $(3^2)^3$.
   a. 9^6 b. $3^2 \cdot 3^2 \cdot 3^2$
   c. 3^5 d. 3^8
   e. None of these

4. Select the equivalent expression for $2(3)^2$.
   a. 6^2 b. $2^2 \cdot 3^2$
   c. $2 \cdot 3^2$ d. $(2 \cdot 3)^2$
   e. None of these

5. Select the equivalent expression for $3^4 - 2^2$.
   a. $(3)(3)(3)(3) - (2)(2)$
   b. 1^2 c. $(3 - 2)^2$
   d. $(3 + 3 + 3 + 3) - (2 + 2)$
   e. None of these

6. Select the equivalent expression for $\frac{4^5}{4^2}$.
   a. 3
   b. $(4 + 4 + 4 + 4 + 4) \div (4 + 4)$
   c. 1^3 d. 4^3
   e. None of these

7. Select the equivalent expression for $\frac{4^4}{2^2}$.
   a. 2^2
   b. $(4 \cdot 4 \cdot 4 \cdot 4) \div (2 \cdot 2)$
   c. $(4 \cdot 4 \cdot 4 \cdot 4) - (2 \cdot 2)$
   d. 4^2 e. None of these

8. Select the equivalent expression for $(5 - 3)^2$.
   a. $5^2 - 3^2$ b. $(5)(2) - (3)(2)$
   c. 2^2 d. $(5)(5)(3)(3)$
   e. None of these

9. Use scientific notation to compute $0.00250 \div 1,250,000$.
   a. $2 \cdot 10^{-9}$ b. $2 \cdot 10^3$
   c. $2 \cdot 10^9$ d. $2 \cdot 10^{-2}$
   e. None of these

10. Use scientific notation to compute $(2.1 \cdot 10^5)(1.3 \cdot 10^{-6})$.
    a. 0.273 b. 0.0273
    c. 2.73 d. -0.273
    e. None of these

11. Express the answer to $82,000 + 6,000$ in scientific notation.
    (Hint: Perform the addition first.)
    a. $8.8 \cdot 10^3$ b. $8.8 \cdot 10^{-3}$
    c. $8.8 \cdot 10^{-4}$ d. $8.8 \cdot 10^4$
    e. None of these

12. Express the answer to $0.0143 - 0.0028$ in scientific notation.
    (Hint: Perform the subtraction first.)
    a. $1.15 \cdot 10^2$ b. $1.15 \cdot 10^{-2}$
    c. $11.15 \cdot 10^3$ d. $11.15 \cdot 10^{-3}$
    e. None of these

13. Select the place value associated with the underlined digit in 3.01$\underline{5}$6 (base ten).
    a. $\frac{1}{10^0}$ b. $\frac{1}{10^1}$
    c. $\frac{1}{10^2}$ d. $\frac{1}{10^3}$
    e. None of these

Chapter 4

14. Select the base ten equivalent of 2013_5 (a base five numeral).
    - a. 258
    - b. 283
    - c. 270
    - d. 133
    - e. None of these

15. Select the place value of the underlined digit in $2\underline{4}310_5$ (a base five numeral).
    - a. 5^0
    - b. 5^1
    - c. 5^2
    - d. 5^3
    - e. None of these

16. Study the following examples:
    $$a^2 \# a^4 = a^{10} \qquad a^3 \# a^2 = a^7$$
    $$a^5 \# a^3 = a^{11}$$
    Select the statement that is compatible with the data.
    - a. $a^x \# a^y = a^{2x + y}$
    - b. $a^x \# a^y = a^{x + 2y}$
    - c. $a^x \# a^y = a^{x + y + 4}$
    - d. $a^x \# a^y = a^{xy + 2}$

17. Study the following examples:
    $$a^5 * a^3 * a^2 = a^5 \qquad a^3 * a^7 * a^2 = a^6$$
    $$a^2 * a^4 * a^8 = a^7$$
    Select the statement that is compatible with the data.
    - a. $a^x * a^y * a^z = a^{x+y+z}$
    - b. $a^x * a^y * a^z = a^{\frac{xyz}{2}}$
    - c. $a^x * a^y * a^z = a^{\frac{x+y+z}{2}}$
    - d. $a^x * a^y * a^z = a^{\frac{xy}{2} + z}$

18. For r = .03, n = 4, p = 100, I = $(1.03)^4 \cdot 100$
    For r = .05, n = 6, p = 400, I = $(1.05)^6 \cdot 400$
    For r = .07, n = 5, p = 600, I = $(1.07)^5 \cdot 600$
    For r = .06, n = 7, p = 200, I = _____
    - a. $(1.06)^6 \cdot 200$
    - b. $(1.06)^7 \cdot 200$
    - c. $(1.07)^6 \cdot 200$
    - d. $(1.07)^7 \cdot 200$

19. Select the expanded notation for 2007.0005.
    - a. $(2 \cdot 10^4) + (7 \cdot 10) + (5 \cdot 10^{-4})$
    - b. $(2 \cdot 10^3) + (7 \cdot 10) + (5 \cdot 10^{-4})$
    - c. $(2 \cdot 10^3) + (7 \cdot 10^0) + (5 \cdot 10^{-4})$
    - d. $(2 \cdot 10^8) + (7 \cdot 10^5) + (5 \cdot 10^1)$

20. Select the numeral for
    $(5 \cdot 10^3) + (1 \cdot 10) + (7 \cdot 10^{-2})$.
    - a. 51.7
    - b. 5001.007
    - c. 5010.007
    - d. 5010.07

21. Select the numeral for $(2 \times \frac{1}{10}) + (2 \times \frac{1}{10^4})$.
    - a. 0.02002
    - b. 0.20020
    - c. 0.20002
    - d. 2.00002

22. Choose the expression that is equivalent to $6y^3(y^3z^8)$.
    - a. $7y^3z^8$
    - b. $6y^3(z^3y^8)$
    - c. $6y^9z^8$
    - d. $6(y^3z^4)^2$

23. Choose the expression that is equivalent to $4xy^2(5x + y^4)$.
    - a. $20x^2y^2 + 4xy^8$
    - b. $4xy^2(5y^4 + x)$
    - c. $4xy^6 + 20x^2y^2$
    - d. $(4xy^2 + 5x)y^4$

24. Choose the statement that is not true for all non-zero real numbers.
    - a. $a^3 + b^3 = (a + b)^3$
    - b. $(b^2)^3 = b^2 \cdot b^2 \cdot b^2$
    - c. $\frac{(4b^2)^2}{2b} = 8b^3$
    - d. $(7x^3)y^7 = 7(x^3y^7)$

25. $4 + 2 \cdot 5b - \frac{2(3b)^2}{2b} =$
    - a. $b + 4$
    - b. $21b$
    - c. $19b + 4$
    - d. $4 + 7b$

five

Further Topics
In Algebra

Section 1   Statements and Equations

We use sentences to express ideas in the English language. We also use sentences to communicate mathematical ideas. Sentences that can be identified as true or false are called **statements**. Thus, $2 \cdot 3 = 6$ (true), $4 + 5 = 6$ (false), and $2 - 1 = 1$ (true), are examples of statements. In this chapter we will be interested in sentences that cannot be immediately identified as true or false. This leads us to the following definition.

> ### Definition
> An **open sentence** is a sentence which cannot
> be classified as true or false until more information
> is provided.

Examples

1. A sentence such as $x + 1 = 3$ is an open sentence. As it stands, it is neither true nor false until more information is provided about x. If x has a value of 2, the sentence $2 + 1 = 3$ is true. If, however, x has a value of 3, the sentence $3 + 1 = 3$ is false.

2. A sentence such as $x - 3 \neq 4$ is an open sentence.

3. A sentence such as $y + 1 < 2$ is an open sentence.
    ("<" is read "is less than.")

In each of the above examples, letters are used to represent numbers.

> **Definition**
> A **variable** is a symbol (such as x or y) that is used to represent a member of a set of elements.

Example

In the open sentence $x + 1 = 3$, the letter x is a variable. A variable is a placeholder for its replacements. The set of replacements could be the set of counting numbers, the set of integers, the set of real numbers, etc. In the open sentence $x + 1 = 3$, the numbers 1 and 3 are called **constants**.

Statements of equality, particularly open sentences that state equality, are important to us.

> **Definition**
> An **equation** is a sentence, whether true or false or open, that represents a statement of equality.

The expression on the left side of the equality is called the left member of the equation, and the expression on the right side of the equality is called the right member of the equation.

Examples

1. The sentence $2 + 3 = 5$ is an equation.
2. The sentence $3 + 5 = 7$ is an equation.
3. The sentence $x + 5 = 7$ is an equation involving the variable x.
4. The sentence $2y - 4 = 7 - 3y$ is an equation involving the variable y.
5. The sentence $3x = y$ is an equation that involves two variables.
6. The sentence $y + 1 < 2$ is not an equation. It is an inequality.

We are particularly interested in equations that contain at least one variable. Equations are said to be either conditional equations or identities.

> **Definition**
> A **conditional equation** is an equation that is true only for certain replacements of its variable(s).

Example

A sentence such as $x + 4 = 6$ is a conditional equation. This equation is true only when x is replaced by 2, but is false for all other replacements. A conditional equation is also an open sentence.

> ### Definition
> An **identity** is a sentence that is true for all permissible replacements of its variable(s).

Examples

1. A sentence such as $x + 3 = 3 + x$ is true no matter what number is used to replace x.

2. A sentence such as $\frac{3}{x} - \frac{2}{x} = \frac{1}{x}$ is an identity because it is true for all permissible replacements for x. Zero is not a permissible replacement for x because division by zero is not permitted.

Problem Set 1.1

1. Indicate whether each of the following is a conditional equation or an identity:
   a. $x + 6 = 6 + x$ b. $x + 4 = 4x + 8$
   c. $x + 1 + 2 = x + 3$
   d. $(x + 2) + 5 = x + (2 + 5)$
   e. $x \cdot 5 = 5 \cdot x$

2. Are there any whole numbers that can replace the variable x that will make the conditional equation $x - 1 = 4$ a true statement?

3. Are there any negative integers that can replace the variable y that will make the conditional equation $y - 1 = -7$ a true statement?

4. Are there any real numbers that can replace the variable z that will make the equation $z + 4 = 4 + z$ a true statement?

5. Are there any integers that can replace the variable x that will make the equation $x = x + 1$ a true statement?

6. Which of the following statements is true for every replacement of the variable?
   a. $a + 1 \cdot 5 + b - a = 1 \cdot 5 + b$
   b. $4x + 2 = 10$

Review Competencies

7. Evaluate:
   a. 3^3 b. 5^0
   c. 4^{-2} d. $2(3)^2$

8. If a and b are real numbers, state the commutative property of addition.

9. Subtract: $5\frac{1}{7} - 3\frac{5}{7}$

10. Express 15% as a decimal.

11. What is 30% of 60?

12. A car goes 50 miles on $2\frac{1}{2}$ gallons of gas. How many miles per gallon does it travel?

Definitions

The **solution** of a conditional equation is the value that will make the equation a true statement when it replaces the variable. The **solution set** of a conditional equation is the set of all possible solutions.

Finding the solution of a conditional equation is referred to as solving the equation.

Examples

1. If $2y + 4 = 6$, and y is replaced by a real number, then 1 is a solution of this equation because $2(1) + 4 = 6$. The solution set contains only one element, 1, and is expressed as $\{1\}$.

2. If $x + 2 = 2 + x$, and x may be replaced by a real number, then there is not just one value that will make this equation true. The solution set contains all the real numbers since any real number will make the equation a true statement.

Definition

Equivalent equations are equations that have the same solution(s).

Example

Equations such as $x = 3$, $x + 3 = 6$, and $3x = 9$ are equivalent equations because 3 is the solution of each equation.

Problem Set 1.2

1. Is 3 a solution of the equation $10 = 2x + 4$? What is the solution set?

2. Is 2 a solution of the equation $2y - 5 = 1$?

3. Is -4 a solution of equation $3a - 4 = a - 12$? What is the solution set?

4. Is -4 a solution of the equation $0 = y + 4$? What is the solution set?

5. Are the equations $x - 2 = 5$ and $x = 7$ equivalent equations?

6. Are the equations $x + 4 = 1$ and $x = -3$ equivalent equations?

7. Are the equations $\frac{x}{3} = 2$ and $x = 6$ equivalent equations?

8. Are the equations $5x = 30$ and $x = 6$ equivalent equations?

9. Are the equations $3x - 7 = 14$ and $3x = 21$ equivalent equations?

10. Are the equations $5x + 6 = 3x - 4$ and $2x + 6 = -4$ equivalent equations?

Review Competencies

11. Are there any integers that can replace x to make the equation $x - 1 = x$ a true statement?

12. Is $\frac{x}{2} = \frac{3x}{6}$ an identity?

13. Evaluate:
    a. $2^2 \cdot 3^2$ b. $6 \cdot 10^0$ c. 2^{-3} d. $(2 \cdot 3)^2$

14. If a, b, and c are real numbers, state the associative property of multiplication.

15. Express $\frac{28}{5}$ as a mixed number.

16. Subtract: $8.1 - 2.006$

17. Express 2.5 as a percent.

18. Evaluate: $4(2 + 3 \cdot 2 - 1)$

You probably have discovered that certain equations can be solved quickly by inspection. For example, an equation such as $x - 2 = 3$ can be solved mentally. The solution is 5. However, an equation such as $3y + 4 = 2y + 2$ requires more thought to obtain -2 as the solution.

In the discussion that follows, we will discuss four principles for solving equations that will help us solve any equation that we encounter in this chapter. The principles will enable us to transform any equation to a simpler, equivalent equation.

> **Principle 1**
>
> The same real number can be **added** to both members of a conditional equation without changing the solution set. The resulting equation is **equivalent** to the original equation. Thus, if a, b, and c are any real numbers and a = b, then $a + c = b + c$.

Problem 1   Solve $x - 2 = 5$.

Solution

step 1

Add 2 to both members of $x - 2 = 5$ because, by Principle 1, this will not change the solution set. This is expressed in either of two ways:

$$x - 2 + 2 = 5 + 2 \quad \text{or} \quad \begin{array}{c} x - 2 = 5 \\ \underline{+2 \quad +2} \end{array}$$

step 2

Observe that the left member of the equation has $-2 + 2$ which equals 0 and the right member has $5 + 2$ which is 7. Therefore,

$$
\begin{array}{ll}
x - 2 = 5 & \text{becomes} \\
x - 2 + 2 = 5 + 2 & \text{which simplifies to} \\
x + 0 = 7 & \text{which further simplifies to} \\
x = 7 &
\end{array}
$$

7 is the solution of $x - 2 = 5$. Notice that 2 was added to both members of the equation to isolate the term involving x. *An important strategy in solving equations is to isolate the variable in either the left or right member of the equation.* Thus, if x is the variable in the equation, then the final step should be:

$$x = \text{number} \quad \text{or} \quad \text{number} = x$$

step 3

Check the solution by replacing x in the original equation by 7.

$$
\begin{array}{ll}
x - 2 = 5 & \\
(7) - 2 = 5 & \text{replacing x by 7} \\
5 = 5 & \text{because } 7 - 2 = 5
\end{array}
$$

The solution is 7 because it is the value that makes the equation a true statement. The solution set is $\{7\}$.

Notice that adding 2 to both members of the equation $x - 2 = 5$ produced the equivalent equation $x = 7$.

To summarize, the general process of solving an equation requires transforming the equation into successively simpler, equivalent equations.

> ## Principle 2
> The same real number can be **subtracted** from both members of an equation without changing the solution set. The resulting equation is equivalent to the original equation.

Thus, if a, b, and c are real numbers and a = b, then a − c = b − c.

Note
Principle 1 and Principle 2 can be combined and expressed as a biconditional statement to say:

$$a = b \text{ if and only if } a + c = b + c.$$

It says:

If a = b, then a + c = b + c.   (The same real number can be added to both members of the equation.)

It also says:

If a + c = b + c, then a = b.   (The same real number can be subtracted from both members of the equation.)

Problem 2

Solve y + 4 = 1.

Solution

step 1

Subtract 4 from both members of y + 4 = 1 to isolate the variable, y, in the left member of the equation. This is expressed in either of two ways:

$$y + 4 - 4 = 1 - 4 \quad \text{or} \quad \begin{array}{r} y + 4 = 1 \\ \underline{-4 \quad -4} \end{array}$$

step 2

Notice that the left member contains the expression 4 − 4 or 0 and the right member contains 1 − 4 or -3.

$$\begin{aligned} y + 4 - 4 &= 1 - 4 \\ y + 0 &= -3 \\ y &= -3 \end{aligned}$$

The solution of y + 4 = 1 is -3. This should be checked by replacing y with -3 in the original equation and showing that a true statement results. The solution set is {-3}.

Notice that y + 4 = 1 and y = -3 are equivalent equations.

> ### Principle 3
> Any non-zero real number can **multiply** both members of a conditional equation without changing the solution set. The resulting equation is equivalent to the original equation.

Thus, if a, b, and c are real numbers ($c \neq 0$) and a = b, then ac = bc.

Also, if $\frac{a}{c} = \frac{b}{c}$, then a = b.

Problem 3   Solve $\frac{x}{3} = 2$.

Solution

step 1

To isolate x in the left member of $\frac{x}{3} = 2$, both members are multiplied by 3.

$$3 \cdot \frac{x}{3} = 3 \cdot 2$$

step 2

The left member of the equation is simplified to x and the right member to 6 as follows:

$$3 \cdot \frac{x}{3} = 3 \cdot 2$$
$$3 \cdot \frac{1}{3} \cdot x = 6$$
$$1 \cdot x = 6$$
$$x = 6$$

The solution 6 should be checked in the original equation.
The solution set is {6}.

Observe that $\frac{x}{3} = 2$ and x = 6 are equivalent equations.

> ### Principle 4
> Any non-zero real number can **divide** both members of a conditional equation without changing the solution set. The resulting equation is equivalent to the original equation.

Thus, if a, b, and c are real numbers ($c \neq 0$) and a = b, then $\frac{a}{c} = \frac{b}{c}$.

Note
Principle 3 and Principle 4 can be combined and expressed as a biconditional statement to say:

$$a = b \text{ if and only if } ac = bc, c \neq 0$$

Problem 4 Solve $30 = -5a$.

Solution

step 1

Recall that $-5a$ means $-5 \cdot a$. Both members of the equation are divided by -5 to isolate a in the right member of the equation.

$$\frac{30}{-5} = \frac{-5a}{-5}$$

step 2

Both members of the equation are simplified.

$$\frac{30}{-5} = -6 \text{ and } \frac{-5a}{-5} = 1 \cdot a = a$$

$$\frac{30}{-5} = \frac{-5a}{-5}$$

$$-6 = a$$

The solution -6 should be checked in the original equation.
The solution set is $\{-6\}$.

Problem 5 Solve $\frac{x}{3} = \frac{5}{6}$

Solution

step 1

To isolate the variable, x, in the left member of the equation, both members are multiplied by 6, which is the **least common multiple** of the denominators, 3 and 6. This number is also called the **least common denominator**.

$$6 \cdot \frac{x}{3} = 6 \cdot \frac{5}{6}$$

step 2

Both members are simplified:

$$6 \cdot \frac{x}{3} = 2x \text{ and } 6 \cdot \frac{5}{6} = 5$$

step 3

The equation $2x = 5$ is solved by dividing both members by 2.

$$2x = 5$$

$$\frac{2x}{2} = \frac{5}{2}$$

$$x = \frac{5}{2}$$

The solution should be checked in the original equation. The solution set is $\{\frac{5}{2}\}$. Notice that the equations $\frac{x}{3} = \frac{5}{6}$, $2x = 5$, and $x = \frac{5}{2}$ are equivalent equations. Notice also that both Principle 3 and Principle 4 were used to solve this equation.

Problem Set 1.3

Solve each of the following equations using the set of real numbers as replacements for the variable.

1. $x + 9 = 11$ 2. $-5 = -10 + y$

3. $8x = -40$ 4. $-16 = -4a$

5. $-3b = 19$ 6. $\frac{x}{5} = -5$

7. $-4 = \frac{n}{-6}$ 8. $\frac{x}{5} = \frac{12}{20}$

9. $\frac{y}{3} = \frac{-5}{4}$ 10. $\frac{k}{3} = \frac{7}{5}$

11. Select the property that justifies the following statement: If $x - 3 = 7$, then $x = 10$.

   a.   If $a = b$, then $ac = bc$ and $c \neq 0$.

   b.   If $a = b$ and $b = c$, then $a = c$.

   c.   If $a = b$, then $a + c = b + c$.

   d.   If $a = b$, then $\frac{a}{c} = \frac{b}{c}$ and $c \neq 0$.

12. Select the property that justifies the following statement: If $\frac{x}{-4} = -5$, then $x = 20$.

   a.   If $a = b$, then $a + c = b + c$.

   b.   If $a = b$, then $a - c = b - c$.

   c.   If $a = b$, then $ac = bc$ and $c \neq 0$.

   d.   If $a = b$ and $b = c$, then $a = c$.

Review Competencies

13. Is 3 in the solution set of $3x - 10 = 1$?

14. Is 0 in the solution set of $\frac{x}{3} = \frac{4x}{12}$?

15. What real numbers can replace x to make the open sentence $x + 5 = 5 + x$ a true statement?

16. Evaluate:

   a. $2 \cdot 3^2$ b. $2^0 \cdot 2$

   c. 3^{-2} d. $(-2)^{-5}$

17. If a, b, and c are real numbers, state the distributive property.

18. Multiply: $\frac{6}{7} \cdot 1\frac{2}{3}$

19. Divide: $0.500 \div 0.125$

20. Express 45% as a fraction in lowest terms.

21. 40 is what percent of 200?

Some equations involve applying more than one principle in order to obtain a solution.

Problem 6   Solve $3b - 7 = 14$.

Solution

step 1

Isolate 3b in the left member of the equation by adding 7 to both members.

$$3b - 7 + 7 = 14 + 7 \quad \text{or} \quad 3b - 7 = 14$$
$$\underline{ +7 = +7}$$
$$3b + 0 = 21 \qquad\qquad 3b + 0 = 21$$

step 2

Isolate b in the left member of the equation by dividing both members by 3.

$$\frac{3b}{3} = \frac{21}{3}$$
$$b = 7$$

step 3

The solution 7 should be checked in the original equation. The solution set is {7}.

Observe that the equations $3b - 7 = 14$, $3b = 21$, and $b = 7$ are equivalent equations.

Problem Set 1.4

Solve each of the following equations using the set of real numbers as replacements for the variable.

1. $3x - 9 = 6$ 5

2. $60 = 5x + 10$ 10

3. $6 + 3y = 11$ $\frac{5}{3}$

4. $-4x + 18 = 46$ -7

5. $-27 = -2 + 5b$ -5

6. $10x + 40 = 0$ -4

7. $27 + 3a = 0$ -9

8. $3 = y - 10$ 13

9. $-6 - 3k = -12$ 2

Review Competencies

10. Find solutions for:
    a. $-7 + x = 3$ 10
    b. $-5x = 20$ -4
    c. $\frac{x}{-3} = -2$ -6
    d. $\frac{x}{4} = \frac{-3}{5}$ $-\frac{12}{5} = -2.4$

11. Evaluate:
    a. $(-2x^2)(3x^4)$
    b. $\frac{-12x^5}{x^3}$
    c. $5 - (1 - 3)$

12. What real number is the additive inverse of 5?

13. Express 234_5 as a base ten numeral.

14. Add: $\frac{1}{5} + 3\frac{1}{3}$

15. $\frac{3}{5} = \frac{?}{20}$

16. Express $\frac{30}{35}$ as a fraction reduced to lowest terms.

17. Express $\frac{2}{5}$ as a decimal.

18. Express $\frac{3}{4}$ as a percent.

19. Evaluate: $9 \div (6 - 3) + 4 \cdot 3 - 1$

20. Express $7\frac{1}{2}\%$ as a decimal.

Some equations contain terms involving the variable in both the left and right members.

Solve $5x + 6 = 3x - 4$.

The equation $5x + 6 = 3x - 4$ differs from the examples shown earlier because the variable, x, appears in **both** members of the equation.

step 1

The term 3x is eliminated from the right member of the equation by subtracting 3x from both members.

$$
\begin{array}{lll}
5x + 6 - 3x & = 3x - 4 - 3x \\
(5x - 3x) + 6 & = (3x - 3x) - 4 \\
2x + 6 & = \text{-}4
\end{array}
\qquad \text{or} \qquad
\begin{array}{ll}
5x + 6 & = 3x - 4 \\
\underline{\text{-}3x} & = \underline{\text{-}3x} \\
2x + 6 & = \text{-}4
\end{array}
$$

step 2

Both methods used in Step 1 to display the elimination of 3x from the right member of the equation result in the new, equivalent equation $2x + 6 = \text{-}4$. Its solution proceeds in the following steps.

$$
\begin{array}{c}
2x + 6 - 6 = \text{-}4 - 6 \\
2x = \text{-}10 \\
\frac{2x}{2} = \frac{\text{-}10}{2} \\
x = \text{-}5
\end{array}
$$

step 3

To check -5 as the solution for $5x + 6 = 3x - 4$, each member of the equation is separately evaluated. Both members must have the same evaluation when x is replaced by -5.

$$
\begin{array}{ll}
5x + 6 & = 3x - 4 \\
5(\text{-}5) + 6 & = 3(\text{-}5) \text{-}4 \\
\text{-}25 + 6 & = \text{-}15 - 4 \\
\text{-}19 & = \text{-}19
\end{array}
$$

Thus, -5 checks as the solution of $5x + 6 = 3x - 4$, and the solution set is {-5}.

Problem Set 1.5

Solve each of the following equations using the set of real numbers as replacements for the variable.

1. $x + 4 = 3x + 2$

2. $3y - 6 = y - 10$

3. $6b + 3b - 18 = 3b$

4. $12a = 6a - 6$

5. $-4y = 18 + 2y$

6. $x - 9 = 21 - 2x$

7. $11p - 24 = 7p$

8. $-5k = 50 + 5k$

9. $2 = 2x - 10x + 10$

Review Competencies

10. Find solutions for:

    a. $4x - 6 = 6$ b. $2x + 10 = 0$

    c. $3 = 3x - 6$ d. $-5 - 3x = -14$

    e. $\frac{x}{3} = \frac{-3}{4}$

11. Evaluate:

    a. $(-3x^{-3})(4x^{-4})$ b. $\dfrac{10x^4}{-5x^2}$

    c. $7^0 \cdot 0$

    d. $(-2)-5$

12. Is the set of real numbers closed for subtraction?

13. Express 1001_2 as a base 10 numeral.

14. Divide: $3\frac{1}{4} \div \frac{1}{4}$

15. $6 = \frac{?}{3}$

16. Express 0.05% as a decimal.

17. 20 is what percent of 50?

18. Evaluate: $(\frac{1}{3} \div \frac{4}{3}) + (\frac{1}{5} \cdot \frac{5}{4})$

19. What is the place value of the digit 3 in 3204_5 ?

Many equations involve the use of grouping symbols in one or both members.

Chapter 5

Solve $4x + 6 = 3(x + 7) - 2x$.

step 1
First eliminate the parentheses from
$4x + 6 = 3(x + 7) - 2x$ using the distributive property.

$4x + 6 = 3x + 21 - 2x$ because $3(x + 7) = 3x + 21$.

step 2
Combine the similar terms in the right member of the equation.
$$4x + 6 = 3x + 21 - 2x$$
$$4x + 6 = x + 21$$

step 3
Eliminate the x term from the right member of $4x + 6 = x + 21$.
$$4x + 6 - x = x + 21 - x$$
$$3x + 6 = 21$$

step 4
Subtract 6 from both members of $3x + 6 = 21$.
$$3x + 6 - 6 = 21 - 6$$
$$3x = 15$$

step 5
Divide both members of $3x = 15$ by 3.
$$\frac{3x}{3} = \frac{15}{3}$$
$$x = 5$$

step 6
Evaluate each member of the equation using 5 as the replacement for x.
$$4x + 6 = 3(x + 7) - 2x$$
$$4(5) + 6 = 3(5 + 7) - 2(5)$$
$$20 + 6 = 3(12) - 10$$
$$26 = 36 - 10$$
$$26 = 26$$

Thus, 5 is the solution, and the solution set is $\{5\}$.

Problem 9   Solve $2x - (4x + 2) = 5x + 12$.

Solution

step 1

Eliminate the parentheses in the equation using the distributive property and multiplying each term of $(4x + 2)$ by -1.

$$2x - (4x + 2) \;=\; 5x + 12 \text{ becomes}$$
$$2x - 4x - 2 \;=\; 5x + 12$$

step 2

Combine similar terms.

$$-2x - 2 \;=\; 5x + 12$$

step 3

Subtract 5x from both members.

$$-2x - 2 - 5x \;=\; 5x + 12 - 5x$$
$$-7x - 2 \;=\; 12$$

step 4

Add 2 to both members.

$$-7x - 2 + 2 \;=\; 12 + 2$$
$$-7x \;=\; 14$$

step 5

Divide both members by -7.

$$\frac{-7x}{-7} \;=\; \frac{14}{-7}$$

$$x \;=\; -2$$

step 6

Check -2 as the solution.

$$\begin{aligned}
2x - (4x + 2) &= 5x + 12 \\
2(-2) - [4(-2) + 2] &= 5(-2) + 12 \\
-4 - [-8 + 2] &= -10 + 12 \\
-4 - [-6] &= 2 \\
-4 + 6 &= 2 \\
2 &= 2
\end{aligned}$$

Thus, -2 is the solution, and {-2} is the solution set.

Chapter 5

Solve $2(x + 3) = 3[x - (1 - x)]$.

step 1

Eliminate the innermost grouping symbols (parentheses) first.

$$2(x + 3) = 3[x - (1 - x)] \text{ becomes}$$
$$2(x + 3) = 3[x - 1 + x]$$

step 2

Combine similar terms within the brackets.

$$2(x + 3) = 3[x - 1 + x] \text{ becomes}$$
$$2(x + 3) = 3[2x - 1]$$

step 3

Eliminate the remaining grouping symbols using the distributive property.

$$2(x + 3) = 3[2x - 1] \text{ becomes}$$
$$2x + 6 = 6x - 3$$

step 4

Isolate all the terms with the variable, x, in one member of the equation and all the terms without the variable in the other member.

$$-4x = -9 \text{ or } 9 = 4x$$

step 5

The two equations obtained in Step 4 are equivalent. In either case, divide both members by the number multiplying x.

$$x = \frac{9}{4}$$

step 6

Check $\frac{9}{4}$ as the solution.

$$2(x + 3) = 3[x - (1 - x)]$$
$$2(\tfrac{9}{4} + 3) = 3[\tfrac{9}{4} - (1 - \tfrac{9}{4})]$$
$$2(\tfrac{21}{4}) = 3[\tfrac{9}{4} - (-\tfrac{5}{4})]$$
$$\frac{42}{4} = 3[\tfrac{7}{2}]$$
$$\frac{21}{2} = \frac{21}{2}$$

Thus, $\frac{9}{4}$ is the solution, and $\{\frac{9}{4}\}$ is the solution set.

Problem Set 1.6

Solve each of the following equations using the set of real numbers as replacements for the variable.

1. $x + 6x + 5 = 2(x + 5) + 4x - 1$

2. $7y - (3y + 5) = 6y + 8$

3. $3(a + 4) + 5(2a - 1) = 7a - 9$

4. $8 - 4(p - 1) = 2 + 3(4 - p)$

5. $(b - 1) - (b + 2) - (b - 3) = -2$

6. $3(k + 1) = 2[k + 3(k - 2)]$

7. $3[x - 2(2x - 1)] = x - 4$

Review Competencies

8. Find solution sets for:
    a. $x - 5 = 4x + 7$ b. $6x + 3 - 4x = -5$
    c. $0 = 7x - 5 - 2x$ d. $\frac{x}{4} = \frac{-6}{9}$

9. Evaluate:
    a. $x^4 \cdot x^{-2}$ b. $\frac{-8x^6}{-4x^4}$
    c. $-7 - (-1 - 4)$

10. What real number is the multiplicative inverse of 3?

11. What real number is the multiplicative identity?

12. Express 312_6 as a base ten numeral.

13. Subtract: $3\frac{1}{2} - 1\frac{2}{3}$

14. Express 65% as a fraction.

15. Evaluate: $10 - 2 + 1 - 3 + 2$

16. Express $3\frac{3}{4}\%$ as a decimal.

17. Express the fraction $\frac{5}{8}$ as a decimal.

18. Find the smallest positive multiple of 6 which yields a remainder of 2 when divided by 7.

We now discuss another important type of equation.

> ### Definition
> A **quadratic equation** is an equation of the form $ax^2 + bx + c = 0$, where $a \neq 0$ and a,b, and c are real numbers.

Examples

1. $3x^2 + 5x - 2 = 0$ is a quadratic equation in which $a = 3$, $b = 5$, and $c = -2$.

2. $4x^2 - 13 = 0$ is a quadratic equation in which $a = 4$, $b = 0$, and $c = -13$.

3. $7x^2 + 4x = 0$ is a quadratic equation in which $a = 7$, $b = 4$, and $c = 0$.

Chapter 5

The ability to solve some quadratic equations is related to **factoring** expressions in the form $ax^2 + bx + c$ where, again, $a \neq 0$. Since factoring is the reverse of multiplication, we will begin by multiplying **binomials** (expressions containing two terms, such as $2x + 3$ or $3x - 1$) to produce $ax^2 + bx + c$. We call $ax^2 + bx + c$ a **quadratic** expression or a **second degree** expression.

Problem 11 Find the product of $2x + 3$ and $4x + 7$.

Solution The multiplication can be done in a horizontal format using the distributive property. We will multiply each term in the second binomial $(4x + 7)$ by each term in the first binomial $(2x + 3)$.

$$
\begin{aligned}
(2x + 3)(4x + 7) &= 2x(4x + 7) + 3(4x + 7) \\
&= (2x)(4x) + (2x)(7) + (3)(4x) + (3)(7) \\
&= 8x^2 \quad + \quad 14x \quad + \quad 12x \quad + \quad 21 \\
&= 8x^2 + 26x + 21
\end{aligned}
$$

Because the multiplication of binomials occurs so frequently in algebra, we need to further examine the multiplication process used in Problem 11. In general, we have $(A + B)(C + D)$.

We distribute A and then B to each term in the second factor.

$$
\begin{aligned}
(A + B)(C + D) &= A(C + D) + B(C + D) \\
&= AC + AD + BC + BD
\end{aligned}
$$

The product $(AC + AD + BC + BD)$ can be computed by:

a. multiplying **FIRST** terms of each binomial: AC
b. multiplying **OUTSIDE** terms of each binomial: AD
c. multiplying **INSIDE** terms of each binomial: BC
d. multiplying **LAST** terms of each binomial: BD

The word "FOIL" is a memory device for the multiplication of binomials. The word contains the first letter of the words FIRST, OUTSIDE, INSIDE and LAST. The FOIL method is often a fast way to quickly multiply two binomials.

Problem 12   Find the product of (3x + 4) and (2x + 1) using the FOIL method.

Solution   **step 1**
Multiply **F**irst terms. $(3x)(2x) = 6x^2$

step 2
Multiply **O**utside terms $(3x)(1) = 3x$

step 3
Multiply **I**nside terms $(4)(2x) = 8x$

step 4
Multiply **L**ast terms $(4)(1) = 4$

Thus, $(3x + 4)(2x + 1) = 6x^2 + 3x + 8x + 4$.
Combining like terms, the product becomes $6x^2 + 11x + 4$.

Problem Set 1.7

Multiply the binomials in problems 1-10 using the FOIL method.

1. $(5x + 3)(7x + 1)$
2. $(9x - 4)(3x + 5)$
3. $(8x - 3)(5x - 2)$
4. $(5x + 1)(3x - 4)$
5. $(9x - 5)(7 + 4x)$
6. $(2x + 7)(2x - 7)$
7. $(8x + 7)(2x + 3)$
8. $(8x - 5)(2x + 11)$
9. $(7x + 4)(7x - 4)$
10. $(5x + 7)^2$

Review Competencies

11. If 25 is reduced to 15, what is the percent decrease?

12. If you increase 92 by 25%, what is the result?

13. The equality $f(x) = 3x^2 - 5x + 2$ is read "f at x equals $3x^2 - 5x + 2$" and gives the value of f for each specified value of x. Substitute -4 for x to find f(-4). We read f(-4) as "f evaluated at -4 or f at -4."

14. Write the negation of the following statement: *If it snows, then all people ski.*

We will now discuss a general method for factoring a quadratic expression. We will begin with $ax^2 + bx + c$ and arrive at the equivalent **product.** This is the factoring process, and the distributive property is the basis for it.

In our previous work we multiplied (3x + 4) and (2x + 1), obtaining $6x^2 + 11x + 4$. We will now employ the reverse process, starting with $6x^2 + 11x + 4$ and obtaining (3x + 4) and (2x + 1). We call (3x + 4) and (2x + 1) the factors of $6x^2 + 11x + 4$. Since each factor involves only x raised to the first power, these factors are referred to as **linear factors.**

Chapter 5

Since factoring is the reverse of multiplication, we know that
$$6x^2 + 11x + 4 = (3x + 4)(2x + 1)$$
or that
$$6x^2 + 11x + 4 \text{ factors into } (3x + 4)(2x + 1).$$

Observe the following relationships:

1. The first terms in each binomial factor (3x and 2x) are factors of the first term of the quadratic expression ($6x^2$). Thus, $(3x)(2x) = 6x^2$. In a product like $6x^2$, the number 6 is called the **numerical coefficient.**

2. The last terms in each binomial factor (4 and 1) are factors of the last term of the quadratic expression. Thus, $(4)(1) = 4$.

3. The sum of the outer product (3x) and the inner product (8x) equals the middle term of the quadratic expression (11x).

These observations lead us to a general procedure for factoring $ax^2 + bx + c$.

> To factor a quadratic expression $ax^2 + bx + c$:
>
> 1. Find the pairs of factors of the numerical coefficients of the first term and the last term of the quadratic expression.
>
> 2. By trial and error, position the pairs of factors in such a way that the sum of the outer product and the inner product will equal the middle term of the quadratic expression.

Problem 13   Factor $2x^2 + 7x + 3$.

Solution

step 1

The pairs of factors of the coefficient of the first term (2) are
                2 and 1 or -2 and -1.
The pairs of factors of the last term (3) are
                3 and 1 or -3 and -1.

The first term in each factor must contain x so that the product (x • x) will result in x^2, the letter portion of the first term $2x^2$. Let us agree that when the first term in the quadratic expression is positive we will use only positive factors of that first term. Thus, we will not use -2x and -x even though $(-2x)(-x) = 2x^2$. Thus, we have: $(2x\quad)(x\quad)$.

step 2
Since the signs in this quadratic expression are all positive, the sign in each factor is positive.

The quadratic expression has the possible factors:
$\quad(2x + 3)(x + 1) = 2x^2 + 5x + 3$ (wrong middle term)
$\quad(2x + 1)(x + 3) = 2x^2 + 7x + 3$ (correct)

Since only $(2x + 1)(x + 3) = 2x^2 + 7x + 3$,
then $2x^2 + 7x + 3$ factors as $(2x + 1)$ $(x + 3)$ or $(x + 3)$ $(2x + 1)$.
Observe that in the expression $(2x + 1)(x + 3)$
the sum of the outer product (6x) and the inner product (x) equals the middle term of the quadratic expression (7x).
We can check using the FOIL method, showing that
$\quad(2x + 1)(x + 3) = 2x^2 + 7x + 3$.

Problem 14   Factor $15x^2 - 17x - 4$.

Solution

step 1
The pairs of positive factors of 15 are 15 and 1 or 5 and 3.
We have the possibilities
$$(15x\quad)(x\quad) \text{ or } (5x\quad)(3x\quad).$$
The pairs of factors of -4 are 4 and -1, -4 and 1, 2 and -2, or -2 and 2.

step 2
We must position these factors in such a way that the sum of the outer product and inner product equals -17x, the middle term.

The list below shows all possible factorizations:

$(15x + 4)(x - 1)$	$(5x + 4)(3x - 1)$
$(15x - 4)(x + 1)$	$(5x - 4)(3x + 1)$
$(15x + 1)(x - 4)$	$(5x - 1)(3x + 4)$
$(15x - 1)(x + 4)$	$(5x + 1)(3x - 4)$ (correct answer)
$(15x - 2)(x + 2)$	$(5x + 2)(3x - 2)$
$(15x + 2)(x - 2)$	$(5x - 2)(3x + 2)$

Since only $(5x + 1)(3x - 4)$ has a sum of outer and inner products equal to -17x, then $15x^2 - 17x - 4$ factors as $(5x + 1)(3x - 4)$.
Thus, $5x + 1$ and $3x - 4$ are the linear factors of the quadratic expression $15x^2 - 17x - 4$.

Problem Set 1.8

Factor the quadratic expressions in problems 1-15.

1. $3x^2 + 8x + 5$ 2. $2x^2 + 9x + 7$

3. $5x^2 + 56x + 11$ 4. $4x^2 + 9x + 2$

5. $8x^2 + 10x + 3$ 6. $6x^2 - 23x + 15$

7. $16x^2 - 6x - 27$ 8. $8x^2 - 18x + 9$

9. $4x^2 - 27x + 18$ 10. $12x^2 - 19x - 21$

11. $4x^2 - x - 18$ 12. $6x^2 + 11x + 3$

13. $4x^2 - 12x + 9$ 14. $8x^2 - 26x + 21$

15. $6x^2 + 19x - 7$

Review Competencies

16. If 50 is increased to 60, what is the percent increase?

17. If you decrease 32 by $12\frac{1}{2}\%$, what is the result?

18. Find the sum of 3,200 and 960, writing the answer in scientific notation.

19. If $f(x) = 3x^2 + 4x - 3$, substitute -5 for x to find f(-5).

20. Select the incorrect negation of the statement: *Both Pensacola and Key West are cities.*
    a. Pensacola is not a city or Key West is not a city.
    b. Pensacola is not a city and Key West is not a city.
    c. If Pensacola is a city, then Key West is not a city.
    d. If Key West is a city, then Pensacola is not a city.

We now turn our attention to solving quadratic equations. We will use a property of real numbers that states **if a product is zero, then at least one of the factors must be zero.** Symbolically, we can write:

$$\text{If } AB = 0, \text{ then } A = 0 \text{ or } B = 0.$$

We can use this property and factoring to solve many quadratic equations.

Problem 15   Solve $x^2 - 2x - 15 = 0$.

Solution

step 1
Factor $x^2 - 2x - 15$. We obtain $(x + 3)(x - 5) = 0$.
We now have the product of two factors equal to zero.
Consequently, at least one factor is zero.

step 2
Set each factor equal to zero.
$$x + 3 = 0 \quad \text{or} \quad x - 5 = 0$$

step 3
Solve the equations in step 2.
$$x + 3 = 0 \quad \text{or} \quad x - 5 = 0$$
$$x = -3 \qquad\qquad x = 5$$
These two values are solutions of $x^2 - 2x - 15 = 0$.

step 4

We can check each solution in the original equation.

If x = -3, we obtain
$$x^2 - 2x - 15 = (-3)^2 - 2(-3) - 15$$
$$= 9 + 6 - 15 = 0$$

If x = 5, we have
$$x^2 - 2x - 15 = 5^2 - 2(5) - 15$$
$$= 25 - 10 - 15 = 0$$

We can also say that -3 and 5 are **roots** of the equation
$x^2 - 2x - 15 = 0$.

Problem 16   Solve $x(3x - 7) = 6$.

Solution   We must begin by writing the equation in the form $ax^2 + bx + c = 0$.
Thus,
$$x(3x - 7) = 6$$
$$3x^2 - 7x = 6$$
$$3x^2 - 7x - 6 = 0$$

step 1
Factor.
$$3x^2 - 7x - 6 = 0$$
$$(3x + 2)(x - 3) = 0$$

step 2
Set each factor equal to zero.
$$3x + 2 = 0 \quad \text{or} \quad x - 3 = 0$$

step 3
Solve for x.
$$3x + 2 = 0 \quad \text{or} \quad x - 3 = 0$$
$$3x = -2 \qquad\qquad x = 3$$
$$x = \frac{-2}{3}$$

step 4
We can easily check the two values in the original equation.

Thus, $\frac{-2}{3}$ and 3 are roots of the quadratic equation.

Problem Set 1.9

Find the roots of the quadratic equations in problems 1-12 using factoring.

1. $3x^2 + 10x - 8 = 0$

2. $2x^2 - 5x - 3 = 0$

3. $5x^2 - 8x + 3 = 0$

4. $7x^2 - 30x + 8 = 0$

5. $x^2 - x = 2$

6. $x^2 + 8x = -15$

7. $3x^2 - 17x = -10$

8. $4x^2 - 11x = -6$

9. $x(x - 3) = 54$

10. $x(2x - 5) = -3$

11. $x(2x + 1) = 3$

12. $x(x - 6) = 16$

13. The product of a number and the number decreased by 8 is -15. Find the number(s).

14. The product of a number and 2 less than 7 times the number is 5. Find the number(s).

Review Competencies

15. If 80 is increased to 100, what is the percent increase?

16. If $f(x) = 2x^3 - 4x^2 + x - 3$, find f(-1) by substituting -1 for x.

17. What property is illustrated by $7x + (-7x) = 0$?

18. What is the place value for the underlined digit in the given base two numeral?

101011

There are many quadratic equations that cannot be solved by factoring. Indeed, not every quadratic expression $ax^2 + bx + c$ is factorable. Consequently, a second technique is needed to solve quadratic equations. This method will work whether or not the equation can be factored.

This second method involves a formula that is traditionally proved in intermediate algebra using a process called completing the square. The derivation of the formula can be found in any intermediate algebra textbook. We will consider this important result without its derivation.

> If $ax^2 + bx + c = 0$, then $x = \dfrac{-b + \sqrt{b^2 - 4ac}}{2a}$ or $x = \dfrac{-b - \sqrt{b^2 - 4ac}}{2a}$.
>
> We can abbreviate by writing $x = \dfrac{-b \pm \sqrt{b^2 - 4ac}}{2a}$.
>
> This formula is called the **quadratic formula**.

Problem 17   Use the quadratic formula to solve $12x^2 - 5x - 3 = 0$.

Solution   Since the equation is in the form $ax^2 + bx + c = 0$, $a = 12$, $b = -5$, and $c = -3$. Observe that a is the coefficient of x^2, b is the coefficient of x, and c is the constant term.

We now substitute these values in the formula. $x = \dfrac{-b \pm \sqrt{b^2 - 4ac}}{2a}$

$$x = \dfrac{-(-5) \pm \sqrt{(-5)^2 - 4(12)(-3)}}{2(12)}$$

$$x = \dfrac{5 \pm \sqrt{25 + 144}}{24}$$

$$x = \dfrac{5 \pm \sqrt{169}}{24}$$

$$x = \dfrac{5 \pm 13}{24}$$

$$x = \dfrac{18}{24} = \dfrac{3}{4} \text{ or } x = \dfrac{-8}{24} = \dfrac{-1}{3}$$

The roots of the equation are $\dfrac{3}{4}$ and $\dfrac{-1}{3}$.

The following problem illustrates a quadratic equation with irrational roots.

Problem 18   Use the quadratic formula to solve $3x^2 = 2 - 4x$.

Solution   We begin by writing the equation in the form $ax^2 + bx + c = 0$.

Thus, $3x^2 + 4x - 2 = 0$. We now see that $a = 3$, $b = 4$ and $c = -2$.

Since $x = \dfrac{-b \pm \sqrt{b^2 - 4ac}}{2a}$, we obtain:

$$x = \dfrac{-4 \pm \sqrt{4^2 - 4(3)(-2)}}{2(3)}$$

$$x = \dfrac{-4 \pm \sqrt{16 + 24}}{6}$$

$$x = \dfrac{-4 \pm \sqrt{40}}{6}$$

$$x = \dfrac{-4 \pm 2\sqrt{10}}{6} = \dfrac{2(-2 \pm \sqrt{10})}{6} = \dfrac{-2 \pm \sqrt{10}}{3}$$

Thus, the roots of the equation are $\dfrac{-2 + \sqrt{10}}{3}$ and $\dfrac{-2 - \sqrt{10}}{3}$.

Problem Set 1.10

Find the roots of the quadratic equations in problems 1-10 using the quadratic formula.

1. $5x^2 + 9x + 4 = 0$

2. $12x^2 + x - 1 = 0$

3. $x^2 + 2x - 4 = 0$

4. $3x^2 - 2x - 2 = 0$

5. $4x^2 - 3x - 2 = 0$

6. $x^2 = 1 - 7x$

7. $x^2 = 1 - x$

8. $x(x - 6) = 7$

9. $x(x - 4) = 45$

10. $25x^2 + 10x = -1$

11. If you decrease 120 by $33\frac{1}{3}\%$, what is the result?

12. Write the contrapositive of the statement:
    If $|x| > 5$, then $x > 5$ or $x < -5$.

13. Simplify: $3\sqrt{2} + 5\sqrt{8}$

14. Simplify: $30 \div 5 \cdot 2 + 4 \cdot 3 \div 6$

15. Perform the indicated operation and write in scientific notation:
    $$(8.3 \times 10^{17})(3 \times 10^{-6})$$

Section 2   Inequalities

In this section we consider statements and sentences in which one member is greater than, less than, or not equal to another member. Such statements and sentences are called **inequalities**.

The symbols used to express inequalities are:

$\neq$ "is not equal to"
$>$ "is greater than"
$\geq$ "is greater than or equal to"
$<$ "is less than"
$\leq$ "is less than or equal to"

Examples

1. The inequality $2 \neq 3$ is read "2 is not equal to 3." The left member of the inequality is 2, and the right member is 3.

2. The inequality $2 > 1$ is read "2 is greater than 1."

3. The inequality $5 \geq 5$ is read "5 is greater than or equal to 5."

4. The inequality $4 < 6$ is read "4 is less than 6."

5. The inequality $2 \leq 3$ is read "2 is less than or equal to 3."

A helpful way to distinguish between the inequality symbols $>$ and $<$ is to remember that the point of the symbols is directed toward the smaller of the two members of the inequality.

Inequalities using the symbols $>$ and $<$ are called **strict** inequalities. The symbol $>$ is said to be the **reverse** of the symbol $<$.

The order of real numbers can be considered on the real number line. Recall that a number x is **greater** than a number y on the number line if x is to the right of y. Also, a number x is **less** than a number y on the number line if x is to the left of y.

Example

On the number line, the following order relations are true.

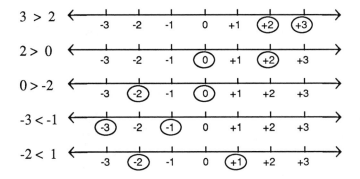

From the previous discussion, it should be obvious that if a, b, and c are real numbers and if $a > b$ and $b > c$, then $a > c$. For example, if $3 > 2$ and $2 > 1$, then $3 > 1$. Also, if $a < b$ and $b < c$, then $a < c$. For example, if $-2 < 1$ and $1 < 4$, then $-2 < 4$.

Chapter 5

It should be apparent that if x and y are any pair of real numbers, then exactly one of the following is a true statement:

$$x < y \qquad \text{or} \qquad x > y \qquad \text{or} \qquad x = y$$

This is called the **trichotomy principle** of the real numbers.

Problem Set 2.1

1. If a < b, does this imply that b > a? *yes*

2. On the real number line which of the following are true?
   a. $2 > 0$ ✓
   b. $4 > -3$ ✓
   c. $-4 > -1$ ✗
   d. $-3 < -4$ ✗
   e. $-1 < 0$ ✓
   f. $-4 > 3$ ✗

3. Order the real numbers using the symbol >.
   Example: $3 > 0 > -2$
   a. $4, -12, -1$ $4 > -1 > -12$
   b. $-4, -5, -2$ $-2 > -4 > -5$
   c. $\frac{-1}{2}, \frac{-1}{3}, \frac{-2}{3}$ $\frac{-1}{3} > \frac{-1}{2} > \frac{-2}{3}$
   -0.5, 0.333, 0.666

4. If a < b, where a and b are positive integers, how does $\frac{1}{a}$ compare with $\frac{1}{b}$? $\frac{1}{a} < \frac{1}{b}$

5. If a < b, where a and b are negative integers, how does $\frac{1}{a}$ compare with $\frac{1}{b}$? $\frac{1}{a} > \frac{1}{b}$

6. Insert the correct inequality symbol (< or >) in the space to make each of the following true.
   a. $\frac{3}{4} \boxed{>} \frac{2}{3}$
   b. $-\frac{4}{8} \boxed{<} -\frac{3}{10}$
   c. $-\frac{5}{8} \boxed{<} -\frac{2}{3}$
   d. $-\frac{5}{2} \boxed{<} -2$
   e. $.70 \boxed{<} \frac{3}{4}$
   f. $4.306 \boxed{<} 4.31$
   g. $5.\overline{3} \boxed{<} 5.31$

7. Which of the following statements is true for every integer x?
   a. $\frac{1}{x^2} < 0$
   b. $(x + 1)^2 = 0$ ✓
   c. $(x + 1)^2 > x^2$ ✓
   d. $x^2 + 1 > 0$ ✓

Review Competencies

8. Find the solution set for each of the following:
   a. $3 = -3x - 5 - 5x$ $-3x - 5 - 5x = 3 / -8x = 8 \Rightarrow x$
   b. $\frac{-3}{4} = \frac{x}{3}$ $\frac{x}{3} = \frac{-3}{4}$ $x = \frac{-9}{4}$

9. Is $\sqrt{24}$ a rational number or an irrational number?   *irrational*

10. Express 3122_4 as a base ten numeral.

11. Evaluate:
    a. $2(5)^2$
    b. 2^{-1}
    c. $-1 - (-1 - 1)$

12. Divide: $24 \div 0.006$

13. Multiply: $2\frac{1}{5} \cdot 5$

14. Express 0.56 as a percent.

15. Express $\frac{5}{4}$ as a decimal.

There are six basic principles, related to inequalities, that are similar in nature to the principles for solving equations.

> **Principle 1**
>
> If the same real number is **added** to both members of an inequality, the direction of the inequality is not changed.

272

If a < b and c is added to both members, then a + c < b + c. This principle also applies to inequalities using the other symbols of inequality.

Examples

1. If 2 < 3, then 2 + 4 < 3 + 4 or 6 < 7.
2. If 4 ≥ 1, then 4 + 2 ≥ 1 + 2 or 6 ≥ 3.
3. If -3 < 1, then -3 + 2 < 1 + 2 or -1 < 3.
4. If 7 < 12 < 15, then 7 + 2 < 12 + 2 < 15 + 2 or 9 < 14 < 17.

> ### Principle 2
> If the same real number is **subtracted** from both members of an inequality, the direction of the inequality is not changed.

If a > b and c is subtracted from both members, then a − c > b − c. This principle also applies to inequalities using the other symbols of inequality.

Examples

1. If 3 > 2, then 3 − 1 > 2 − 1 or 2 > 1.
2. If 3 ≤ 5, then 3 − 2 ≤ 5 − 2 or 1 ≤ 3.
3. If -4 < -2, then -4 − 1 < -2 − 1 or -5 < -3.
4. If 7 < 12 < 15, then 7 − 3 < 12 − 3 < 15 − 3 or 4 < 9 < 12.

Note
Principle 1 and Principle 2 can be combined and expressed as a biconditional statement to say:
$$a > b \text{ if and only if } a + c > b + c.$$

> ### Principle 3
> If both members of an inequality are **multiplied** by the same **positive** real number, the direction of the inequality is not changed.

If a > b and c is a positive real number (c > 0), then ac > bc. The important qualification in Principle 3 is that the multiplier must be a positive number. The principle also holds for inequalities involving other symbols of inequality.

Examples

1. If $2 < 3$, then $2 \cdot 4 < 3 \cdot 4$ or $8 < 12$.
2. If $-1 \geq -3$, then $-1 \cdot 2 \geq -3 \cdot 2$ or $-2 \geq -6$.
3. If $7 < 12 < 15$, then $7 \cdot 3 < 12 \cdot 3 < 15 \cdot 3$ or $21 < 36 < 45$.

> ### Principle 4
> If both members of an inequality are **divided** by the same **positive** real number, the direction of the inequality is not changed.

If $a > b$ and c is a positive real number ($c > 0$), then $\frac{a}{c} > \frac{b}{c}$. This principle also applies to inequalities using other symbols.

Examples

1. If $4 < 6$, then $\frac{4}{2} < \frac{6}{2}$ or $2 < 3$.   2. If $-4 \leq 8$, then $-\frac{4}{4} \leq \frac{8}{4}$ or $-1 \leq 2$.

Note
Principle 3 and Principle 4 can be combined and expressed as a biconditional statement to say: **$a > b$ if and only if $ac > bc$ and $c > 0$.**

> ### Principle 5
> If both members of an inequality are **multiplied** by the same negative real number, the direction of the inequality is **reversed**.

If $a < b$ and c is a negative real number ($c < 0$) then $ac > bc$. Recall that $<$ and $>$ are reverse symbols of inequality. Notice how they were used in applying Principle 5.

Examples

1. If $4 > 2$, then $4 \cdot (-1) < 2 \cdot (-1)$ or $-4 < -2$.

2. If $2 \leq 3$, then $2 \cdot (-2) \geq 3 \cdot (-2)$ or $-4 \geq -6$.

> **Principle 6**
> If both members of an inequality are **divided** by the same **negative** real number, the direction of the inequality is **reversed**.

If $a > b$ and c is a negative real number ($c < 0$), then $\frac{a}{c} < \frac{b}{c}$.

Examples

1. If $6 > 4$, then $\frac{6}{-2} < \frac{4}{-2}$ or $-3 < -2$.
2. If $-2 \le 4$, then $\frac{-2}{-2} - \ge \frac{4}{-2}$ or $1 \ge -2$.

Example

If $a < 0$, then we can divide both members of $a^2 > ab + a$ by a and reverse the direction of the inequality.

$$\frac{a^2}{a} < \frac{ab}{a} + \frac{a}{a}$$

$$a < b + 1$$

Note
Principle 5 and Principle 6 can be combined and expressed as a biconditional statement to say:

$$a > b \text{ if and only if } ac < bc \text{ and } c < 0.$$

Problem Set 2.2

1. If $a + c < b + c$ where c is any real number, is $a < b$? *True*

2. If $a - c > b - c$ where c is any real number, is $a > b$? *True*

3. If $ac < bc$ where c is any negative real number, is $a < b$? *False*

4. If $ac > bc$ where c is any positive real number, is $a > b$? *True*

5. Complete each of the following by inserting the correct symbol of inequality.
   a. If $4 - 2 > 3 - 2$ then 4 __>__ 3.
   b. If $2 \cdot 4 < 3 \cdot 4$ then 2 __<__ 3.
   c. If $3 + 5 < 4 + 5$ then 3 __>__ 4.
   d. If $6 < 12$ then 3 __<__ 6.
   e. If $6 < 12$ then -1 __>__ -2.

6. Select the property that justifies the following statement: If $-3x > 6$, then $x < -2$.
   a. If $a < b$, then $a + c < b + c$.
   b. If $a < b$, then $a - c < b - c$.
   c. If $ac < bc$ and $c > 0$, then $a < b$
   d. If $ac < bc$ and $c < 0$, then $a > b$.

275

7. If $b < 0$, then $b^2 < ab + b$ is equivalent to which of the following?

    a. $b < a + 1$ b. $b > a + 1$
    c. $b < -a - 1$ d. $b > -a - 1$

8. If $b < 0$, then $ab^3 > ab^4 + b^2$ is equivalent to which of the following?

    a. $ab^2 < ab^3 + b^2$

    b. $ab^2 > ab^3 + b^2$

    c. $ab^2 < ab^3 + b$

    d. $ab^2 < -ab^3 - b$

9. If $x > 0$, then $x^2 > x + xy$ is equivalent to which of the following?

    a. $x < x + y$ b. $x < 1 + y$
    c. $x > 1 + y$ d. $x > x + y$

Review Competencies

10. True or False?

     a. $-1 > 0$ b. $-3 < -2$
     c. $-1 < -3$ d. $2 > -4$

11. Order the following numbers using the symbol $<$.   3, -1, -3

12. Find the solution for
     $2(x + 2) = 13 - (2x - 5) - 3x$.

13. Express 0.35 as a fraction.

14. What is 15% of 70?

15. Evaluate: $12 - (3 \cdot 2) + 2(8 \div 4) - 1$

16. Each day a movie theater sells 80 reduced tickets at $5 each and 120 tickets at $7 each. If expenditures for the theater come to $950 per day, what is the theater's profit for a ten-day period?

17. If 160 is increased by 75% of itself, what is the result?

18. Roy paints $\frac{1}{4}$ of a house and Cora paints $\frac{1}{8}$ of the remainder. What fraction of the house is left unpainted?

Now let us see how to solve an inequality involving one variable and to graph the solution set on the real number line.

Problem 1   Solve $x + 2 < 0$ and graph the solution set.

Solution   The sequence of steps used to solve inequalities is similar to those used to solve equations. First, isolate the variable, x, in one member of the inequality.

step 1
Subtract 2 from both members of $x + 2 < 0$.
$$x + 2 < 0 \text{ becomes}$$
$$x + 2 - 2 < 0 - 2$$

step 2
Simplify.
$$x + 2 - 2 < 0 - 2 \text{ becomes}$$
$$x < -2$$

The inequality $x < -2$ states that the solution set is any real number less than -2.

step 3
The graph of the solution set of x < - 2 consists of all points on the number line to the left of -2, but **not** including -2. The graph would appear as follows:

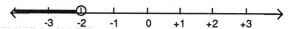

Recall that the solution set of x = -2 contains only one real number, -2. The solution set of x < - 2 contains infinitely many real numbers.

Notice that -3 is in the solution set of x + 2 < 0 because -3 + 2 < 0 or -1 < 0. Replacing x by -3 produces a true statement.

Notice also that the inequalities x + 2 < 0 and x < -2 have the same solution set. They are called **equivalent** inequalities.

Problem 2

Solution

Solve 3 ≤ y - 1 and graph the solution set.

In the inequality 3 ≤ y - 1, the variable, y, is in the **right** member and will be most easily isolated there.

step 1
Add 1 to both members of the inequality 3 ≤ y - 1.

$$3 \le y - 1$$

becomes

$$3 + 1 \le y - 1 + 1$$

step 2
Simplify.

$$3 + 1 \le y - 1 + 1$$

becomes

$$4 \le y$$

Notice that 4 ≤ y is equivalent to y ≥ 4.
Thus the solution set is any real number greater than or equal to 4.

step 3
Graph y ≥ 4 on the number line by including all points to the right of 4 and including 4. The graph is shown below.

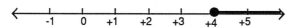

Notice that 5 is in the solution set of 3 ≤ y − 1 because 3 ≤ 5 − 1 or 3 ≤ 4. Replacing y by 5 produces a true statement.

Problem Set 2.3

Find and graph on the real number line the solution set for each inequality.

1. $x + 4 < 0$

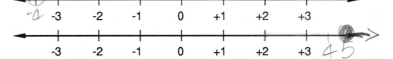

2. $y - 3 \geq 2$

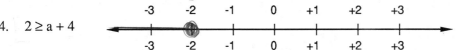

3. $5 + x > 3$

4. $2 \geq a + 4$

5. True or False?

   a. If $-6x < -6$, then $x > 1$.

   b. If $4 > 8y$, then $\frac{1}{2} < y$.

   c. If $bx < c$ and $b > 0$, then $x < \frac{c}{b}$.

   d. If $bx < c$ and $b < 0$, then $x > \frac{c}{b}$.

6. If $(x - p)(x - q) < 0$, then $x - p < 0$ and $x - q > 0$ or $x - p > 0$ and $x - q < 0$. Is this statement true or false?

7. Is 2 in the solution set of $5a < a + 4$?

8. Is -3 in the solution set of $(x + 4)(x - 1) \leq -3$?

9. Is -1 in the solution set of $x^2 + 3x + 5 < 4$?

Review Competencies

10. Insert the correct inequality symbol in each of the following:

    a. If $3 \cdot 5 > 2 \cdot 5$, then 3 _____2.

    b. If $4 < 12$, then 2 _____6.

    c. If $6 < 18$, then -2_____-6.

11. True or false?

    a. $-2 > -3$ b. $2 < -1$ c. $-1 < 0$

    d. $-\frac{3}{4} > -\frac{1}{2}$ e. $-\frac{5}{2} < -2$ f. $-\frac{2}{3} > -\frac{5}{6}$

12. Find solution sets for:

    a. $10x + 6 = 4x - 6$

    b. $3x - 7 = 4x + 14$

    c. $2x + 4 = 3x + 5$

    d. $\frac{x}{4} = -2$ e. $\frac{x}{2} = -\frac{1}{5}$ f. $\frac{x}{6} = -\frac{2}{9}$

13. Express 0.000045 in scientific notation.

14. Express 321_5 as a base 10 numeral.

15. What percent of 40 is 15?

16. Evaluate: $12 - (5 \cdot 2 + 6 \div 2)$

17. How many positive factors of 35 are odd and yield a remainder of 2 when divided by 3?

Denise Gonzalez

The problems that follow involve the application of two or more principles in order to solve the inequalities.

Problem 3   Solve $2x + 4 < 8$ and graph the solution set.

Solution

step 1
Isolate the variable, x, in the left member of the inequality by subtracting 4 from both members of the inequality and simplifying.

$$2x + 4 < 8 \qquad \text{or} \qquad 2x + 4 < 8$$
$$2x + 4 - 4 < 8 - 4 \qquad \qquad \underline{-4 \quad -4}$$
$$2x < 4 \qquad \qquad 2x \quad < \quad 4$$

step 2
Divide both members of $2x < 4$ by 2.
$$\frac{2x}{2} < \frac{4}{2}$$
$$x < 2$$

Notice that the direction of the inequality remained the same because the divisor was a **positive** number.

step 3
Graph $x < 2$ by including all points on the number line to the left of 2, but not including 2.

Problem 4   Solve $1 - 2x > 5$ and graph the solution set.

Solution

step 1
Isolate x in the left member of the inequality by subtracting 1 and simplifying.

$$1 - 2x > 5$$
$$1 - 2x - 1 > 5 - 1$$
$$-2x > 4$$

step 2
Divide both members of $-2x > 4$ by -2.

$$\frac{-2x}{-2} < \frac{4}{-2} \text{ becomes}$$
$$x < -2$$

Notice that the direction of the inequality was **reversed** because the divisor was a **negative** number. Step 3 on the next page completes the graph.

step 3

Graph the solution set of $x < -2$, which is equivalent to the solution set of $1 - 2x > 5$, by using all points on the number line to the left of -2.

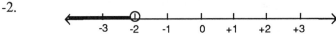

An inequality such as $-1 < x < 4$ is called a **continued inequality.** Such an inequality implies that $-1 < x$ and $x < 4$. The graph of $-1 < x < 4$ is:

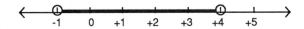

The graph of the continued inequality $-1 \leq x \leq 4$ is:

Problem 5

Solve the continued inequality $4 < 3x + 7 < 13$ and graph the solution set.

Solution

We extend our properties to cover this situation. If a number is subtracted from the middle expression, then the same number must be subtracted from the outside expressions. If the middle expression is divided by a number, then the same must be done to the outside expressions.

step 1

Subtract 7 from all three members of the inequality.
$$4 - 7 < 3x + 7 - 7 < 13 - 7 \text{ becomes}$$
$$-3 < 3x < 6$$

step 2

Divide all three members by 3.
$$-\frac{3}{3} < \frac{3x}{3} < \frac{6}{3} \text{ becomes}$$
$$-1 < x < 2$$

The graph of the solution set is:

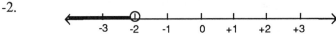

Problem Set 2.4

Find and graph solution sets for:

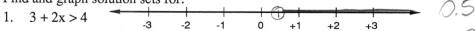

1. $3 + 2x > 4$ 0.5

2. $1 - 2x \geq 7 - 1$ -3

3. $4 - 3x \leq 10$ -2

4. $-2 > 6 - 4a$ 2

5. Select the property that justifies the following statement: If $-9 < 3x < 6$, then $-3 < x < 2$.

   a. If $a + c > b + c$, than $a > b$.
   b. If $a - c > b - c$, then $a > b$.
   c. If $ac < bc$ and $c < 0$, then $a > b$.
   d. If $ac < bc$ and $c > 0$, then $a < b$.

6. Select the property that justifies the following statement: If $-5 < -5x < 10$, then $-2 < x < 1$.

   a. If $a + c < b + c$, then $a < b$.
   b. If $a - c < b - c$, then $a < b$.
   c. If $ac < bc$ and $c < 0$, then $a > b$.
   d. If $ac < bc$ and $c > 0$, then $a < b$.

7. Solve the continued inequality $-1 \leq 5x - 6 \leq 14$.

Review competencies

8. Find and graph solution sets for:

   a. $x - 3 > 0$

   b. $y - 2 \leq -1$

   c. $-3 + x > 2$

   d. $3 < y + 2$

9. Is 3 in the solution set of $2x < x + 5$?

10. Is -1 in the solution set of $(x - 1)(x + 2) \leq 2$?

11. True or false? $-1 > 0$

12. Does $a > b$ imply $b < a$?

13. Find solution sets for:

   a. $10x - 5x - 5 = 0$
   b. $-2x + 4 = -3x + 5$
   c. $1 = 2x - 5x - 5$
   d. $\frac{x}{3} = -7$
   e. $\frac{x}{4} = \frac{-5}{3}$

14. Simplify:

   a. $\frac{-15x}{-5}$
   b. $-4 - (4 - 5)$
   c. $-(-1)^2$

15. Is the set of real numbers closed for multiplication?

16. Express 925,000,000 in scientific notation.

17. Multiply: $2\frac{2}{5} \cdot 1\frac{3}{4}$

18. Express 72% as a decimal.

19. 12% of what number is 36?

20. Evaluate: $\frac{1}{2} \cdot 8 + \frac{1}{4} \div \frac{1}{4} + \frac{3}{5} \cdot \frac{5}{6}$

21. True or false? If $a > b$ and $b > c$, then $a > c$.

22. What is the next term in the arithmetic progression below? (The definition for an arithmetic progression is in problem set 5.6, Chp. 1, #9.)
   $-17, -14, -11, \_\_\_\_$

Chapter 5

Some inequalities involve variables in both members.

Problem 6   Solve $3a - 2 > a + 2$ and graph the solution set.

Solution   **step 1**
The inequality $3a - 2 > a + 2$ has the variable a in *both* members.
Either 3a or a must be eliminated. In the display below, a has been
eliminated from the right member of the inequality.

$$3a - 2 > a + 2 \qquad \text{or} \qquad 3a - 2 > a + 2$$
$$3a - 2 - a > a + 2 - a \qquad\qquad \underline{-a} \qquad \underline{-a}$$
$$2a - 2 > 2 \qquad\qquad\qquad 2a - 2 > 2$$

$2a - 2 > 2$ is equivalent to $3a - 2 > a + 2$, and the solution of $2a - 2 > 2$
proceeds as it was done in previous problems.

step 2
$$2a - 2 \quad > \quad 2$$
$$2a - 2 + 2 \quad > \quad 2 + 2$$
$$2a \quad > \quad 4$$
$$\frac{2a}{2} \quad > \quad \frac{4}{2}$$
$$a \quad > \quad 2$$

step 3
The graph of the solution set of $a > 2$ or $3a - 2 > a + 2$ is shown below.

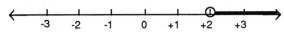

Problem 7   Solve $2(x + 5) \geq 6(x + 1)$.

Solution   **step 1**
First eliminate the $2(x + 5) \quad \geq \quad 6(x + 1)$ becomes
parentheses using
the distributive $2x + 10 \quad \geq \quad 6x + 6$
property.

step 2
Eliminate 6x $2x + 10 - 6x \quad \geq \quad 6x + 6 - 6x$ becomes
from the right
member of the $-4x + 10 \quad \geq \quad 6$
inequality.

step 3

Isolate -4x in the left member of the inequality.

$-4x + 10 - 10 \geq 6 - 10$ becomes

$-4x \geq -4$

step 4

Divide both members by -4 and reverse the direction of the inequality.

$\frac{-4x}{4} \leq \frac{-4}{-4}$ becomes

$x \leq 1$

Problem Set 2.5

Find and graph solution sets for:

1. $4x + 3 \geq 2x - 1$
2. $3x + 3 < 5x - 5$
3. $2x + 5 < 4x + 1$
4. $3 - 2a < 4a + 15$
5. $3(x + 5) \leq 2 - (x - 1)$

6. Choose the inequality that is equivalent to $4 - 2y > 12$.
   a. $-2y > 8$ ✓
   b. $-2y < 8$
   c. $2y > 8$
   d. $2y < 8$

7. Choose the inequality that is equivalent to $-5x > 25$.
   a. $x < 5$
   b. $x < -5$
   c. $x > -5$ ✓
   d. $x > 5$

8. Choose the inequality that is equivalent to $7 < x + 3 < 12$.
   a. $12 < x < 15$
   b. $4 < x < 9$
   c. $21 < x < 36$
   d. $4 > x > 9$

9. Choose the inequality that is equivalent to $14 < 2x < 20$.
   a. $12 < x < 18$
   b. $16 < x < 22$
   c. $7 < x < 10$
   d. $28 < x < 40$

10. Choose the inequality that is equivalent to $18 < -3x < 24$.
    a. $-8 < x < -6$
    b. $-6 < x < -8$
    c. $21 < x < 27$
    d. $6 > x > 8$

11. Which of the following has/have -4 as a solution?
    i. $x - 4 = 0$
    ii. $y + 5 \geq 0$
    iii. $z^2 + 3z - 4 = 0$
    a. i only
    b. ii only
    c. i and iii only
    d. ii and iii only

12. Which of the following has/have $\frac{1}{4}$ as a solution?
    i. $y^2 = \frac{1}{8}$
    ii. $2x - \frac{1}{2} = 0$
    iii. $8y - 2 \leq 0$
    a. i only
    b. ii only
    c. iii only
    d. ii and iii only

Review Competencies

13. Find and graph solution sets for:
    a. $2 - 3x > -4$
    b. $4 + 2x \geq 2$
    c. $-1 \geq 5 - 2a$
    d. $5 - x > 1$
    e. $-x \leq 2$

14. Is 3 in the solution set of $x > 2x - 1$?

15. Is 2 in the solution set of $(x + 3)(x - 2) \leq 0$?

16. Is -2 in the solution set of $y^2 - 4y - 10 = 5$?

17. True or false? $-\frac{5}{3} < -2$

283

18. Find solution sets for:

    a. $-x + 15 = 3x - 1$ b. $\frac{x}{7} = \frac{-3}{14}$

19. Simplify:

    a. $\frac{21x^2}{7x}$ b. $(-7) + 4$

    c. $3 \cdot 10^3$ d. $3^2 \cdot 3^2$

20. Which one of the following is not true for all real numbers?

    a. $7x^2(y + x) = 7x^2y + 7x^3$
    b. $(2x + 3y^2)(4x + y) = (y + 4x)(3y + 2x^2)$
    c. $5(xy) = (5x)y$
    d. $5x - 10y = -5(2y - x)$

21. If F = 7 and and M = 5, then W = $\frac{49}{5}$. If F = 8 and M = 2, then W = 32. Select the formula that is compatible with the data.

    a. $W = \frac{7F}{M}$ b. $W = 2FM$

    c. $W = \frac{8F}{M}$ d. $W = \frac{F^2}{M}$

22. If $f(x) = 5 - 3x$, find $f(-\frac{5}{4})$.

Section 3   Verbal Problems

This section describes the solution of problems stated in verbal form. Certain words imply various operations. Addition is implied by such clue words as sum, more than, increased by, and exceeded by. Subtraction is implied by difference, decreased by, less than, and diminished by. Product and times imply multiplication. The answer to a division problem is its quotient.

An important skill is the translation of English statements into algebraic expressions.

Problem 1   Translate the following into an algebraic expression: Five times a number (x) increased by 3.

Solution   Since the term "increased" implies addition and "times" implies multiplication, the translation is $5x + 3$.

Problem 2   Translate the following into an algebraic expression: The difference of twice a number and 10.

Solution   **Twice** a number means multiplying by 2 and **difference** implies subtraction. Thus the translation is $2x - 10$.

Problem Set 3.1

Translate each of the following into an algebraic expression.

1. The sum of a number (x) and 5

2. 12 more than a number (x)

3. 10 less than a number (x)

4. The difference between a number (x) and 6

5. Twice a number (x) decreased by 4

6. The product of 3 and a number (x)

7. A number (x) multiplied by 4

8. The quotient of a number (x) and 5

Review Competencies

9. $8 - 2(a + 2b) =$
   a. $8 - 2a + 4b$ b. $8 - 2a - 4b$
   c. $6a + 2b$ d. $6a + 12b$

10. $\sqrt{5} \cdot \sqrt{5} =$
    a. 5   b. 25   c. $\sqrt{10}$ d. $2\sqrt{5}$

11. Identify the symbol that should be placed in the box to form a true statement. $-13 \,\square\, -51$
    a. $=$ b. $<$ c. $>$

12. Is -2 in the solution set of $2a \geq a - 1$?

13. Find and graph on the number line the solution set of $-2x - 2 < 0$.

14. Write the number represented by $4.2 \cdot 10^{-5}$.

15. Insert the correct inequality symbol in $-\frac{3}{2}$ [ ] -2 to make it true.

16. Express 1111_2 as a base ten numeral.

17. How many prime factors of 30 are also factors of 60?

English phrases such as **a number increased by 2** translate as algebraic expressions. English statements such as **A number increased by 2 is 5** translate as equations.

Problem 3   Translate the following into an equation:
A number (x) decreased by 7 is 25.

Solution   The word "is" translates as "equals." Thus "A number (x) decreased by 7 is 25" translates as: $x - 7 = 25$.

Problem Set 3.2

Translate 1-4 into equations.

1. 10 added to twice a number (x) is 32.

2. A number (x) divided by 6 is 4.

3. 4 times a number (x) increased by 3 is 2 times the same number increased by 11.

4. A number (x) increased by 10 is equal to 3 times the number decreased by 16.

5. One number is 3 times a second number. If x is the second number, the first number may be expressed as ____.

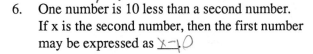

6. One number is 10 less than a second number. If x is the second number, then the first number may be expressed as $x - 10$

7. The sum of two numbers is 20. If x is the first number, the second number may be expressed as $20 - x$

8. Joe is half as old as Bill. If x represents Bill's age, then Joe's age may be expressed as $\frac{1}{2}x$.

Review Competencies

9. Translate:
   a. The sum of twice a number (x) and 3
   b. The product of 3 and the quotient of a number (x) and 4 $3 \times \frac{x}{4}$

10. 2,600,000 • 0.00014 =
    a. $3.64 \cdot 10^2$ b. $3.64 \cdot 10^{10}$
    c. $3.64 \cdot 10^{-24}$ d. $3.64 \cdot 10^{-2}$

11. Evaluate $-2x + xy - y^2$ when x = 2 and y = -1.

12. Simplify:
    a. $(-3x^5)(2x^{-3})$ b. $-2 - 1$ c. $-(-2)^2$

13. Find and graph on the number line the solution set of $-3 < x - 1$.

14. Is $n^2 + 4n$ a prime number when n is 3?

To solve verbal problems, it is important to read the problem very carefully to determine what unknown quantities are being sought. If there is only one unknown quantity, represent it by some variable, usually x. If there are two or more unknown quantities, represent one of the unknowns by a variable and then express the other unknown quantities in terms of that same variable. Finally, formulate an equation expressing the given relationships and solve the equation.

Problem 4   One number is 3 times a second number. The sum of the two numbers is 40. Find the numbers.

Solution   Let x = the second number
Then 3x = the first number
Therefore: x + 3x = 40 (The sum of the two numbers is 40.)
                    x + 3x = 40   simplifies as 4x = 40 or x = 10.
Since x = 10 and x represents the second number, the second number is 10.
3x represents the first number, and x = 10. Therefore, the first number is 30.

Problem 5   8% of what number is 64?

Solution   Let x represent the number.

Therefore: $0.08 \cdot x = 64$ or $x = \frac{64}{0.08}$ or x = 800.

Thus, 8% of 800 is 64.

Problem 6   What percent of 60 is 15?

Solution   Let x represent the desired percent.

Therefore $x \cdot 60 = 15$ or $x = \frac{15}{60} = \frac{1}{4} = 0.25 = 25\%$

Thus, 25% of 60 is 15.

Problem 7   John spends 35% of his income on rent. One month he spent $420 on rent. What was his income that month?

Solution   Let x represent John's income. 35% of his income = money spent on rent. Therefore,

$$0.35x = 420 \quad \text{or} \quad x = \frac{420}{0.35} \quad \text{or} \quad x = 1,200$$

John's income was $1,200.

Problem 8   A certain number is unknown. If 1 is added to the unknown number, the result is 9 less than 2 times the number. Find the number.

Solution   Let x = unknown number.
Then x + 1 represents **1 added to the unknown number**
and 2x – 9 represents "9 less than two times the number"
Therefore:

If 1 is added to the number	the result is	9 less than twice the number
x + 1	=	2x – 9

Solving $x + 1 = 2x - 9$ gives $x = 10$. Therefore, the unknown number is 10.

When the sum of two numbers is 10 and one number is x, then the other is $10 - x$. Similarly, if two numbers have a sum of 19 and one number is x, then the other will be expressed as $19 - x$.

Problem 9   Two numbers have a sum of 16. If twice the second is subtracted from 3 times the first, the result is 28. Find the numbers.

Solution   Let x = first number
Then $16 - x$ = second number (since their sum is 16)
Therefore:

| 3 times the first | $-$ | twice the second number | | is | 28. |

$$3x \quad - \quad 2(16 - x) \quad = \quad 28$$

Solving $3x - 2(16 - x) = 28$ gives $3x - 32 + 2x = 28$
or $5x - 32 \quad = 28$
or $5x \qquad = 60$
or $x \qquad = 12$

The first number is 12 and the second is $(16 - 12)$ or 4.

Problem Set 3.3

1. If 3 is added to 5 times a number, the result is 43. What is the number?

2. 3 less than a number equals 42. What is the number?

3. In a mathematics class there are 4 more men than women. If there are 30 students in the class, how many men are there?

4. If a certain number is multiplied by 3 and then 7 is added, the result is 11 less than 5 times the original number. What is the original number?

5. If 4 times a number is increased by 30, the result is the same as when 9 times the number is decreased by 10. What is the number?

6. If a second number is twice a first number and 3 times the first equals 12 increased by the second number, what are the numbers?

7. If Jim is 3 years younger than Ken and twice Jim's age decreased by 5 years equals Ken's age increased by two years, what are their ages? Jim 3   Ken 5.

8. 15% of what number is 75?

9. What percent of 50 is 20?

10. One number is 4 times another number. If the larger number is diminished by 3, the result is 6 more than the smaller number. Find the numbers.

11. Yoko's monthly budget allows 25% for housing, 35% for rent, 10% for clothing, and the rest on miscellaneous. One particular month Yoko spent $615 in the miscellaneous category. What was her income that month?

12. The difference between two numbers is 7 and their product is 44. If x represents the greater number what equation could be used to find x?
    a. $x(x - 7) = 44$ b. $x(7 - x) = 44$
    c. $x - (x - 7) = 44$ d. $x + (x - 7) = 44$

Review Competencies

13. Find the solution set of
    $y - 2(1 - 3y) = 6 + 3(4 - y)$.

14. Find the solution set of $b < b - (8 - 2b)$.

15. The formula for finding the value (V) of an investment (P) after one year at a given rate (r) is $V = P(1 + r)$. If $12,500 is invested at 9%, determine the value after one year.

We now consider problems involving **consecutive integers**, such as 8, 9, and 10 or 23, 24, and 25. If we let x represent the first integer in a series of consecutive integers, the accompanying table should be helpful in solving consecutive integer problems.

Consecutive Integers

English Phrase	Algebraic Expression	Example
Two consecutive integers	x, x + 1	13, 14
Three consecutive integers	x, x + 1, x + 2	-8, -7, -6
Two consecutive even integers	x, x + 2	40, 42
Two consecutive odd integers	x, x + 2	-37, -35
Three consecutive even integers	x, x + 2, x + 4	30, 32, 34
Three consecutive odd integers	x, x + 2, x + 4	9, 11, 13

Problem 10 Find three consecutive even integers such that the sum of the first and third, increased by 4, is 20 more than the second.

Solution Let

x = the first even integer
$x + 2$ = the second consecutive even integer
$x + 4$ = the third consecutive even integer

Then

Sum of the first and third	increased by 4	is	20 more than the second.
$x + (x + 4)$	$+4$	$=$	$(x + 2) + 20$

$$x + x + 4 + 4 = x + 2 + 20$$
$$2x + 8 = x + 22$$
$$x + 8 = 22$$
$$x = 14$$

The integers are 14, 16, and 18.

Check

Sum of the first and third	increased by 4	is	20 more than the second.
$14 + 18$	$+ 4$	$\overset{?}{=}$	$16 + 20$
		$36 = 36$	

Problem Set 3.4

1. The sum of three consecutive integers is 30. What are the integers?

2. The sum of three consecutive integers is 234. Find the integers.

3. The sum of three consecutive even integers is 198. Find the integers.

4. The sum of three consecutive odd integers is 51. Find the integers.

5. The sum of four consecutive odd integers is 216. Find the integers.

6. The sum of four consecutive even integers is 180. Find the integers.

7. The sum of three consecutive integers is 25. Find the integers.

8. The sum of three consecutive odd integers is 186. Find the integers.

9. The largest of three consecutive even integers is 6 less than twice the smallest. Find the integers.

10. Find three consecutive odd integers such that the largest is 13 less than twice the smallest.

11. Find three consecutive even integers such that the sum of the first and the third exceeds one-half of the second by 15.

12. Find three consecutive odd integers such that the sum of three times the first and twice the second exceeds three times the third by 26.

13. Find three consecutive even integers such that twice the second is equal to the sum of the first and the third.

14. Find three consecutive odd integers such that twice the second is equal to the sum of the first and the third increased by one.

Review Competencies

15. The square of a number, decreased by 6 times the number, is 18 more than the number. Which equation should be used to find x, the number?
    a. $(x-6)^2 = x + 18$ b. $x^2 - 6x^2 = x + 18$
    c. $x^2 - 6x + 18 = x$ d. $x^2 - 7x = 18$

16. A molecule contains 1 more atom of carbon than twice the number of atoms of oxygen and 1 less atom of hydrogen than carbon. If the molecule contains a total of 21 atoms, how many atoms of carbon are there?

17. The library fees for a borrowed book are $.70 for the first day and $.30 for each additional day. If the loan of a book came to $4.30, for how many days was the book borrowed?

Many number problems focus on the digits of a two-digit number. If t represents the tens digit and u the units digit, the accompanying table should be helpful in solving digit problems.

Algebraic Translations of English Phrases

English Phrase	Algebraic Translation	Example
Tens digit	t	$83; t = 8$
Units digit	u	$u = 3$
Sum of digits	$t + u$	$8 + 3$
The number	$10t + u$	$10(8) + 3$ (or 83)

Problem 11 The tens digit of a two-digit number is 1 more than two times the units digit. If the sum of the digits is 13, what is the number?

Solution Let u = the units digit.
Thus, $2u + 1$ = the tens digit (1 more than two times the units digit).
The sum of the digits is 13.

$$u + (2u + 1) = 13$$
$$3u + 1 = 13$$
$$3u = 12$$
$$u = 4$$

The units digit is 4 and the tens digit is $2u + 1$ or $2(4) + 1 = 9$.
The number is 94.

Problem 12 The tens digit of a two-digit number is 3 more than the units digit. If the number itself is 17 times the units digit, what is the number?

Solution Let u = the units digit.
Thus, $u + 3$ = the tens digit (3 more than the units digit).
The number $= 10t + u$
$= 10 \text{ (tens digit)} + \text{units digit}$
$= 10 (u + 3) + u$

We now translate:

The number	is	17 times the units digit.
$10(u + 3) + u$	$=$	$17 \cdot u$

We now solve the equation.

$$10(u + 3) + u = 17u$$
$$10u + 30 + u = 17u$$
$$11u + 30 = 17u$$
$$30 = 6u$$
$$5 = u$$

The units digit is 5 and the tens digit is $u + 3$ or $5 + 3 = 8$.
The number is 85.

Problem Set 3.5

1. The tens digit of a two-digit number is 1 more than three times the units digit. If the sum of the digits is 9, what is the number?

2. The tens digit of a two-digit number is 4 less than the units digit. If the sum of the digits is 14, what is the number?

3. The tens digit of a two-digit number is 2 more than the units digit. If the number itself is 16 times the units digit, what is the number?

4. The tens digit of a two-digit number is 6 more than the units digit. If the number itself is 31 times the units digit, what is the number?

5. A two-digit integer is equal to 4 times the sum of its digits. Which equation should be used to find the number if t represents its tens digit and u represents its units digit?
   a. $4(10t + u) = t + u$
   b. $tu = 4(t + u)$
   c. $10t + u = 4(t + u)$
   d. $4tu = t + u$

6. The tens digit of a two-digit number is 3 more than the units digit. If the number is 16 times its units digit, what equation can be used to find t, the tens digit?
   a. $10(t + 3) + (t - 3) = 16t$
   b. $10(t + 3) + (t - 3) = 16$
   c. $10t + (t - 3) = 16$
   d. $10t + (t - 3) = 16(t - 3)$

Review Competencies

7. In a community of 600 people, candidates A and B ran for mayor. Candidate B received 121 more votes than candidate A. Both candidates together received 493 votes. How many votes were received by candidate A?

8. Choose the algebraic description that equivalently translates the following verbal situation: For any three consecutive positive odd integers, the square of the middle integer is greater than the product of the smallest integer, x, and the largest integer.
   a. $(x + 1)^2 > x(x + 2)$
   b. $x(x + 4) < (x + 2)^2$
   c. $x^2 > (x + 2)(x + 4)$
   d. $x^2 + 4 > x(x + 4)$

9. The sum of 5 more than a certain number and 10 more than twice the number is equal to the product of 2 and the number increased by 8. Find the number.

10. The difference between two numbers is 2 and the sum of their reciprocals is $\frac{5}{12}$. If x represents the greater number, what equation could be used to find x?
    a. $\frac{1}{x} + \frac{1}{2 - x} = \frac{5}{12}$
    b. $\frac{1}{x + (x - 2)} = \frac{5}{12}$
    c. $x + (x - 2) = \frac{12}{5}$
    d. $\frac{1}{x} + \frac{1}{x - 2} = \frac{5}{12}$

Section 4   Ratio and Proportion

> ### Definition
> A **ratio** of two numbers, x and y, is a comparison of x to y by division. The ratio of x to y is written as $\frac{x}{y}$ or x:y (y ≠ 0).

A ratio expressed as a fraction should be reduced to lowest terms.

Example

A class contains 20 female students and 24 male students. The ratio of female students to male students may be expressed as:

$$\frac{\text{female students}}{\text{male students}} = \frac{20}{24} = \frac{5}{6} \text{ or } 5:6$$

Problem Set 4.1

1. Express the following as ratios in simplest form.
    a. 7 to 14
    b. 25 to 35
    c. 6 to 21
    d. 20 to 30
    e. 15 to 12
    f. 28 to 21

2. Determine the ratio of an 18-tooth gear to a 24-tooth gear.

3. A football team won 7 games and lost 4 games. What is the ratio of games won to games lost? What is the ratio of games won to games played?

4. The compression ratio is the ratio of the total volume of a cylinder to its clearance volume. If a given cylinder has a total volume of 75 cubic inches and a clearance volume of 25 cubic inches, find its compression ratio.

Review Competencies

5. Find three consecutive integers such that the sum of the second and the third is four less than three times the first.

6. If Sean spends 35% of his income on mortgage payments, and one year his mortgage payments came to $6,300, what was his yearly income?

7. Select a factor of $4x^2 - 12x + 9$.
    a. 2x − 3   b. 4x − 3
    c. 4x − 1   d. x + 9

8. 120 is what percent of 80?

9. Solve for the variable: $6x^2 - x - 2 = 0$.

> ### Definition
> A **proportion** is a statement that one ratio equals another.

A proportion appears symbolically in the following form:

$$\frac{a}{b} = \frac{c}{d} \ (b \neq 0 \text{ and } d \neq 0)$$

This proportion is read "a is to b as c is to d."

When one ratio equals another, the four quantities are said to be **in proportion** or **proportional.**

Example

The following equality is a proportion: $\frac{8}{14} = \frac{4}{7}$

The four numbers 8, 14, 4, and 7 are proportional. The equality is read as "8 is to 14 as 4 is to 7." Notice also that:

$$8 \cdot 7 = 14 \cdot 4.$$

These cross products will always be equal in a proportion.

A most important principle for working with proportions is:

> If $\frac{a}{b} = \frac{c}{d}$, then $a \cdot d = b \cdot c$. $(b \neq 0, d \neq 0)$

The equality of cross products can be used to check the truth of a proportion.

Example

Is $\frac{2}{3} = \frac{9}{12}$ a true proportion? To answer the question, compare the cross products. Since $2 \cdot 12 \neq 3 \cdot 9$, it is not a true proportion.

Example

To find the missing number in the proportion $\frac{3}{5} = \frac{x}{20}$, use the cross products to write the equation

$$3 \cdot 20 = 5 \cdot x \qquad \text{or} \qquad 60 = 5x.$$

Solving the equation gives $12 = x$. The value, 12, should be checked in the original proportion.

Earlier in this chapter, equations like $\frac{3}{5} = \frac{x}{20}$ were solved using the least common multiple of the denominators. For review, it would be valuable to try that method and show that the same result is achieved as with the cross products method.

Problem 1   Solve the proportion $\frac{4}{7} = \frac{3}{x}$.

Solution   First multiply the cross products of $\frac{4}{7} = \frac{3}{x}$.

$$4 \cdot x = 7 \cdot 3 \quad \text{or} \quad 4x = 21$$

Solving $4x = 21$ gives the correct result: $\frac{21}{4}$.

Problem 2   On a map $\frac{1}{4}$ inch represents 60 miles. How many miles are represented by $1\frac{1}{2}$ inches?

Solution   Setting up a proportion comparing inches to miles gives

$$\frac{\text{inches per 60 miles}}{60 \text{ miles}} = \frac{\text{inches per ? miles}}{? \text{ miles}}$$

or
$$\frac{\frac{1}{4}}{60} = \frac{1\frac{1}{2}}{x}$$

This gives
$$\frac{1}{4} \cdot x = 60 \cdot 1\frac{1}{2}$$

or
$$\frac{1}{4}x = 90.$$

Multiplying both members by 4 gives $x = 360$.
Since x stands for unknown miles, the answer is 360 miles.

Problem 3   The ratio of Jan's weight to Bob's weight is 4 to 3. If Jan's weight is 120 pounds, find Bob's weight.

Solution   The ratio of Jan's weight to Bob's weight is

$$\frac{\text{Jan's weight}}{\text{Bob's weight}} = \frac{4}{3}$$

This ratio is equal to $\frac{120}{x}$, where x represents Bob's weight.
Using the cross products of the proportion $\frac{4}{3} = \frac{120}{x}$ gives

$$4x = 360 \quad \text{or} \quad x = 90.$$

Thus, Bob weighs 90 pounds.

Chapter 5

Problem 4

Two machines can complete 5 tasks every 8 days. Let t represent the number of tasks these machines can complete in 30 days. Which of the following proportions accurately represent this situation?

a. $\frac{8}{5} = \frac{30}{t}$ b. $\frac{8}{5} = \frac{t}{30}$ c. $\frac{t}{5} = \frac{8}{30}$

Solution

It is not necessary to write all correct proportions. If we write one proportion that is correct and multiply the cross products, any other correct choice must give us an identical result when the cross products are multiplied.

Let us compare tasks to days.
(Observe that we can also compare days to tasks.)

$$\frac{\text{Tasks (in 8 days)}}{8 \text{ days}} = \frac{\text{Tasks (in 30 days)}}{30 \text{ days}}$$

or $$\frac{5}{8} = \frac{t}{30}$$

We now multiply cross products. Since this proportion is correct, any given choice that results in $8t = 5 \cdot 30$ is also correct.

Choice a, $\frac{8}{5} = \frac{30}{t}$, becomes $8t = 5 \cdot 30$, so a is correct.

Choice b, $\frac{8}{5} = \frac{t}{30}$, becomes $5t = 8 \cdot 30$, so b is incorrect.

Choice c, $\frac{t}{5} = \frac{8}{30}$, becomes $30t = 5 \cdot 8$, so c is incorrect.

Problem 5

A forest service catches, tags and then releases 72 deer back into a park. Two weeks later they select a sample of 336 deer, 21 or which were found to be tagged. Assuming that the ratio of tagged deer in the sample holds for all deer in the park, approximately how many deer are in the park?

Solution

Set up a proportion comparing tagged deer with the total number of deer.

$$\frac{\text{tagged deer}}{\text{total number}} : \frac{72}{x} = \frac{21}{336}$$

Reducing gives:
$$\frac{72}{x} = \frac{1}{16}$$
$$1x = (72)(16)$$
$$x = 1{,}152$$

Since x stands for the total number of deer in the park, there are approximately 1,152 deer.

Problem Set 4.2

1. True or false proportions?
   a. $\frac{5}{20} = \frac{1}{4}$ b. $\frac{5}{9} = \frac{15}{18}$ c. $\frac{7}{4} = \frac{21}{12}$

2. The ratio of a mother's age to her son's age is 6 to 1. If the son is 7 years old, how old is the mother?

3. The ratio of men to women in a class is 9 to 4. If there are 27 men in the class, how many women are there?

4. If the ratio of dogs to cats in a pet store is 5 to 2, determine how many cats there are if there are 20 dogs in the store.

5. If there are 36 inches in 3 feet, set up a proportion and determine how many inches are in 12 feet.

6. If there are 100 centimeters in 1 meter, set up a proportion and determine how many centimeters are in 4.3 meters.

7. A snake consumes 7 mice every 5 days. How many mice will the snake consume in 30 days?

8. A blood count records the ratio of white blood cells to red blood cells as 4 to 2500. How many white blood cells are in a sample that contains 25,000 red blood cells?

9. A photograph that measures 4 inches wide and 7 inches high is to be enlarged so that the height will be 10 inches. What will be the width of the enlargement?

10. The counter on a tape recorder registers 420 after the recorder has been running for 20 minutes. What will the counter register after half an hour?

11. If 13 pounds of fertilizer will cover 2000 square feet of lawn, how much fertilizer is needed to cover 3200 square feet?

12. Suppose that one pound of hamburger makes spaghetti for three people. Let x represent the amount of hamburger needed to make spaghetti for eight people. Select the correct statement of the given conditions.
    a. $\frac{1}{3} = \frac{x}{8}$ b. $\frac{1}{3} = \frac{8}{x}$ c. $\frac{1}{x} = \frac{8}{3}$

13. Suppose 3 long distance runners cover 51 miles. Let x represent how many runners are needed to cover 85 miles. Select the correct statement of the given conditions.
    a. $\frac{3}{51} = \frac{85}{x}$ b. $\frac{3}{x} = \frac{51}{85}$ c. $\frac{3}{x} = \frac{85}{51}$

Review Competencies

14. Evaluate:
    a. $2x^2 - 3x - 1$; $x = -1$
    b. $C = \frac{5}{9}(F - 32)$; $F = 59$

15. Find and graph the solution set of $2x - 4 \geq 3x + 2$.

16. Is 3 in the solution set of $-3x \geq x - 2$?

17. What is the place value of 2 in 234_5?

18. Express 3% as a decimal.

19. Multiply: $\frac{1}{4} \cdot 2\frac{2}{3}$

20. Find the smallest positive multiple of 8 which leaves a remainder of 4 when divided by 5 and a remainder of 3 when divided by 7.

21. Evaluate $n^2 + n + 11$ when n is 3. Is the result a prime number?

Section 5   Variation

In an equation containing two variables, the value of one variable depends on the value given the other variable. In this section, we discuss three important relationships between variables.

If y is a variable that is equal to a constant k times another variable x (y = kx), then we say that:

y varies directly as x

or

y varies as x

or

y is directly proportional to x

or

y is proportional to x.

The fixed constant, k, is called the **constant of variation** or the **constant of proportionality.**

Problem 1   y varies directly as x. When y = 6, then x = 3. Find the constant of proportionality (constant of variation).

Solution   The words "varies directly as" means the variables y and x are related by the equation y = kx.

When y = 6 and x = 3, y = kx becomes 6 = k • 3.

6 = 3k solves as k = 2. Therefore, the constant of proportionality is 2.

Notice that when the constant of proportionality is 2, then the equation relating the variables is y = 2x. Once k is known, any value for x may be substituted to find a matching value for y. For example,
    If y = 2x and x is 3, then y = 2 • 3 = 6.

Problem 2   y varies directly as x. When y = 24, then x = 64.
Find y when x = 32.

Solution   Since y "varies directly as" x, the variables are related by the
equation y = kx. This equation is used to find a value for k.

$$y = kx, \text{ y = 24 and x = 64 gives } 24 = k \cdot 64 \text{ or } k = \frac{24}{64} = \frac{3}{8}.$$

Therefore, the equation is $y = \frac{3}{8} \cdot x$. When x = 32, this gives:
$$y = \frac{3}{8} \cdot 32 = 12.$$

The preceding problem provides a step-by-step procedure for solving variation problems.

> **Procedure for Solving a Variation Problem**
> 1. Translate the English statement into an equation.
> 2. Find the value for the constant k by substituting paired values for the variables.
> 3. Use the equation of step (1) with the value of k found in step (2).
> 4. Substitute the value of the unmatched variable and solve the equation.

Problem 3   The force, F, required to stretch a spring is proportional to the
elongation, E. If 24 pounds stretches a spring 3 inches, find the
force required to stretch the spring $4\frac{1}{2}$ inches.

Solution   The steps shown in this solution are identical to those listed above in
the procedure for solving this type of problem.

step 1
The words "is proportional to" gives the equation F = kE.

step 2
F = kE when F = 24 and E = 3 becomes 24 = k · 3 or k = 8.

step 3
Since k = 8, F = kE becomes F = 8E.

step 4
F = 8E and E = $4\frac{1}{2}$ becomes F = 8 · $4\frac{1}{2}$ = 36.

Thus, the required force is 36 pounds.

Problem 4   y varies directly as x squared. When y = 5, then x = 10.
Find y when x = 5.

Solution   "x squared" means x^2.

step 1
"y varies directly as x^2" gives the equation $y = kx^2$.

step 2
$y = kx^2$, y = 5 and x = 10 becomes $5 = k \cdot 10^2$

or   5 = 100k or $\frac{5}{100} = k$ or $k = \frac{1}{20}$

step 3
$y = kx^2$ becomes $y = \frac{1}{20}x^2$.

step 4
When x = 5, $y = \frac{1}{20}x^2$ becomes

$$y = \frac{1}{20} \cdot (5)^2 = \frac{1}{20} \cdot 25 = \frac{5}{4}$$

Problem Set 5.1

1. y varies directly as x. When y = 5, then x = 10.
    Find y when x = 35.

2. y varies directly as x. When y = 9, then x = 27.
    Find y when x = 17.

3. y varies directly as x squared. When x = 4, then
    y = 64. Find y when x = 10.

4. y varies directly as x^3. When x = 2, then y = 4.
    Find y when x = 6.

5. The amount of paint required to paint a circular
    floor varies as the square of the radius. If it
    takes 4 liters of paint for a floor of radius 10
    meters, find the amount of paint required for a
    floor of radius 16 meters.

Review Competencies

6. Simplify or evaluate:
    a. $(2x^2)(-3x^4)$ b. $a^{-3} \cdot a^{-2}$
    c. $3 \cdot 5^{-2}$ d. $(3 \cdot 3)^2$
    e.   - (-5)   f. $3 - (-5)$

7. Evaluate: $4(4 \div 2 + 3 - 1 + 2)$

8. A football team won 8 games and lost 3 games.
    What is the ratio of games lost to games
    played?

9. What number is the multiplicative identity?

10. True or false? -2 > -1

11. Express 0.456 as a percent.

12. What is 20% of 75?

Let us now discuss the second type of variation.

If y is a variable that is equal to a constant k times the reciprocal of x ($y = \frac{k}{x}$), then we say that:

y varies inversely as x

or

y is inversely proportional to x.

Problem 5 y varies inversely as x. When y = 10, then x = 5. Find y when x = 2.

Solution

step 1

"y varies inversely as x" implies the equation $y = \frac{k}{x}$.

step 2

When y = 10 and x = 5, then $y = \frac{k}{x}$ becomes $10 = \frac{k}{5}$ or k = 50.

step 3

$y = \frac{k}{x}$ becomes $y = \frac{50}{x}$.

step 4

When x = 2, $y = \frac{50}{x}$ becomes $y = \frac{50}{2} = 25$.

Problem 6 y varies inversely as x squared. When x = 5, then y = 3. Find y when x = 10.

Solution

step 1

"y varies inversely as x squared" implies $y = \frac{k}{x^2}$.

step 2

When x = 5 and y = 3, then $y = \frac{k}{x^2}$ becomes $3 = \frac{k}{5^2}$ or $3 = \frac{k}{25}$ or k = 75.

step 3

$y = \frac{k}{x^2}$ becomes $y = \frac{75}{x^2}$.

step 4

When x = 10, $y = \frac{75}{x^2}$ becomes $y = \frac{75}{10^2} = \frac{75}{100} = \frac{3}{4}$.

Problem 7   z varies directly as x and inversely as y. When $x = 10$ and $y = 2$,

then $z = \frac{1}{3}$. Find z when $x = 30$ and $y = 5$.

Solution

step 1

"z varies directly as x and inversely as y" means $z = \frac{kx}{y}$

step 2

When $x = 10$, $y = 2$, and $z = \frac{1}{3}$, then $z = \frac{kx}{y}$ becomes $\frac{1}{3} = k \cdot \frac{10}{2}$ or $\frac{1}{3} = 5k$
or $k = \frac{1}{15}$.

step 3

$z = \frac{kx}{y}$ becomes $z = \frac{\frac{1}{15}x}{y}$

step 4

When $x = 30$, $y = 5$, then $z = \frac{\frac{1}{15}x}{y}$ becomes $z = \frac{\frac{1}{15} \cdot 30}{5} = \frac{2}{5}$.

Finally, let us discuss the third type of variation.

If y is a variation that is equal to a constant k times the product x and z
$(y = kxz)$ then we say that:

y varies jointly as x and z

or

y varies directly as x and z.

Problem 8   y varies jointly as x and z. When $x = 2$ and $z = 5$, then $y = 25$.
Find y when $x = 8$ and $z = 12$.

Solution

step 1
"y varies jointly as x and z" implies $y = kxz$.

step 2
When $x = 2$, $z = 5$, and $y = 25$, $y = kxz$ becomes

$25 = k(2)(5)$ or $25 = k(10)$ or $k = \frac{25}{10}$ or $k = \frac{5}{2}$.

step 3
$y = kxz$ becomes $y = \frac{5}{2}xz$.

step 4
When $x = 8$ and $z = 12$, $y = \frac{5}{2}xz$ becomes $y = \frac{5}{2} \cdot 8 \cdot 12 = 240$.

Problem 9   y varies jointly as p and x squared and inversely as D. When
y = 14, p = 3, and x = 2, then D = 6.
Find y when p = 1, x = 4 and D = 2.

Solution

step 1

"y varies jointly as p and x squared and inversely as D" implies $y = \frac{kpx^2}{D}$.

step 2

When y = 14, p = 3, x = 2 and D = 6, then $y = \frac{kpx^2}{D}$

becomes $14 = \frac{k \cdot 3 \cdot 2^2}{6}$ or $14 = k \cdot \frac{12}{6}$ or $14 = k \cdot 2$ or $k = 7$.

step 3

$y = \frac{kpx^2}{D}$ becomes $y = \frac{7px^2}{D}$.

step 4

When p = 1, x = 4 and D = 2, $y = \frac{7px^2}{D}$ becomes $y = \frac{7 \cdot 1 \cdot 4^2}{2} = 56$.

Problem Set 5.2

1. Translate each of the following into an
   equation:
   a. The area A of a circle varies directly as the
   square of the radius r.
   b. The area A of a triangle varies jointly as the
   base b and the altitude h.
   c. The attraction F between two bodies varies
   inversely as the square of the distance s
   between them.
   d. The illumination I produced on a page by a
   lamp is directly proportional to the wattage
   W of the lamp and inversely proportional to
   the square of the distance D between the
   lamp and the page.
   e. The volume V of a right circular cone varies
   jointly as the square of the radius r of the
   base and the altitude h.
   f. y varies jointly as x squared and the square
   root of z and inversely as the square of R.

   g. The pressure P of a gas varies directly as its
   absolute temperature T and inversely as its
   volume V.

2. y varies inversely as x. When x = 3, then y = 5.
   Find y when x = 9.

3. y varies directly as x and inversely as z. When
   x = 6 and z = 3, then y = 20. Find y when x = 5
   and z = $\frac{1}{3}$.

4. y varies inversely as x cubed. When y = 6, then
   x = 2. Find y when x = $\frac{1}{2}$.

5. y varies directly as x and inversely as the
   square of z. When y = 20 and x = 50, then
   z = 5. Find y when x = 3 and z = 6.

6. y varies jointly as x and z. When y = 10 and x = 5, then z = $\frac{1}{4}$. Find y when x = $\frac{1}{16}$ and z = 24.

7. y varies jointly as x squared and z and inversely as R cubed. When y = $\frac{1}{3}$, x = 2 and z = 6, then R = 2. Find y when x = 3, z = 2, and R = 1.

8. The current I flowing in an electrical circuit varies inversely as the resistance R in the circuit. When R = 4 ohms, then I = 24 amperes. What is the current I when the resistance is 6 ohms?

9. The time (T) required to do a job is inversely proportional to the product of the number of workers (W) and the number of hours they work per day (H). 9 workers working 10 hours a day can do a job in 14 days. (When W = 9 and H = 10, then T = 14.) Find how long it would take 12 workers working 7 hours per day to do the job.

10. The ratio of men to women in a class is 2 to 5. If there are 25 women in the class, how many men are there?

11. The ratio of a father's age to his son's age is 8 to 3. If the son is 15 years old, how old is the father?

12. Two machines can complete 6 tasks every 5 days. How many tasks can these machines complete in 30 days?

13. y varies directly as x. When y = 4, then x = 12. Find y when x = 21.

14. How many prime factors of 630 are also factors of 792?

Section 6   Principles of Graphing

In this section we will consider equations that contain **two** variables. Such equations will have the form Ax + By = C, where A, B, and C are constants (A and B not both zero) and x and y are variables. We will see that a pair of numbers, one for x and one for y, is required to make the equation a true statement. This pair of numbers is written as (x,y). Such a pair is said to be an **ordered pair** to indicate that the order in which x and y appear makes a difference. Thus, the ordered pair (1,2) is not the same as the ordered pair (2,1) because the order of the numbers is not the same.

The equation 2x + y = 3 is an equation with two variables. **Two** numbers, one for x and one for y, must be found to make 2x + y = 3 a true statement. The two numbers can be considered an ordered pair (x,y). If x = 0 and y = 3, then 2x + y = 3 becomes the true statement 2 • 0 + 3 = 3, and the ordered pair (0,3) is a solution. But, if x = 3 and y = 0, then 2x + y = 3 becomes the false statement 2 • 3 + 0 = 3, and the ordered pair (3,0) is not a solution.

Problem 1   Find the set of ordered pairs (x,y) that are solutions for $2x + y = 3$ if the variables can only be replaced by elements of $\{-1,0,1,2,3\}$.

Solution   Only the numbers -1, 0, 1, 2, and 3 are allowed as replacements for x and y in $2x + y = 3$. Begin by replacing x by -1 and solve the equation to find the matching value for y. When $x = -1$, $2x + y = 3$ becomes $2(-1) + y = 3$ or $-2 + y = 3$ or $y = 5$. But 5 is not an allowed replacement for y, so the ordered pair (-1,5) is *not* an acceptable solution.

Try replacing x by 0. When $x = 0$, then $2x + y = 3$ gives
$$2 \cdot 0 + y = 3 \quad \text{or} \quad 0 + y = 3 \quad \text{or} \quad y = 3.$$

Both 0 and 3 are allowed as replacements for the variables. Therefore, (0,3) is an acceptable solution.

Try replacing x by 1. When $x = 1$, then $2x + y = 3$ gives
$$2 \cdot 1 + y = 3 \quad \text{or} \quad 2 + y = 3 \quad \text{or} \quad y = 1.$$

1 is an acceptable replacement for both variables, so (1,1) is an acceptable solution.

Try replacing x by 2. When $x = 2$, then $2x + y = 3$ gives
$$2 \cdot 2 + y = 3 \quad \text{or} \quad 4 + y = 3 \quad \text{or} \quad y = -1.$$

Both 2 and -1 are acceptable replacements for x and y. Therefore, (2,-1) is an acceptable solution.

Finally, try replacing x by 3. When $x = 3$, then $2x + y = 3$ gives
$$2 \cdot 3 + y = 3 \quad \text{or} \quad 6 + y = 3 \quad \text{or} \quad y = -3.$$

But -3 is not an acceptable replacement for y. Therefore, (3,-3) is not an acceptable solution.

A **table of values** can be used to record the acceptable replacements for x and y as follows:

x	0	1	2
y	3	1	-1

Using elements of $\{-1,0,1,2,3\}$ as replacements for x and y, the complete set of ordered pair solutions of $2x + y = 3$ is
$$\{(0,3),(1,1),(2,-1)\}.$$

The ability to find ordered pair solutions is a skill that is basic to the graphing process.

Problem Set 6.1

1. Find the set of ordered pair solutions (x,y) for each of the following with the restriction that the variables must be replaced by whole numbers.

   a. $2x + y = 4$ b. $x + 3y < 3$ c. $x + y \leq 2$

2. Find the set of ordered pair solutions (x,y) for each of the following using only 0, 1, 2, and 3 as replacements for the variables.

   a. $y = x + 1$ b. $x + y = 2$ c. $x + 2y = 3$
   d. $x + y < 3$ e. $2x + y \leq 3$

3. Find the set of ordered pair solutions (x,y) for each of the following using only -2, -1, 0, 1 and 2 as replacements for the variables.

   a. $y = x$ b. $x - y = 1$
   c. $x + y = 2$ d. $2x - y = 2$

4. Find the set of ordered pair solutions (x,y) for each of the following using only -1, 0, 1 and 2 as replacements for the variables.

   a. $y < x$ b. $y \leq x - 1$ c. $y \geq x + 1$

Review Competencies

5. The ratio of Joan's age to Bill's age is 3 to 2. If Joan is 36, find Bill's age.

6. y varies inversely as x. When y = 6, then x = 3. Find y when x = 2.

7. Find the solution set for $\frac{x}{3} = -\frac{2}{5}$.

8. Evaluate $F = \frac{9}{5}C + 32$ when C = 10.

9. Is 4 in the solution set of $-x + 2 > 2x - 1$?

10. Write the number expressed by $3.5 \cdot 10^5$.

11. Find and graph the solution set of $-3 + 2x \leq 4x - 5$.

12. Evaluate: $-5 + 6 \div 2 - 2 \cdot 4 + 3$

Recall that to graph an inequality with one variable, like x < 5, one number line is used. To graph an equation with two variables, like x + y = 4, or an inequality with two variables, like x + y < 4, two number lines are used, like those shown below.

The rectangular (or Cartesian) coordinate system consists of two number lines that are perpendicular and intersect at the point 0 on both lines. Each number line is called an **axis.** The horizontal line is called the **x-axis** and the vertical line is called the **y-axis.** The point of intersection of the two axes is called the **origin.** The axes divide the plane into four regions, called **quadrants,** which are numbered I through IV in a counter-clockwise direction.

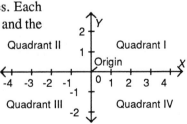

The x and y axes can be used to associate ordered pairs of numbers with points in the plane. Each point in the plane will have an ordered pair of real numbers associated with it. Locating a point in a plane is called **plotting the point.** The ability to correctly plot points is essential to the graphing process.

In an ordered pair (x,y) the **x-number indicates horizontal movement right or left**. If the x-number is positive, the point is to the right of the origin. If the x-number is negative, the point is to the left of the origin. The **y-number indicates vertical movement up or down**. If the y-number is positive, the point is above the origin. If the y-number is negative, the point is below the origin.

Problem 2 Plot the point represented by the ordered pair (3,2).

Solution The point (3,2) is found by:

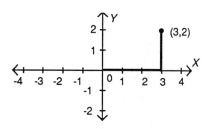

1) Starting at the origin,
2) Moving 3 units right, and
3) Moving 2 units up.

The diagram at the right shows the process used to plot (3,2). The point lies in Quadrant 1.

Problem 3 Plot the point represented by the ordered pair (-1,-2).

Solution The point (-1,-2) is plotted by:

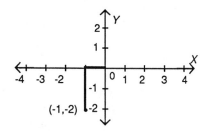

1) Starting at the origin,
2) Moving 1 unit left, and
3) Moving 2 units down.

The graph at the right shows the process used to plot (-1,-2). The point lies in Quadrant III.

Observe that every point on the x-axis has an ordered pair (x,0). Examples are (2,0) and (-1,0). Every point on the y-axis has an ordered pair (0,y). Examples are (0,4) and (0,-2). The ordered pair associated with the origin is (0,0). A point on either axis does not lie in any quadrant. Also observe that for every ordered pair there is precisely **one** point in the rectangular coordinate system.

Problem Set 6.2

Plot the following ordered pair on the axes shown at the right and indicate the quadrant in which each lies.

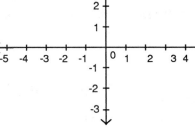

1. (0,0)
2. (3,5)
3. (-1,4)
4. (-3,-3)
5. (3,0)
6. (3,-2)
7. (-3,0)
8. (0,2)
9. (0,-1)

Review Competencies

10. Find the solution set of ordered pairs for $x + y \leq 2$ using only whole numbers as replacements for the variables.

11. Evaluate $P = 2L + 2W$ when $L = 8$ and $W = 5$.

12. Find the solution set for $-5 = -x + 5 - 4x$.

13. Simplify or evaluate:
   a. $2^2 \cdot 2^2$ b. $5 \cdot 0 \cdot 5^0$ c. $3 \cdot 4^{-1}$
   d. $\frac{x^4}{x^2}$ e. $a^5 \cdot a^{-6}$

14. If a, b, and c are real numbers, state the distributive property of multiplication over addition.

15. There are 14 boys and 20 girls in a class. What is the ratio of boys to the entire class?

16. Divide: $1.2 \div 0.002$

17. 10% of what number is 24?

Let us now discuss the process for graphing an equation of the form $Ax + By = C$. Such an equation is called a **first degree equation** because the exponent of each variable is 1. Such an equation is also called a **linear equation in two variables** or a **linear equality.**

Examples

1. $2x + 3y = 7$ is a linear equation with two variables, x and y.

2. $x^2 + 2y = 9$ is not a linear equation. This is not a first degree equation because not all the exponents of the variables are 1.

3. $x = 7$ is a linear equation, but only one variable appears. To write $x = 7$ with two variables, use the fact that $0y = 0$. Thus, $x = 7$ is equivalent to $x + 0y = 7$.

The graph of a linear equation in two variables is always a straight line. Since a straight line is determined by any two points on it, it is possible to find the graph by finding any two ordered pair solutions of the equation. As a check, it is generally wise to find **three** ordered pairs of the solution set. If they lie on the same straight line, then in all likelihood the work has been done correctly, and the line is the correct graph of the equation.

In the next two problems we will see that it is possible to graph a linear equation in either of two ways.

Problem 4 Graph x + 2y = 6 by plotting points. The variables may be replaced by real numbers.

Solution Three ordered pair solutions need to be found, but **any three** solutions will work. Consequently, any three numbers can be chosen to replace x, and the matching numbers for y will provide the three ordered pair solutions.

Let x = 0. Then, x + 2y = 6 gives 0 + 2y = 6 or y = 3. Thus, (0,3) is an ordered pair solution.

Let x = 2. Then, x + 2y = 6 gives 2 + 2y = 6. Solving for y leads successively to 2y = 4 and then to y = 2. Therefore, (2,2) is another ordered pair solution.

Let x = 4. Then x + 2y = 6 gives 4 + 2y = 6. Solving for y leads successively to 2y = 2 and then to y = 1. (4,1) is a third ordered pair solution for the equation.

Notice that we have completed a table of values as follows:

x	0	2	4
y	3	2	1

The three solutions, (0,3), (2,2) and (4,1), are plotted on the graph shown at the right. Notice that all three points are on the same straight line. This line is the graph of all the solutions of x + 2y = 6 and represents the solution set for the equation.

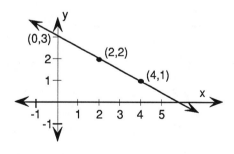

It is also possible to graph a linear equation by finding only **two** points and drawing the line that contains them. The two points where the line crosses the axes are most easily found.

In the previous problem, the graph crossed the y-axis at (0,3). Thus, 3 is called the **y-intercept**. The graph crosses the x-axis at (x,0). Thus, x is called the **x-intercept**. The intercepts are useful for graphing linear equations.

To graph a linear equation using the intercepts:

1. Find where the graph crosses the x-axis by letting y = 0 and solving the equation to find (x,0).

2. Find where the graph crosses the y-axis by letting x = 0 and solving the equation to find (0,y).

3. Draw the straight line through the two points.

Problem 5   Graph 2x + 3y = 6 by first finding the intercepts.

Solution

Let y = 0.

$$2x + 3 \cdot 0 = 6$$
$$2x + 0 = 6$$
$$2x = 6$$
$$x = 3$$

The x-intercept is 3, so the graph crosses the x-axis at the point (3,0).

Let x = 0.

$$2 \cdot 0 + 3y = 6$$
$$0 + 3y = 6$$
$$3y = 6$$
$$y = 2$$

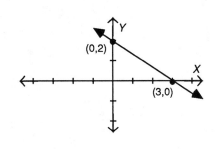

The y-intercept is 2, so the graph crosses the y-axis at the point (0,2). The graph is correctly drawn at the right.

Problem 6   Graph y = -x by finding the intercepts.

Solution

Let y = 0.   Then   0 = -x or x = 0.
The x-intercept is 0, so the graph crosses the x-axis at the point (0,0).

Let x = 0.   Then   y = -0 or   y = 0.

The y-intercept is also 0, so the graph crosses the y-axis also at the point (0,0).

Since only one point, (0,0), of the line has been found, a second, different, point is needed. Select any value of x.

    Let x = 1. Then y = -1.

The point (1,-1) is a second point on the line.

The graph of the line appears at the right.

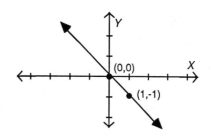

Problem 7   Graph x = -2.

Solution   The graph of x = -2 is a vertical line parallel to the y-axis and 2 units to the left of it. The graph appears at the right.

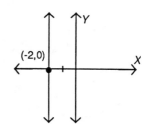

In general, any equation of the form x = C has a vertical line as its graph. The equation x = 0 graphs as the y-axis.

Problem 8   Graph y = 3

Solution   The graph of y = 3 is a horizontal line parallel to the x-axis and 3 units above it. The graph appears at the right.

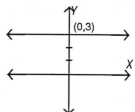

In general, any equation of the form y = C has a horizontal line as its graph. The equation y = 0 graphs as the x axis.

Problem Set 6.3

In problems 1-7 graph the linear equations using either method.

1. x = 2y   2. y + 2x = 2   3. 2x − y = -6
4. x + 2y = 5   5. x − 2y = -4   6. x = 3
7. y = -1

Review Competencies

8. y varies inversely as x. When x = 4, y = 7. Find y when x = 12.

9. Two eagles eat an average of 8 mice every 6 days. How many mice will these eagles eat in a 30 day period?

10. Evaluate L = a + (n − 1)d when a = 10, n = 4 and d = 2.

11. Find and graph the solution set of -8 ≤ -3x − x.

12. Is -2 in the solution set of (x − 2)(x + 2) ≤ 1?

13. Subtract: 5 − 1.002

14. Express $\frac{4}{25}$ as a percent.

15. True or false? The only prime number that is also even is 2.

Chapter 5

It is possible to graph the solution set for an inequality in two variables such as x < y + 1. We will see that the solution set is a region of the plane.

Problem 9 Graph the solution set for x < y + 1.

Solution **step 1**
Graph the linear **equation** x = y + 1. The equation is the **boundary** for the graph and is obtained from the inequality by replacing the inequality symbol with an equal sign.

Any three ordered pair solutions of x = y + 1 (or simply the two intercepts) are needed. Three solutions are (0,-1), (1,0) and (2,1) because their x and y values make x = y + 1 true.

For (0,-1) 0 = -1 + 1 For (1,0), 1 = 0 + 1 For (2,1), 2 = 1 + 1

step 2
Use the three solutions, (0,-1), (1,0), and (2,1) to graph the straight line of x = y + 1, but use a **dashed line** to indicate < rather than =. A dashed line is used because the points on the boundary line do not belong to the solution set of the given inequality. Inequalities with the symbols ≤ and ≥ would use a **solid line** because the points on the boundary line would be part of the solution set.

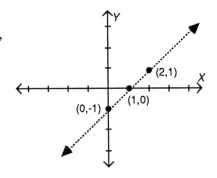

step 3
The graph of x < y + 1 is the region above the dashed line or it is the region below it. To decide which is correct, select any ordered pair that does not lie on the straight line. Frequently (0,0) is a selection that will make this step simple. Use the selected ordered pair to test whether the given inequality is true or false.

For (0,0), x < y + 1 is 0 < 0 + 1 or 0 < 1 which is true.

step 4
Since the point (0, 0) tested in Step 3 made the inequality true, all points on the same side of the dashed line graphed in Step 2 will also make the inequality true. The solution set of the inequality x < y + 1 is, therefore, the shaded region shown at the right.

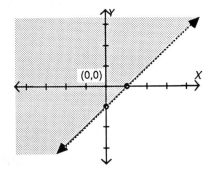

Problem 10   Graph the solution set for x ≥ y.

Solution

step 1
Find any three solutions of the equation
x = y. Three such solutions
are (0,0), (1,1), and (2,2).

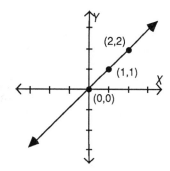

step 2
Graph the boundary line of x = y using
the three solutions found in Step 1. A
solid line is used because solutions of
x = y are also solutions of x ≥ y.

step 3
The graph of x ≥ y is the region above the line or the region below the line.
To decide which region is correct, select any ordered pair for a point not on
the line. The origin, (0,0), cannot be used in this case because (0,0) is a
point on the line.

The point (2,3) is selected and tested in x ≥ y, the result is 2 ≥ 3 and false.
Therefore, (2,3) and all other points on the same side of the line of x = y
make the inequality x ≥ y false.

step 4
The testing of (2,3) in x ≥ y indicates
that the solution set of the inequality is
on the opposite side of the line from (2,3).
The shaded portion of the graph at the
right represents the solution set. The
solid line is part of this solution set.

Problem 11   Graph the solution set for x < -1.

Solution

The graph of x = -1 is parallel to the
y-axis and 1 unit to the left of it.
x < -1 is the region to the left of the
line, but does not include the line.

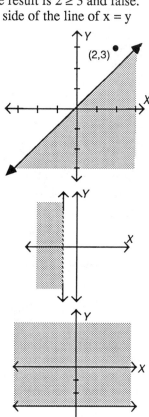

Problem 12   Graph the solution set for y ≥ -3.

Solution

The graph of y = -3 is the line parallel to the
x-axis and 3 units below it. y ≥ -3 includes
the line of y = -3 and the region above it.

Problem Set 6.4

Graph the solution sets for:

1. $y > 2$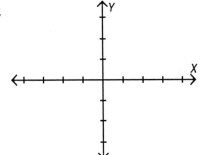

2. $x < y$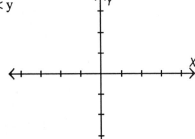

3. $2x - 4y \leq 4$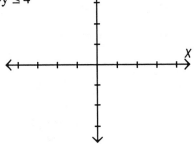

4. $y > x + 2$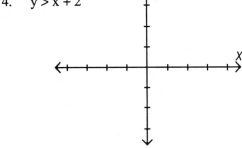

Review Competencies

5. Graph the solution set of $x - y = 3$.

6. What is the ratio of 4 nickels to 3 dimes?

7. The ratio of Jane's weight to Bob's weight is 2 to 3. If Jane's weight is 140 pounds, find Bob's weight.

8. Evaluate $(2x)^2$ when $x = -1$.

9. Is 1 in the solution set of $-3a > a + 1$?

10. Insert the correct symbol of inequality to make $-\frac{3}{4} [\ ] - \frac{2}{3}$ true.

11. If a and b are real numbers, state the commutative property of multiplication.

12. What is the additive inverse of -10?

13. Find the solution set for the equation $\frac{x}{2} = -\frac{3}{5}$.

14. Express 175% as a decimal

15. Subtract: $4\frac{1}{5} - 2\frac{3}{5}$

A pair of inequalities such as $y \geq 0$ and $x \geq 0$ is called a **system of inequalities.** We will see that the solution of a system of inequalities is the **intersection** of the separate solution sets of the individual inequalities. The solution set will be a collection of points on the plane.

Problem 13   Graph the solution of y ≥ 0 and x ≥ 0.

Solution

step 1

Graph y ≥ 0. Notice
that the graph at the
right consists of all
the points on or above
the x-axis. The region
is represented by a
series of vertical
lines.

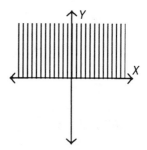

step 2

Graph x ≥ 0. Notice that
the graph at the right consists
of all the points on or to
the right of the y-axis.
The region is represented
by a series of horizontal
lines and is imposed over
the graph of y ≥ 0.

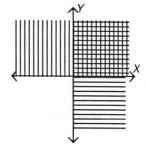

step 3

The solution of the system y ≥ 0
and x ≥ 0 is the region that is
common to the separate graphs
of the two inequalities. It is the
intersection of the graphs as shown
by the area where the lines overlap.

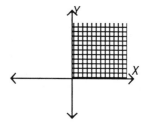

Problem 14   Graph the solution of the system of y ≥ x + 1 and x ≥ 0.

Solution

step 1

To graph y ≥ x + 1 we must first graph
y = x + 1. Two solutions, (-1,0)
and (0,1) are found and plotted. The
solid line through these two points
is shown at the right.

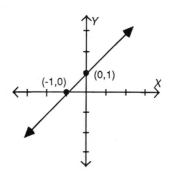

315

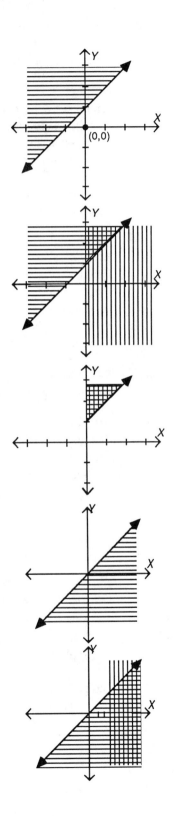

step 2
The point (0,0) is tested in $y \geq x + 1$. Since $0 \geq 0 + 1$ or $0 \geq 1$ is false, the region representing the solution of $y \geq x + 1$ appears as horizontal lines as shown at the right.

step 3
The graph of $x \geq 0$ is imposed by the use of vertical lines.

step 4
The set of points such that $y \geq x + 1$ and $x \geq 0$ is the region shown at the right.

Problem 15 Graph the solution of $x \geq y$, $x \geq 3$ and $y \geq 0$.

Solution

step 1
Graph $x \geq y$. In the figure at the right, the solution set is indicated by horizontal lines.

step 2
Graph $x \geq 3$. In the figure at the right, the solution set of $x \geq 3$ is indicated by the vertical lines.

step 3
Graph y ≥ 0. The
solution set is
shown by slanted
lines in the figure
at the right.

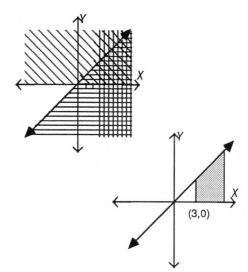

step 4
The solution of the
system is the set of
points common to all
three regions.

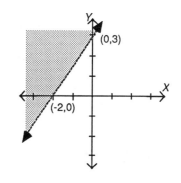

Problem 16   Which of the condition(s) listed below
corresponds to the graph at the right?

a.   x < 2 and y > 3

b.   y = -x

c.   x < 2, y > 3, and y > x

d. $y > \frac{3}{2}x + 3$

Solution   Try to eliminate each choice.

choice a   The graph of x < 2 would include all points
to the left of the vertical line x = 2.
Therefore, choice a is incorrect.

choice b   y = -x is an equation. Its graph is a straight line —
not a region. Choice b is incorrect.

choice c   Again, x < 2 eliminates this option just as it did in
choice a. Choice c is incorrect.

choice d   Let us confirm that this is the correct option. The
dashed line crosses the axes at (-2,0) and (0,3), and
both pairs should be checked in the equation

$$y = \frac{3}{2}x + 3$$

$$0 = \frac{3}{2} \cdot -2 + 3 \text{ is true.}$$

$$3 = \frac{3}{2} \cdot 0 + 3 \text{ is true.}$$

Finally, test a point in the shaded region to see if it satisfies the inequality. The point (0,4) is tested in the steps below.

$$y > \frac{3}{2}x + 3$$

$$4 > \frac{3}{2} \cdot 0 + 3 \text{ or } 4 > 3 \text{ is true.}$$

The test confirms the fact that choice d is correct.

Problem Set 6.5

Graph solutions for the following systems of inequalities:

1. $x + y \leq 2$ and $x \geq 0$

2. $x \geq y, x \leq 2$ and $y \geq 0$

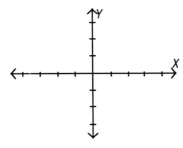

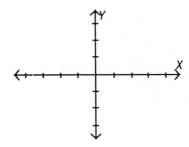

3. $x - 2y \leq 6, 2x + y \geq 6,$
    and $y \leq 4$

4. $x \geq 2y, x + y \leq 8,$
    $x \geq 3,$ and $y \geq 0$

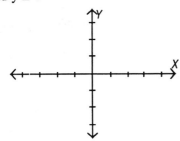

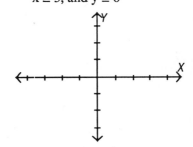

5. Identify the condition(s) that correspond to each shaded region.

a.   b.   c.   d.   e.

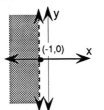

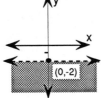

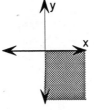

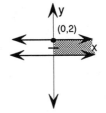

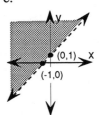

6. Which option gives the condition(s) that correspond to the shaded region of the plane shown below?

    a. $3x + y > 6$
    b. $3x + y < 6$
    c. $y \geq 6$ and $x \geq 2$
    d. $y > 6$ and $x > 2$

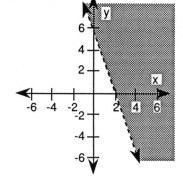

7. Which option gives the condition(s) that correspond to the shaded region of the plane shown below?

    a. $2x + 2y \geq 2$ and $y < 4$
    b. $x \geq 2$ and $y \geq 2$
    c. $x + y \geq 2$ and $x < 4$
    d. $x + y \geq 2$ and $y < 4$

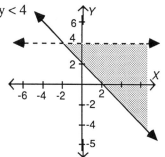

Review Competencies

8. Graph the solution set for $y < x - 1$.

9. Evaluate $y = (x - 1)^2$ when $x = 3$.

10. Is 4 in the solution set of $3x - 4 = x - 12$?

11. Express 2001_4 as a base ten numeral.

12. Use scientific notation to calculate $(8,000,000)(0.0004)$.

13. If a, b, and c are real numbers, state the associative property of addition.

14. True or false? Every odd counting number greater than 1 is a prime number.

Section 7 Solving Two Equations in Two Variables

We now turn our attention to solving a system of two equations in two variables. An example of such a system is:

$$2x - y = 4$$
$$3x + y = 6$$

> The solution set of a system of equations in two variables is the set of all ordered pairs of values (a, b) that satisfy every equation in the system.

319

Chapter 5

Problem 1   Show that $(2, 0)$ is a solution of the system $2x - y = 4$ and $3x + y = 6$.

Solution   Using substitution we can show that $(2, 0)$ satisfies both equations.

$$
\begin{array}{rcl}
2x - y & = & 4 \\
2(2) - 0 & = & 4 \\
4 & = & 4
\end{array}
\qquad
\begin{array}{rcl}
3x + y & = & 6 \\
3(2) + 0 & = & 6 \\
6 & = & 6
\end{array}
$$

In set notation, the solution set to this system is $\{(2, 0)\}$ – that is, the set consisting of the ordered pair $(2, 0)$.

> The solution to a system of equations in two variables corresponds to the point(s) of intersection of their graphs.

In the diagram at the right, we see that the graphs of $2x - y = 4$ and $3x + y = 6$ intersect at $(2, 0)$, which was precisely the solution of the system.

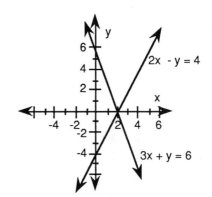

Now let us consider a strictly algebraic approach. To solve a system of equations algebraically, we want to transform the system into an equivalent system that can be solved by inspection. **Equivalent systems** of equations are systems that have the same solution set.

The following systems of equations

$$
\begin{cases} 2x - y = 4 \\ 3x + y = 6 \end{cases}
\quad \text{and} \quad
\begin{cases} x = 2 \\ y = 0 \end{cases}
$$

are equivalent because they both have the same solution set, $\{(2, 0)\}$. The system on the right can be solved by inspection.

When solving a linear system, we will begin with

$$
a_1 x + b_1 y = c_1
$$
$$
a_2 x + b_2 y = c_2
$$

and end with an equivalent system in the form

$$
x = x_1
$$
$$
y = y_1
$$

which gives the solution set explicitly as $\{(x_1, y_1)\}$. One way to transform the original system into $x = x_1$ and $y = y_1$ involves **substitution**. We can substitute for one variable in any equation either an equivalent expression for that variable obtained from another equation in the system or the actual value for the variable.

Let's see exactly what this means. In the original system that we considered,

$$2x - y = 4$$
$$3x + y = 6$$

we can work with either equation, obtain an expression for y in terms of x (or x in terms of y) and then substitute this expression for y (or for x) into the other equation. This means that we are faced with four possibilities: Solve the first equation for x in terms of y, solve the first equation for y in terms of x, solve the second equation for x in terms of y, or solve the second equation for y in terms of x. When possible, **solve for the variable that has a coefficient of 1 or -1**. In the example,

$$2x - y = 4$$
$$3x + y = 6$$

we will solve for y in terms of x in the second equation. We obtain

$$y = -3x + 6$$

We now substitute $-3x + 6$ for y in the other equation. Thus $2x - y = 4$ becomes

$$2x - (-3x + 6) = 4$$

and we have transformed the system to one equation in one variable. We now solve this equation.

$$
\begin{aligned}
2x - (-3x + 6) &= 4 \\
2x + 3x - 6 &= 4 \\
5x &= 10 \\
x &= 2
\end{aligned}
$$

Since we must still find y, we should ideally substitute 2 into either of the original equations. However, since we initially solved the second equation for y in terms of x, many students prefer to use this transformation of the second equation. Thus,

$$
\begin{aligned}
y &= -3x + 6 \\
y &= -3(2) + 6 \\
y &= 0
\end{aligned}
$$

Our original system has now been transformed to

$$x = 2 \qquad y = 0$$

and we see that $(2, 0)$ appears to be the solution of the system.

To check that $(2, 0)$ is indeed a solution to both equations, we substitute 2 for x and 0 for y as we did in Problem 1.

Since (2, 0) satisfies both equations, we say that (2, 0) is the solution to the system and consequently the solution set is $\{(2, 0)\}$. You should get into the habit of checking ordered pair solutions to both equations of the system.

Before considering additional illustrative examples, let us summarize the steps used in the substitution method.

Solving Systems of Two Equations in Two Variables (x and y) by the Substitution Method

1. Solve one equation for x in terms of y or for y in terms of x.
2. Substitute this expression for that variable into the other equation.
3. Solve the resulting equation in one variable.
4. Substitute the solution from step 3 into either original equation to find the value of the other variable.
5. Check the solution in both of the given equations.

Problem 2

Solve by the substitution method.
$$3x - 2y = -5$$
$$4x + y = 8$$

Solution

step 1
Solve for y in the second equation, obtaining
$$y = 8 - 4x$$

step 2
Substitute $(8 - 4x)$ for y in the first equation.

$$3x - 2y = -5$$
$$3x - 2(8 - 4x) = -5$$

step 3
Solve this equation in one variable.

$$3x - 2(8 - 4x) = -5$$
$$3x - 16 + 8x = -5$$
$$11x - 16 = -5$$
$$11x = 11$$
$$x = 1$$

step 4

Substitute 1 for x in either original equation.

$$4x + y = 8$$
$$4(1) + y = 8$$
$$y = 4$$

step 5

Check that (1, 4) is a solution by substituting 1 for x and 4 for y in both original equations.

$3x - 2y = -5$	$4x + y = 8$
$3(1) - 2(4) \overset{?}{=} -5$	$4(1) + 4 \overset{?}{=} 8$
$3 - 8 \overset{?}{=} -5$	$4 + 4 \overset{?}{=} 8$
$-5 = -5$	$8 = 8$

The solution set is {(1, 4)}.

Problem Set 7.1

Solve each system by the substitution method.

1. $x = 2y - 5$
    $x - 3y = 8$

2. $x = 2y - 2$
    $2x - 2y = 1$

3. $4x + y = 5$
    $2x - 3y = 13$

4. $x - y = 4$
    $2x - 5y = 8$

5. $x + y = 0$
    $3x + 2y = 5$

6. $3x - 2y = 4$
    $2x - y = 1$

7. $7x - 3y = 23$
    $x + 2y = 13$

Review Competencies

8. Solve for x. $x > x - (12 - x)$

9. $1{,}700{,}000 \cdot 0.00012 =$
    a. $2.04 \cdot 10^2$ b. $2.04 \cdot 10^{10}$
    c. $2.04 \cdot 10^{-24}$ d. $2.04 \cdot 10^{-2}$

10. Find four consecutive odd integers such that the sum of the least integer and greatest integer is 40.

11. A resistor having a voltage of 60 volts produces a current of 35 amps. How many amps of current are produced by a resistor of 72 volts?

The second algebraic method we will consider for solving systems of equations is the **elimination method,** sometimes called the addition method. This method is particularly useful for linear systems in which none of the variables has a coefficient of 1 or -1, such as

$$3x + 2y = 2$$
$$8x - 5y = -5$$

In the preceding section we saw that our aim in solving a system of linear equations was to transform the system into equivalent equations of the form

$$x = x_1$$
$$y = y_1$$

explicitly giving a solution set of $\{(x_1, y_1)\}$. Consider the system

$$3x + 2y = 2 \qquad (1)$$
$$8x - 5y = -5 \qquad (2)$$

Using properties of equality, we can multiply both sides of equation (1) by the same nonzero number, say 5, and similarly multiply both sides of equation (2) by the same nonzero number, say 2, obtaining the **equivalent system:**

$$15x + 10y = 10 \qquad (3)$$
$$16x - 10y = -10 \qquad (4)$$

Equations (1) and (3) are equivalent, and so are equations (2) and (4). The sum of equations (3) and (4),

$$31x = 0 \qquad (3) + (4)$$

obtained by adding left and right sides of the equations, is called a linear combination of the two equations. Notice that the linear combination $31x = 0$ is an equation that is free of the variable y, making it possible to solve for x, resulting in $x = 0$.

The concept of a linear combination can be used to solve a system by choosing appropriate multipliers so that the coefficients of one of the variables, x or y, are additive inverses. This will result in an equation free of one of the variables. Let's see exactly what this means by considering an illustrative problem.

Problem 3 Solve the system

$$3x + 2y = 48 \qquad (1)$$
$$9x - 8y = -24 \qquad (2)$$

Solution

step 1

We must rewrite one or both equations in equivalent forms so that the coefficients of the same variable (either x or y) will be additive inverses of one another. We can accomplish this in a number of ways. If we wish to eliminate x, we can multiply each term of the first equation by -3 and then add the equations.

$$3x + 2y = 48 \quad \text{multiply by -3}$$
$$9x - 8y = -24$$

We obtain the equivalent system

$$-9x - 6y = -144 \quad (3)$$
$$9x - 8y = -24 \quad (4)$$

Add the corresponding members of (3) and (4). We get

$$-14y = -168 \quad \text{(a linear combination of the original equations)}$$
$$y = 12$$

which contains in its solution set any solution common to the solution sets of the original equations in the system. The solution is in the form (x, 12).

step 2

Substitute 12 for y in either (1) or (2) to determine the x-component for the ordered pair (x, 12) that satisfies both equations. If (1) is used, we obtain

$$3x + 2y = 48$$
$$3x + 2(12) = 48$$
$$3x + 24 = 48$$
$$3x = 24$$
$$x = 8$$

If (2) is used we obtain

$$9x - 8y = -24$$
$$9x - 8(12) = -24$$
$$9x - 96 = -24$$
$$9x = 72$$
$$x = 8$$

The ordered pair (8, 12) satisfies both (1) and (2), so the solution set is {(8, 12)}.

Notice that we could have made the decision to eliminate y instead of x. To accomplish this it would have been necessary to multiply the first equation by 4 and then add the equations.

The goal of this method, then, is to transform the original system into an equivalent system so that the coefficients of one of the variables are additive inverses. We then add the equations to eliminate this variable. For this reason, we call the method the **elimination** or **addition** method.

As we noted in problem 3, we can usually make a number of decisions regarding how to eliminate a variable. This is illustrated below.

Eliminate x:

$$
\begin{array}{lll}
2x + y = 2 & \text{multiply by -1} & -2x - y = -2 \\
2x - 6y = 30 & \text{no change} & 2x - 6y = 30 \\
& \text{Add:} & -7y = 28
\end{array}
$$

Eliminate y:

$$
\begin{array}{lll}
2x + y = 2 & \text{multiply by 6} & 12x + 6y = 12 \\
2x - 6y = 30 & \text{no change} & 2x - 6y = 30 \\
& \text{Add:} & 14x = 42
\end{array}
$$

$$
\begin{array}{lll}
x + y = 3 & \text{no change} & x + y = 3 \\
x - y = 1 & \text{no change} & x - y = 1 \\
& \text{Add:} & 2x = 4
\end{array}
$$

Eliminate x:

$$
\begin{array}{lll}
x + y = 3 & \text{no change} & x + y = 3 \\
x - y = 1 & \text{multiply by -1} & -x + y = -1 \\
& \text{Add:} & 2y = 2
\end{array}
$$

Problem 4   Solve by the elimination method.

$$
\begin{array}{l}
7x = 5 - 2y \\
3y = 16 - 2x
\end{array}
$$

Solution   **step 1**

We first arrange the system so that variable terms appear on the left and constants appear on the right. We obtain:

$$
\begin{array}{l}
7x + 2y = 5 \\
2x + 3y = 16
\end{array}
$$

We can eliminate x or y. Let us eliminate y by multiplying the first equation by 3 and the second equation by -2.

$$7x + 2y = 5 \qquad \text{multiply by 3}$$
$$2x + 3y = 16 \qquad \text{multiply by -2}$$

$$21x + 6y = 15$$
$$\underline{-4x - 6y = -32}$$
$$\text{Add:} \quad 17x \qquad = -17$$
$$x = -1$$

(This is the x-coordinate of the solution to our system.)

step 2

Substitute -1 for x in either original equation to find the y-coordinate.

$$3y = 16 - 2x$$
$$3y = 16 - 2(-1)$$
$$3y = 16 + 2$$
$$3y = 18$$
$$y = 6$$

The solution (-1, 6) can be shown to satisfy both equations in the system. Consequently, the solution set is $\{(-1, 6)\}$.

Before considering additional illustrative examples, let us summarize the steps involved in the solution of a system of two equations in two variables by the elimination method.

Solving Systems by Elimination

1. Write the system in the form

$$a_1x + b_1y = c_1$$
$$a_2x + b_2y = c_2$$

2. If necessary, multiply either equation or both equations by appropriate numbers so that the coefficients of x or y will have a sum of zero.
3. Add the equations in step 2. The sum is an equation in one variable.
4. Solve the equation from step 3.
5. Substitute the value obtained in step 4 into either of the given equations and solve for the other variable.
6. Check the solution in both of the original equations.

Problem 5 Solve by the elimination method.

$$x - 10y = 64$$
$$3x - 14y = 90$$

Solution **step 1**

We can eliminate x by multiplying the first equation by -3 and leaving the second unchanged.

$x - 10y = 64$	multiply by -3	$-3x + 30y = -192$
$3x - 14y = 90$	no change	$3x - 14y = 90$
	Add:	$16y = -102$
		$y = -\dfrac{102}{16}$

Reducing the fraction, the y-coordinate of our solution is $-\dfrac{51}{8}$.

step 2

Substitution of this value back into either original equation in the system results in cumbersome arithmetic. Another option is to go back to the equations with integral coefficients and this time eliminate y instead of x.

$x - 10y = 64$	multiply by -7	$-7x + 70y = -448$
$3x - 14y = 90$	multiply by 5	$15x - 70y = 450$
	Add:	$8x = 2$
		$x = \dfrac{2}{8} = \dfrac{1}{4}$

The solution to our system is $\left(\frac{1}{4}, \frac{-51}{8}\right)$ and its solution set is $\left\{\left(\frac{1}{4}, \frac{-51}{8}\right)\right\}$.

Problem Set 7.2

Solve each system by the elimination method.

1. $x + y = 7$
   $x - y = 3$

2. $2x + y = 3$
   $x - y = 3$

3. $12x + 3y = 15$
   $2x - 3y = 13$

4. $2x + y = 3$
   $2x - 3y = -41$

5. $x - 2y = 5$
   $5x - y = -2$

6. $4x - 5y = 17$
   $2x + 3y = 3$

7. $2x - 9y = 5$
   $3x - 3y = 11$

8. $3x - 4y = 4$
   $2x + 2y = 12$

9. $3x - 7y = 1$
   $2x - 3y = -1$

10. $2x - 3y = 2$
    $5x + 4y = 51$

11. $4x + y = 2$
    $2x - 3y = 8$

12. $3x + 4y = 16$
    $5x + 3y = 12$

Review Competencies

13. The tens digit of a two-digit number is 3 more than the units digit. If the number itself is 17 times the units digit, what is the number?

14. Select the numeral for $\left(3 \cdot \frac{1}{10^2}\right) + \left(5 \cdot \frac{1}{10^4}\right)$.

    a. 0.0305   b. 0.3005

    c. 3.00005   d. 0.35

15. Identify the symbol that should be placed in the box to form a true statement.

$$\frac{1}{3} \;\Box\; 0.3$$

a. =   b. <   c. >

16. Choose the expression that is equivalent to $8x^2 + 2y^3$

a. $2x^2 + 8y^3$ b. $2(y^3 + 4x^2)$

c. $2x^3 + 8y^2$ d. $(2x)^2 + 2y^3$

17. Choose the inequality equivalent to $6 - 2y < 6$.

a. $-2y > 0$ b. $-2y < 0$

c. $-2y > 12$ d. $-2y < 12$

As we shall now see, some systems of equations have no solution and others have infinitely many solutions.

Problem 6   Solve the system.

$$2x + 3y = 5$$
$$4x + 6y = 11$$

Solution   We will use the elimination method, eliminating x by multiplying the first equation by -2. We obtain the equivalent system

$$-4x - 6y = -10$$
$$4x + 6y = 11$$

Adding, we have $0 = 1$. Since this is a **contradiction**, the false statement $0 = 1$ indicates there is no solution to the system. The solution set for the system is the empty set, Ø. If we graph the two equations, the resulting lines would be parallel. Whenever both variables have been eliminated and the resulting statement is false, the solution set for the system is Ø. Such a system is called an **inconsistent** system.

Problem 7   Solve the system.

$$y = 3 - 2x \quad (1)$$
$$4x + 2y = 6 \quad (2)$$

Solution   Using the substitution method and substituting the expression $3 - 2x$ for y in the second equation, we obtain

$$4x + 2(3 - 2x) = 6$$
$$4x + 6 - 4x = 6$$
$$6 = 6$$

Both variables have been eliminated and the resulting statement $6 = 6$ is true. This identity indicates that the system has infinitely many solutions. Such a system is called a **dependent system**. The lines representing the graphs of the equations coincide. Any ordered pair that satisfies the first equation also satisfies the second equation. Such ordered pairs include (0, 3), (1, 1), (2, -1), and so on. Thus the solution set consists of all ordered pairs that satisfy either equation. We write this as $\{(x, y) \mid y = 3 - 2x\}$ [the set of all ordered pairs (x, y) such that $y = 3 - 2x$] or $\{(x, y) \mid 4x + 2y = 6\}$ [the set of all ordered pairs (x, y) such that $4x + 2y = 6$].

Problem Set 7.3

Classify each system as inconsistent or dependent.

1. $x + 2y - 3 = 0$
   $12 = 8y + 4x$

2. $3x + 3y = 2$
   $2x + 2y = 3$

3. $0.2x - 0.5y = 0.1$
   $0.4x = y - 0.2$

4. $2x - y = -2$
   $4x - 2y = 5$

5. $3x - 2y = -2$
   $6x - 4y = 2$

6. $\frac{x}{3} + \frac{y}{5} = 15$
   $10x + 6y = 5$

7. $x + 2y = 3$
   $4x + 8y = 12$

Review Competencies

8. Find three consecutive integers such that the sum of the first and three times the second is five less than the third.

9. How many positive factors of 24 are even, less than 24, and also divisible by 4?

10. How much change should one receive from $1.00 after the purchase of $(30 - a)$ three-cent items?

    a. $10 - 3a$ b. $10 + 3a$
    c. $10 - a$ d. $70 + a$

Chapter 5 Summary

1. ## Principles Used to Solve Equations
    a. $a = b$ if and only if $a + c = b + c$.
        The same real number can be added to or subtracted from both members of an equation.
    b. $a = b$ if and only if $ac = bc$, $c \neq 0$.
        The same non-zero real number can multiply or divide both members of an equation.
    c. Some quadratic equations (equations of the form $ax^2 + bx + c = 0$, where $a \neq 0$ and a, b, and c are real numbers) can be solved by factoring using the property that if a product is zero, then at least one of the factors is zero. If $AB = 0$, then $A = 0$ or $B = 0$.
    d. All quadratic equations can be solved using the quadratic formula in which
        $$x = \frac{-b \pm \sqrt{b^2 - 4ac}}{2a}.$$

2. ## Principles Used to Solve Inequalities
    a. $a > b$ if and only if $a + c > b + c$.
        The same real number can be added to or subtracted from both members of an inequality.
    b. $a > b$ if and only if $ac > bc$ and $c > 0$.
        The same positive real number can multiply or divide both members of an inequality without affecting the direction of the inequality symbol.
    c. $a > b$ if and only if $ac < bc$ and $c < 0$.
        The same negative real number can multiply or divide both members of an inequality, reversing the direction of the inequality symbol.

3. ## Solving Verbal Problems
    a. Represent the unknown quantity by some variable, usually x.
    b. If there are additional unknown quantities, if possible represent these in terms of x.
    c. Formulate an equation expressing the given verbal relationships.
    d. Solve the equation.
    e. Consecutive integer problems can be solved by representing the integers by x, x + 1, x + 2, and so on. Consecutive even or consecutive odd integers are represented by x, x + 2, x + 4, and so on.
    f. If t represents the tens digit and u the units digit of a two-digit number, t + u is the sum of the digits and 10t + u is the number.

Chapter 5

4. Principle For Working With Proportions

If $\frac{a}{b} = \frac{c}{d}$, $b \neq 0$, $d \neq 0$, then $a \cdot d = b \cdot c$.

5. Variation

    a. $y = kx$ translates as y varies directly as x or y is proportional to x.

    b. $y = \frac{k}{x}$ translates as y varies inversely as x.

    c. $z = kxy$ translates as z varies jointly as x and y or z varies directly as x and y.

6. Graphing

    a.   The graph of $Ax + By = C$ ($A \neq 0$ and $B \neq 0$) is a line that is neither horizontal nor vertical. The equation can be graphed by finding the x-intercept (set $y = 0$) and the y-intercept (set $x = 0$) and then drawing the straight line through these two points. If these points are both the origin, another point is necessary to graph the equation.

    b. $x = C$ has a vertical line as its graph.

    c. $x = 0$ has the y-axis as its graph.

    d. $y = C$ has a horizontal line as its graph.

    e. $y = 0$ has the x-axis as its graph.

7. Solving Two Equations in Two Variables (x and y)

    a.   The substitution method works well when one of the variables has a coefficient of 1. Solve for this variable in terms of the other variable and substitute this expression into the other equation.

    b.   The elimination method involves adding the equations in order to eliminate a variable. It is often necessary to multiply one or both equations by a nonzero number to create the situation where coefficients of a variable are additive inverses.

    c.   If either method results in a contradiction, such as $0 = 3$, the system has no solution and is called inconsistent. The solution set is $\emptyset$.

    d.   If either method results in a statement of the form $a = a$, such as $3 = 3$, the system has infinitely many solutions and is called dependent.

Chapter 5 Self-Test

1. If $21 - (13 - 2x) = 16 - (13x - 7)$, then
   a. $x = \frac{1}{11}$
   b. $x = -\frac{15}{11}$
   c. $x = \frac{15}{11}$
   d. $x = 1$

2. If $3(2b - 1) \geq 9b - 15$, then
   a. $b \geq 4$
   b. $b \leq 4$
   c. $b \leq \frac{14}{3}$
   d. $b \leq \frac{6}{5}$

3. If $2[x - 3(2x - 1)] = x - 2$, then
   a. $x = -\frac{4}{11}$
   b. $x = -\frac{8}{11}$
   c. $x = \frac{5}{11}$
   d. $x = \frac{8}{11}$

4. If $y \geq y - (6 - 2y)$, then
   a. $y \leq 3$
   b. $y \geq 3$
   c. $y \leq -3$
   d. $y \geq -3$

5. Choose the inequality equivalent to $-4y - 4 > 6$.
   a. $-4y > 10$
   b. $-4y < 10$
   c. $-4y > 2$
   d. $-4y < 2$

6. Choose the inequality equivalent to $-3x < -12$.
   a. $x < -4$
   b. $x > -4$
   c. $x < 4$
   d. $x > 4$

7. Choose the equation equivalent to $4x - 7 = 3x - 9$.
   a. $4x = 3x - 16$
   b. $7x - 7 = -9$
   c. $x - 7 = -9$
   d. $4x + 9 = 3x - 7$

8. Choose the inequality equivalent to $-12 < x - 5 < 9$.
   a. $-17 < x < 14$
   b. $-7 > x > 14$
   c. $-7 < x < 14$
   d. $-17 < x < 4$

9. If $y < 0$, then $y^3 < xy^2 + y$ is equivalent to which of the following?
   a. $y^2 > xy + 1$
   b. $y^2 < xy + 1$
   c. $y^2 > xy + y$
   d. $y^2 < xy + y$

10. Identify the symbol that should be placed in the box to form a true statement.
    $$-\frac{5}{6} \ \square \ -0.1$$
    a. $=$
    b. $>$
    c. $<$

11. Identify the symbol that should be placed in the box to form a true statement.
    $$4.12 \ \square \ \sqrt{15}$$
    a. $=$
    b. $>$
    c. $<$

12. Identify the symbol that should be placed in the box to form a true statement.
    $$2.6\overline{8} \ \square \ 2.\overline{68}$$
    a. $=$
    b. $>$
    c. $<$

13. Which is a linear factor of the following expression?
    $$8x^2 - 22x + 15$$
    a. $2x - 5$
    b. $4x - 5$
    c. $8x - 3$
    d. $4x + 3$

14. Find the correct solutions to this equation:
    $4x^2 = 8x - 1$
    a. $1 + \sqrt{3}$ and $1 - \sqrt{3}$
    b. $\frac{2+\sqrt{3}}{2}$ and $\frac{2-\sqrt{3}}{2}$
    c. $1 + 4\sqrt{3}$ and $1 - 4\sqrt{3}$
    d. $\frac{-2+\sqrt{3}}{2}$ and $\frac{-2-\sqrt{3}}{2}$

15. Find the correct solutions to this equation:
    $6y^2 = 7y + 3$
    a. $\frac{7+\sqrt{120}}{12}$ and $\frac{7-\sqrt{120}}{12}$
    b. $\frac{-7+\sqrt{120}}{12}$ and $\frac{-7-\sqrt{120}}{12}$
    c. $\frac{3}{2}$ and $-\frac{1}{3}$
    d. $-\frac{3}{2}$ and $\frac{1}{3}$

16. Which of the following has/have $\frac{1}{4}$ as a solution?

    i. $y^2 = \frac{1}{8}$ ii. $5x - \frac{5}{4} = 0$ iii. $8y - 2 \ \ 0$
    a. i only   b. ii only
    c. iii only   d. ii and iii only

17. Which of the following has/have (-2) as a solution?
    i. $2x - 4 = 0$ ii. $(x + 4)(x + 2) < 0$
    iii. $3x^2 + 5x + 6 = 8$
    a. i only   b. ii only
    c. iii only   d. ii and iii only

18. The difference between two numbers is 7 and their product is 120. What equation can be used to find one of the numbers x?
    a. $x(x - 7) = 120$ b. $x(7 - x) = 120$
    c. $x - (x - 7) = 120$ d. $x + (x - 7) = 120$

19. The sum of three consecutive odd integers is five more than twice the middle integer. Select the equaton that can be used to find x, the smallest integer.
    a. $x + (x + 2) + (x + 4) = 2(x + 2) + 5$
    b. $x + (x + 2) + (x + 4) + 5 = 2(x + 2)$
    c. $x + (x + 1) + (x + 2) = 2(x + 1) + 5$
    d. $x + (x + 1) + (x + 2) + 5 = 2(x + 1)$

20. Choose the algebraic description that is equivalent to the verbal description: For any three consecutive positive integers, the cube of the middle integer is greater than the product of the smallest integer, x, and the largest integer.
    a. $(x + 2)^3 > x(x + 1)$
    b. $x^3 + 1 > x(x + 2)$
    c. $(x + 1)^3 > x^2 + 2x$
    d. $(x + 1)(x + 2) > x^3$

21. A two-digit positive integer is equal to 5 times the sum of its digits. If t represents the tens digit and u the units digit, which equation is equivalent to this verbal description?
    a. $5(10t + u) = t + u$

    b. $tu = 5(t + u)$
    c. $10t + u = 5(t + u)$
    d. $5tu = t + u$

22. The tens digit of a two-digit number is 3 more than the units digit. If the number itself is 21 times the units digit, what equation can be used to find t, the tens digit?
    a. $10(t + 3) + (t - 3) = 21t$
    b. $10t + (t - 3) = 21(t - 3)$
    c. $10(t + 3) + (t - 3) = 21$
    d. $10t + (t - 3) = 21$

23. Jerry spends 45% of his monthly income on housing. One particular month he spent $3,150 on housing. What was his income that month?
    a. $1,417.50   b. $8,000.00
    c. $10,150.00
    d. The correct answer is not given.

24. 32 is 60% of what number?
    a. 1.92   b. 19.2
    c. $53.\overline{3}$ d. 192

25. Two machines can complete 9 tasks every 6 days. Let t represent the number of tasks these machines can complete in a 30-day month. Select the correct statement of the given condition.
    a. $\frac{6}{9} = \frac{t}{30}$ b. $\frac{t}{9} = \frac{6}{30}$
    c. $\frac{t}{6} = \frac{9}{30}$ d. $\frac{9}{6} = \frac{t}{30}$

26. A seamstress can sew 3 dresses in 2 days. An assistant works half as fast. In how many days can the assistant sew 9 dresses?
    a. 3   b. 6   c. 9   d. 12

27. A forest service catches, tags, and then releases 19 bears back into a park. Two weeks later they select a sample of 156 bears and find that 13 of them are tagged. Assuming that the ratio of tagged bears in the sample is proportional to the actual ratio in the park, approximately how

many bears are in the park?
a. 114   b. 200   c. 228   d. 456

28. The lift of an airplane wing varies directly as the square of its width. A wing whose width is 5 inches has a lift of 35 pounds. Find the lift when the wing's width is 20 inches.
a. 70 pounds   b. 140 pounds
c. 540 pounds   d. 560 pounds

29. The number of pens sold varies inversely as the price per pen. If 4,000 pens are sold at a price of $1.50 each, how many pens will be sold at a price of $1.20?
a. 2,000   b. 3,000
c. 5,000   d. 6,000

30. Which shaded region identifies the portion of the plane in which $x \geq 2$ and $y \leq 0$?

a.
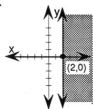
(2,0)

b.
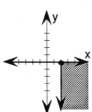

c.
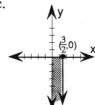
$(\frac{3}{2},0)$

d.
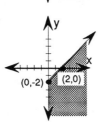
(0,-2)   (2,0)

31. Which option gives the condition(s) that correspond(s) to the shaded region of the plane illustrated at right?
a. $x \leq -2$ and $y \leq -3$
b. $x \geq -2$ and $y \leq 3$
c. $x \leq -2$ and $y \leq 3$
d. $x \geq -2$ and $y \leq -3$
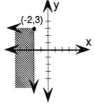
(-2,3)

32. Which option gives the condition(s) that correspond(s) to the shaded region of the plane illustrated at the right?
a. $y < x + 3$
b. $y \leq x + 3$
c. $x < 3$ and $y > 3$
d. $x < 3$, $y > 3$, and $y > x$

(0,3)
(0,-3)

33. Identify the conditions which correspond to the shaded region of the plane.
a. $x \geq -4$ and $x \geq 2$
b. $y \geq -4$ and $y \geq 2$
c. $-4 \leq x \leq 2$
d. $-4 \leq y \leq 2$
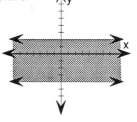

34. Choose the correct solution set for the system of linear equations. $y = 4x - 5$
$5x - y = 11$
a. {(-6, -41)}   b. {(6, 19)}
c. $\{(\frac{2}{3}, \frac{-23}{3})\}$ d. Ø

35. Choose the correct solution set for the system of linear equations. $3x - 2y = -7$
$2x - 5y = -12$
a. {(1, -2)}   b. {(2, -1)}
c. {(-1, 2)}   d. Ø

36. Choose the correct solution set for the system of linear equations. $3x + y = -7$
$2y = -14 - 6x$
a. {(1, -10)}   b. {(2, -13)}
c. $\{(x, y) \mid y = \frac{1}{2}(-14 - 6x)\}$ d. Ø

37. Choose the correct solution set for the system of linear equations. $2x - 3y = 1$
$-4x + 6y = 5$
a. $\{(1, \frac{1}{3})\}$ b. $\{(\frac{1}{2}, \frac{7}{6})\}$
c. $\{(x, y) \mid 2x - 3y = 1)\}$ d. Ø

six

Informal Geometry

The principles of geometry presented today in many high schools are basically those originally established by the Greek mathematician, Euclid, more than 2,000 years ago. These principles comprise what has come to be called Euclidean geometry. This formal geometry is approached in a logical manner. Specific statements (theorems) are proved by reasoning while other statements (axioms or postulates) are accepted without proof.

Another type of geometry will be of interest to us in this chapter. The approach is based on observation and intuition. This geometry may be called **informal** as opposed to the formal geometry presented in the secondary grades. In this chapter, we will consider various geometric figures and the properties they possess.

Section 1   Points, Lines, and their Properties

The basic terms used in geometry include **point, line,** and **plane.** It is tempting to try to define these terms, but, in fact, it is impossible to define such basic words. In any mathematical system, an attempt to define every word takes one in circles. For example, Euclid defined **point** as that which has no part, but never defined what was meant by part.

In mathematics today, it is recognized that some basic, primitive notions must be accepted without definition. Then, these basic terms can be used as building blocks in the development of a mathematics vocabulary. These notions are the **undefined terms** of our informal geometry.

Point is an undefined term, but a point may be thought of as a precise location or some fixed position. A point has no size or dimension. It is invisible. However, a point can be represented on paper by drawing a dot. The representation for a point shown at the right and labeled A can be seen, but the point itself has no size. Points are frequently denoted by capital letters, and the representation above could be referred to as point A.

The notion of **line** will be another undefined term which depends upon the intuitive ideas of line. The figure at the right is a representation of a line, and the small letter m indicates it could be referred to as line m. The arrowheads indicate that the line continues even though the drawing stops. In general, the word **line means straight line** and should be interpreted that way unless specifically directed otherwise.

A line is a one-dimensional figure. It has length, but no width because its width is the same as the width of a point. Since our attempts to draw other geometric figures involve points and lines, it must be recognized that every such drawing is only a representation for its figure.

Lines possess certain properties even though the term **line** is not defined.

> ### Property 1
> A line is a set of points.

Each point of the line is **on** the line. Consider the line at the right. The point A is on the line.

A line is an infinite set of points and appears as the **real number line**. The real number line is useful in algebra. Every point on the line has a real number associated with it. Every real number has a corresponding point on the line.

> ### Property 2
> Exactly one line may be determined by any two distinct points.

Two distinct points are represented at the right. This paper could be creased in exactly **one** straight line so that the crease passes through both points A and B. This straight line is symbolized as $\overleftrightarrow{AB}$ and read as **line AB.** Thus, a line may be named by any two of its points. Notice that $\overleftrightarrow{AB}$ is the same line as $\overleftrightarrow{BA}$.

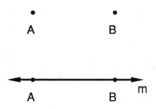

> ### Property 3
> For any three distinct points on a line, exactly one of the three points is between the other two points.

Consider the line at the right with the three distinct points A, B, and C. Notice that point B is between the other two points A and C. This property is often called the **betweenness property** for points on a line. The idea of betweenness is an intuitive one and can be considered an undefined term. Certainly there are points on a line that are between any other two points no matter how closely they appear to each other.

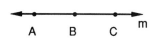

Note

It is customary to speak of three points that lie on one line as **collinear** points. **Noncollinear** points do not lie on the same line. Notice that the line through the three collinear points above may be expressed as:

$\overleftrightarrow{AB}$, $\overleftrightarrow{BA}$, $\overleftrightarrow{AC}$, $\overleftrightarrow{CA}$, $\overleftrightarrow{BC}$, or $\overleftrightarrow{CB}$.

Property 4

Any point on a line separates the line into three separate parts: the point itself and two **half-lines**, one on each side of the point.

Consider the line at the right in which point C is between points A and B. This property (known as the **separation property** for lines) says that point C separates the line into **three** separate and distinct parts. The point C may be thought of as the boundary between the two half-lines though it is **not** part of either half-line. Observe that the direction of one half-line is opposite of the direction of the other half-line.

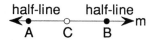

In the next section we discuss half-lines more fully.

Property 5

Two distinct lines are said to **intersect** if they share exactly one point.

Examples

1. Observe that the distinct lines m_1 and m_2 at the right intersect in exactly one point. Point 0 is the one and only point shared by the two lines.

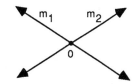

2. Observe that the lines l_1 and l_2 at the right are said to intersect. Remember that a line continues indefinitely. The two lines will intersect somewhere.

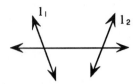

Problem Set 1.1

1. Why is it necessary that any two points be distinct to determine exactly one line?

2. In the representation below,

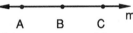

   a. Is $\overleftrightarrow{AB}$ the same as $\overleftrightarrow{AC}$?
   b. Is $\overleftrightarrow{AB}$ the same as $\overleftrightarrow{BA}$? Is $\overleftrightarrow{AB}$ the same as $\overleftrightarrow{BC}$?
   c. How many different names may be formed for this line using the points A, B, and C?
   d. Point B separates the line. Do the two resulting half-lines have any points in common?
   e. Since point B separates the line, is the set of points of the two half-lines equivalent to the set of points in the original line?

3. How many lines are determined by any three noncollinear points?

4. How many lines are determined by any four points? Assume that no three of the points are collinear.

5. Is it true that any line has two endpoints?

6. Is it possible that two lines may have more than one point in common?

Review Competencies

7. Add: $\frac{5}{12} + \frac{5}{18}$

8. Multiply: $2\frac{1}{2} \cdot \frac{1}{5}$

9. Divide: $3\frac{1}{3} \div 6$

10. Convert 123_5 to a base ten numeral. Consult Chapter 4, Section 4 for help if needed.

Section 2   Planes and their Properties

What is a **plane**? Intuitively we might describe it as a smooth surface. But what do these terms mean? We will let the word plane be another of our undefined terms. We can draw a representation of a plane and denote it by the letter P (read: plane P) as shown at the right. There are obvious limitations to our drawing. A plane has no thickness, and it has no boundaries. It is a two-dimensional surface that extends indefinitely. The boundaries in our drawing do not indicate this. Actually we have only represented a portion of a plane. It is not possible to reach the edge of an idealized plane.

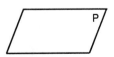

Planes possess certain properties even though the term **plane** is not defined.

> ## Property 1
> A plane is a set of noncollinear points.

Every point in the plane is said to be **on** the plane. Obviously, a plane contains an infinite set of points.

> ## Property 2
> Exactly one plane may be determined by any three distinct, noncollinear points.

There is only one plane that passes through the three points A, B, and C shown at the right. Each of the points lies on the plane.

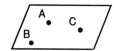

> ## Note
> Have you ever noticed that a chair with only three legs will always rest firmly on the floor while one with four legs might wobble? The ends of the three legs all lie on the plane of the floor.

> ## Property 3
> Any line on a plane separates the plane into three separate parts: the line itself and two **half-planes**, one on each side of the line.

Consider the line m on plane P shown at the right. The **separation property** for planes says that line m separates the plane into three separate parts. The line m may be considered the boundary (called the edge) between the two half-planes, but is **not** part of either half-plane.

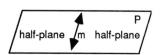

> ## Property 4
> Two distinct planes have either a line in common or no points in common.

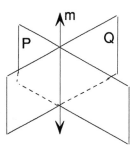

Observe the intersection of planes P and Q at the right. The points shared by these two planes may be represented by all the points on the line m.

In the figure at the right plane P does not intersect plane Q. Plane P is said to be **parallel** to plane Q.

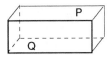

Consider the intersection of line m and plane P shown at the right. The **dashed line** indicates the portion of the line on the other side of the plane. Notice that point C is shared by both the line and the plane.

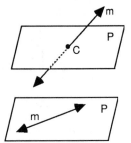

Another possibility is that the intersection of line m and plane P is the entire line itself. All points of the line are shared with the plane.

Property 6

A line that does not intersect a plane is said to be **parallel** to that plane.

Example

Line m does not intersect plane P at any point in the drawing shown at the right. Thus, line m is parallel to plane P.

Property 7

Lines in the same plane that do not intersect are called **parallel lines**.

Line m_1 is parallel to line m_2 in plane P since they do not intersect at any point no matter how far the lines are extended. Only one plane may contain these parallel lines.

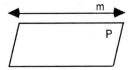

Note

Lines that lie in the same plane are said to be **coplanar**. Lines that are noncoplanar are **skew** lines. Skew lines are lines that do not lie in the same plane; they are not parallel even though they never intersect each other. Consider the figure at the right and the skew lines m_1 and m_2.

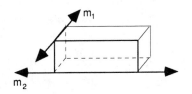

Problem Set 2.1

1. Why is it necessary that three points that determine a plane should not be on the same straight line?

2. Is it possible that two planes could intersect in only one point?

3. Is it true that if any two points lie in a plane, then the line determined by those points also lies in the plane?

4. Is it true that any two parallel lines must be coplanar, but any two coplanar lines need not be parallel?

5. Can skew lines be coplanar?

6. Can a point separate a plane?

7. How many planes may contain the same point?

8. How many planes may contain the same line?

Review Competencies

9. In the figure below is $\overleftrightarrow{XY}$ the same as $\overleftrightarrow{YZ}$?

10. How many lines are determined by any five points, no three of them collinear?

11. Subtract: $5 - 1.003$

12. Divide: $3.6 \div 0.12$

13. Express 35% as a fraction in simplest form.

14. Express $4\frac{1}{2}$ % as a decimal.

15. Express 0.6 as a fraction in lowest terms.

Section 3   Half-Lines, Rays, Line Segments, and Angles

In this section, we discuss some important sets of points and their symbolic representations.

We know that point C on line m at the right separates the line into half-lines. Point C, the separation point, is not a part of either half-line, yet is called the **endpoint** of either half-line. The symbol for the half-line to the **right** of point C is $\overset{\circ}{C}\overrightarrow{B}$ (read **half-line CB**), and the symbol for the half-line to the **left** of point C is $\overset{\circ}{C}\overleftarrow{A}$. Notice that the little circle at the end of the arrow indicates that point C is **not** contained in either half-line. Notice also that $\overset{\circ}{C}\overrightarrow{B}$ does **not** represent the same set of points on line m as $\overset{\circ}{B}\overrightarrow{C}$.

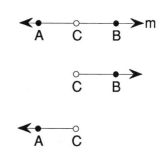

We now consider a set of points in which the separation point is included with the half-line. Such a set of points is called a **ray**.

> ### Definition
> A **ray** is the union of a half-line and its endpoint.

In the figure at the right, the ray whose endpoint is C and extends to the **right** is read ray CB and is denoted by $\overrightarrow{CB}$. The ray whose endpoint is C and extends to the **left** is read **ray CA** and is denoted by $\overrightarrow{CA}$. The first letter in the symbol indicates the endpoint and the second letter indicates any point on the ray. Notice that $\overrightarrow{CB}$ and $\overrightarrow{CA}$ name different sets of points as do $\overrightarrow{CB}$ and $\overrightarrow{BC}$.

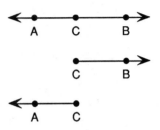

Note

A ray has an endpoint and proceeds indefinitely along the line in one direction. Any point on a line may be an endpoint and there are two different rays emanating in either direction from that endpoint.

Problem 1   Given a line with four points marked A, B, C, and D, find $\overrightarrow{AB} \cap \overrightarrow{CD}$. ($\cap$, called the **intersection**, is the set of the points that lie on both $\overrightarrow{AB}$ and $\overrightarrow{CD}$.)

Solution   **step 1**
Represent $\overrightarrow{AB}$.

step 2
Represent $\overrightarrow{CD}$.

step 3
$\overrightarrow{AB} \cap \overrightarrow{CD}$ is represented by $\overrightarrow{CD}$.

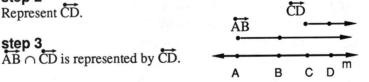

Thus, $\overrightarrow{AB} \cap \overrightarrow{CD} = \overrightarrow{CD}$. The points in $\overrightarrow{CD}$ are common to both $\overrightarrow{AB}$ and $\overrightarrow{CD}$.

Problem Set 3.1

1. Use the line m and its marked points to complete the following:

   a. $\overleftrightarrow{FK} \cap \overleftrightarrow{NK}$
   b. $\overleftrightarrow{NF} \cup \overleftrightarrow{NK}$ ($\cup$, called the **union**, represents the points that lie either on $\overleftrightarrow{NF}$ or on $\overleftrightarrow{NK}$.)
   c. $\overleftrightarrow{KN} \cap \overleftrightarrow{NF}$
   d. $\overleftrightarrow{FI} \cup \overleftrightarrow{FI}$
   e. $\overleftrightarrow{NF} \cap \overleftrightarrow{NK}$

2. If $\overrightarrow{AB} = \overrightarrow{AC}$, is it necessarily true that point B coincides with point C?

Review Competencies

3. Use the diagram below to determine which of the following is true:

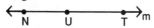

   a. Point U separates $\overleftrightarrow{NT}$ into two rays.
   b. Point T is part of $\overleftrightarrow{UN}$.

4. How many planes may contain the same line?

5. Can skew lines be coplanar?

6. Does a line have endpoints?

7. How many lines are determined by any four points, no three of them collinear?

8. Express 0.09 as a percent.

9. Subtract: $3\frac{1}{4} - 1\frac{1}{3}$

10. What is 7% of 30?

11. Express 0.12 as a fraction in lowest terms.

12. Multiply: $\sqrt{3} \cdot \sqrt{6}$

13. Simplify: $\sqrt{12} + \sqrt{27}$

Problem 2 Consider now the line m at the right.

Find $\overleftrightarrow{AB} \cap \overleftrightarrow{BA}$.

Solution **step 1**
Represent $\overleftrightarrow{AB}$.

step 2
Represent $\overleftrightarrow{BA}$.

step 3
$\overleftrightarrow{AB} \cap \overleftrightarrow{BA}$ is the portion of the line containing points A and B and the set of points between them. Such a set of points is called a **line segment**.

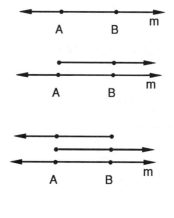

345

Definition

A **line segment** is a portion of a line that consists of
two endpoints and all the points between them.

Example

Observe the geometric
representation of line
segment RS at the right:

We use the symbol $\overline{RS}$ to represent the line segment RS. Points R and S are
endpoints. Notice that $\overline{RS} = \overline{SR}$ since both represent the same set of points.
Notice that $\overset{\circ}{RS}$ represents the set of points between points R and S but not
including points R and S. Observe also that a line segment has a definite
length. It does not extend indefinitely in one direction or the other.

Problem 3 Use the line at the right
and find $\overline{BC} \cap \overline{CD}$.

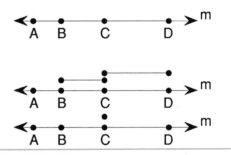

Solution The set of points common to
these two line segments is the
set containing only the point C.
Thus, $\overline{BC} \cap \overline{CD}$ = point C.

Problem Set 3.2

1. Does $\overleftrightarrow{AB}$ include the points which $\overrightarrow{AB}$ and $\overrightarrow{BA}$
   have in common?

2. Determine which of the following is true or
   false given the line below.

   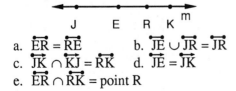

   a. $\overline{ER} = \overline{RE}$ b. $\overrightarrow{JE} \cup \overrightarrow{JR} = \overrightarrow{JR}$
   c. $\overrightarrow{JK} \cap \overrightarrow{KJ} = \overline{RK}$ d. $\overrightarrow{JE} = \overrightarrow{JK}$
   e. $\overrightarrow{ER} \cap \overrightarrow{RK}$ = point R

3. Is it true that a line segment is an infinite set of
   points?

Review Competencies

4. Use the line below to find:

   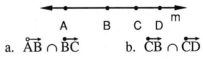

   a. $\overline{AB} \cap \overline{BC}$ b. $\overleftrightarrow{CB} \cap \overleftrightarrow{CD}$

5. Must two parallel lines also be coplanar?

6. If a point separates a line, do the resulting half-
   lines have any common points?

7. $\frac{6}{10} = \frac{?}{30}$

8. Evaluate: $5 \cdot 2 \div 2 + 4 \cdot 3 - 1$

9. Find the smallest positive multiple of 6 which leaves a remainder of 2 when divided by 8 and a remainder of 4 when divided by 7.

Observe the geometric figures below in which two distinct rays ($\overrightarrow{DP}$ and $\overrightarrow{DQ}$) share a common endpoint D. Each of these figures is an example of a **plane angle**.

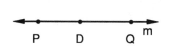

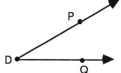

> ### Definition
> A **plane angle** is the set of points formed by the union of two rays that have a common endpoint. The common endpoint is called the **vertex** of the angle.

Examples

1. The angle at the right is **angle PDQ** (denoted ∠PDQ) or **angle QDP** (denoted ∠QDP) or simply ∠D. The letter representing the vertex of the angle is the **middle** letter when three letters name the angle. Each ray is a side of the angle.

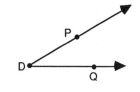

2. The angle with vertex D, shown at the right, has its rays (sides) on the same straight line. Thus, ∠PDQ is called a **straight angle**.

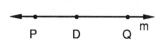

There is a **separation property** that relates to angles. Notice in the figure at the right that an angle separates a plane into three separate and distinct parts. There is the set of points outside the angle which is the **exterior** of the angle, the set of points inside the angle which is the **interior** of the angle, and there is the set of points on the angle itself. Remember that there is no point that lies in the exterior or the interior that also lies on the angle itself. A straight angle has no interior or exterior.

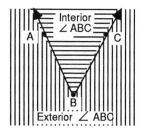

Problem 4
Consider the plane figure at the right and determine the set of points represented by:
(Exterior ∠BCD) ∩ B⃗A

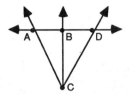

Solution

step 1
Shade the area representing exterior ∠BCD remembering that this region does not touch any points of ∠BCD.

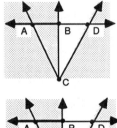

step 2
Represent B⃗A.

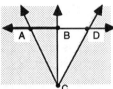

step 3
Observe that all the points of the ray are in common with points in exterior ∠BCD except point B which is part of ∠BCD.
(Exterior ∠BCD) ∩ B⃗A = B⃗A.

Problem Set 3.3

1. Determine each of the following sets of points using the accompanying figure.

   a. ∠ZXY ∩ Z⃡Y
   b. X⃡Y ∪ X⃡Z
   c. ∠XZY ∩ ∠ZYX
   d. X⃡Z ∩ Y⃡Z

2. Consider the figure below and determine the set of points in each of the following.

   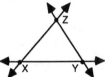

   a. (Exterior ∠HAT) ∩ (Exterior ∠MAT)
   b. (Interior ∠HAM) ∩ A⃡T

3. In each of the following identify the set of points using the accompanying figure.
   a. ∠DCJ ∩ P⃡U
   b. (Interior ∠DCJ) ∪ P⃡U

4. Is the union of the interior of an angle and the exterior of an angle the same identical set of points as the entire plane?

5. Consider the accompanying line and determine:
   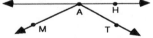
   a. A⃡B ∩ C⃡D   b. A⃡D ∪ A⃗D

6. Can two planes intersect in just one point?

7. How many lines are determined by any three noncollinear points?

8. Express 120% as a decimal.

9. Evaluate: 4(2 • 3 ÷ 6 + 2)

10. Divide: 16 ÷ 0.004

11. Express $2\frac{3}{4}$% as a decimal.

12. Express 0.025 as a fraction in lowest terms.

Section 4 Curves and Polygons

If one is asked to take a pencil and draw a **curve** on paper, one would probably draw some line that is "curvy" or not straight, such as the one shown at the right. Notice that the pencil was not lifted from the paper as the curve was drawn. One continuous path was traced with no breaks. Can you see that this curve consists of a set of points that begins at a certain point and ends at a certain point? We may use the word curve in a more general mathematical sense.

> ## Definition
> A **curve** is a set of points that forms a continuous path on a plane.

Examples

1. The curve at the right intersects itself at some point other than the starting point. Some point is used more than once.

2. We may think of a straight line segment also as a curve. Thus, curves may contain straight portions. Any curve that consists of a series of connected line segments is called a **broken line**. Figures a, b, and d are broken lines.

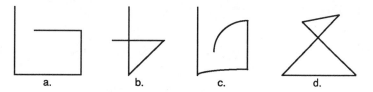

a.   b.   c.   d.

As you see, there are some curves that intersect themselves and there are others that do not.

> ## Definition
> A curve that does not intersect itself is a **simple curve**. A simple curve does not pass through any point more than once. A curve that does intersect itself is **not simple**.

Examples

1. The curves at the right are simple.

2. The curves at the right are not simple.

 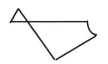

 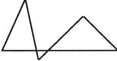

It is possible to begin tracing a curve at a starting point and eventually return to that very same point without lifting the pencil from the paper. Such a curve begins and ends at the same point. A curve in which the endpoints coincide is called a **closed curve**.

Examples

1. The curves at the right are closed curves.

2. The curves at the right are not closed.

> **Definition**
> A **simple closed curve** is one whose endpoints coincide and which does not intersect itself.

Examples

The figures at the right are simple closed curves. The simple closed curves in figures a and b are closed broken lines.

a.

b.

c.

Note

A simple closed curve does not cross its own path. No point, other than the starting point, is used more than once.

Problem Set 4.1

1. Is the accompanying figure a curve? Why?

2. Is every closed curve also a simple curve?

3. Is every simple curve also a closed curve?

4. Why is the curve shown not a simple closed curve?

5. Why is the curve shown not a simple closed curve?

Review Competencies

6. Determine the following using the accompanying figure.

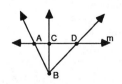

   a. ∠CBD ∩ $\overrightarrow{AD}$
   b. Interior ∠ABD ∩ Interior ∠ABC

7. Consider the accompanying line and determine:

   a. $\overrightarrow{QR} \cap \overrightarrow{RQ}$ b. $\overleftrightarrow{QS} \cap \overrightarrow{QR}$

8. What is the intersection of two distinct planes?

9. 10 is what percent of 40?

10. Express $\frac{19}{5}$ as a mixed number.

11. Multiply: $\frac{5}{6} \cdot 5$

12. Express 0.125 as a fraction in lowest terms.

13. Express 0.00041 in scientific notation.

Chapter 6

We are particularly interested in simple closed broken lines.

Examples

Each figure shown below is a polygon:

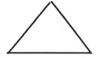  

There are names for polygons based upon the number of sides they contain. Below are some of these different polygons and the shapes they could assume. There are, of course, other possible shapes.

triangle	quadrilateral	pentagon	hexagon	octagon
(three sides)	(four sides)	(five sides)	(six sides)	(eight sides)

The following figures are examples of regular polygons. In each figure the sides are of equal length and the angles are of equal size.

 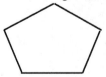

The vertices of a polygon are often labeled by capital letters. Thus, the accompanying polygon could be called **quadrilateral ABCD**. In the quadrilateral, ∠A and ∠B are said to be **consecutive angles**. Another pair of consecutive angles would be ∠B and∠C. In quadrilateral ABCD, ∠A and ∠C are **opposite angles**. Another pair of opposite angles would be ∠B and ∠D. In the same way, side AB and side DC are **opposite sides**.

A **diagonal** of a polygon is a line segment joining any two non-adjacent vertices of the polygon.

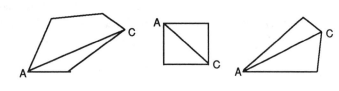

Observe the diagonal AC in each polygon above.

Let us now consider some special kinds of quadrilaterals.

1. A **parallelogram** is a quadrilateral containing two pairs of equal and parallel sides. The figure shown is an example of a parallelogram with diagonals. Any parallelogram has the following properties:
    a. a quadrilateral
    b. opposite sides are equal in length
    c. opposite sides are parallel
    d. all sides are not necessarily equal in length
    e. opposite angles are equal in measure
    f. all angles are not necessarily equal in measure
    g. diagonals are not necessarily equal in length
    h. diagonals bisect each other (intersect each other at their midpoints)

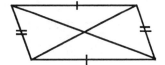

2. A **rectangle** is a parallelogram containing four angles that are equal in measure. These four angles are right angles. A right angle is indicated by ⊾. The figure shown is an example of a rectangle with diagonals. Any rectangle has the following properties:
    a. a quadrilateral
    b. a parallelogram
    c. opposite sides are parallel and equal in length
    d. all sides are not necessarily equal in length
    e. opposite angles are equal in measure
    f. all angles are equal in measure, and each angle is a right angle
    g. diagonals are equal in length and bisect each other

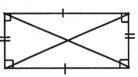

Notice that a rectangle includes the properties of a parallelogram.

3. A **square** is a rectangle with four equal sides. The square is the only regular polygon (all sides and angles equal) among the quadrilaterals. The figure shown is an example of a square with diagonals. Any square has the following properties.

   a. a quadrilateral
   b. a parallelogram
   c. a rectangle
   d. opposite sides are parallel and equal in length
   e. all sides are equal in length
   f. opposite angles are equal in measure
   g. all angles are right angles
   h. diagonals are equal in length and bisect each other
   i. diagonals form right angles with each other

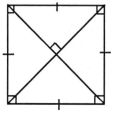

A square has the properties of a rectangle and a parallelogram and some additional properties.

4. A **rhombus** is a parallelogram containing four equal sides. The figure shown is a rhombus with diagonals. (A square can be defined as a rhombus with four right angles.) Any rhombus has the following properties:

   a. a quadrilateral
   b. a parallelogram
   c. all sides are equal in length
   d. opposite sides are parallel and equal in length
   e. all angles are not necessarily equal in measure
   f. diagonals are not necessarily equal in length
   g. diagonals form right angles with each other

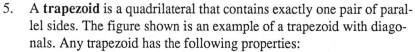

5. A **trapezoid** is a quadrilateral that contains exactly one pair of parallel sides. The figure shown is an example of a trapezoid with diagonals. Any trapezoid has the following properties:

   a. a quadrilateral
   b. exactly one pair of parallel sides
   c. non-parallel sides not necessarily equal in length
   d. parallel sides not equal in length
   e. opposite angles not equal in measure
   f. diagonals not necessarily equal in length

Problem 1   For each of the figures shown, describe it by using two or more names from the list: quadrilateral, parallelogram, rectangle, rhombus, square, and trapezoid.

Solutions

1. Quadrilateral
   Trapezoid

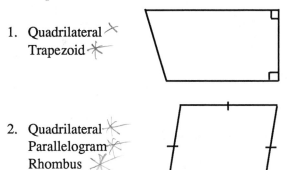

2. Quadrilateral
   Parallelogram
   Rhombus

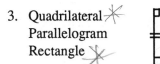

3. Quadrilateral
   Parallelogram
   Rectangle

Problem 2   Select the geometric figure that does not possess all of the characteristics of:
1. quadrilateral
2. opposite sides equal in length
3. opposite angles equal in measure
   a. square
   b. parallelogram
   c. pentagon
   d. rhombus

Solution   Letter c is the correct choice. The pentagon contains five sides and is not a quadrilateral. The polygons in choices a, b, and d possess all of the characteristics.

Chapter 6

Problem Set 4.2

1. Should a polygon have as many sides as vertices? *yes*

2. Why is it necessary that a polygon contain at least three sides? *cant with 2 sides it won't be a close plane*

3. Every polygon is a simple closed curve. Sketch a figure that illustrates that not every simple closed curve is a polygon.

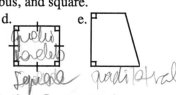

4. Is the curve shown a polygon? Why? *because no clos plane*

5. Describe each figure shown by selecting as many names as possible from the list: quadrilateral, parallelogram, rectangle, rhombus, and square.

a. *quadrilatral*
b. *quadilatral parallelogram*
c. *rectangle parallelogram rect*
d. *rhombus parallelogram square*
e. *quadrilatral*

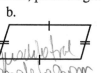

6. Which of the geometric figures is a quadrilateral with opposite sides equal and diagonals that form right angles?
   a. parallelogram   b. rectangle
   c. rhombus   d. triangle

7. Which of the geometric figures is a quadrilateral with opposite angles equal in measure and diagonals equal in length?
   a. hexagon   b. rhombus
   c. pentagon   d. rectangle

8. Which of the geometric figures does not possess both of the following characteristics?
   i. quadrilateral
   ii. diagonals bisect each other
   a. rectangle   b. square
   c. parallelogram   d. octagon

9. Select the geometric figure that possesses all of the following characteristics.
   i.   quadrilateral
   ii. all sides equal in length
   iii. opposite angles equal in measure
   a. square   b. hexagon
   c. rectangle   d. parallelogram

Review Competencies

10. Is the curve at right a simple curve?

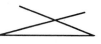

11. Is the curve at right a closed cuve?

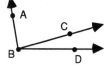

12. Use the figure at the right to find:
    a. Exterior $\angle CBD \cap \overrightarrow{BA}$
    b. $\overrightarrow{BC} \cup \overrightarrow{BD}$

13. How many planes may contain the same point?

14. How many distinct points are necessary to determine a plane?

15. Express $4\frac{1}{6}$ as an improper fraction.

16. Divide: $0.50 \div 0.025$

17. Express 65% as a fraction in lowest terms.

18. If a and b represent real numbers, state the commutative property for multiplication.

Definitions

The union of a simple closed curve and its interior is called a **closed region**. The curve itself is called the **boundary** of the region. The union of a polygon and its interior is a **polygonal region**.

The figures at the right are examples of closed regions.

The figures at the right are examples of polygonal regions.

Note

Observe that a polygon separates the entire plane into **three** separate parts. There is the set of points within the polygon which is known as the **interior** of the polygon. There is the set of points outside the polygon which is known as the **exterior** of the polygon. Finally, there is the set of points of the polygon itself. There is no point that is common to any two of these three sets of points.

Definitions

A simple closed curve is said to be **convex** if, for any two points A and B in the interior of the curve, the entire line segment AB lies in the interior. Any polygon that is not convex is said to be **concave**.

Examples

1. The curves at the right are convex.
   Figure b is a convex polygon.

a   b.

2. The following curves are **not** convex.
   Notice that the line segments do not
   always lie within the curve. These
   curves are said to be concave. Figure
   b is a concave polygon.

a   b.

The **triangle** is a convex polygon that is of special interest to us.
A triangle is a polygon formed by three line segments joining three
noncollinear points. Each segment is called a side of the triangle,
and each point of intersection of the segments is a **vertex**. We
generally name a triangle by its vertices. We may name the
pictured triangle at the right triangle ABC (denoted △ABC).

A triangle separates the points of the plane into three separate parts:

1. the set of points in the
   interior of the triangle.
2. the set of points in the
   exterior of the triangle.
3. the set of points of the
   triangle itself.

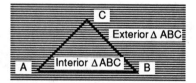

A **triangular region** is the union of the triangle and the interior of the
triangle.

Problem 3 Use the figure at the
right to find the set of
points representing:

(Interior △ACE) ∩ (Interior △BCD)

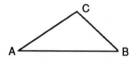

Solution **step 1**
Mark the set of
points representing
Interior △ACE using
vertical markings.

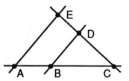

step 2

Mark the set of points representing
Interior ΔBCD using horizontal
markings.

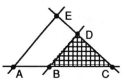

step 3

The answer is the intersection of the
two areas marked in Steps 1 and 2.
This common area is the same as
the interior of ΔBCD.

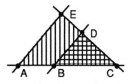

(Interior ΔACE) ∩ (Interior ΔBCD) = Interior ΔBCD.

Problem Set 4.3

1. Indicate which of the following is a polygonal
   region.

   a.   b.   c.

   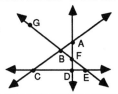

2. Indicate which of the following curves is not
   convex.

   a.   b.   c.

3. Is it possible to sketch a triangle that is not
   convex?

4. Consider the given
   figure and identify
   the set of points for:

   a. (Interior ΔACD) ∩ (Interior ΔABD)
   b. ΔADC ∩ $\overleftrightarrow{BC}$
   c. Interior ΔBCD ∪ Exterior ΔBCD ∪ ΔBCD

5. Consider the accompanying figure and identify
   the set of points for:

   a. (Interior ΔABF) ∪ (Interior ΔACD)
   b. ∠ACD ∩ ∠CDA
   c. (Interior ∠ACD) ∩ (Interior ΔABF)

Review Competencies

6. What is the name of a polygon containing
   five sides?

7. Is the figure shown
   a polygon?

8. Use the figure
   shown to
   determine:

   (Interior ∠CAB) ∩ (Interior ∠ACB)

9. What are parallel lines?

10. 20% of what number is 10?

11. Subtract: $4\frac{1}{5} - 2\frac{3}{5}$

12. Evaluate: $-3(2 + 6 - 3 \div 3)$

13. If a, b, and c represent real numbers, state the
    associative property for addition.

Chapter 6

Section 5   Solid Figures

Previous sections of this chapter have dealt with figures which lie entirely on one plane. Such figures are **plane figures**.

It is obvious that there are other figures which do not lie on one plane. Such figures are **solid** or **space figures**. Space figures have three dimensions usually referred to as length, width, and depth.

Because space figures have three dimensions, drawings to represent them on a piece of paper usually use dashed lines to create the illusion of depth. Four drawings of space figures are shown below.

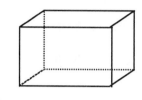  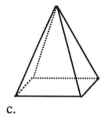 

a.   b.   c.   d.

The surface of each of the space figures above are closed, but not all of the surfaces in every figure are polygonal regions. Consider figure d. In this section, we are concerned with space figures whose surfaces are polygonal regions. Figures a, b and c all have surfaces that are polygonal regions.

> ## Definitions
> Any closed surface formed by the union of polygonal regions is called a **polyhedron**. Each polygonal region is called a **face** of the polyhedron. A **regular polyhedron** is one whose faces are all made up of one kind of regular polygon. The line segment formed by the intersection of two faces is called an **edge**. Each endpoint of the line segment is called a **vertex**.

In any polyhedron the following is true:

$$v - e + f = 2$$

in which: v = the number of vertices in the polyhedron

                    e = the number of edges in the polyhedron

                    f = the number of faces in the polyhedron

Example

Consider the polyhedron at the right. This polyhedron has eight vertices (A,B,C,D,E,F,G,H), twelve edges, and six faces. Can you identify them? We call this polyhedron a **rectangular solid** or a **rectangular parallelepiped**.

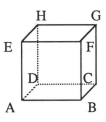

To verify that $v - e + f = 2$,
we see that $v = 8$, $e = 12$, and $f = 6$.
Thus: $8 - 12 + 6 = 2$.

Note

A polyhedron separates the set of points in space into three separate parts. There is the set of points on the surface of the polyhedron. There is also the set of points within the polyhedron (the interior) and the set of points outside the surface (the exterior).

The following polyhedrons are also important:

1. The **hexahedron (cube)**
   (All the faces are squares.)

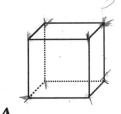

2. The **tetrahedron (triangular pyramid)**
   (All the faces are triangles.)

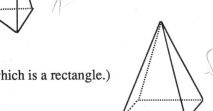

3. The **right pyramid (rectangular pyramid)**
   (All the faces are triangles except the base, which is a rectangle.)

4. The **right triangular prism**
   (Two faces are triangles.
   The remaining faces are rectangles.)

5. The **octahedron**
   (All the faces are triangles.)

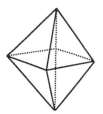

Problem Set 5.1

1. Consider the accom-
   panying tetrahedron:
   a. How many vertices are contained
      in this polyhedron? 4 How many of these
      vertices may all lie on the same plane? 3
   b. How many edges are contained in this
      polyhedron? 6
   c. How many faces are contained in this
      polyhedron? 4 Each face is formed by what
      kind of region? triangular region
   d. Verify that $v - e + f = 2$.
      $4 - 6 + 4 = 2$

2. Consider the accom-
   panying right pyramid:
   a. How many vertices are
      contained in this
      polyhedron? 5
   b. How many edges are contained in this
      polyhedron? 8
   c. How many faces are contained in this
      polyhedron? 5
   d. Verify that $v - e + f = 2$. ✓

3. Consider the accompanying
   right triangular prism:
   a. How many vertices are
      contained in this polyhedron? 6
   b. How many edges are contained in this
      polyhedron? 9
   c. How many faces are contained in this
      polyhedron? 5
   d. Verify that $v - e + f = 2$. ✓

4. Consider the accompanying
   octahedron:
   a. How many vertices are
      contained in this polyhedron? 6
   b. How many edges are contained
      in this polyhedron? 12
   c. How many faces are contained in this
      polyhedron? 8
   d. Verify that $v - e + f = 2$.

5. What is the smallest number of vertices that
   may be contained in a polyhedron? 4

6. What is the smallest number of edges that may
   be contained in a polyhedron? 6

7. What is the smallest number of faces that may
   be contained in a polyhedron? 4

Review Competencies

8. Is the curve shown below concave?

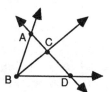

9. Is the figure below a closed curve?

10. Consider the
    accompanying
    figure and
    determine:

    Interior $\triangle ABC \cup$ Interior $\angle ABC$

11. How many sides does a hexagon contain?

12. Can two lines have more than one point in
    common?

13. Can two parallel lines be noncoplanar?

14. Express 1.5 as a percent.

15. Multiply: $1\frac{1}{5} \cdot 2\frac{2}{3}$

16. Express $\frac{35}{50}$ in simplest form.

17. Express $5\frac{3}{4}\%$ as a decimal.

18. Express 1.8 as an improper fraction.

Section 6   Measurement

When a set of distinct objects is counted, the result, assuming the count is accurately taken, is **exact**. However, when any object is measured, even assuming the accuracy of the measurer, the result is **approximate**. Counting results in exact answers; measurement results in approximate answers.

Every measurement consists of two parts: a numerical expression and a **unit of measure**. In the measurement 123 feet, the numerical expression is 123, and the unit of measure is one foot.

The smaller the unit of measure, the more accurate will be the measurement (again assuming that the measurer is equally careful). When measuring a line segment with a ruler marked off in $\frac{1}{4}$ inches, the unit of measure is $\frac{1}{4}$ inch, and the result is given to the nearest $\frac{1}{4}$ inch. A more accurate measurement could be made with a ruler marked off in $\frac{1}{8}$ inches. Then the unit of measure would be $\frac{1}{8}$ inch, and the result would be to the nearest $\frac{1}{8}$ inch.

The result obtained from measuring a line segment is called a **linear measurement** and is stated in **linear units**. A linear measurement is used to find a length.

Measurements are always approximations, and the degree of accuracy is dependent upon the unit of measure. Since the degree of accuracy of a measurement is the unit of measure, it is necessary to round the measurement and state it in terms of its unit of measure. This means that if a measurement is **less than half** of the unit, we simply round it to the next smaller unit. If a measurement is **one half or more** of the unit, we simply round it to the next larger unit. Thus, a measurement of $21\frac{1}{3}$ pounds (unit of measure: one pound) would be rounded to 21 pounds because $\frac{1}{3}$ is less than half of the unit.

Similarly, a measurement of 21.6 centimeters (unit of measure: one centimeter) would be rounded to 22 centimeters because 0.6 is more than half of the unit.

To round 6.635 centimeters to the nearest hundredth centimeter, look first at the hundredths digit, which is 3, and then look at the first digit to the right. If that digit is 0, 1, 2, 3, or 4, leave the hundredths digit unchanged. If the digit is 5, 6, 7, 8, or 9, increase the hundredths digit by 1. Since the digit to the right of the hundredths digit is 5, 6.635 rounded to the nearest hundredth is 6.64 centimeters.

Chapter 6

Two line segments are shown next to a ruler in the figure below.
Use the figure to find:

a. The measure of AB to the
   nearest $\frac{1}{2}$ inch.

b. The measure of CD to the
   nearest $\frac{1}{4}$ inch.

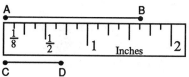

Solutions   a. $1\frac{1}{2}$ inches.   b. $\frac{3}{4}$ inches.

Problem Set 6.1

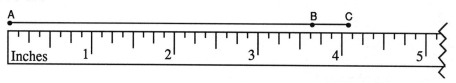

1. Round each of the following.
    a. 9 yd. 1 ft. (to the nearest yard)
        (Recall: 3 ft. = 1 yd.)
    b. 4 yd. 2 ft. (to the nearest yard)
        (Recall: 12 in. = 1 ft.)
    c. 3 ft. 7 in. (to the nearest foot)
    d. $3\frac{2}{3}$ yd. (to the nearest yard)
    e. 6.5 in. (to the nearest inch)
    f. 37.5 cm. (to the nearest centimeter)
    g. 5.6874 km. (to the nearest kilometer)
    h. 2,500.5 km. (to the nearest kilometer)
    i. 2.735 km. (to the nearest tenth of a
        kilometer)
    j. 24.2739 cm. (to the nearest hundredth of
        a centimeter)
    k. 62.3019 m. (to the nearest thousandth of
        a meter)
    l. 5,475 ft. (to the nearest hundred feet)

2. Use the ruler shown above. Find the
    measure of:
    a. AB to the nearest inch
    b. AB to the nearest $\frac{1}{2}$ inch
    c. AB to the nearest $\frac{1}{4}$ inch
    d. AB to the nearest $\frac{1}{8}$ inch
    e. AC to the nearest inch
    f. AC to the nearest $\frac{1}{2}$ inch
    g. AC to the nearest $\frac{1}{4}$ inch
    h. AC to the nearest $\frac{1}{8}$ inch

Review Competencies

3. How many faces
    are contained in
    the polyhedron
    below?

4. Is the figure
    shown below a
    convex curve?

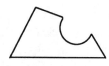

5. Should a polygon contain as many vertices
    as sides?

6. Consider the line below and find $\overleftrightarrow{BC} \cap \overleftrightarrow{BC}$.

7. Is the figure at the right a simple closed curve?

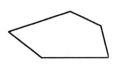

8. How many sides does an octagon contain?

9. Evaluate: $5 \cdot 3 + 4 \div 2 + 5 - 2$

10. Express 8% as a decimal.

11. Subtract: $10 - 2\frac{3}{5}$

12. Determine the common relationship between numbers in each pair. Then identify the missing term.

$(16, 4) \qquad (\frac{1}{9}, \frac{1}{3}) \qquad (1, 1) \qquad (0.25, \underline{\quad})$

13. Juan has $250 to purchase new tires. Tire prices are:
   Brand A: $42 each + 5% tax
   Brand B: $65 each + 5% tax
If Juan purchases two Brand A tires and two Brand B tires, how much change will he receive?

Section 7   Perimeter and Circumference

We are often interested in measuring the distance around a polygon. This distance is known as the **perimeter** of the polygon. The perimeter of a polygon is simply the sum of the lengths of its sides and is always stated in **linear** units.

Examples

1. The perimeter of the accompanying rectangle is:
   $5 + 5 + 3 + 3$ or 16 ft.

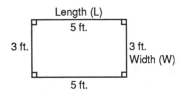

   If L = length and W = width, the perimeter (P) of a rectangle is:
   $P = L + L + W + W$
   or
   $P = 2L + 2W$

2. The perimeter of the accompanying triangle is:
   $10 + 9 + 4$ or 23 in.

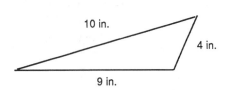

365

Problem 1   The length of a rectangle is five more than three times its width. If the perimeter of the rectangle is 42 centimeters, find the length and width.

Solution   Let x = width. Then "5 more than 3 times the width" is 3x + 5. Thus, 3x + 5 = length.

The rectangle is pictured at the right. Since its perimeter is 42 centimeters, the following equation may be written and solved.

Length = 3x + 5

Width = x

$$P = 2 \cdot L + 2 \cdot W$$
$$42 = 2(3x + 5) + 2x$$
$$42 = 6x + 10 + 2x$$
$$42 = 8x + 10$$
$$32 = 8x$$
$$4 = x$$

The width (x) is 4 cm. The length (3x + 5) is 3 • 4 + 5 or 12 + 5 or 17 cm.

The **circle** is an important plane figure. A circle is a set of points in the plane equally distant from a given point called its **center**. The **radius** of a circle is a line segment that joins the center of the circle to a point on the circle. The **diameter** of a circle is a line segment that passes through the center of a circle and joins two points on the circle. The diameter (d) is twice the length of the radius (r). Thus, d = 2 • r.

The distance around a circle is the **circumference**. To determine the circumference of a circle, its diameter or radius must be known. The ratio of the circumference of a circle to its diameter is always the same regardless of the size of the circle. This result is the number π (the Greek letter "pi"), an irrational number approximated by 3.14 or $\frac{22}{7}$, but closer to 3.14159.

Thus $\pi = \frac{\text{circumference (C)}}{\text{diameter (d)}}$ or $\pi = \frac{C}{d}$.

This relationship implies that C = πd.

> The circumference (C) of a circle is the product of π and the diameter. (C = πd) The circumference (C) of a circle is the product of 2π and the radius. (C = 2πr)

The two formulas given previously are equivalent because the diameter is twice the radius (d = 2 • r).

Problem 2 Find the circumference of a circular flower bed with diameter 21 feet. Use $\frac{22}{7}$ for π.

Solution We use the formula C = πd

and obtain $C = \frac{22}{7} \cdot 21^{3}$

$C = 66$

Thus, the circumference is approximately 66 feet. The circumference may also be expressed in terms of π as 21π feet. Notice that a circumference is a **linear** measurement and is always stated in terms of linear units.

Problem 3 Find the circumference of the circle with radius 4 yards. Use 3.14 for π.

Solution We use the formula C = 2πr

and obtain C = 2 • 3.14 • 4

C = 25.12

Thus, the circumference is approximately 25.12 yards. The circumference may also be expressed in terms of π as 8π yards.

Problem Set 7.1

1. What is the perimeter of a quadrilateral whose sides are 10.7 yards each? *42.8 yards*

2. Find the perimeter of a pentagon whose sides are all 6 inches long. *24 inch*

3. If a hexagon has 6 equal sides and its perimeter is 42 inches, how long is each side? *7 inch*

4. If the perimeter of a quadrilateral is 50 feet and the lengths of three sides are 10 feet, 15 feet, and 20 feet, what is the length of the fourth side? *5*

5. Find the length of a fence around a rectangular field that is 50 feet long and 40 feet wide. *180*

6. A rectangular field is 70 feet long and 30 feet wide. If fencing costs $2.00 per yard, how much will it cost to enclose the field? *$133.40*

7. One side of a square flower bed is 12 feet long. How many plants would be needed if they are to be spaced 6 inches apart around the outside of the bed? *48 feet   576 inches   96 plants*

8. The length of a rectangle is twice the width. If the perimeter is 60 feet, find the dimensions of the rectangle.

9. The length of a rectangle is three more than two times the width. If the perimeter is 30 inches, find the dimensions of the rectangle.

10. If a circular pool has a diameter of 10 feet, what is the circumference of the pool using 3.14 for π?

11. If a circle has a radius of 7 inches, what is the circumference of the circle using $\frac{22}{7}$ for π?

12. If the circumference of a circle is 10π units, what is the measure of the radius and the diameter?

13. The minute hand of a clock is 5 inches long. What is the distance that the tip of the minute hand moves in one hour? Use 3.14 for π.

14. How many yards of fencing would be needed to enclose a circular yard that has a diameter of 30 feet? Use 3.14 for π.

Review Competencies

15. The perimeter of a regular octagon is 48 inches. How long is each side?

16. 2 yd. 5 ft. = ___ ft.

17. Round each of the following:
    a. $5\frac{3}{5}$ ft. to the nearest foot.
    b. 4.255 cm to the nearest tenth of a centimeter.
    c. 5.5 m to the nearest meter.

18. What is the smallest number of faces that may be contained in a polyhedron?

19. Is this figure a regular polygon?

20. Is every simple curve also a closed curve?

21. Is every rectangle also a parallelogram?

22. Express $5\frac{1}{6}$ as an improper fraction.

23. $\frac{7}{5} = \frac{21}{?}$

24. Divide: $9 \div 0.3$

25. Express 0.06 as a percent.

Section 8   Angle Measurement

An angle is generated by a ray that begins at a certain initial position and rotates about a point (the vertex) until it stops at some terminal position. In the diagram, the arrow indicates the direction of the counterclockwise rotation of the angle.

If an angle generates one complete revolution, it would be represented by the figure at the right.

It is often necessary to measure the size of an angle. The measure of an angle indicates the number of **unit angles** it contains. For example, consider ∠ABC at the right along with a given basic unit angle. Observe that the measure of ∠ABC is 3 units. Thus, the size of an angle is the measure together with the unit.

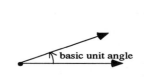

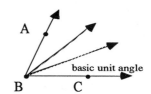

The most common basic unit angle is a one degree (1°) angle. A **degree** is defined to be $\frac{1}{360}$ of a complete rotation. The instrument used to measure an angle is a **protractor**. It contains many unit angles (degrees) placed side by side so that measuring can be done easily. Observe that the accompanying protractor contains 180 degrees (denoted 180°).

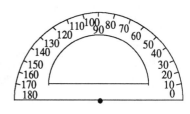

Consider ∠ABC at the right that is superimposed upon a protractor. Can you see that the size of this angle is 30 degrees (30°)? In notation, we write

m∠ABC = 30°.

The notation "m∠ABC" indicates the **measure** of ∠ABC to distinguish it from the rays that form the angle itself.

The angle with measure 30° is shown at the right.

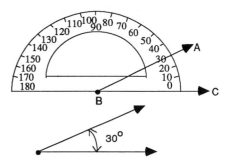

369

An angle whose measure is 90 degrees is a **right angle**. When two rays meet to form a right angle, they are **perpendicular** to each other. Notice in the accompanying figure that ∠B is a right angle or that m ∠B = 90°. Ray BA ($\overrightarrow{BA}$) is perpendicular to ray BC ($\overrightarrow{BC}$). We may express this as: $\overrightarrow{BA} \perp \overrightarrow{BC}$. In the diagram we agree that $\overrightarrow{BC}$ occupies a **horizontal** position since it can be said to follow the direction of the horizon. We also agree that $\overrightarrow{BA}$ occupies a **vertical** position. This position has an "up and down" sense.

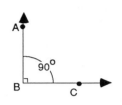

Vertical lines and horizontal lines are perpendicular to each other regardless of the orientation of the lines. For example, consider the accompanying figure. If you are told that line l_1 is vertical, then you can conclude that line l_2 is horizontal because l_2 is perpendicular to l_1.

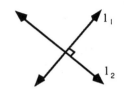

Problem 1   Use the figure at the right to determine whether the following statement is true or false.

If line m_3 is vertical, the lines m_1 and m_2 are both horizontal and perpendicular to line m_3.

Solution   Lines m_1 and m_2 are both horizontal because they are perpendicular to line m_3. Thus, the statement is true.

Angles can be classified according to the number of degrees they contain. An angle whose measure is less than 90 degrees is an **acute angle**. In the accompanying figure, ∠ABC is an acute angle.

An angle whose measure is greater than 90 degrees and less than 180 degrees is an **obtuse angle**. In the accompanying figure, ∠ABC is an obtuse angle.

An angle whose measure is exactly 180 degrees is a **straight angle**. In the accompanying figure ∠ABC is a straight angle.

Problem Set 8.1

Use the protractor pictured below to find the measure of the following angles.
Indicate whether they are right angles, acute angles, obtuse angles, or straight angles:

1. ∠BAC
2. ∠DAC
3. ∠EAD
4. ∠FAC
5. ∠GAE
6. ∠HAD
7. ∠HAE
8. ∠HAC

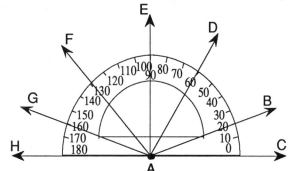

9. Is a degree equal to $\frac{1}{180}$ of a straight angle?

10. How many right angles are generated by two perpendicular lines?

Review Competencies

11. If the circumference of a circle is 20π units, find the measure of the diameter.

12. If a circle has a radius of 14 inches, what is the circumference of the circle? Use $\frac{22}{7}$ for π.

13. Indicate the perimeter of a triangle whose sides are 6.3 feet, 2.8 feet, and 4.1 feet.

14. Round 0.836 m to the nearest meter.

15. 97 ft. = _____ yd.

16. Is every square also a rectangle?

17. Subtract: $7\frac{1}{9} - 4\frac{8}{9}$

18. Express $\frac{36}{48}$ as a fraction in lowest terms.

19. What is 40% of 60?

Pairs of angles can form special relationships. In the figure at the right, line m_1 is perpendicular to line m_2 at point E. Notice that ∠CEB and ∠AEC share the same vertex (point E) and the same ray ($\overrightarrow{EC}$). These angles are **adjacent angles**. Can you find other pairs of adjacent angles?

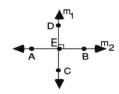

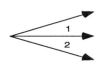

Intersecting lines need not be perpendicular to form adjacent angles. In the figure at the left, two more adjacent angles are shown. The adjacent angles are ∠1 and ∠2.

In the figure at the right, the intersection of two lines has formed four angles. A relationship exists between certain non-adjacent pairs of these angles. We say that $\angle 1$ and $\angle 3$ are **vertical** angles. Also, $\angle 2$ and $\angle 4$ are vertical angles.

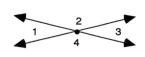

It can be shown that **vertical angles have the same measure**. Angles that have the same measure are said to be **congruent**. We use the symbol $\cong$ to denote congruence of two angles. Since $\angle 1$ and $\angle 3$ are vertical angles, then $\angle 1 \cong \angle 3$. Also, $\angle 2 \cong \angle 4$.

Pairs of angles can form other special relationships. Two acute angles are **complementary angles** if the sum of their measures is 90°. Each angle is the **complement** of the other. Complementary angles need not be adjacent angles. Thus, $\angle ABC$ and $\angle XYZ$, in the figures at the right, are complementary.

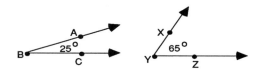

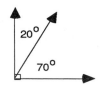

Complementary angles can be adjacent angles. The figure at the left illustrates two acute angles whose sum is 90°. The acute angles are both adjacent and complementary.

Two angles are **supplementary angles** if the sum of their measures is 180°. Each angle is called the **supplement** of the other. Supplementary angles need not be adjacent. The pairs of angles at the right are supplementary, but not adjacent.

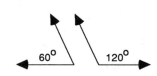

The figure at the right illustrates two adjacent supplementary angles. Notice that when two angles are supplementary, then generally one angle is acute and the other angle is obtuse. What is the one exception to this generality?

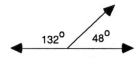

Problem 2   Find x (in degrees), and then find the measure of the two angles.

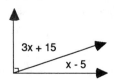

Solution   From the sketch we can see that the two angles have measures whose sum is 90°. (The angles are complementary.) Thus, we solve the following equation.

$$x - 5 + 3x + 15 = 90$$
$$4x + 10 = 90$$
$$4x = 80$$
$$x = 20 \quad (20°)$$

The angles measure: $x - 5 = 20 - 5 = 15°$ and $3x + 15 = 3(20) + 15 = 75°$

Problem 3 Find x (in degrees), and then find the measure of the two angles.

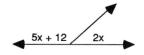

Solution From the sketch we can see that the two angles have measures whose sum is 180°. (The angles are supplementary.) Thus, we solve the following equation.

$$5x + 12 + 2x = 180$$
$$7x + 12 = 180$$
$$7x = 168$$
$$x = 24 \quad (24°)$$

The angles measure:
$$5x + 12 = 5(24) + 12 = 120 + 12 = 132° \text{ and } 2x = 2(24) = 48°$$

Problem Set 8.2

1. Use the accompanying figure to answer:

    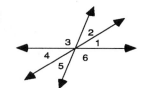

    a. Which angle(s) is/are adjacent to ∠4?

    b. Which angle(s) is/are vertical to ∠6?

2. Indicate the measure of the angle that is complementary to an angle measuring:
    a. 73°   b. 24°

3. Indicate the measure of the angle that is supplementary to an angle measuring:
    a. 117°   b. 89°

4. If two supplementary angles are congruent, what is their measure?

5. If two complementary angles are congruent, what is their measure?

6. Indicate the measure of an angle whose measure is twice the measure of its complementary angle.

7. If two lines intersect, what can be said about the adjacent angles formed?

8. Two lines intersect so as to form congruent adjacent angles. How large is each angle?

9. Which is larger: the supplement of an acute angle or its complement? How much larger?

10. Find the measurement of the angles in each of the following.

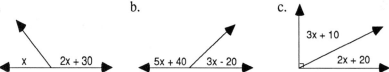

a.   b.   c.   d.

Review Competencies

11. Consider the accompanying figure:
    a. If $\overrightarrow{AB}$ is vertical, is $\overrightarrow{AC}$ horizontal?
    b. Is $\overrightarrow{AC}$ perpendicular to $\overrightarrow{AB}$?

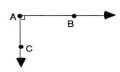

12. Describe the following angles as obtuse, right, straight, or acute.

    a.   b.   c.

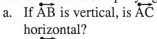

13. If a circle has a diameter of 20 feet, what is the circumference of the circle? Use 3.14 for π.

14. Indicate the perimeter of a regular octagon if one side is 10 inches.

15. Round 4.075 km to the nearest hundredth of a kilometer.

16. Is every rhombus also a parallelogram?

17. Divide: $1 \div 0.005$

18. Express $4\frac{3}{5}\%$ as a decimal.

19. What is the next term in the following geometric progression?

    $1, -\frac{1}{2}, \frac{1}{4}, \underline{\hspace{1cm}}$

20. A television selling for $390 was purchased for 35% off the original price. What was the original price?

Section 9   Parallel Lines and their Properties

In the figure at the right, line m_1 is parallel to line m_2. Observe that line m_3 intersects the parallel lines at point A on line m_1 and point B on line m_2. Line m_3 is a **transversal**. Generally speaking, a transversal is a line which intersects any two other lines in two distinct points.

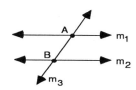

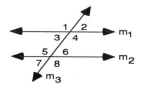

Several angles are formed by two parallel lines and a transversal and are given special names. In the figure at the left, line m_1 is parallel to line m_2. **Alternate interior angles** are pairs of angles whose interiors lie between the parallel lines but on opposite sides of the transversal. Thus, $\angle 3$ and $\angle 6$ are alternate interior angles as are $\angle 4$ and $\angle 5$.

It can be shown that, **when two lines are parallel, alternate interior angles have the same measure**. Thus, $\angle 3 \cong \angle 6$ and $\angle 4 \cong \angle 5$. Two lines are parallel if and only if each pair of alternate interior angles is congruent.

In the preceding figure, $\angle 1$ and $\angle 8$ are said to be **alternate exterior angles**. These angles lie outside the parallel lines but on opposite sides of the transversal. Another pair of alternate exterior angles is $\angle 2$ and $\angle 7$. It can be shown that **when two lines are parallel, alternate exterior angles have the same measure**. Thus, $\angle 1 \cong \angle 8$ and $\angle 2 \cong \angle 7$. Two lines are parallel if and only if each pair of alternate exterior angles is congruent.

Corresponding angles are two non-adjacent angles whose interiors lie on the same side of the transversal such that one angle lies between the parallel lines and the other lies outside the parallel lines. In the preceding figure, four pairs of corresponding angles are $\angle 1$ and $\angle 5$, $\angle 2$ and $\angle 6$, $\angle 3$ and $\angle 7$, and $\angle 4$ and $\angle 8$.

It can be shown that **when two lines are parallel, corrresponding angles have the same measure**. Thus $\angle 1 \cong \angle 5$, $\angle 2 \cong \angle 6$, $\angle 3 \cong \angle 7$, and $\angle 4 \cong \angle 8$. Two lines are parallel if and only if each pair of corresponding angles is congruent.

Notice, in the preceding figure, that $\angle 3$ and $\angle 5$ is a pair of interior angles on the same side of the transversal whose interiors lie between the parallel lines. Another pair of angles is $\angle 4$ and $\angle 6$. It can be shown that, **when two lines are parallel**, interior angles on the same side of the transversal are supplementary. Observe that $\angle 1$ and $\angle 7$ is a pair of exterior angles on the same side of the transversal as are $\angle 2$ and $\angle 8$. These pairs of angles are also supplementary.

Chapter 6

Problem 1   Find the measure of all
the angles in the accompanying
figure if m_1 is parallel to m_2.

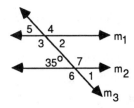

Solution $m\angle 1 = 35°$ since vertical angles are congruent.
$m\angle 6 = 180 - 35 = 145°$ since $\angle 1$ and $\angle 6$ are supplementary.
$m\angle 7 = 180 - 35 = 145°$ (or $\angle 7 \cong \angle 6$ since these are vertical angles)
$m\angle 3 = 145°$
$\angle 3 \cong \angle 7$ These are alternate interior angles (on opposite sides of
the transversal and within the parallel lines).
$m\angle 2 = 180 - 145 = 35°$
$m\angle 4 = 145°$
$\angle 4 \cong \angle 7$ These are corresponding angles (above the parallel lines
and to the right of the transversal).
(Or $\angle 4 \cong \angle 3$ since these are vertical angles)
$m\angle 5 = 180 - 145 = 35°$ since $\angle 5$ and $\angle 4$ are supplementary.

Notice that if the size of one angle is known when a transversal intersects
parallel lines, then all other seven angles can be found. Three of the other
angles will have equal measure as vertical, corresponding, alternate interior
or alternate exterior angles. The other four angles will be supplementary.

Problem 2   In the accompanying figure, m_1
is parallel to m_2. Find x,
and then find the measure of the
angles.

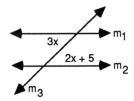

Solution   Since the lines are parallel, the alternate interior angles are equal in
measure. Thus, we solve the following equation.

$$3x = 2x + 5$$
$$x = 5 \ (5°)$$

The angles measure $3x = 3(5) = 15°$ and $2x + 5 = 2(5) + 5 = 10 + 5 = 15°$.

Denise Gonzalez

Problem 3 Use the figure at the right to determine
the truth or falsity of each statement.
a. If m_4 is horizontal, then lines
   m_1 and m_2 are both vertical
   and parallel.
b. Line m_4 intersects m_3.

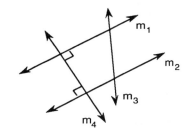

Solution Answer a is true. m_1 and m_2 are vertical (they are perpendicular to a horizontal line) and are parallel (alternate interior angles are congruent because
they are right angles).

Answer b is also true. A line extends indefinitely in both directions. The
two lines will intersect somewhere in the plane.

Consider the accompanying figure in which the line through points A
and B (line AB) is parallel to the line through points C and D (line CD).
Furthermore, the transversal through B, D, and E (line BD, DE, or BE)
is perpendicular to both parallel lines. The following notations can be
used:

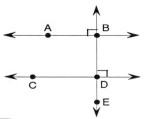

> **Notations for Parallel and Perpendicualr Lines**
> 1. $\overline{AB} \parallel \overline{CD}$ means line AB is parallel to line CD.
>
> 2. $\overline{BE} \perp \overline{AB}$ means line BE is perpendicular to line
>   AB (that is, the lines intersect at right angles).
>   Similarly, $\overline{BE} \perp \overline{CD}$.

Problem Set 9.1

Consider the accompanying figure in which m_1 is parallel to m_2 and m_3 is a transversal:

1. Name the angles that have the same measure as $\angle 3$.

2. Name the angles that are supplementary to $\angle 5$.

3. Name the angles that have the same measure as $\angle 2$.

4   Name the angles that are supplementary to $\angle 7$.

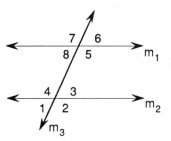

377

5. If m_1 is parallel to m_2, find the measure of the remaining angles.

6. If m_1 is parallel to m_2 and m_3 is a transversal, find $m\angle x$.

7. If m_1 is parallel to m_2, find $m\angle x$.

8. Find the value of x, and then find the measure of the angles if m_1 and m_2 are parallel lines.

9. The diagram below shows five lines in the same plane. Line m_1 is horizontal. Two congruent angles are indicated. Which of the following statements is/are true of the lines in the diagram?
   i. Lines m_2 and m_3 form the only pair of parallel lines.
   ii. Lines m_1 and m_4 intersect.
   iii. Line m_2 is the only vertical line.

   a. i only
   b. ii only
   c. iii only
   d. i and ii only

10. Which of the following statement(s) is/are true based upon the sketch?

    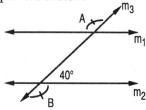

    a. $m\angle A = 40°$
    b. m_1 is parallel to m_2.
    c. m_1 and m_3 intersect.
    d. m_3 is a vertical line if m_1 is a horizontal line.

Review Competencies

11. Consider the accompanying figure.

    a. Which angle(s) is/are adjacent to $\angle 1$?
    b. Which angle(s) is/are vertical to $\angle 2$?
    c. Which angle(s) is/are supplementary to $\angle 3$?

12. Find the value of x and then find the measure of the two angles in the accompanying figure.

13. Find the value of x, and then find the measure of either vertical angle in the accompanying figure.

    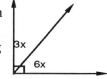

14. Find the circumference of a circular vegetable garden whose diameter is 70 feet. Use $\frac{22}{7}$ for π.

15. Find the perimeter of a regular pentagon having one side equal to $3\frac{1}{3}$ feet.

Section 10   Triangles

We know that a triangle contains three sides and three interior angles. It can be proved that **the sum of the measures of the three interior angles is 180°.** This fact helps us to solve different types of problems involving triangles.

Problem 1   Find x (in degrees), and then find the measures of the three angles in the accompanying figure.

Solution   The sum of the measures of the angles is 180°. Thus, we solve the following equation.

$$x + 4x + 5x = 180$$
$$10x = 180$$
$$x = 18 \ (18°)$$

The angles measure $x = 18°$; $4x = 4(18) = 72°$; and $5x = 5(18) = 90°$.

Problem 2   Find the measure of all missing angles in the following figure.

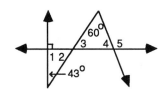

Solution $m\angle 1 = 90°$ since it is supplementary to the right angle.

$m\angle 2$ can be found using the idea that the sum of the measures of the angles of a triangle is 180°.
$m\angle 1 + 43° = 90 + 43 = 133°$ and $m\angle 2 = 180 - 133 = 47°$.

$m\angle 3$ can be found using the idea that vertical angles are congruent.
$\angle 3 \cong \angle 2$. Thus, $m\angle 3 = 47°$.

To find $m\angle 4$, we can add $m\angle 3 + 60°$ and then subtract from 180°.
$m\angle 3 + 60° = 47 + 60 = 107°$ and $m\angle 4 = 180 - 107 = 73°$.

$\angle 5$ is supplementary to $\angle 4$.
Thus, $m\angle 5 = 180 - m\angle 4 = 180 - 73 = 107°$.

Problem 3   In the accompanying figure, m_1 is parallel to m_2.

Find $m\angle 1$.

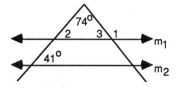

Solution   We can immediately find $m\angle 2$: $m\angle 2 = 41°$ since, when lines are parallel, the corresponding angles are congruent.

Since a triangle has an angle sum of 180°, we can find $m\angle 3$ by adding $m\angle 2$ to 74°($41° + 74° = 115°$) and subtracting this result from 180°. Thus, $m\angle 3 = 180 - 115 = 65°$.

$\angle 1$ is supplementary to $\angle 3$. Thus, $m\angle 1 = 180 - 65 = 115°$.

When a side of a triangle is extended, an **exterior angle** is created. Since this exterior angle is supplementary to the adjacent interior angle of the triangle, its measure is related to the measures of the other two angles (**remote interior angles**) of the triangle. Consider the figure at the right.

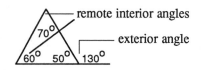

Notice that $130° = 60° + 70°$.

The measure of an exterior angle of a triangle equals the sum of the measures of the two remote interior angles.

Problem 4   Find the measure of $\angle 1$ in the figure at the right.

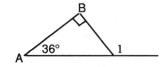

Solution   Since $\angle 1$ is an exterior angle,

$$m\angle 1 = m\angle A + m\angle B$$
$$= 36 + 90$$
$$= 126 \quad \text{Thus, } m\angle 1 = 126°.$$

Triangles are classified according to the kinds of angles they contain. An **acute triangle** contains three acute angles.

A **right triangle** is one which contains one right angle. In the figure at the right, the side of the right triangle that lies opposite the right angle is the **hypotenuse**. The hypotenuse is the longest side of the right triangle. The other two sides of the right triangle are **legs**. A right triangle can only contain one right angle and two acute angles. The sum of the measures of the two acute angles is 90°. Thus, the acute angles of a right triangle are complementary.

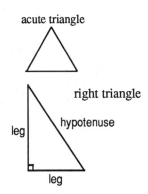

An **obtuse triangle** is one which contains one obtuse angle. The figure on the right shows an obtuse triangle. The longest side of an obtuse triangle always lies opposite the obtuse angle. An obtuse triangle cannot contain more than one obtuse angle.

A triangle can also be classified according to relationships between the lengths of its sides. One triangle in the figure at the right has three equal sides and is an **equilateral triangle**. Each of the angles has a measure of 60°. The triangle is also **equiangular**.

The other triangle at the right has two equal sides and is an **isosceles triangle**.

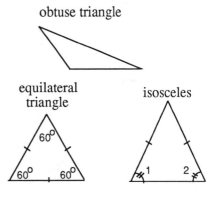

obtuse triangle

equilateral triangle

isosceles

Note

It can be proved that a triangle has two equal sides if and only if the angles opposite those sides are congruent. In the preceding isosceles triangle figure, the two base angles have equal measure (denoted by two strokes) because the opposite sides have equal length. Thus, ∠1 ≅ ∠2.

According to these classifications, an equilateral triangle is also isosceles, but an isosceles triangle is not necessarily equilateral.

A **scalene triangle** is one in which no two sides are equal.

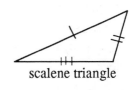

scalene triangle

Problem Set 10.1

1. Can a right triangle also be an isosceles triangle?

2. Can a right triangle also be an equilateral triangle?

3. Can a scalene triangle also be an obtuse triangle?

4. Can an isosceles triangle also be an obtuse triangle?

5. Is an equilateral triangle also an acute triangle?

6. Classify each triangle shown below as equilateral, isosceles, or scalene. Then classify the same triangle as acute, right, or obtuse.

a.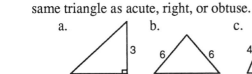

3
3

b.

6   6
9

c.

4   6
7

d.

5   5
5

e.

19   14
10

7. Find m∠x in the figure at right.

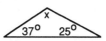

8. Find x and the measure of each of the angles of the triangle.

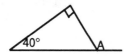

9. Find m∠x for the figure at right.

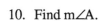

10. Find m∠A.

11. Find m∠1.

12. Find the measure of angle y in terms of x degrees.

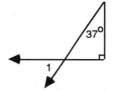

13. Determine if the following are true or false for the figure shown.
    a. $\overline{AB} \perp \overline{BC}$
    b. $m\angle 2 + m\angle 3 = 90°$
    c. $m\angle 2 + m\angle 3 + m\angle 4 = 180°$
    d. $\angle 1$ and $\angle 2$ are complementary.
    e. $m\angle 1 = m\angle 3 + m\angle 4$

14. Which of the following statements is/are true for the accompanying figure?
    i. $\angle A \cong \angle E$ ii. $m\angle D = 140°$
    iii. $\angle A$ is supplementary to $\angle C$
    a. i only   b. iii only
    c. i and iii only d. ii and iii only

Review Competencies

15. Consider the accompanying figure in which line m_1 is parallel to line m_2 and line m_3 is a transversal.

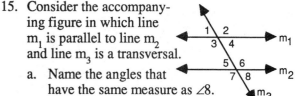

    a. Name the angles that have the same measure as $\angle 8$.
    b. Name the angles that are supplementary to $\angle 6$.
    c. Name the angles that have the same measure as $\angle 4$.
    d. Name the angles that are supplementary to $\angle 1$.

16. Consider the accompanying figure in which line m_1 is parallel to line m_2 and line m_3 is a transversal. Find the value of x and then find the measure of either angle.

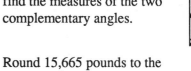

17. Find the value of x and then find the measures of the two complementary angles.

18. Round 15,665 pounds to the nearest thousand pounds.

19. The perimeter of a regular hexagon is 30 feet. How long is each side?

20. Express $\frac{27}{45}$ in lowest terms.

21. Add: $\frac{5}{6} + \frac{3}{4}$

22. Subtract: $5 - 2.015$

23. Express 80% as a fraction.

24. What is 15% of 30?

25. Evaluate: $5(7 \cdot 2 - 5 \div 5 - 10)$

26. Express 2.1 as an improper fraction.

Section 11   Sum of the Angle Measures in Any Convex Polygon

We know that the sum of the three interior angles of any triangle is 180°. What is the sum of the measures of the four interior angles in a rectangle, or indeed, any convex quadrilateral?

Let us consider a convex quadrilateral and draw a diagonal from any vertex, as shown in the figure at the right. Observe that we obtain two triangles. The sum of the angle measures of the two triangles (360°) should equal the sum of the angle measures of the quadrilateral. Let us express this sum as 2 • 180° or (4 – 2) • 180°.

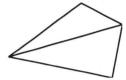

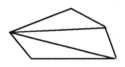

Now let us consider the convex pentagon shown at the left. Notice that we can draw two diagonals from one vertex. When this is done, we obtain three triangles. Thus, the sum of the measures of the interior angles of the convex pentagon is 540° or 3 • 180° or (5 – 2) • 180°.

We can find the sum of the measures of the interior angles of a convex polygon by drawing all the diagonals from any one vertex of the polygon. Consider the pattern below.

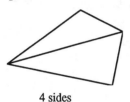 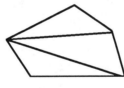 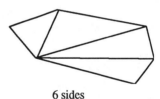

4 sides	5 sides	6 sides
2 triangles	3 triangles	4 triangles
Angle sum: 2(180°) = 360°	Angle sum: 3(180°) = 540°	Angle sum: 4(180°) = 720°

This pattern leads to a generalization.

> The sum of the measures of the interior angles of a convex polygon of n sides is (n – 2) • 180°.

Observe that in the case of a triangle (n = 3), the formula gives us:
$$(3 - 2) • 180° \text{ or } 1 • 180° \text{ or } 180°$$
Thus, this relationship verifies that the sum of the measures of the interior angles of a triangle is 180°.

The sum of the measures of the interior angles of a convex hexagon is:
$$(6 - 2) • 180° \text{ or } 4 • 180° \text{ or } 720°$$

Consider the accompanying rectangle ABCD. In this rectangle, we know that $\angle A$ and $\angle B$ are consecutive angles and that $\angle A \cong \angle B$ since both angles are right angles. Since $m\angle A = 90°$ and $m\angle B = 90°$, the sum of $m\angle A$ and $m\angle B$ is 180°. Thus, $\angle A$ and $\angle B$ are supplementary. In the same way, $\angle B$ and $\angle C$ are supplementary. Thus, **the consecutive angles of a rectangle are supplementary.**

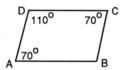

It can be shown that **the consecutive angles of a parallelogram are supplementary.** Thus, in the accompanying parallelogram ABCD, if $m\angle D = 110°$, then $m\angle A = 70°$ and $m\angle C = 70°$.

Problem Set 11.1

1. What is the sum of the measures of the angles of a convex hexagon? A convex octagon?

2. Three angles of a quadrilateral measure 64°, 121°, and 157°. What is the measure of the fourth angle?

3. Find $m\angle x$ in the diagram.

4. A plane geometric figure has 12 sides. How many degrees are there in the sum of the angles?

5. Consider the accompanying parallelogram. Find $m\angle B$ and $m\angle C$.

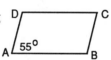

Review Competencies

6. Find $m\angle x$ in each of the following figures.

   a.

   b.

   c.

   d.

7. In the diagram m_1 is parallel to m_2. Find $m\angle 1$.

8. In the diagram m_1 is parallel to m_2. Find $m\angle x$.

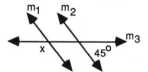

9. The perimeter of a regular pentagon is 60 feet. How long is each side?

10. Find the circumference of a circle whose radius is 5 inches. Express your answer in terms of π.

11. Round 13.0177 cm. to the nearest thousandth of a centimeter.

12. Express $\frac{31}{8}$ as a mixed number.

13. Subtract: $3\frac{1}{2} - 1\frac{3}{4}$

14. Multiply: $5.01 \cdot 2.6$

15. Express $\frac{3}{8}$ as a decimal.

16. 20% of what number is 40?

17. Evaluate: $(3 \cdot 5 - 8 \div 4) - (12 + 1 - 4 \cdot 3)$

18. Express 0.214 as a fraction in lowest terms.

19. Find the solution set for $3(x - 2) - 5(x - 3) = 17$.

20. For the figure at the right, which of the following is/are true?
   i. $m\angle C = 60°$
   ii. $\angle A \cong \angle B$
   iii. $\angle D \cong \angle E \cong \angle F$

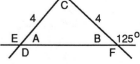

Section 12   The Pythagorean Theorem

One of the most powerful and useful theorems of geometry is the **Pythagorean Theorem** which states:

> In any right triangle, the sum of the squares of the legs is equal to the square of the hypotenuse.

In the accompanying figure the Pythagorean Theorem may be stated symbolically as:

$$a^2 + b^2 = c^2$$

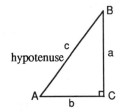

Note

In the right triangle ABC in which $\angle C$ is the right angle, the leg opposite $\angle A$ is called "a." The leg opposite $\angle B$ is called "b." The hypotenuse is called "c."

Problem 1   Use the Pythagorean Theorem to determine the missing side (c) in the right triangle in which $\angle C$ is the right angle, side a is 3 feet, and side b is 4 feet.

Solution   Substitute the values into the statement of the theorem.

Thus, $a^2 + b^2 = c^2$ when a = 3 and b = 4
becomes: $3^2 + 4^2 = c^2$
which gives: $9 + 16 = c^2$
                $25 = c^2$
                 $5 = c$ because $\sqrt{25} = 5$

The missing side, c, measures 5 feet.

Chapter 6

Problem 2

Use the Pythagorean Theorem and the principles for solving equations to determine the missing side in the right triangle in which $\angle C$ is the right angle, side b is 5 inches, and side c is 13 inches.

Solution

The missing side to be determined is side a.

$$\text{Thus,} \quad a^2 + b^2 = c^2 \text{ when } b = 5 \text{ and } c = 13$$
$$\text{becomes:} \quad a^2 + 5^2 = 13^2$$
$$a^2 + 25 = 169$$
$$a^2 + 25 - 25 = 169 - 25$$
$$a^2 = 144$$
$$a = 12 \text{ because } \sqrt{144} = 12$$

The missing side, a, measures 12 inches.

Problem 3

Find side b in the right triangle with hypotenuse 4 and one side (leg) $\sqrt{7}$.

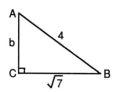

Solution

$$a^2 + b^2 = c^2$$
$$(\sqrt{7})^2 + b^2 = 4^2$$
$$7 + b^2 = 16$$
$$b^2 = 9$$
$$b = 3$$

Note

If a is positive, $(\sqrt{a})^2 = a$; thus, $(\sqrt{7})^2 = 7$.

Problem 4

A 26 foot ladder is placed against a wall. The bottom of the ladder is 24 feet away from the wall. How high up does the ladder reach?

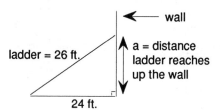

Solution

$$a^2 + b^2 = c^2$$
$$a^2 + 24^2 = 26^2$$
$$a^2 + 576 = 676$$
$$a^2 = 100$$
$$a = 10$$

Thus, the ladder reaches 10 feet up the wall.

Problem Set 12.1

1. Use the accompanying right triangle in which ∠C is the right angle.

   a. Find c if b = 8 inches and a = 15 inches.

   b. Find b if a = 24 feet and c = 25 feet.

   c. Find a if c = 10 yards and b = 6 yards.

   d. Find c if b = √3 centimeters and a = 1 centimeter.

   e. Find b if c = √170 meters and a = 11 meters.

   f. Find a if c = √2 inches and b = 1 inch.

2. Use the Pythagorean Theorem to solve each of the following. (Hint: Sketch a right triangle in each problem and assign values to the appropriate parts.)

   a. A sail boat leaves a dock sailing 9 miles due west and then 12 miles due north. How far away is the boat from the dock?

   b. How high up on a wall does a 26 foot ladder reach if the foot of the ladder is 10 feet from the wall?

   c. What must be the length of a guy wire on a 24 foot flagpole if the wire is to be secured on the ground 7 feet from the base of the pole?

   d. An empty lot is 40 feet long and 30 feet wide. How many feet does a person save by walking diagonally across the lot instead of walking the length and width of the lot?

   e. Two flag poles are 42 feet and 49 feet high respectively and are 24 feet apart. How long is a wire from the top of one pole to the top of the second pole?

   f. A builder wishes to test whether the walls at a corner of a building form a right angle. The builder measures 8 feet from the corner along one wall and 6 feet from the corner along the other wall, discovering the distance between the ends of these lines is 10 feet. Can the builder conclude that the walls form a right angle? (Hint: Show that the distances satisfy the Pythagorean theorem.)

3. In the sketch that is shown, find the cost of constructing the new street if construction costs $100 per linear foot. One mile equals 5,280 linear feet.

4. If a bicyclist averages 10 miles per hour, how long will it take to cover the distance represented by AB?

   a. 1 hour

   b. 1½ hours

   c. 2½ hours

   d. 3 hours

Review Competencies

5. What is the sum of the measures of the angles of a convex pentagon?

6. Find m∠x.

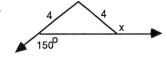

7. One of the equal sides of an isosceles triangle is 3a long. The unequal side is 7b long. What is the perimeter of the triangle?

8. Find the measure of each of the angles numbered 1-7. The lines m₁ and m₂ are parallel.

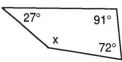

9. Find m∠x.

387

10. In the figure at the right, $\overline{CD} \perp \overline{AD}$

    If $m\angle BDC = 35°$, find $m\angle ADB$.

11. Which of the following statements is/are true for the figure shown at the right?

    i. $m\angle 1 + m\angle 2 = 90°$
    ii. $m\angle 3 = m\angle 2 + 90°$
    iii. $\angle 1$ and $\angle 3$ are complementary.

    a. i only
    b. ii only
    c. i and ii only
    d. all of the statements

12. Round 6285 feet to the nearest hundred feet.

13. $\frac{5}{6} = \frac{?}{30}$

14. Add: $\frac{3}{5} + 1\frac{2}{3}$

15. Add: $0.015 + 43 + 3.02$

16. Multiply: $\frac{2}{3} \cdot \frac{6}{7} \cdot \frac{5}{2}$

17. Express 27.5% as a decimal.

18. 50 is 25% of what number?

19. Evaluate: $6 \div (4 - 2) + 3 \cdot (1 + 2)$

20. Express 0.65 as a fraction in lowest terms.

21. Solve the inequality $2(x - 7) > 5(x - 1)$.

Section 13   Similar Triangles

Recall that the **ratio** of two numbers a and b (b ≠ 0) compares a to b by division and can be represented as the fraction $\frac{a}{b}$.

Examples

1. The ratio of 20 to 15 is $\frac{20}{15}$ or $\frac{4}{3}$.

2. The ratio of 3 to 2 is equal to the ratio of 9 to 6 since $\frac{3}{2} = \frac{9}{6}$.
   A statement that ratios are equal is called a **proportion**. Notice that $3 \cdot 6 = 2 \cdot 9$.

> If $\frac{a}{b} = \frac{c}{d}$ then ad = bc.

Consider the two triangles shown at the right. Suppose they both have the same exact shape, even though they are not identical in size. We say that $\triangle ABC$ and $\triangle DEF$ are **similar triangles**. We write:

$$\triangle ABC \sim \triangle DEF.$$

If two triangles are similar, the corresponding angles are congruent. That is, $\angle A \cong \angle D$, $\angle B \cong \angle E$, and $\angle C \cong \angle F$. If two angles of one triangle are congruent, respectively, to two angles of another triangle, then the third angles are also congruent.

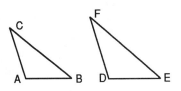

If two triangles are similar, the corresponding sides are proportional. Let us return to the similar triangles shown on the right. If we place the sides of the larger triangle in the denominator of each ratio, we obtain:

$$\frac{AB}{DE} = \frac{BC}{EF} = \frac{AC}{DF}$$

Observe that this proportion involves **three** ratios. In our problems, we will generally use two ratios.

If two triangles are similar, two corresponding pairs of sides are proportional, and the angles included between these sides are congruent. In the two similar triangles we have been discussing, it is true that $\frac{AC}{DF} = \frac{AB}{DE}$ and the respective included angles, $\angle A$ and $\angle D$, are congruent.

> Note
> Two triangles are similar if and only if:
> 1. two angles of one triangle are congruent, respectively, to two angles of the other triangle, or
> 2. all three pairs of corresponding sides are proportional, or
> 3. two corresponding pairs of sides are proportional and the angles included between these sides are congruent.

Let us again consider the same similar triangles and assign values to represent the lengths of certain sides.

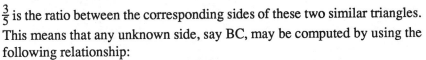

$$AB = 3, DE = 5, \text{ and } EF = 10.$$

Since the sides of these similar triangles are proportional, we know that:

$$\frac{AB}{DE} = \frac{3}{5} \quad \text{Thus, } \frac{BC}{EF} = \frac{AC}{DF} = \frac{3}{5}.$$

$\frac{3}{5}$ is the ratio between the corresponding sides of these two similar triangles. This means that any unknown side, say BC, may be computed by using the following relationship:

$$\frac{BC}{EF} = \frac{3}{5}$$

Since EF = 10 we have: $\frac{BC}{10} = \frac{3}{5}$, and this proportion is solved as:

$$5 \cdot BC = 10 \cdot 3 \text{ or } 5 \cdot BC = 30 \text{ or } BC = 6$$

Chapter 6

Problem 1 Consider the accompanying triangles, and find the length of side x.

 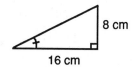

Solution The two triangles are similar since two angles of one are congruent to two angles of the other as indicated by the angle markings. Hence, the corresponding sides are proportional.

$$\frac{x}{8} = \frac{12}{16} \text{ or } \frac{x}{8} = \frac{3}{4} \text{ gives } 4 \cdot x = 8 \cdot 3 \text{ or } 4x = 24 \text{ or } x = 6 \text{ cm.}$$

Problem 2 Consider the following figure in which $\overline{AB} \parallel \overline{CD}$. Find AB.

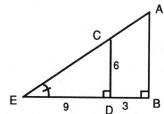

Solution $\triangle ABE$ is similar to $\triangle CDE$ because two angles of the smaller triangle are congruent to two angles of the larger triangle. Both triangles share $\angle E$, and both triangles contain right angles.

$$\frac{AB}{CD} = \frac{BE}{DE} \text{ or } \frac{AB}{6} = \frac{12}{9} \text{ or } \frac{AB}{6} = \frac{4}{3}$$

Thus, we obtain: $3 \cdot AB = 6 \cdot 4$ or $3 \cdot AB = 24$ or $AB = 8$.

Problem 3 The sun is shining in a way that a 5 foot tall pole casts a shadow of 2 feet. Find the height of a pole that casts a shadow of 50 feet.

Solution A drawing for this type of problem makes the similarity of the triangles easier to see and the pairs of matching sides easier to select.

$$\frac{x}{5} = \frac{50}{2} \text{ or } \frac{x}{5} = \frac{25}{1}$$

Solving gives $x \cdot 1 = 5 \cdot 25$ or $x = 125$. Thus, the height of the pole is 125 feet.

 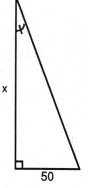

Problem 4   Consider the similar triangles shown below.
Which of the following
proportions is incorrect?

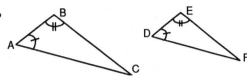

1) $\dfrac{AB}{DE} = \dfrac{AC}{DF}$

2) $\dfrac{AB}{AC} = \dfrac{DE}{DF}$

3) $\dfrac{AB}{DF} = \dfrac{AC}{DE}$

Solution   To be correct, both ratios of a proportion must make the **same**
comparison of corresponding sides.

1) $\dfrac{\text{AB (corresponds to DE)}}{\text{DE}} = \dfrac{\text{AC (corresponds to DF)}}{\text{DF}}$ is a correct proportion.

Notice that the ratio on the left of the correct proportion above compares
corresponding sides of ΔABC to ΔDEF, and the ratio on the right also
compares corresponding sides of ΔABC to ΔDEF.

2) $\dfrac{\text{AB (corresponds to DE)}}{\text{AC (corresponds to DF)}} = \dfrac{\text{DE}}{\text{DF}}$ is a correct proportion.

Notice that the ratio on the left of the correct proportion above compares
two sides of ΔABC, and the ratio on the right compares the corresponding
sides of ΔDEF.

3) $\dfrac{\text{AB (corresponds to DE)}}{\text{DF}} = \dfrac{\text{AC (corresponds to DF)}}{\text{DE}}$ is not correct.

Notice that the ratio on the left makes a different comparison than the ratio
on the right. To be correct, the two ratios must follow exactly the same
pattern.

Problem Set 13.1

1. Find the lengths of the unknown
    sides in the following pairs
    of similar triangles.

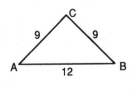

2. Find the lengths of the unknown
    sides in the following pairs of
    similar triangles.

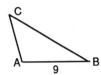

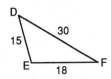

3. In the figure, △ABC is similar to △DEC. Find DE.

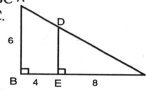

4. In the figure, △ABC is similar to △DEC. Find AB.

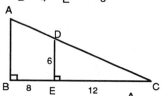

5. In the accompanying figure, △ABC is similar to △DEB. Find BC and EC.

6. Which of the statements is/are true for the pictured triangles?

   i. $\angle B \cong \angle D$

   ii. $\dfrac{BC}{CD} = \dfrac{ED}{AB}$

   iii. $\dfrac{BC}{AC} = \dfrac{CD}{CE}$

7. Which statement(s) listed below is/are true for the pictured triangles?

   i. $\angle BCA \cong \angle BED$

   ii. $\angle BDE \cong \angle BCA$

   iii. $\dfrac{EB}{CB} = \dfrac{AC}{ED}$ iv. $\dfrac{AB}{BD} = \dfrac{BC}{EB}$

8. Which pairs of triangles are similar?

   i.   ii.

   iii.   iv.

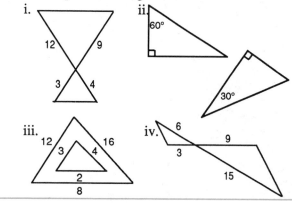

9. The ratio between corresponding sides of two similar triangles is $\frac{2}{3}$. The lengths of the sides of the larger triangle are 12, 15, and 18. What are the lengths of the sides of the smaller triangle?

10. The sides of a triangle measure 4, 6, and 9 inches respectively. The shortest side of a similar triangle measures $6\frac{2}{3}$ inches. Determine the measures of the other two sides.

11. A flagpole casts a shadow 56 feet long at the same time that a fence post casts a shadow 8 feet long. If the post is 3 feet high, how tall is the flagpole?

12. Find the height of a tower that casts a shadow of 50 feet while a person that is 5 feet tall casts a shadow of 2 feet.

Review Competencies

13. Find the perimeter of a right triangle with legs of 12 and 16 inches.

14. A seven foot ladder is placed against a wall. The bottom of the ladder is $\sqrt{24}$ feet from the wall. How far up the wall does the ladder reach?

15. Use the diagram to find the value of x and the measures of the two angles.

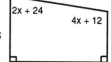

16. Round 37.49 centimeters to the nearest centimeter.

17. Express $\frac{2}{5}$ as a percent.

18. Express 0.01% as a decimal.

19. Express $2\frac{1}{5}\%$ as a decimal.

20. Subtract: $12 - 3\frac{4}{7}$

21. Divide: $0.42 \div 0.021$

22. Evaluate: $\frac{1}{2}(4-2) \cdot 2(10 \div 2)$

Section 14   Area

Often we are interested in measuring the interior surface that is enclosed by a region. In this section, we consider a very important concept in measurement. We use the word **area** to refer to the measure of the interior surface of a closed curve.

We must select a basic unit of measure to determine the area of a region. We want to be able to completely cover the region by placing the units so they touch but do not overlap.

The most conveniently shaped unit is a **square unit**. A square unit is a square with sides that measure one unit in length.

square unit

1 unit

Our task is to determine how many square units are contained in a region. Thus, **area is always measured in terms of square units**.

Consider the rectangular region ABCD at the right. Our unit of measure will be the square unit of measure shown in the figure.

square unit of measure

3 rows (width)

4 in each row (length)

square unit of measure

We simply count the square units to find the area. Observe that it takes 12 square units to cover this region. Using the basic unit, we say that the area of the region is 12 square units.

Notice there are 3 rows with 4 units in each row. We call the 3 rows the **width** of the region and the 4 units in each row the **length** of the region. The length and width of the rectangular region are its **dimensions**. Can you see that the area may simply be determined by finding the product of the length and the width? Thus, Area = 4 • 3 or A = 12 square units.

Note

People frequently refer to the area of a rectangle when what they really mean is the area of a rectangular region. We will continue this practice also.

The area (A) of a rectangle is the product of the length (L) and the width (W). (A = L • W).

Chapter 6

There are various standard square units used to measure area. The **square inch** (in^2) is a square with sides that are one inch in length. A square with one foot sides is called a **square foot** (ft^2). A square with sides that are one yard in length has an area of one **square yard** (yd^2).

Problem 1 Determine the area of the rectangle with length of 10 feet and width of 6 feet.

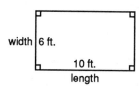

Solution Using the formula: $A = L \cdot W$
$$A = 10 \cdot 6$$
$$A = 60$$

Thus, the area is 60 square feet (60 ft^2).

Some important relationships exist among the square units:

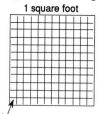

144 square inches is equivalent to 1 square foot.

$144 \text{ in}^2 = 1 \text{ ft}^2$

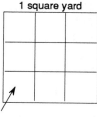

9 square feet is equivalent to 1 square yard.

$9 \text{ ft}^2 = 1 \text{ yd}^2$

To find the area of a rectangle, the length and width must be stated in the same units. Thus, we state the dimensions of a rectangle that measures 3 ft. by 24 in. as 3 ft. by 2 ft. or 36 in. by 24 in. We will follow this procedure when we state the dimensions of any polygonal region.

Problem 2 An 8 ft. by 11 ft. rectangular wall can be painted for $13. Find the cost to paint a rectangular wall 12 ft. by 11 ft.

Solution Set up a proportion as:
$$\frac{\text{original area}}{\text{new area}} = \frac{\text{original cost}}{\text{new cost}}$$

original area $\dfrac{8 \cdot 11}{12 \cdot 11} = \dfrac{13.00}{x}$ ($) original cost
new area   ($) new cost

Reduce before solving for x. $\dfrac{8 \cdot \cancel{11}}{12 \cdot \cancel{11}} = \dfrac{13.00}{x}$ becomes $\dfrac{2}{3} = \dfrac{13}{x}$

The proportion is solved as:

$$2x = 3 \cdot 13$$

Recall: If $\frac{a}{b} = \frac{c}{d}$ then $ad = bc$.

$$2x = 39$$
$$x = \frac{39}{2} = 19.5$$

Since x stands for dollars, the cost is $19.50.

It is possible to find the **surface area** of a rectangular solid. The surface area refers to the area of all the polygons that make up the solid figure. The surface area is always measured in square units.

Problem 3   Find the surface area of the accompanying rectangular solid.

Solution   To find the surface area, we must find the areas of the six rectangles that make up the solid and add them.

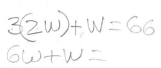

4 cm
3 cm
5 cm

Area of base	$= 5 \cdot 3 = 15$ cm^2
Area of top	$= 15$ cm^2
Area of front	$= 5 \cdot 4 = 20$ cm^2
Area of back	$= 20$ cm^2
Area of side	$= 4 \cdot 3 = 12$ cm^2
Area of opposite side	$= 12$ cm^2

Surface area $= 15 + 15 + 20 + 20 + 12 + 12 = 94$ cm^2

$3(2w) + W = 66$
$6w + W =$
7

Problem Set 14.1

$16\text{^{?}}$

$35 += \$ 14$

1. Find the areas of rectangles with the following dimensions.
    2 feet
    a. 6 ft. by 36 in. (in ft^2 or in^2) $18\ ft^2$
    b. 2 yds. by 9 ft. (in yd^2 or ft^2) $6\ yd^2$
    c. 2 yds. by 36 in. (in yd^2 or in^2) $2\ yd^2$
    139 feet

2. A room is 9 feet by 21 feet. What is the cost of carpeting the room at $12.98 per square yard? $21 sy$
    272.58

3. If the smaller rectangle in the diagram has dimensions 10 ft. by 30 ft. and the larger rectangle has dimensions 20 ft. by 60 ft., find the area of the shaded region. $900 ft^2$

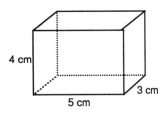

60
20   10   30
1200

4. The length of a rectangle is 3 more than 2 times the width. If the perimeter of the rectangle is 66 feet, what is the area?

5. A 5 ft. by 7 ft. rectangular wall can be painted for $14. Find the cost to paint a 10 ft. by 21 ft. rectangular wall. $210 = 34$
    $72 = $

6. A 6 ft. by 12 ft. rectangular wall can be painted for $20. Find the cost to paint a 9 ft. by 12 ft. rectangular wall. $108 = 30$

7. What is the surface area of a rectangular solid that is 10 feet by 8 feet by 6 feet?

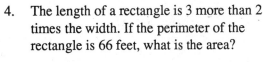

395

8. What is the surface area of a rectangular solid that is 4 inches long, 4 inches wide, and 4 inches tall?

9. Consider the accompanying diagram and determine if the following statement is true or false:

    $$\frac{5}{10} = \frac{3}{BE}$$

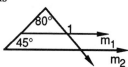

10. A 24' vertical tower is to be anchored by a wire from the ground to the top of the tower. The wire is to be anchored 18' from the base of the tower. What should the length of the wire be?

11. In the diagram, m_1 is parallel to m_2. Find $m\angle 1$.

12. The non-equal side of an isosceles triangle is one foot less than three times either equal side. If the perimeter of the triangle is 14 feet, find each side.

13. This figure can be described as a:
    a) quadrilateral
    b) parallelogram
    c) rectangle
    d) rhombus
    e) all of these
    f) a, b, and c
    g) a, b, and d

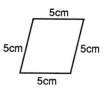

14. Round 6.75 meters to the nearest tenth of a meter.

15. Express $\frac{32}{48}$ in lowest terms.

16. Express $\frac{5}{8}$ as a decimal.

17. Divide: $2\frac{1}{5} \div 5$

18. Express $2\frac{1}{4}\%$ as a decimal.

19. $\frac{12}{8} = \frac{24}{?}$

20. Express 2.007 as an improper fraction.

21. The sum of two numbers is 11. Their product is 24. What is the sum of their reciprocals?

We now discuss the formulas for areas of other polygons.

Consider the square region ABCD shown in which the letter s represents the length of any of the four equal sides. Since a square is a rectangle, the area can be determined by using the formula for the area of a rectangle.

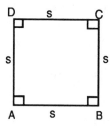

$A = L \bullet W$. Thus, we obtain: $A = L \bullet W$ or $A = s \bullet s$ or $A = s^2$

> The area (A) of a square is the square of the length of a side (s). $(A = s^2)$

Problem 4   Consider the square ABCD
shown in which the measure
of a side is 5 feet. Find its
area.

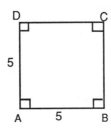

Solution   Using the formula $A = s^2$ and substituting 5 for s gives:
$$A = s^2 = 5^2 = 25$$

Thus, the area is 25 square feet (25 ft^2).

Consider the parallelogram ABCD shown at the right. The line
segment DE, drawn from a vertex and perpendicular to the base, is
called the **height** or the **altitude**.

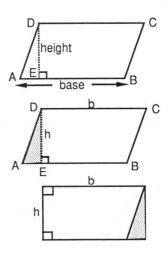

It can be demonstrated that the area of the parallelogram is the same
as the area of a rectangle. In the figure at the right, if $\triangle AED$ is removed
from the left of the parallelogram and attached to the right, the result
is seen to be a rectangle with the same base (length) and height (width)
as the original parallelogram. Since the length of the rectangle equals
the base (b) of the parallelogram and the width equals the height (h),
the formula for the area of the rectangle can be transformed into a
formula for the area of the parallelogram.
$$A = L \bullet W \text{ becomes } A = b \bullet h \text{ or } A = bh$$

> The area (A) of a parallelogram equals the product
> of its base (b) and height (h).   ($A = b \bullet h$)

Problem 5   Find the area of the parallelogram shown.

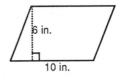

Solution   Since the base (b) is 10 in. and the height (h) is 6 in., the formula
$$A = bh \text{ becomes } A = 10 \bullet 6 = 60.$$

Thus, the area of the parallelogram is 60 square in. (60 in^2).

Chapter 6

Consider the accompanying parallelogram
ABCD in which DE is the height (h) and
AB is the base (b). The diagonal BD
forms △ABD.

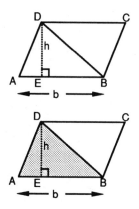

It can be demonstrated that the area of a
triangle is one-half the area of a
parallelogram. Since the area of a
parallelogram is $A = b \cdot h$ or $A = bh$,

the area of △ABD is $\frac{1}{2}$ this area or

$A = \frac{1}{2} \cdot b \cdot h$ or $A = \frac{1}{2}bh$.

> The area (A) of a triangle equals half the product
> of the base (b) and the height (h). ($A = \frac{1}{2}bh$)

There is only one formula for the area of a triangle, but three figures are shown
below to aid in applying the formula. In all cases the formula is:

$$A = \frac{1}{2}bh$$

where b is the length of
the side called the base
and h is the height (the
perpendicular distance
from the vertex to the
base or extended base).

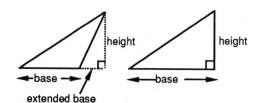

Problem 6   Consider the triangle shown in which
              b = 10 cm. and h = 8 cm.
              Find the area of the triangle.

Solution   Since b = 10 cm. and h = 8 cm.

  $A = \frac{1}{2} \cdot b \cdot h = \frac{1}{2} \cdot 10 \cdot 8 = 40$

  Thus, the area is 40 square cm (40 cm^2).

Problem 7   Find the area of the accompanying figure.

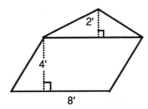

Solution   The figure above needs to be seen as a parallelogram and a triangle which do not overlap. The area of the entire figure is found by adding the area of the parallelogram to the area of the triangle.

Area of parallelogram = $b \cdot h = 8 \cdot 4 = 32$ ft^2

Area of triangle = $\frac{1}{2} \cdot b \cdot h = \frac{1}{2} \cdot 8 \cdot 2 = 8$ ft^2

Total area = $32 + 8 = 40$ ft^2

Consider the accompanying trapezoid ABCD with height DE (indicated as h). If the lower base AB is a and the upper base DC is b, it is possible to develop the formula for the area of the trapezoid.

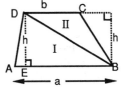

Draw a diagonal BD and form two triangles. Thus,

area (A) of trapezoid	= area of triangle I	+ area of triangle II
A	= $\frac{1}{2}$ah	+ $\frac{1}{2}$bh
A	= $\frac{1}{2}$h(a + b)	

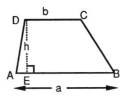

> The area (A) of a trapezoid equals half the product of the height (h) and the sum of the bases (a and b).
>
> $A = \frac{1}{2}h(a + b)$

Problem 8   Find the area of the accompanying trapezoid ABCD in which the lower base is 10 feet, the upper base is 6 feet, and the height is 5 feet.

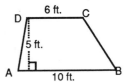

Solution   Since a = 10, b = 6, and h = 5,

$$A = \tfrac{1}{2}h(a + b) = \tfrac{1}{2} \cdot 5 \cdot (10 + 6) = \tfrac{1}{2} \cdot 5 \cdot (16) = 40$$

Thus, the area is 40 square ft. (40 ft^2)

Chapter 6

To find the area of plane regions having unusual shapes, it is necessary to break the region up into figures that are familiar. The total area of the region is then determined by adding the areas of the known figures.

Problem 9   Find the area of the accompanying figure.

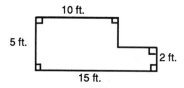

Solution   The figure is seen as two non-overlapping rectangles.

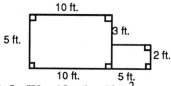

Area of larger rectangle is L •W = 10 • 5 = 50 ft²
Area of smaller rectangle is L • W = 5 • 2 = 10 ft²
Total Area = 50 + 10 = 60 ft²

Problem Set 14.2

1. Find the area of a square if one side measures 9 meters.

2. Find the area of a parallelogram if the base measures 3 feet and the height measures 5 inches, and express the result in square feet and square inches.

3. Find the area of a triangle if the base measures 2 yards and the height measures 4 feet, and express the result in square yards and square feet.

4. Find the area of a trapezoid if the lower base is 5 inches, the upper base is 2 inches, and the height is 4 inches.

In problems 5-8 find the area of each figure.

5.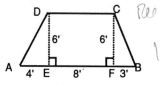

6.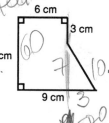

7.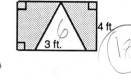

8.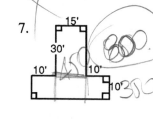

9. Find the area of the shaded region.

10. If the sides of a rectangle are doubled, the area is increased how many times?

11. A person needs to buy wall-to-wall carpeting for a living room which is 18 feet long and 12 feet wide. If carpet costs $6 a square yard, what is the total cost?

400

Review Competencies

12. Are these triangles similar?

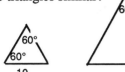

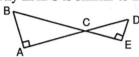

13. Explain why ΔABC is similar to ΔEDC.

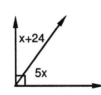

14. An eight foot ladder is placed against a wall. The bottom of the ladder is $\sqrt{39}$ feet away from the wall. How high up the wall does the ladder reach?

15. Find m ∠1 if m₁ is parallel to m₂.

16. Find the value of x and the measure of the two complementary angles.

17. The figure below can be described as a:
    a) quadrilateral
    b) parallelogram
    c) rectangle
    d) rhombus
    e) square
    f) all of these

18. Round 47,261 feet to the nearest hundred feet.

19. Add: $3\frac{2}{5} + 2\frac{1}{4}$

20. Subtract: $10\frac{1}{3} - 7\frac{3}{4}$

21. Multiply: $0.004 \cdot 12.1$

22. Divide: $0.15 \div 0.03$

23. What is 30% of 70?

24. Express 0.04% as a decimal.

25. Express 1.23 as an improper fraction.

Up to this point, we have considered the areas of polygons. We now turn our attention to the area of a circle.

It can be demonstrated that the area of a circle is slightly greater than $3r^2$ where r is the radius of the circle. The exact area of a circle is πr^2.

> The area (A) of a circle is the product of π and the square of the radius (r).   ($A = \pi r^2$)

401

Problem 10 Using 3.14 for π, find the area of the circle whose radius is 10 inches.

Solution Using the formula $A = \pi r^2$ where $r = 10$,

$$A = 3.14 \cdot 10^2$$

$$A = 3.14 \cdot 100$$

$$A = 314$$

Thus, the area of the circle is approximately 314 square in. (314 in^2).

Note

If it is required that the result be left in terms of π units, then the area would be expressed exactly as 100π in^2.

Problem 11 Using $\frac{22}{7}$ for π, find the area of the circle with radius 7 feet.

Solution Using the formula $A = \pi r^2$ where $r = 7$,

$$A = \frac{22}{7} \cdot 7^2$$

$$A = \frac{22}{\cancel{7}} \cdot \cancel{49}^{7}$$

$$A = 154$$

Thus, the area of the circle is approximately 154 ft^2 (or 49π ft^2).

Problem 12 Find the area of the figure shown.

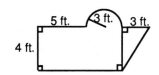

Solution The figure is composed of:
1) a semi-circle with radius 3 ft.
2) a rectangle with $L = 5 + 6 = 11$ ft. and $W = 4$ ft.
3) a triangle with base 3 ft. and height 4 ft.

The area of the entire figure is found by separately finding the areas of these non-overlapping figures and then adding them. Using 3.14 for π:

area of the semi-circle:

$\frac{1}{2} \cdot \pi \cdot r^2 = \frac{1}{2} \cdot 3.14 \cdot 3^2 = \frac{1}{2} \cdot 3.14 \cdot 9 = 14.13 \ \text{ft}^2$

area of the rectangle:
$L \cdot W = 4 \cdot 11 = 44 \ \text{ft}^2$

area of the triangle:

$\frac{1}{2} \cdot b \cdot h = \frac{1}{2} \cdot 3 \cdot 4 = 6 \ \text{ft}^2$

Thus, the total area is: $44 + 6 + 14.13 = 64.13 \ \text{ft}^2$

Problem Set 14.3

1. Using 3.14 for π, find the area of a circular pool whose diameter is 60 feet.

2. Using $\frac{22}{7}$ for π, find the area of the circle whose radius is $3\frac{1}{2}$ inches.

Find the total areas for the figures in problems 3 and 4. Use 3.14 for π.

3.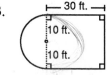

4.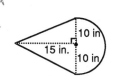

Find the areas of the shaded regions for the figures in problems 5 and 6. Leave the answers in terms of π.

5.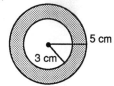

6.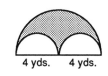

7. Find the area of the figure consisting of a square and four semi-circles. Use 3.14 for π.

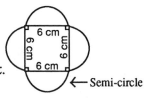

← Semi-circle

8. Find the shaded area using 3.14 for π.

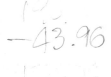

9. If the radius of a circle is doubled, what happens to the area?

10. The lookout in the "crow's nest" of a ship can see for a distance of 10 miles. How many square miles can be observed? Use 3.14 for π.

11. How many times as much water will flow through a pipe 20 inches in diameter as through a pipe 10 inches in diameter if the water pressure is the same for both pipes?

Review Competencies

12. What combination of the following formulas is needed to calculate the area of the shaded region?

   a. $A = \pi r^2$ b. $A = L \cdot W$

   c. $A = \frac{1}{2}bh$ d. $A = \frac{1}{2}h(a + b)$

i. a and b   ii. c and d

iii. a and d   iv. b and d

v. b and c

13. What combination of the following formulas is needed to calculate the area of the shaded region?
    a. $A = \pi r^2$ b. $A = L \cdot W$
    c. $A = \frac{1}{2}h(a + b)$ d. $A = \frac{1}{2}bh$
    i.   a and b   ii. b and c
    iii. a and c   iv. a and d
    v.   b and d

14. If $\overline{AC} \parallel \overline{DE}$, explain why $\triangle ABC$ and $\triangle DEF$ are similar.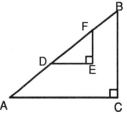

15. In the figure shown, DE is parallel to AB. If AC = 12, CD = 4, CE = 8, find BC.

16. Which of the sets of pictured triangles contain(s) similar triangles?
    i.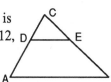

    ii. 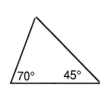   iii.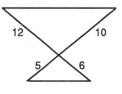

17. This triangle is:
    a. equilateral and right
    b. isosceles and acute
    c. scalene and obtuse
    d. isosceles and obtuse
    e. none of these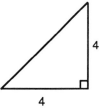

18. If $\overline{AB} \parallel \overline{CD}$, express the measure of $\angle CHE$.
    i. $a°$
    ii. $180° - a°$
    iii. $a° - 180°$
    iv. $90° - a°$
    v. $a° - 90°$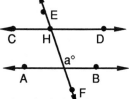

19. If the circumference of a circle is 16π meters, find the radius of the circle.

20. Express $\frac{7}{4}$ as a decimal.

21. Express 0.55 as a fraction.

22. Multiply: $\frac{3}{4} \cdot \frac{5}{6} \cdot \frac{8}{5}$

23. Divide: $6 \div 2\frac{1}{3}$

24. 30 is what percent of 15?

25. Evaluate: $4 \cdot 2 - 3 + 1 \cdot 5 \div 5 - 2$

26. Express 270,000 in scientific notation.

Section 15 Volume

It is often necessary to measure the amount of space enclosed by a solid figure. This amount of space is the **volume** of the solid figure.

The basic unit for the measurement of volume is the **unit cube**. Recall that a cube is a polyhedron. A cube is also a rectangular solid composed of six square faces. A unit cube is a cube with an edge that measures 1 unit in length. Notice in the figure at the right that a cube has three dimensions: length, width, and height.

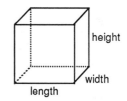

Cubic units must be counted to obtain the volume of a solid. Thus, **volume is always measured in terms of cubic units**. The figure at the right shows a rectangular solid which is next to the basic unit cube. Notice that the solid occupies as much space as 12 units. Thus, the volume of this solid is 12 cubic units. Notice also that the volume is represented by a number.

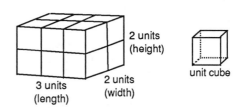

Observe in the solid above that the length is 3 units, the width is 2 units, and the height is 2 units. The volume may be determined by finding the product of the length, the width, and the height: $3 \cdot 2 \cdot 2 = 12$.

> The volume (V) of a rectangular solid is the product of the length (L) and width (W) and height (H).
> $$(V = L \cdot W \cdot H)$$

A **cubic inch** (in^3) is a very common unit of measurement. It is the volume of a cube that is one inch long, one inch wide, and one inch high. The **cubic foot** (ft^3) and the **cubic yard** (yd^3) are other common basic units of measurement.

Problem 1   Find the volume of the accompanying rectangular solid.

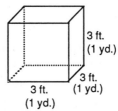

Solution   Notice that this rectangular solid is a cube with an edge that measures 3 feet or 1 yard. Thus,

$$V = L \cdot W \cdot H = 3 \cdot 3 \cdot 3 = 27 \text{ cubic ft } (27 \text{ ft}^3)$$
$$V = L \cdot W \cdot H = 1 \cdot 1 \cdot 1 = 1 \text{ cubic yd } (1 \text{ yd}^3)$$

Chapter 6

Problem 2   Find the volume of the accompanying
rectangular solid.

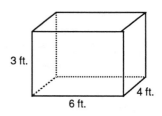

3 ft.

4 ft.

6 ft.

Solution $V = L \cdot W \cdot H = 6 \cdot 4 \cdot 3 = 72$ cubic ft (72 ft^3)

Problem Set 15.1

1. Consider the accompanying
    rectangular solid and
    answer the following
    questions:

    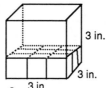

    3 in.

    3 in.

    3 in.

    a. How many cubes are in
        each layer of 1 inch cubes?
    b. How many layers will fit inside the solid?
    c. What is the volume of this solid in
        cubic inches?

2. Find the volumes of boxes having the follow-
    ing dimensions. Express the volume in cubic
    inches, cubic feet, or cubic yards. Remember to
    convert all measurements to the same unit
    before finding the volume.
    a. 2 ft. by 4 ft. by 48 in.
    b. 2 yd. by 9 ft. by 144 in.
    c. 24 in. by 12 in. by 3 ft.

3. A family is moving to Florida. The mover
    estimates their possessions will take up 750
    cubic feet of space. Would a truck with a
    container 8 feet wide, 6 feet high, and 16 feet
    long hold all the possessions?

4. Compare the volumes of three 4-inch cubes
    and four 3-inch cubes, and indicate which is
    greater.

5. What is the volume of a rectangular solid that
    is 4 yards long, 4 yards wide, and 4 yards high?

6. A rectangular solid of dimensions 3 by 5 by 7
    (cm) can be filled with 100 jellybeans. How
    many jellybeans will fill a rectangular solid 3
    by 10 by 21 (cm)?

7. Find the volume of a rectangular solid with
    $L = \frac{5}{24}$ feet, $W = \frac{8}{3}$ feet, and $H = \frac{21}{16}$ feet.

Review Competencies

8. A square has the same area as a triangle with
    base 16 cm. and height 2 cm. Find the length
    of the side of the square.

9. Find the ratio of the areas of a smaller circle to
    a larger circle if one has a diameter of 12 feet
    and the other a diameter of 16 feet.

10. Find the length of x.
    $\overline{DE} \parallel \overline{AB}$.

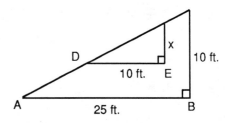

D

10 ft.   E

x

10 ft.

A

25 ft.

B

11. What combination of the formulas below is needed for calculating the area of the figure shown? (The figure consists of a triangle and a semi-circle.)

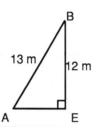

  a. $A = \frac{1}{2}\pi r^2$

  b. $A = L \cdot W$

  c. $A = \frac{1}{2}h(a + b)$

  d. $A = \frac{1}{2}bh$

  i.   All of these   ii. a, b, c

  iii. b, c   iv. a, c

  v. c, d   vi. a, d

12. Find the area of the accompanying right triangle.

13. Express $5\frac{3}{8}$ as an improper fraction.

14. Add: $26.014 + 0.03 + 1.0005$

15. Express 0.250 as a fraction.

16. Express $5\frac{1}{4}\%$ as a decimal.

17. Insert < or > in the space to make the following true: $3.\overline{4}$ [   ] $3.\overline{41}$

We now discuss the formulas for the volumes of some other important polyhedrons.

Recall that the polyhedron at the right is called a **rectangular pyramid** or a **right pyramid**. Notice that there are four triangular faces and one rectangular face. The base of the rectangular pyramid is a rectangle.

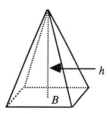

> The volume (V) of a rectangular pyramid is $\frac{1}{3}$ the product of the area of the base (B) and the height (h).
>
> $(V = \frac{1}{3} \cdot B \cdot h)$

Problem 3   Find the volume of the pyramid
at the right with the following
dimensions: the rectangular base
is 10 feet by 3 feet, and the height
is 6 feet.

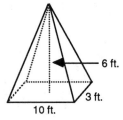

— 6 ft.

3 ft.

10 ft.

Solution   Since the base is a rectangle, the area of the base (B) is:

$$B = L \cdot W$$
$$B = 10 \cdot 3$$

Thus, B = 30 sq. ft.

The formula for the volume of the pyramid is now applied:

$$V = \frac{1}{3} \cdot B \cdot h$$
$$V = \frac{1}{3} \cdot 30 \cdot 6$$
$$V = 60$$

Thus, the volume of the pyramid is 60 cubic ft (60 ft³).

Recall that the polyhedron shown at the right is
called a **right triangular prism**.

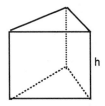

h

> The volume (V) of a right triangular prism is the
> product of the area of the base (B) and the height (h).
> (V = B • h)

The base of the right triangular prism is a **triangle**.

Problem 4   Find the volume of the right
triangular prism in which the
triangular base (B) has a base
(b) of 6 inches and a height (h)
of 3 inches. The height of the
prism itself is 10 inches.

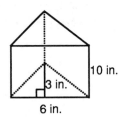

10 in.

3 in.

6 in.

Solution   Since the base is a triangle, the area (B) of the triangle is:

$$B = \frac{1}{2} \cdot b \cdot h$$

$$B = \frac{1}{2} \cdot 6 \cdot 3$$

$$B = 9$$

Thus, the area of the base is 9 square inches. To find the volume of the right triangular prism, we have:

$$V = B \cdot h$$

$$V = 9 \cdot 10$$

$$V = 90$$

Thus, the volume is 90 cubic in. (90 in³).

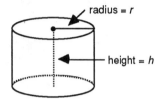

Recall that the solid figure at the right is a **right circular cylinder**.

> The volume (V) of a right circular cylinder is the product of the area of the circular base and the height (h).   (V = πr²h)

Note

Sometimes the formula for the volume of the right circular cylinder is stated as V = B • h, in which B denotes the area of the circular base.

Problem 5   Find the volume of a right circular cylinder if the radius of the circular base is 10 meters, and the height of the cylinder is 20 meters. Use 3.14 for π.

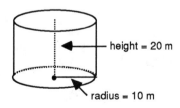

Solution $$V = \pi \cdot r^2 \cdot h$$

$$V = 3.14 \cdot 10^2 \cdot 20$$

$$V = 3.14 \cdot 100 \cdot 20$$

$$V = 3.14 \cdot 2000$$

$$V = 6,280$$

Thus, the volume of the right circular cylinder is approximately 6,280 cubic meters (6280 m³) or 2,000πm³.

Chapter 6

Recall that the solid figure shown at the right is a **right circular cone**.

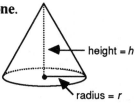

The volume (V) of a right circular cone is $\frac{1}{3}$ the product of the area of the circular base and the height (h). $(V = \frac{1}{3} \cdot \pi r^2 h)$

Note

Sometimes the formula for the volume of the right circular cone is stated as $V = \frac{1}{3} \cdot B \cdot h$, in which B denotes the area of the circular base.

Problem 6 Find the volume of a right circular cone if the radius of the circular base is 10 feet, and the height of the cone is 15 feet. Use 3.14 for π.

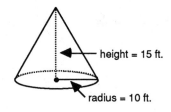

Solution

$V = \frac{1}{3} \cdot \pi \cdot r^2 \cdot h$

$V = \frac{1}{3} \cdot 3.14 \cdot 10^2 \cdot 15$

$V = \frac{1}{3} \cdot 3.14 \cdot 100 \cdot 15$

$V = 3.14 \cdot 500$

$V = 1,570$

Thus, the volume of the right circular cone is approximately 1,570 cubic ft. $(1,570 \text{ ft}^3)$ or $500\pi \text{ ft}^3$.

Definition

A **sphere** is a set of points in space equally distant from a given point, called the center. A sphere may be represented as shown at the right.

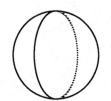

> The volume (V) of a sphere is $\frac{4}{3}$ the product of π and the cube of the radius (r). $(V = \frac{4}{3} \cdot \pi \cdot r^3)$

Problem 7

Find the volume of a sphere with radius 3 feet. Use 3.14 for π.

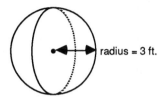

radius = 3 ft.

Solution

$V = \frac{4}{3} \cdot \pi \cdot r^3 = \frac{4}{3} \cdot 3.14 \cdot 3^3$

$V = \frac{4}{3} \cdot 3.14 \cdot \frac{27}{1} = 113.04$

Thus, the volume is approximately 113.04 ft^3 or 36π ft^3.

Problem 8

A spherical tank of radius 1 meter can be filled with a liquid for $12. How much would it cost to fill a spherical tank of radius 3 meters with the same liquid?

Solution

We set up a proportion: $\dfrac{\text{original volume}}{\text{new volume}} = \dfrac{\text{original cost}}{\text{new cost}}$

Volume $= \frac{4}{3}\pi r^3$

$\dfrac{\text{original volume } \frac{4}{3}\pi 1^3}{\text{new volume } \frac{4}{3}\pi 3^3} = \dfrac{12}{x} \quad \begin{array}{l}\text{original cost}\\ \text{new cost}\end{array}$

$\dfrac{\frac{4}{3}\pi 1}{\frac{4}{3}\pi 27} = \dfrac{12}{x}$

We obtain: $\dfrac{1}{27} = \dfrac{12}{x}$ or $x = 27 \cdot 12$ or $x = 324$ or $324

Thus, the new cost is $324.

Chapter 6

Use the figure of the right circular cone
at the right. Identify the type of
measure needed for each of the following:

    a. the radius of the circular base
    b. the area of the circular base
    c. the interior region of the cone

Solution   a. The length of a quantity is always measured in **linear** units.
    b. The area of any region is always measured in **square** units.
    c. The space (volume) inside a solid figure is always measured
       in **cubic** units.

Problem Set 15.2

1. Find volumes for the following rectangular
   pyramids.
   a. The dimensions of the base are 4 inches
      by 3 inches and the height is 8 inches.
   b. Each side of the base measures 6 feet and
      the height is 12 feet.

2. Find volumes for the following right triangular
   prisms.
   a. The triangle has a base of 9 yards and
      height of 4 yards and the height of the
      prism is 20 yards.
   b. The triangle has a base of 2 feet and a
      height of 6 inches and the height of the
      prism is 1 yard. (Express the volume in
      terms of cubic inches.)

3. Find volumes for the following right circular
   cylinders.
   a. The radius of the base is 2 inches and the
      height of the cylinder is 10 inches.
      Use 3.14 for π.
   b. The radius of the base is 5 feet and the
      height of the cylinder is 20 feet.
      Use 3.14 for π.

4. Find volumes for the following right circular
   cones.
   a. The radius of the base is 5 meters and the
      height of the cone is 6 meters.
      Use 3.14 for π.
   b. The diameter of the base is 6 feet and the
      height of the cone is 5 feet. Use 3.14 for π.

5. Find the volume of a sphere with diameter 20
   feet. Use 3.14 for π.

6. Find the volume of a sphere with radius 2 feet.
   Use 3.14 for π.

7. A spherical tank of radius 2 meters can be
   filled with liquid for $6.00. How much would
   it cost to fill a spherical tank of radius 10
   meters with the same liquid?

8. A spherical tank of radius 2 meters can be
   filled with a liquid for $16. How much would
   it cost to fill a spherical tank of radius 4 meters
   with the same liquid?

9. A right circular cylinder of radius 2 yd. and
   height 1 yd. can be filled with a liquid for $6.
   How much would it cost to fill a cylinder of
   radius 4 yd. and height 3 yd.?

10. For each figure below, indicate the appropriate type of measure needed. Mark L for linear, S for square, C for cubic.

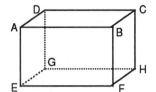

  a. line segment AB, one of the edges of the rectangular solid
  b. region ABCD, one of the faces of the rectangular solid
  c. region N, the interior of the pyramid

Review Competencies

11. Find the area of right triangle ABC.

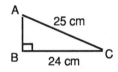

12. Find the area of the trapezoid. Each unit on the graph paper represents meters.

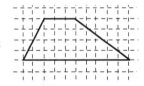

13. A pizza with radius 4 in. sells for $1.50. Find the charge for a pizza of radius 12 inches.

14. The sun is shining in a way that a 7 foot pole casts a shadow of 4 feet. Find the height of a pole that casts a shadow of 12 feet.

15. The length of a rectangle is 4 times the width. If the perimeter of the rectangle is 120 yards, find the length and width.

16. A large rectangle has dimensions 4 ft. by 3 ft. A smaller rectangle has dimensions $3\frac{1}{3}$ ft. by $2\frac{1}{2}$ ft. How much larger is the area of the large rectangle?

17. If $\overleftrightarrow{AD} = 35$ in., find $\overleftrightarrow{AB}$.

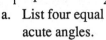

18. Use the diagram in which m_1 is parallel to m_2:

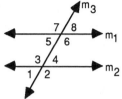

  a. List four equal acute angles.
  b. If $m\angle 6 = 100°$, find the measure of $\angle 4$.

19. Add: $1\frac{1}{3} + 1\frac{3}{4}$

20. Divide: $0.246 \div 0.2$

21. Express 37.5% as a decimal.

22. Express $\frac{7}{8}$ as a percent.

23. What is 40% of 50?

24. Express 0.444 as a fraction in lowest terms.

25. What is the next term in the following arithmetic progression?
$$10, 6, 2, -2, \underline{\quad}$$

26. A **harmonic progression** is a sequence of numbers whose reciprocals form an arithmetic progression. What is the next term in the following harmonic progression?
$$\frac{1}{3}, \frac{1}{6}, \frac{1}{9}, \underline{\quad}$$

27. Determine the formula for computing the surface area of the accompanying solid figure which is composed of three attached cubes whose edges each measure x.

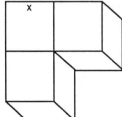

  a. $10x^2$
  b. $12x^2$
  c. $14x^2$
  d. $16x^2$

Section 16   The Metric System

Although the English system of measurement (inch, foot, yard, etc.) is most commonly used in the United States, most industrialized countries use the **metric system** of measurement. One of the advantages of the metric system of measurement is that the units are based on powers of ten, making it much easier than the English system to change from one unit of measure to another.

The basic unit for **linear measure** in the metric system is the **meter** (abbreviated as m). A meter is slightly larger than a yard (approximately 39.37 inches). The most commonly used units related to the meter are shown in the following table.

1 millimeter (mm) = $\frac{1}{1000}$ of a meter

(1 mm = 0.001 m and 1 m = 1000 mm)

1 centimeter (cm) = $\frac{1}{100}$ of a meter

(1 cm = 0.01 m and 1 m = 100 cm)

1 decimeter (dm) = $\frac{1}{10}$ of a meter

(1 dm = 0.1 m and 1 m = 10 dm)

1 kilometer (km) = 1000 meters

(1 km = 1000 m and 1 m = 0.001 km)

(1 kilometer is approximately $\frac{3}{5}$ miles)

Examples

1. 7.2 m = 7.2 (1000 mm) = 7,200 mm
2. 7.2 m = 7.2 (100 cm) = 720 cm
3. 7.2 m = 7.2 (10 dm) = 72 dm
4. 7.2 m = 7.2 (0.001 km) = 0.0072 km

Problem 1   Packages are wrapped with tape that costs 40 cents per meter. If each package requires 25 centimeters of tape, find the cost of wrapping 500 packages.

Solution   25 cm • 500 packages = 12,500 cm

= (12,500) (.01 m) = 125 meters

The cost is (125)(.4) = $50

The basic metric unit for measuring weight (mass) is the **gram** (g). The prefixes (milli, centi, deci, and kilo) are used with this basic unit. A gram is much smaller than a pound. A kilogram (kg = 1000 grams) is approximately 2.2 pounds.

Similarly, the basic metric unit for measuring volume is the **liter** (l). One liter is approximately 1.05 quarts. Here again, the metric prefixes apply, so that 1 kiloliter (kl) = 1000 liters. Since volume is also measured in cubic units (see section 15, Chapter 6), there is a relationship between liters and cubic centimeters given by: 1 liter = 1000 cubic centimeters.

Problem Set 16.1

1. Convert:
   a. 3.7 m = ____ mm
   b. 5.2 m = ____ cm
   c. 9.1 m = ____ dm
   d. 4.8 m = ____ km

2. What is the perimeter, in kilometers, of a four-sided figure having sides measuring 720 meters, 300 meters, 450 meters, and 200 meters?

3. Pipe, having a diameter of 6 mm., is to be placed along a path measuring 10 km. If pipe costs 20 cents per meter, find the total expense (in dollars) of laying the pipe.

4. Convert:
   a. 10 g = ____ kg
   b. 200 kg = ____ g

5. What is the volume, in cubic centimeters, of a 1.37 liter flask?

7. Round the reading on the gram scale shown below to the nearest ten grams.

8. For each item, identify the appropriate measure (cm, cm², liter) needed for the figure.
   a. segment BF
   b. the area of plane region BFGC
   c. the interior of the solid region.

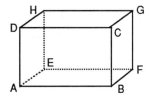

Review Competencies

6. Find the measure of AB to:
   a. the nearest centimeter.
   b. the nearest $\frac{1}{2}$ centimeter.

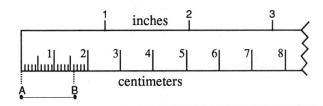

Chapter 6 Summary

1. Quadrilaterals

   a. Trapezoid: Exactly one pair of parallel sides.
   b. Parallelogram: Opposite sides are equal and parallel. Consecutive angles are supplementary. Diagonals bisect each other.
   c. Rectangle: A parallelogram with four right angles. Diagonals are equal.
   d. Rhombus: A parallelogram with four equal sides. Diagonals are perpendicular.
   e. Square: A rectangle with four equal sides or a rhombus with four right angles. Diagonals are equal, perpendicular, and bisect.

2. Measurement

   a. Line segments are measured in linear units, area in square units, and volume in cubic units or liters. 1 liter = 1,000 cm^3
   b. Common metric units for length include:

$$\text{millimeter (mm)} = \tfrac{1}{1000} \text{ of a meter}$$

$$\text{centimeter (cm)} = \tfrac{1}{100} \text{ of a meter}$$

$$\text{decimeter (cm)} = \tfrac{1}{10} \text{ of a meter}$$

$$\text{kilometer (km)} = 1,000 \text{ meters}$$

3. Angles

   a. Angles are measured in degrees. Acute angles measure less than 90°, right angles measure exactly 90°, obtuse angles measure more than 90° but less than 180°, and straight angles measure exactly 180°.
   b. Complementary angles have a sum of their measures of 90°, and supplementary angles have a sum of their measures of 180°.
   c. $\angle A \cong \angle B$ means that m$\angle A$ = m$\angle B$.
   d. Vertical angles are congruent.

4. Parallel lines

   a. Parallel lines lie in the same plane and do not intersect.
   b. Two lines are parallel if and only if alternate interior or alternate exterior or corresponding angles are congruent.

5. Triangles

   a. The sum of the measures of the interior angles is 180°.
   b. Any exterior angle has a measure equal to the sum of the measures of the two remote interior angles.
   c. Equilateral triangles have three equal sides (each angle measures 60°), isosceles triangles have two equal sides (and two congruent angles opposite these sides), and scalene triangles have no equal sides.

d. The Pythagorean Theorem states that in a right triangle, the sum of the squares of the legs is equal to the square of the hypotenuse.

6. Polygons

a. Regular polygons have equal sides and congruent angles.
b. The sum of the measures of the interior angles of a convex polygon of n sides is $(n-2) \cdot 180°$.

7. Similar Triangles

a. Similar triangles have identical shapes but not necessarily the same size.
b. Two triangles are similar if and only if:
1. two angles of one triangle are congruent to two angles of the other triangle, or
2. all three pairs of corresponding sides are proportional, or
3. two corresponding pairs of sides are proportional and the angles included between these sides are congruent.

8. Geometric Formulas for Area

a. Rectangle $\qquad A = LW$

b. Square $\qquad A = s^2$

c. Parallelogram $\qquad A = bh$

d. Triangle $\qquad A = \frac{1}{2}bh$

e. Trapezoid $\qquad A = \frac{1}{2}h(a + b)$

f. Circle $\qquad A = \pi r^2$ (Also: $C = \pi d = 2\pi r$)

g. Rectangular Solid   The surface area is found by finding the area of the six rectangles that form the solid and taking the sum of these six numbers.

9. Geometric Formulas for Volume

a. Rectangular Solid $\qquad V = LWH$

b. Rectangular Pyramid $\qquad V = \frac{1}{3}Bh$

c. Right Triangular Prism $\qquad V = Bh$

d. Right Circular Cylinder $\qquad V = \pi r^2 h$

e. Right Circular Cone $\qquad V = \frac{1}{3}\pi r^2 h$

f. Sphere $\qquad V = \frac{4}{3}\pi r^3$

Chapter 6 Self-Test

1. Round 78.349 cubic feet to the nearest tenth of a cubic foot.
   a. 78.3 ft^3 b. 78.35 ft^3
   c. 78.4 ft^3 d. 80 ft^3

2. Use the protractor to find m∠ABC. Then classify the angle by its measure.

   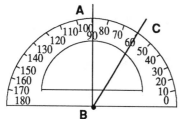

   a. 90°, right   b. 60°, obtuse
   c. 30°, obtuse   d. 30°, acute

3. Round 14.645 cubic yards to the nearest ten cubic yards.
   a. 10 yd^3 b. 14.6 yd^3
   c. 14.7 yd^3 d. 20 yd^3

4. What is the perimeter of a rectangle that is 500 meters wide and 700 meters long?
   a. 2.4 km   b. 350 km
   c. 350,000 m^2 d. 2,400,000 km

5. What is the formula for the total surface area of the rectangular solid shown below with a base measuring A feet by 3A feet and a height measuring A feet?

   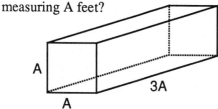

   a. 3A^3 ft^3 b. 3A^3 ft^2
   c. 13A^2 ft^2 d. 14A^2 ft^2

6. The two circles shown below have the same center. The smaller, unshaded circle has a diameter of 4 inches. The larger circle has diameter of 2 feet. What is the area of the shaded ring?

   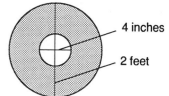

   4 inches

   2 feet

   a. 20π in^2
   b. 140π in^2
   c. 160π in^2
   d. 560π in^2

7. Which one of the following names a polygon having no two sides equal in length?
   a. parallelogram   b. scalene triangle
   c. quadrilateral   d. isosceles triangle

8. The solid shown at right is made up by attaching 3 identical cubes whose edges each measure 12 feet.

   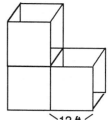

   12 ft.

   The solid is completely filled with dirt. The dirt is to be removed by trucks that can carry 8 cubic yards per trip. If there is a charge of $10 for each truckload to haul the dirt away, what will it cost to remove all the dirt from the solid?
   a. $240   b. $1,920
   c. $1,940   d. $2,160

9. How much fencing is needed to enclose a circular garden that measures 20 yards across?
   a. 400π yd.   b. 100π yd.^{2}
   c. 100π yd.   d. 20π yd.

10. What is the volume of a cone that is 15 meters high if the circular base is 10 meters in diameter?
    a. 500π m^3 b. 187.5π m^3
    c. 125π m^3 d. 50π m^3

11. Select the formula that correctly expresses the area inside the trapezoid and outside the square in the accompanying diagram. The trapezoid has bases of 10A and 6A and a height of 2A. One side of the square is represented by A.

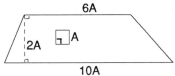

a. $9A^2$ b. $15A^2$ c. $19A^2$ d. $31A^2$

12. Find the volume of the solid shown below consisting of a cylinder and a cone. The circular top of the cylinder is the circular base of the cone.

a. $213\frac{1}{3}\pi$ ft^3
b. 208π ft^3
c. 176π ft^3
d. $69\frac{1}{3}\pi$ ft^3

13. Which statement listed below is true of polygons?
a. Diagonals of a parallelogram are equal.
b. Every rhombus is a square.
c. Each interior angle of a regular octagon measures 135°.
d. Opposite angles of a trapezoid are congruent.

14. What is the cost to cover a triangular plot with a base of 9 feet and a height of 4 feet with shrubs if shrubs cost $25 per square foot?
a. $900   b. $450   c. $420   d. $325

15. Round 7 ft. 5 in. to the nearest foot.
a. 7 ft.   b. 7.5 ft.   c. 8 ft.   d. 9 ft.

16. What is the measure of an interior angle of a regular pentagon, a five-sided polygon?

a. 108°   b. 135°   c. 144°   d. 180°

17. Find the perimeter of the figure shown below composed of a square and four identical equilateral triangles. One side of the square measures 2k and one side of each triangle measures K.

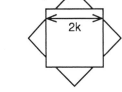

a. $6k^2$ b. 8k   c. 12k   d. 16k

18. An empty rectangular lot measures 12 miles long and 16 miles wide. How long will it take a person to walk diagonally across the lot if the person averages 2 miles per hour?

a. 10 hours   b. 14 hours
c. 28 hours   d. 40 hours

19. Which one of the following is true for the accompanying figure?

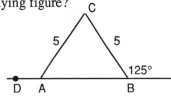

a. Triangle ABC is an isosceles right triangle.
b. m∠DAC = 55°
c. ∠ABC is the complement of the 125° angle.
d. m∠C = 70°

20. What will it cost to cover a rectangular floor measuring 40 feet by 50 feet with square tiles that measure 2 feet on each side if a package of 10 tiles costs $13 per package?

a. $13,000   b. $6,500
c. $1,300   d. $650

21. The figure below shows a metal casting 10 cm square and 2 cm thick with a hole in the center 5 cm square. What is the volume of the metal casting?

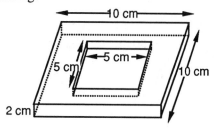

a. 50 cm^3 b. 75 cm^3
c. 150 cm^3 d. 175 cm^3

22. If one sphere has a radius of 2 cm and a second sphere has a radius of 3 cm, what fraction will result if the volume of the smaller sphere is divided by the volume of the larger sphere?

a. $\frac{2}{3}$ b. $\frac{4}{9}$ c. $\frac{4}{9\pi}$ d. $\frac{8}{27}$

23. Round the measure of AB to the nearest $\frac{1}{4}$ inch.

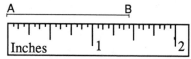

a. 1 in.   b. $1\frac{1}{4}$ in.   c. $1\frac{1}{2}$ in.   d. $1\frac{7}{8}$ in.

24. What is the perimeter of an isosceles triangle whose equal sides measure 6 meters each and whose third side measures 6 centimeters?
a. 12.6 m   b. 18 m
c. 126 cm   d. 1,206 cm

25. How many cubic feet of space is available in the tent, a prism, shown below? The triangles at both ends have a base of 3 ft and height measuring 2 ft. The tent is 2 yards long.

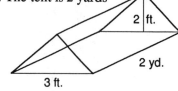

a. 6 ft^3
b. 12 ft^3
c. 18 ft^3
d. 36 ft^3

26. Sue travels 16 miles south and then 12 miles east. Her travel costs $0.15 per mile. What would Sue save if she had traveled directly from her starting point to her point of destination?
a. $1.20
b. $3.00
c. $4.20
d. None of these

27. The accompanying figure consists of a square with an equilateral triangle and a semicircle attached. One side of the square measures 2A linear units. What is the distance around this figure?

a. 8A + πA   b. 12A + πA
c. 8A + 2πA   d. 12A + 2πA

28. Which statement is true for the pictured triangles?

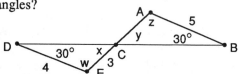

a. m∠z ≠ m∠w   b. $\frac{CE}{CA} = \frac{CB}{CD}$
c. AC = 3.75   d. m∠x = 30°

29. The outside dimensions of a picture frame are 2 feet by 30 inches. If its inside dimensions are $1\frac{1}{4}$ feet by 25 inches, what is the area of the frame?

a. 28.75 ft^2 b. 50.125 ft^2
c. 300 in^2 d. 345 in^2

30. The figure below shows a hemisphere (a half-sphere) on top of a cylinder. What is the volume of the resulting solid?

a. 738π ft^3
b. 864π ft^3
c. $1,008\pi$ ft^3
d. None of these

20 ft.

12 ft.

31. Which statement listed below is true of quadrilaterals?
    a. Not all quadrilaterals have a sum of measures of the four interior angles equal to 360°.
    b. Some rhombuses are not parallelograms.
    c. Consecutive angles of a parallelogram are supplementary.
    d. Opposite sides of a trapezoid are equal.

32. Which of the following names a geometric figure that must contain at least one angle measuring 45°?
    a. equilateral triangle
    b. rhombus
    c. isosceles right triangle
    d. parallelogram

33. Which statement is true for the figure shown, given that $\overline{AB} \parallel \overline{CD}$?

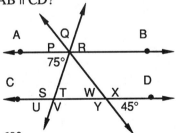

a. $m\angle Q = 60°$
b. $m\angle V = m\angle R$
c. $\angle T \cong \angle W$
d. None of these statements is true.

34. Which statement is true for the pictured triangles?

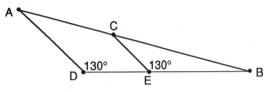

a. $m\angle DAB = m\angle DBC$
b. $\dfrac{CE}{AD} = \dfrac{AB}{CB}$
c. $m\angle DAB = m\angle BCE$
d. None of these statements is true.

35. Which of the following pairs of angles are complementary?

a. C and D
b. A and D
c. A and E
d. C and F

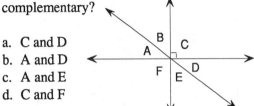

36. Which of the following could be used to report the distance around a circular swimming pool?
    a. meters   b. degrees
    c. square meters   d. cubic feet

37. Which one of the following would not be used to report the amount of water in an aquarium?
    a. cubic feet   b. liters
    c. gallons   d. meters

38. Which statement is true for the pictured triangles below?
    a. $CB = 10\sqrt{13}$ b. $\dfrac{CB}{FE} = \dfrac{3}{2}$
    c. $\dfrac{AB}{DE} = \dfrac{2}{3}$ d. $m\angle ACB \neq m\angle DEF$

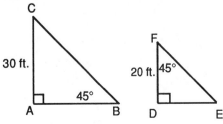

39. Consider the figure shown below

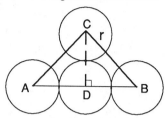

There are four circles. Each circle has a radius of r. The three outside circles each touch the circle whose center is D at exactly one point. The outside circles have centers at A, B, and C, which also serve as vertices for △ABC.

Which statement is true for this figure?
a. The perimeter of △ABC is 8r.
b. The area of △ABC is $8r^2$.
c. The area of △ABC is $4r^2$.
d. AC = 3r.

40. Which statement is true for the figure shown, given that quadrilateral PQRS is a parallelogram?

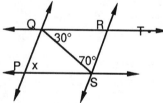

a. m∠x = 70°   b. m∠x = m∠SRT
c. m∠SRT = 100°
d. None of the above is true.

41. What type of triangle is shown below?

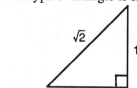

a. scalene   b. isosceles
c. equilateral   d. acute

42. Study the information given showing the areas of three figures of the same type.

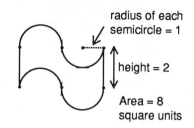

radius of each semicircle = 1
height = 2
Area = 8 square units

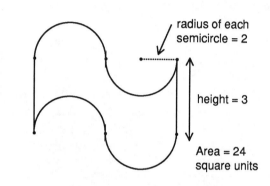

radius of each semicircle = 2
height = 3
Area = 24 square units

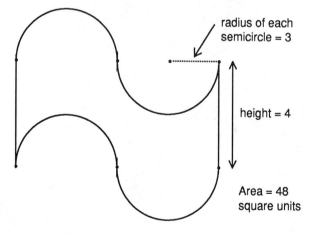

radius of each semicircle = 3
height = 4
Area = 48 square units

What is the area of the same type of figure with a radius of 9 and a height of 10?
a. 80   b. 360   c. 810   d. 900

43. Select the geometric figure that does not possess all of the following characteristics:

    i.   quadrilateral
    ii.   opposite sides parallel
    iii.   all angles equal in measure

  a. rhombus   b. square
  c. rectangle   d. none of these

44. Which of the sets of pictured triangles contain(s) similar triangles?

i.

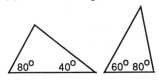

ii. 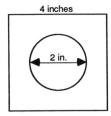   iii.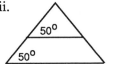

  a. i only   b. ii only
  c. i and ii only   d. ii and iii only
  e. All of the sets contain similar triangles.

45. Figure (1) below shows one face of a cube which contains a circular hole. The hole goes completely through the cube. The resulting solid is shown in figure (2). Find the volume of the solid in terms of π.

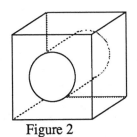

      Figure 1   Figure 2

  a. $(32 - 4\pi)\text{in}^3$ b. $(64 - 2\pi)\text{in}^3$
  c. $(64 - 4\pi)\text{in}^3$ d. None of these

46. What is the volume in centiliters of a 4.35 liter bottle?

  a. 4.35 cl   b. 43.5 cl
  c. 435 cl   d. 4350 cl

47. In the accompanying figure $\angle\text{TOP} \cap \overrightarrow{\text{NR}}$ is:

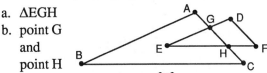

  a. $\overrightarrow{\text{NR}}$ b. $\varnothing$
  c. $\angle\text{TOP}$ d. $\overrightarrow{\text{OR}}$
  e. $\overleftrightarrow{\text{TR}}$

48. In the accompanying figure $\triangle\text{ABC} \cap \triangle\text{DEF}$ is:

  a. $\triangle\text{EGH}$
  b. point G and point H
  c. Interior of $\triangle\text{EGH}$ d. $\overleftrightarrow{\text{GH}}$
  e. None of these

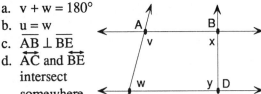

49. Given that $\overline{\text{AB}} \parallel \overline{\text{CD}}$, which one of the statements is not true for the figure shown? (The measure of angle ABD is represented by x, with the lower-case letters representing the measures of the various angles.)

  a. $v + w = 180°$
  b. $u = w$
  c. $\overline{\text{AB}} \perp \overline{\text{BE}}$
  d. $\overleftrightarrow{\text{AC}}$ and $\overleftrightarrow{\text{BE}}$ intersect somewhere in the plane.

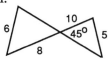

50. A rectangular wall measures 14 feet by 10 feet and has a triangular opening with base 10 feet and height 4 feet. The wall, except for the triangular opening, is to be covered with wallpaper. The wallpaper comes in rolls measuring 40 feet by $1\frac{1}{2}$ feet and sells for $12 per roll. The paper hanger will charge $20 for the labor. Find the total cost of papering the wall.

  a. $44   b. $60   c. $88
  d. None of these

seven

Probability

Many things in life have uncertainty associated with them. Regrettably, only taxes and death appear certain! Statements such as: "I will probably pass the math test," or "There is an 80% chance of rain today," or "The odds are 3 to 5 that the Dolphins will win the game," or "If I park in the faculty parking lot, I will most likely get a ticket," are typical instances in which the outcome is not certain. However, some confidence can be expressed that the prediction will eventually be verified.

All of us are required to make decisions based only on what is likely rather than that which is certain. In this chapter we will consider the theory of **probability** which enables us to study uncertainties mathematically. We will be applying knowledge of what occurs **in the long run** to some specific instance. If the weather report, for example, says that the probability of rain today is 80 percent, the implication is that there has been rain on 80 percent of the days in which the weather has been similar.

The first important work in the area of probability was done by two mathematicians, Blaise Pascal and Pierre Fermat, in the seventeenth century. These men were really more interested in finding methods for winning at the gambling table than in the creation of a theory.

Probability theory plays a role in such games of chance as cards, dice, and roulette. The speculation involved in odds, expectations, and the chances of events occurring has always intrigued some people. For example, one can only speculate on the outcome of the toss of dice. It is not possible to control the outcome (if the dice are fair, that is). Probabilities have applications in everyday life also. Probabilities play a role in opinion polls, elections, quality control and genetics. Insurance companies use them to calculate life expectancy or the likelihood of automobile accidents.

Of course, some people get carried away with the theory of probability. One man figured he could protect himself on a plane trip by taking a harmless bomb along with his baggage. He reasoned that the odds against any one person taking a bomb aboard were high, but the odds against two people doing it were astronomical. What do you think?

Before beginning this chapter, it might be helpful to review sections 1 and 2 of the Set Theory chapter.

Section 1   The Fundamental Counting Principle

The process of **counting** is an important part of the study of probability. In order for a probability to be calculated, it is necessary to count all the possible outcomes that can occur. In this section we will consider some ways of counting possible outcomes.

Consider a women's track meet at Suntan University with the following four runners in the mile race:Maria, Aretha, Thelma, and Debbie. Points are to be awarded only to the women finishing first or second. In how many different ways can these four runners finish first or second? There is a need to count this number.

Observe that if Maria finishes first, then each of the other three runners could finish second. This observation might help us construct a **tree diagram** in which each of the ways that the runners can finish first or second is determined by reading along the branches of the diagram. The possible ways are shown below:

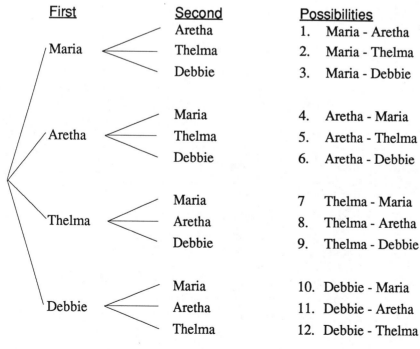

First	Second	Possibilities	
	Aretha	1.	Maria - Aretha
Maria	Thelma	2.	Maria - Thelma
	Debbie	3.	Maria - Debbie
	Maria	4.	Aretha - Maria
Aretha	Thelma	5.	Aretha - Thelma
	Debbie	6.	Aretha - Debbie
	Maria	7	Thelma - Maria
Thelma	Aretha	8.	Thelma - Aretha
	Debbie	9.	Thelma - Debbie
	Maria	10.	Debbie - Maria
Debbie	Aretha	11.	Debbie - Aretha
	Thelma	12.	Debbie - Thelma

The tree diagram shows there are 12 possible ways (arrangements) for the runners to finish first and second in the race. Each of four runners could finish first and, for each of these possibilities, each of three runners could finish second.

Because constructing a tree diagram may be a tedious task, it is fortunate that there is a method of quickly and accurately counting the number of possibilities even in situations more complex than the example with the runners. This method is appropriately called **The Fundamental Counting Principle.**

> ## The Fundamental Counting Principle
> If a first task can be performed in m different ways, and a second task can be performed in n different ways, then the two tasks can be performed in that order in $m \cdot n$ different ways. If there are additional tasks, with a third task performed in p ways, a fourth task in q ways, etc., the principle can be extended as follows: $m \cdot n \cdot p \cdot q \cdot \ldots$

The following problems illustrate the use of **The Fundamental Counting Principle.**

Problem 1

In how many ways can first place and second place be taken in a race involving four runners?

Solution

Let m be the number of ways that the task of finishing first can be performed. Since the task of finishing first can be performed in 4 different ways, m = 4. The task of finishing second (n) can be performed in 3 different ways. Thus, n = 3. If one runner finished first, then 3 runners remain that could possibly finish second. **The Fundamental Counting Principle** says that the two numbers, 4 and 3, should be multiplied.

$$m \cdot n = 4 \cdot 3 = 12$$

There are 12 ways that the first place and second place can be taken in a race involving four runners. This result agrees with the number of possibilities obtained in the tree diagram.

Problem 2   How many arrangements of three letters can be formed from the letters of the word MATH if any letter may not be repeated? (Examples of arrangements of three letters are AMT, HMA, etc.)

Solution   This problem is an extension of **The Fundamental Counting Principle.** The task of filling the first position in an arrangement may be performed in any of **four** ways. This means that we can initially choose any one of the four letters — M, A, T, or H. Since we are not permitted to repeat the letters, the task of filling the second position may be performed in **three** ways. For example, if A had been chosen for the first position, we could now choose any one of the three remaining letters - M,T, or H for the second position. Finally, the task of filling the third position may be performed in **two** ways, choosing any one of the two remaining letters. To find the total number of arrangements of these letters that can be formed, we use **The Fundamental Counting Principle.**

$$4 \cdot 3 \cdot 2 = 24$$

Thus, 24 arrangements of three letters can be formed. AMT and HMA are two of these 24 arrangements.

Problem 3   A man has 2 different jackets, 5 different shirts and 3 different ties in his wardrobe. How many possible outfits can this man select to wear?

Solution   The task of selecting a jacket can be performed in **two** ways. The task of selecting a shirt can be performed in **five** different ways. The task of selecting a tie can be performed in **three** different ways. To find the total number of outfits this man could wear, we use **The Fundamental Counting Principle**:

$$2 \cdot 5 \cdot 3 = 30$$

Thus, the man can select 30 possible outfits.

Problem Set 1.1

Determine the solution of each of the following by using **The Fundamental Counting Principle.**

1.   From a committee of 10 members, a chairperson, vice-chairperson and secretary are to be elected. Each person can fill only one position. In how many ways can it be done?   720

2.   How many arrangements of 5 letters can be formed from the letters of the word LOGIC if no letter may be repeated?   120

3.   How many arrangements of four letters may be formed using the vowels a, e, i, o, u if any letter may be repeated?   625

4.   A person can order a new car with a choice of 6 possible colors, with or without air-conditioning, with or without automatic transmission, with or without power windows, and with or without a radio. In how many different ways can a new car be ordered in terms of these options? $6 \times 2 \times 2 \times 2 \times 2 = 96$

5. At the Greasy Spoon Restaurant a person can order a dinner with or without salad, with or without beverage, with or without dessert, have the meat broiled rare, medium, or well done, and have the potato baked, mashed, or french fried. In how many different ways can a person order a dinner? *[handwritten: $2 \times 2 \times 2 \times 3 \times 3 = 72$]*

6. How many 7-digit telephone numbers can be formed using the decimal digits (0-9) if zero is not allowed as the first digit only, and any digit may be repeated? *[handwritten: 9,000,000]*

7. Consider the digits 3, 6, 7, 8, 9.
   a. How many 4-digit numbers may be formed using these digits if the first digit must be 3, and any digit may be repeated? *[handwritten: 125.]*
   b. How many 3-digit numbers may be formed using these digits if the last digit must be 9, and any digit may be repeated? *[handwritten: 25.]*

8. How many student numbers can be issued at Ivy League University if each number consists of two letters of the alphabet followed by four decimal digits (0-9)? Any letter and any digit may not be repeated. (A calculator would be helpful.)

9. Consider the digits 1, 2, 3, 4, 5, 6.
   a. How many two-digit even numbers can be formed if any digit may be repeated?
   b. How many two-digit even numbers can be formed if any digit may not be repeated?

10. If a 3-digit number is formed from the digits 1, 2, 3, 4, how many arrangements give numbers that are greater than 300? No digit in an arrangement may be repeated. *[handwritten: 12, 1 4 3]*

11. Subtract: $3\frac{1}{5} - 1\frac{2}{3}$

12. Express 35% as a decimal.

13. Express $2\frac{1}{4}\%$ as a decimal.

14. Evaluate: $6 \div 2 - 3 + 4 \cdot 2 - 6$

Section 2   Permutations

Let us return again to our race consisting of the runners Maria, Aretha, Thelma and Debbie. Since we are only concerned with first and second place finishers, there are 4 runners being considered 2 at a time. We saw in Section 1 that there are 12 different arrangements of 4 runners taken 2 at a time.

Notice that the possibility in which Maria finishes first and Aretha finishes second is different from the possibility in which Aretha finishes first and Maria finishes second. Even though the same two runners are involved in each possibility, the **order** in which they are considered makes a difference. This discussion leads us to the following definition.

Definition

A **permutation** is a sequential arrangement of n distinct objects taken r at a time, denoted by $_nP_r$, in which the order makes a difference.

Alternate notations for a permutation are $P(n,r)$ and P_r^n.

Example

Referring to our race in which 4 runners are being considered 2 at a time, n would represent 4 and r would represent 2. We could express the number of permutations of 4 runners taken 2 at a time as: $_4P_2$.

The expression $_nP_r$, in which n and r are positive integers and $n \geq r$, may be evaluated using the following "short cut" formula:

$$_nP_r = n \cdot (n-1) \cdot (n-2) \cdot \ldots \cdot (n-r+1)$$

Notice that we are multiplying a series of factors in which n is the first factor and the quantity $(n-r+1)$ is the last factor. Notice also that in each factor the quantity is decreasing by 1.

Where does this "short-cut" formula come from? In this formula we are concerned with filling r positions with n different objects. We will apply an extension of **The Fundamental Counting Principle** to the formula. Let us examine the following relationship:

r positions	1	2	3	...	r
n choices	n	$(n-1)$	$(n-2)$	...	$n-(r-1)$

We can fill the first of the r positions in any one of n different ways. After choosing the first, the second can be chosen in $(n-1)$ different ways, since there is one less object. Notice the relationship of the r position (2) and the number subtracted from n (1). This number is one less than the number representing the r position, or $r-1$. The third position can be filled in $(n-2)$ different ways. Again, observe the relationship of the r position (3) and the number subtracted from n (2). This number is one less than the number representing the r position, or $r-1$. Finally, the r position can be filled in $n-(r-1)$ different ways. Here we see clearly the relationship of the r position and the number subtracted from n. The number is indeed one less than the number representing the r position, or $r-1$. It is possible to express $n-(r-1)$ as $n-r+1$. Notice that when $n=r$, the last factor in the product is 1. The reason this formula represents a "short-cut" is stated later.

Problem 1 Use the "short-cut" formula to evaluate $_4P_2$.

Solution We observe the following:

$$n = 4 \ \text{(our first factor)}$$
$$r = 2$$
$$n - r + 1 = 4 - 2 + 1 = 3 \ \text{(our last factor)}$$

Using the formula $n \cdot (n - 1) \cdot (n-2) \cdot \ldots \cdot (n - r + 1)$ we have:

$$_4P_2 = 4 \cdot 3$$
$$= 12$$

This result agrees with the number of possible ways in which the four runners could finish first or second in the race. Some counting problems that may be solved by **The Fundamental Counting Principle** are also problems involving permutations and may be solved by using the "short-cut" formula.

Notice that $_4P_2 = 4 \cdot 3$ (**two** factors in the product.)
Observe the pattern in each of the following:

a. $_8P_3 = 8 \cdot 7 \cdot 6$ (**three** factors in the product)

b. $_5P_4 = 5 \cdot 4 \cdot 3 \cdot 2$ (**four** factors in the product)

c. $_7P_5 = 7 \cdot 6 \cdot 5 \cdot 4 \cdot 3$ (**five** factors in the product)

Problem Set 2.1

1. Is it possible to evaluate an expression such as $_5P_6$? Why? No because the rate element can't be greater than the choices

2. Evaluate each of the following by using the "short-cut" formula:

   a. $_4P_1$ b. $_7P_2$ c. $_{10}P_3$ d. $_6P_4$

3. Is it possible to evaluate $_4P_0$ using the "short-cut" formula?

Review Competencies

4. In how many ways can 7 people fill the positions of president, vice-president, and treasurer in a club election?

5. In how many ways can 5 different books be arranged on a shelf?

6. Consider all the arrangements made up with letters J, A, C, K with each letter used only once (such as JCKA, ACJK, etc.). How many arrangements begin with the letter A?

7. If a three digit number is formed from the digits 3, 4, 5, and 8, how many arrangements give numbers that are less than 800? Any digit may be repeated.

8. Express $4\frac{3}{5}$ as an improper fraction.

10. Subtract: $8 - 0.652$

9. Multiply: $2\frac{2}{3} \cdot 1\frac{1}{2}$

11. Express 65% as a fraction.

Let us now examine a special situation that can arise using the "short-cut" formula.

Problem 2   Evaluate $_5P_5$ using the "short-cut" formula:

Solution $n = 5$ (our first factor),
$r = 5$, and
$n - r + 1 = 5 - 5 + 1 = 1$ (our last factor).

The expression $n \cdot (n - 1) \cdot (n - 2) \cdot \ldots \cdot (n - r + 1)$ yields:

$$5 \cdot 4 \cdot 3 \cdot 2 \cdot 1 = 120$$

The expression $5 \cdot 4 \cdot 3 \cdot 2 \cdot 1$ is referred to as "5 **factorial**" denoted 5! .
This is the product of the integers from 1 to 5.
In the same way, the expression $3 \cdot 2 \cdot 1$ is referred to as "3 factorial" or 3!.

In general, n factorial, denoted n!, represents the product of the integers from 1 to n. Thus,

$$n! = n \cdot (n - 1) \cdot (n - 2) \cdot \ldots \cdot 3 \cdot 2 \cdot 1$$

It is helpful if factorials are defined for all whole numbers. This leads us to the following definition.

Definition
$0! = 1$

Examples

1. $0! = 1$
2. $1! = 1$
3. $2! = 2 \cdot 1 = 2$
4. $3! = 3 \cdot 2 \cdot 1 = 6$
5. $4! = 4 \cdot 3 \cdot 2 \cdot 1 = 24$
6. $5! = 5 \cdot 4 \cdot 3 \cdot 2 \cdot 1 = 120$

7. $6! = 6 \cdot 5 \cdot 4 \cdot 3 \cdot 2 \cdot 1 = 720$

Notice that the values of the factorials increase rapidly.

8. $7! = 5,040$
9. $8! = 40,320$
10. $10! = 3,628,800$

Problem 3 Evaluate $\frac{4!}{3!}$

Solution

$$\frac{4!}{3!} = \frac{4 \cdot 3 \cdot 2 \cdot 1}{3 \cdot 2 \cdot 1}$$

$$= \frac{4 \cdot \cancel{3} \cdot \cancel{2} \cdot \cancel{1}}{\cancel{3} \cdot \cancel{2} \cdot \cancel{1}}$$

$$= 4$$

Problem 4 Evaluate $2! \cdot 3!$

Solution $2!3! = 2 \cdot 1 \cdot 3 \cdot 2 \cdot 1 = 12$

Problem 5 Evaluate $(4 - 2)!$

Solution Determine the result inside the parentheses first.
Thus, $(4 - 2)! = 2! = 2 \cdot 1 = 2$

Problem 6 Evaluate $\frac{5!}{3! \, 2!}$

Solution $\frac{5!}{3! \, 2!} = \frac{5 \cdot 4 \cdot 3 \cdot 2 \cdot 1}{3 \cdot 2 \cdot 1 \cdot 2 \cdot 1} = \frac{5 \cdot \overset{2}{\cancel{4}} \cdot \cancel{3} \cdot \cancel{2} \cdot \cancel{1}}{\cancel{3} \cdot \cancel{2} \cdot \cancel{1} \cdot \cancel{2} \cdot \cancel{1}} = 10$

Problem Set 2.2

Evaluate each of the following.

1. $\frac{7!}{1!}$

2. $\frac{6!}{4!}$

3. $\frac{10!}{7!}$

4. $\frac{4!}{4!}$

5. $\frac{5!}{0!}$

6. $3! \, 4!$

7. $5!0!$

8. $(5 - 3)!$

9. $(4 - 0)!$

10. $(5 - 5)!$

11. $4! - 2!$

12. $6! - 4!$

13. $\frac{5!}{(5 - 2)!}$

14. $\frac{6!}{(6 - 4)!}$

15. $\frac{5!}{3!(5 - 3)!}$

16. $\dfrac{6!}{4!\,2!}$　　17. $\dfrac{4!}{4!\,0!}$

Review Competencies

18. Evaluate $_6P_4$ using the "short-cut" formula.

19. Evaluate $_4P_4$ using the "short-cut" formula.

20. A man has 7 shirts, 5 pairs of pants, and 3 pairs of shoes. In how many ways can the man select an outfit in terms of shirts, pants, and shoes?

21. How many arrangements can be formed using the letters S, U, and N if any letter can be repeated?

22. If a three digit number is formed using the digits, 1,2,3, and 4, how many give numbers that are even assuming any digit may be repeated?

23. 20% of what number is 40?

24. Divide: $0.142 \div 0.71$

25. Divide: $1\frac{3}{4} \div 1\frac{2}{3}$

26. Evaluate: $7(3 \cdot 2 - 1) + 2(4 \div 2 + 3)$

The "short-cut" formula is so called because it can be used to quickly evaluate the expression $_nP_r$. Let us now consider an alternative "factorial" formula that will provide the same result as the "short-cut" formula.

> The expression $_nP_r$, n and r are nonnegative integers, may be evaluated using factorials as follows:
> $$_nP_r = \frac{n!}{(n - r)!}$$

The "factorial" formula can be generated from the "short-cut" formula as shown below:

Begin with the "short-cut" formula.
$$_nP_r = n(n - 1)(n - 2) \cdot \ldots \cdot (n - r + 1)$$

Multiply the right member by $\dfrac{(n - r)!}{(n - r)!}$
$$_nP_r = n(n - 1)(n - 2) \cdot \ldots \cdot (n - r + 1) \cdot \frac{(n - r)!}{(n - r)!}$$

$$_nP_r = \frac{n(n-1)(n-2) \cdot \ldots \cdot (n - r + 1)(n - r)(n - r - 1) \cdot \ldots \cdot 3 \cdot 2 \cdot 1}{(n - r)!}$$

Simplify the numerator.
$$_nP_r = \frac{n!}{(n - r)!}$$

Thus, the "short-cut" formula is equivalent to the "factorial" formula. Either formula gives the same result. Generally, it is faster and simpler to use the "short-cut" formula for most problems. However, an expression such as $_5P_0$ can only be evaluated using the "factorial" formula.

Problem 7 Evaluate $_4P_2$ using the "factorial" formula.

Solution In the expression $_4P_2$, $n = 4$, $r = 2$, and $n - r = 4 - 2 = 2$.

Since the "factorial" formula is $_nP_r = \dfrac{n!}{(n - r)!}$

$$_4P_2 = \frac{4!}{(4 - 2)!} = \frac{4!}{2!} = \frac{4 \cdot 3 \cdot 2 \cdot 1}{2 \cdot 1} = \frac{4 \cdot 3 \cdot \cancel{2} \cdot \cancel{1}}{\cancel{2} \cdot \cancel{1}} = 4 \cdot 3 = 12$$

The result agrees with that obtained by using the "short-cut" formula.

Problem Set 2.3

Evaluate each of the following by using the "factorial" formula.

1. $_6P_3$

2. $_7P_2$

3. $_6P_6$

4. $_4P_0$

5. Is it true that $_nP_n = n!$?

Review Competencies

6. Evaluate each of the following.

   a. $\dfrac{7!}{4!}$ b. $\dfrac{5!}{5!}$ c. $\dfrac{6!}{0!}$

7. Evaluate $_6P_4$ using the "short-cut" formula.

8. Use the digits 1, 3, 6, and 7. How many three digit numbers can be formed if the number must be greater than 600 and no digit may be repeated?

9. Express $\dfrac{5}{8}$ as a percent.

10. Express 0.01% as a decimal.

11. Add: $\dfrac{3}{7} + \dfrac{6}{10}$

12. Add: $27.041 + 2.86 + 0.345$

Let us see how the permutation formulas help us to find the solution to the following problem.

Problem 8 In how many ways can the letters a, b, c, d be arranged in four-letter arrangements if any letter is not repeated?

Solution Since order is involved, we must find the number of permutations of 4 letters taken 4 at a time. This number can be determined by using either of the two formulas since they both yield identical results. Using the "short-cut" formula,

$$_nP_r = n \cdot (n - 1) \cdot (n - 2) \cdot .. \cdot (n - r + 1),$$

we must evaluate $_4P_4$.

Thus, $n = 4$ (our first factor), $r = 4$, and $n - r + 1 = 1$ (our last factor).
The "short-cut" formula gives: $_4P_4 = 4 \cdot 3 \cdot 2 \cdot 1 = 24$

Using the "factorial" formula, $_nP_r = \dfrac{n!}{(n - r)!}$

$$_4P_4 = \frac{4!}{(4 - 4)!} = \frac{4!}{0!} = \frac{4 \cdot 3 \cdot 2 \cdot 1}{1} = 24$$

Thus, there are 24 ways of arranging the 4 letters. Three examples of these 24 arrangements are acbd, acdb, and dbac.

Notice that **The Fundamental Counting Principle** can also be used to solve this problem. Since there are 4 letters, there are 4 positions to fill. The first position can be filled in 4 ways, the second position in 3 ways, the third position in 2 ways and the fourth position in 1 way. The total number of ways that 4 different letters can be arranged is:
$$4 \cdot 3 \cdot 2 \cdot 1 = 24$$

Problem Set 2.4

The following problems involve permutations. Evaluate each using both formulas and **The Fundamental Counting Principle** if possible.

1. In how many ways can 6 children be seated in 6 chairs in the front row of an auditorium?

2. In how many ways can the 5 players on a basketball team be introduced before the game?

3. There are 10 essay questions on a final examination. In how many different sequences could a student answer any three of these questions?

4. In how many ways can 4 old paintings be hung in each of 4 different places on the walls of a museum?

5. In how many ways can 5 people line up at a single stadium ticket window?

6. How many arrangements can be made using 3 of the letters of the word DOLPHIN if any letter is not repeated?

7. A team has 6 players but only 4 empty lockers in the gymnasium. In how many ways can the coach assign these lockers to the players assuming the lockers may not be used by more than one player?

8. Connie is holding 9 playing cards in her hand. If she randomly plays them one at a time, in how many ways can she play the first four cards?

Review Competencies

9. Evaluate $_6P_4$ using the "factorial" formula.

10. Evaluate $_6P_0$ using the "factorial" formula.

11. Consider all the arrangements consisting of the letters of the word FLORIDA. In how many ways can the letters be arranged if the first letter must be L, the last letter must be F, and no letter may be repeated?

12. 15% of what number is 60?

13. Express 270% as a decimal.

14. Express $\frac{27}{4}$ as a mixed number.

15. Express $\frac{28}{42}$ in lowest terms.

16. Divide: $1 \div 0.005$

17. Express 0.007 in scientific notation.

Section 3   Combinations

Let us recall again our four runners: Maria, Aretha, Thelma and Debbie. Suppose that only two of the runners are allowed to represent the college in the conference track meet. Listed below are all the possible pairs of two runners that could be selected.

1. Maria and Aretha
2. Maria and Thelma
3. Maria and Debbie
4. Aretha and Thelma
5. Aretha and Debbie
6. Thelma and Debbie

There are 6 possible groups of two that can be formed from our four runners. In other words, there are 6 subsets of size 2 that can be formed. The pair consisting of Maria and Aretha is the same pair as the one consisting of Aretha and Maria. In considering these pairs, the order in which the names appear is **not** important.

This discussion leads to the following definition.

> ### Definition
> A **combination** is a group of n distinct objects taken r at a time, denoted by $\binom{n}{r}$, in which the order is not important.

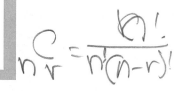

Alternate notations for a combination include $_nC_r$, $C(n,r)$ and C^n_r.

We can now say that there are six combinations that can be formed by taking 4 runners 2 at a time.

Referring to our pair of runners in which 4 are being taken 2 at a time, n would represent 4 and r would represent 2. Thus, we could denote the combinations of 4 runners taken 2 at a time as $\binom{4}{2}$. Since there are 6 combinations that can be formed, $\binom{4}{2} = 6$. Recall

437

there were 12 permutations $\left({_4}P_2\right)$, and now there are 6 combinations. There are twice as many permutations as combinations. Thus,

$$2 \cdot \binom{4}{2} = {_4}P_2$$

$$2! \binom{4}{2} = {_4}P_2 \quad \text{since } 2! = 2$$

$$\text{or } \binom{4}{2} = \frac{{_4}P_2}{2!}$$

$$\text{or } \binom{4}{2} = \frac{4!}{2!\,(4-2)!} \quad \text{since } {_4}P_2 = \frac{4!}{(4-2)!}$$

This discussion motivates us to state the following.

> The expression $\binom{n}{r}$ may be evaluated using the following formula.
> $$\binom{n}{r} = \frac{n!}{r!\,(n-r)!}$$

$$nCr = \frac{n!}{r!\,(n-r)!}$$

Problem 1 Evaluate $\binom{4}{2}$.

Solution In the expression $\binom{4}{2}$, we know that $n = 4$, $r = 2$, and $n - r = 4 - 2 = 2$.

$$\binom{n}{r} = \frac{n!}{r!\,(n-r)!}$$

Thus, $\binom{4}{2} = \dfrac{4!}{2!\,(4-2)!} = \dfrac{4!}{2!\,2!} = \dfrac{4 \cdot 3 \cdot 2 \cdot 1}{2 \cdot 1 \cdot 2 \cdot 1} = \dfrac{\overset{2}{\cancel{4}} \cdot 3 \cdot \cancel{2} \cdot \cancel{1}}{\cancel{2} \cdot 1 \cdot \cancel{2} \cdot \cancel{1}} = 2 \cdot 3 = 6$

Problem 2 Evaluate $\binom{5}{5}$.

Solution Since $n = 5$ and $r = 5$, we obtain:

$$\binom{n}{r} = \frac{n!}{r!\,(n-r)!}$$

$$\binom{5}{5} = \frac{5!}{5!\,(5-5)!} = \frac{5!}{5!\,0!} = \frac{5 \cdot 4 \cdot 3 \cdot 2 \cdot 1}{5 \cdot 4 \cdot 3 \cdot 2 \cdot 1} = \frac{\cancel{5} \cdot \cancel{4} \cdot \cancel{3} \cdot \cancel{2} \cdot \cancel{1}}{\cancel{5} \cdot \cancel{4} \cdot \cancel{3} \cdot \cancel{2} \cdot \cancel{1}} = 1$$

This problem tells us that we can only obtain one combination (or selection) of a group of 5 objects if we take all 5 objects at a time.

Problem Set 3.1

1. Would it be possible to evaluate an expression such as $\binom{3}{4}$? Why?

2. Evaluate each of the following:

   a. $\frac{6!}{4!\,2!}$ b. $\frac{8!}{5!\,3!}$ c. $\frac{10!}{4!\,6!}$

   d. $\frac{7!}{7!\,0!}$ e. $\frac{6!}{0!\,6!}$

3. Evaluate each of the following using the formula for combinations:

   a. $\binom{7}{2}$ b. $\binom{8}{3}$ c. $\binom{6}{6}$

   d. $\binom{7}{0}$ e. $\binom{5}{1}$

Review Competencies

4. In how many ways can 4 people line up at a theatre box office? Check using both formulas and **The Fundamental Counting Principle** if possible.

5. Evaluate $_5P_0$ using the "factorial" formula.

6. Evaluate: $\frac{9!}{5!}$

7. Using the digits 3, 5, 6, and 7, how many three digit numbers may be formed with these digits if the number must be even, and no digit is repeated?

8. Divide: $4\frac{2}{3} \div 1\frac{1}{2}$

9. Express 150% as a decimal.

10. Express $\frac{5}{4}$ as a decimal.

11. Express $5\frac{1}{4}\%$ as a decimal.

12. Evaluate: $7(5 \cdot 2 \div 5 - 1) + 3(6 \div 2 - 3 \cdot 1)$

13. Convert 123_4 to a base ten numeral.

We now see how the combination formula helps us to solve a problem.

Problem 3   Eight women try out for the basketball team. How many teams of 5 women could be formed by the coach?

Solution   Can you see that the order in the groups is not important? A team of 5 women can be tried out in any order. However, it would still be the same team. Thus, this problem deals with the combinations of 8 women taken 5 at a time. We may express this as $\binom{8}{5}$. In the expression $\binom{8}{5}$ we must recognize that n = 8, r = 5, and n − r = 8 − 5 = 3.

We can determine the answer using the formula: $\binom{n}{r} = \frac{n!}{r!\,(n-r)!}$

$$\binom{8}{5} = \frac{8!}{5!\,(8-5)!} = \frac{8!}{5!\,3!} = \frac{8 \cdot 7 \cdot 6 \cdot 5 \cdot 4 \cdot 3 \cdot 2 \cdot 1}{5 \cdot 4 \cdot 3 \cdot 2 \cdot 1 \cdot 3 \cdot 2 \cdot 1}$$

and $\dfrac{8 \cdot 7 \cdot 6 \cdot 5 \cdot 4 \cdot 3 \cdot 2 \cdot 1}{5 \cdot 4 \cdot 3 \cdot 2 \cdot 1 \cdot 3 \cdot 2 \cdot 1} = \dfrac{8 \cdot 7 \cdot \cancel{6} \cdot \cancel{5} \cdot \cancel{4} \cdot \cancel{3} \cdot \cancel{2} \cdot \cancel{1}}{\cancel{5} \cdot \cancel{4} \cdot \cancel{3} \cdot \cancel{2} \cdot \cancel{1} \cdot \cancel{3} \cdot \cancel{2} \cdot \cancel{1}} = 8 \cdot 7 = 56$

The coach can form 56 possible teams of 5 women each.

Problem Set 3.2

The following six problems involve combinations. Evaluate each using the formula.

1. How many choices are there for one to select 4 magazines to read from a collection of 8 magazines?

2. A student located 8 books she needed at the library, but she can only check out three at a time. How many sets of three books can she possibly select?

3. Four out of seven cheerleaders must be selected to accompany the basketball team to the state tournament. How many groups of four could be picked?

4. A vending machine in the student center has 5 different sandwiches available. How many selections can a student make if he wishes to choose two different sandwiches for lunch?

5. There are 7 candidates running for 4 seats on a city council. How many different groups can be elected?

6. An office needs 5 typists, and 8 apply for the jobs. How many different groups of 5 typists can be hired?

7. The triangular array shown below is **Pascal's Triangle.** It may be used to find the number of combinations of n objects taken r at a time. Replace each of the combinations by its equivalent number to obtain an array of numbers. Then determine the numbers in the row where n = 4.

$n = 0$ $\qquad\qquad\qquad \binom{0}{0}$

$n = 1$ $\qquad\qquad \binom{1}{0} \quad \binom{1}{1}$

$n = 2$ $\qquad\quad \binom{2}{0} \quad \binom{2}{1} \quad \binom{2}{2}$

$n = 3$ $\qquad \binom{3}{0} \quad \binom{3}{1} \quad \binom{3}{2} \quad \binom{3}{3}$

$n = 4$ $\quad \binom{4}{0} \quad \binom{4}{1} \quad \binom{4}{2} \quad \binom{4}{3} \quad \binom{4}{4}$

Review Competencies

8. $\dfrac{7!}{5!\,2!}$

9. In how many ways can 6 different books be lined up on a shelf?

10. Evaluate $_5P_3$ using both formulas.

11. Using the digits 1, 2, 3, 4, 5, and 6, how many three digit odd numbers may be formed if no digit may be repeated?

Section 4   Distinguishing Between Permutations and Combinations

The major problem generally encountered in the study of permutations and combinations is the ability to distinguish between each in a given real-world situation. Remember that the question of **order** is important when considering a permutation; it is not important when considering a combination. It is often helpful to associate the word "arrangement" with a permutation and the words "group" and "selection" with a combination.

Problem Set 4.1

Determine whether each of the following is a permutation or a combination. It is not necessary to perform the computation with a formula.

1. There are seven empty seats in the last row of a movie theatre. In how many ways can seven people sit in these seats? (In how many ways can these seats be arranged?) *permutation*

2. A club consists of 8 members. How many committees of 3 members each can be selected? *combination*

3. There are 8 women at a card party. How many tables of 4 women can be chosen? *combination*

4. How many lines are determined by 5 points if no three points are collinear?

5. From a class of 15 students, a president and sergeant-at-arms are to be elected. In how many ways can it be done? *permutation*

6. Five speakers are on the program at a school assembly. In how many ways can the speakers be introduced? *permutation*

7. How many sets of 3 birthday cards can be selected from 10 birthday cards? *comb*

8. In how many ways can 10 different books be arranged on a library shelf? *combination*

9. In how many ways can 5 women be seated around a circular table?

10. A contest has 6 finalists. In how many ways could they be presented for judging?

11. Eight people apply for three identical jobs. How many selections can be made to fill these jobs? *combination*

12. A delegation of 5 students must be chosen from a college to attend a convention. How many delegations can be determined if there are 10 eligible students? *combination*

13. Seven runners compete in a race for which there are first and second prizes. In how many ways may the prizes be won? *permutation*

Review Competencies

14. Evaluate: $\binom{9}{5}$

15. Evaluate: $_6P_3$

16. Multiply: $\frac{4}{5} \cdot \frac{3}{4} \cdot \frac{5}{9}$

17. Divide: $\frac{4}{5} \div 1\frac{3}{4}$

18. Subtract: $0.06 - 0.004$

19. Express $\frac{7}{8}$ as a decimal.

20. Express 0.012% as a decimal.

21. Evaluate: $\frac{1}{3}(\frac{2}{3} \div \frac{1}{3}) - \frac{1}{4}(\frac{3}{4} \cdot \frac{4}{5})$

22. Find the solution set for
$3x + 2(x - 5) = 7 - (x + 3)$.

Section 5   Experiments, Sample Spaces, and Events

In this section, we lay the groundwork for some important notions that will soon follow. Let us begin with the following definition.

> ### Definition
> An **experiment** refers to any act or process that can be performed that yields a collection of outcomes (results). The outcome of the act is not known in advance of the act.

Examples

Most people have an intuitive idea of what the word "experiment" implies in scientific research. The following are examples of the types of experiments that will concern us:

1. the act of tossing a fair coin
2. the act of rolling a fair die (one of a pair of dice) (A die is a cube with the faces containing points numbered from 1 to 6.)
3. the act of drawing a single card from a deck of 52 playing cards
4. the act of drawing a marble out of a box containing several different colored marbles
5. the act of interviewing an individual to determine if he or she is for or against a particular candidate (or undecided)
6. the act of selecting a vowel from the set of letters of the alphabet

> ### Definition
> A **sample space** is the set of all the possible outcomes (results) that can occur in an experiment such that exactly one outcome occurs at a time. An **event** is any subset of the sample space.

We shall use the letter S to represent the sample space. In the following examples notice that the sample space S associated with an experiment can be constructed by either listing the possible outcomes or by constructing a tree diagram.

Examples

1. If a fair coin is tossed, there are two possible, **equally likely** outcomes in the sample space that could occur in this experiment (Heads, H, or Tails, T). We assume that the coin does not land on an edge. One outcome has the same chance of occurrence as the other. Listing the sample space S associated with this experiment yields S = {H,T}. In this experiment, the two individual outcomes may be referred to as events since the sets {H} and{T} are subsets of S = {H,T}. Any outcome of an experiment can be an event. Obviously, just one outcome can occur at a time.

2. If a fair coin is tossed two times in succession, the sample space S could be determined by drawing a tree diagram or by simply listing the possibilities. Listing the sample space S associated with this experiment yields S = {HH, HT, TH, TT}. Observe that the same set of outcomes would have been generated had the experiment consisted of tossing **two** fair coins **simultaneously**. If the coins are, say, both quarters, then it is helpful to distinguish between them by marking one coin as coin #1 and the other as coin #2. There are 4 outcomes in the sample space. Thus, the number of outcomes may be expressed as 2^2 or 4 where the exponent denotes the number of coins tossed. The event of obtaining two tails is {TT}. The event {HT} is the case in which a Head is tossed first and then a Tail. The event {HT, TH} is the event of obtaining exactly one Head. The event {HH, HT, TH} is the event of obtaining at least one Head.

3. If a fair coin is tossed three times in succession, the sample space could be determined by drawing a tree diagram as follows:

First Toss	Second Toss	Third Toss	Outcomes
		H	HHH
	H		
H		T	HHT
		H	HTH
	T		
		T	HTT
		H	THH
	H		
T		T	THT
		H	TTH
	T		
		T	TTT

Listing the sample space S associated with the experiment yields

S = {HHH,HHT,HTH,HTT,THH,THT,TTH,TTT}

Observe that the same set of outcomes would have been generated had the experiment consisted of tossing **three** fair coins **simultaneously**. In this case, it is helpful to distinguish the coins by marking them as coin #1, coin #2, and coin #3. Notice that there are 8 outcomes in the sample space (which may be expressed as 2^3).

The event of tossing at least two Heads is

{HHT, HTH, THH, HHH}.

4. If a fair die is rolled, there are six possible, equally likely outcomes that could occur in the experiment. Listing the sample space S associated with this experiment yields
   S = {1,2,3,4,5,6}.
   Each of these individual outcomes in this experiment is an event because {1}, {2}, {3}, {4}, {5}, and {6} are subsets of S.

   Obviously, no two outcomes can occur at the same time. The event of rolling an even number on the die is {2,4,6}. The event of rolling a 7 on a die is Ø.

   Recall that Ø is a subset of S = {1,2,3,4,5,6}.

5. If a fair die is rolled two times in succession, each outcome can be represented as an ordered pair (a,b) in which a is the number obtained on the first roll, and b is the number obtained on the second roll. Listing the sample space S associated with this experiment yields:

$$
S = \left\{
\begin{array}{l}
(1,1),(1,2),(1,3),(1,4),(1,5),(1,6) \\
(2,1),(2,2),(2,3),(2,4),(2,5),(2,6) \\
(3,1),(3,2),(3,3),(3,4),(3,5),(3,6) \\
(4,1),(4,2),(4,3),(4,4),(4,5),(4,6) \\
(5,1),(5,2),(5,3),(5,4),(5,5),(5,6) \\
(6,1),(6,2),(6,3),(6,4),(6,5),(6,6)
\end{array}
\right\} \quad 36
$$

Notice that there are 36 possible outcomes in the sample space S. The number of outcomes can be obtained by recalling **The Fundamental Counting Principle.** If the number of ways of performing the task of rolling the first die is 6 and the number of ways of performing the task of rolling the second die is 6, then the total number of ways of performing both tasks is 6 • 6 or 36. The event of obtaining the same number on each roll of the die is: {(1,1), (2,2), (3,3), (4,4), (5,5) (6,6)}.

6. Consider the experiment in which a fair coin is tossed, and then a fair die is rolled. The sample space could be determined by drawing a tree diagram as follows.

Coin	Die	Outcomes
	1	H1
	2	H2
H	3	H3
	4	H4
	5	H5
	6	H6
	1	T1
	2	T2
T	3	T3
	4	T4
	5	T5
	6	T6

Listing the sample space S associated with this experiment yields

$$S = \{H1, H2, H3, H4, H5, H6, T1, T2, T3, T4, T5, T6\}.$$

Notice that there are 12 outcomes in the sample space. The number of outcomes can be obtained recalling **The Fundamental Counting Principle.**

If the number of ways of performing the first task of tossing a coin is 2, and the number of ways of performing the second task of rolling a die is 6, then the total number of ways of performing both tasks is 2 • 6 or 12.

The event of tossing a Tail on the coin and then rolling an odd number on the die is: {T1, T3, T5}.

Many of the concepts of probability theory can be related to concepts in set theory. In the Venn diagram shown, the sample space S serves as the universal set. An event is any subset of the sample space. The circle containing A represents the set of outcomes that are in event A. Obviously, there are outcomes of the sample space that are not included in event A.

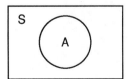

Chapter 7

Problem Set 5.1

1. List the sample space associated with the experiment of selecting a vowel from the set of letters of the alphabet.

2. Two children are born into a family. Let "b" represent a boy and "g" represent a girl. List the sample space associated with this experiment.

3. A box contains three marbles (one orange, one green, one white). Two marbles are drawn in succession with the first marble replaced in the box before the second is drawn. List the sample space associated with this experiment. Represent the event in which both draws would yield a marble of the same color.

4. Box 1 contains a red (R) and a white (W) ball. Box 2 contains a yellow (Y) and a green (G) ball. Consider the experiment in which we first select a box and then select a ball from the box. List the sample space associated with this experiment. $1(R,W) 2(Y,G)$

5. Box A contains three cards numbered 2, 4, and 6. Box B contains two cards numbered 4 and 8. Two cards are to be randomly selected, the first from Box A and the second from Box B. Which set represents an appropriate sample space for this experiment?
   a. {2,4,6,8}
   b. {2,4,4,6,8}
   c. {(2,4), (2,8), (4,8), (6,4), (6,8)}
   d. {(2,4), (2,8), (4,4), (4,8), (6,4), (6,8)}

6. Determine whether each of the following is a permutation or a combination and then evaluate using the appropriate formula:
   a. Ten people want to try out for the tennis team. How many teams of six could be selected?
   b. From a club of 8 members a president and vice-president are to be elected. In how many ways could this be done?

7. Using the vowels a, e, i, o, and u, how many four letter arrangements may be formed using these vowels if any letter may not be repeated and the second letter must be u?

8. Subtract: $6 - 1\frac{3}{5}$

9. Express 45% as a fraction.

10. Rationalize the denominator of $\frac{6}{\sqrt{3}}$.

11. What percent of 30 is 18?

12. Express $4\frac{7}{10}\%$ as a decimal.

13. You are travelling at a constant speed. At 12:45 P.M. you have travelled 340 miles. You have travelled 353 miles by 1:00 P.M. and 366 miles by 1:15 P.M. How many miles will you travel by 3:00 P.M. of the same day?

We now introduce another important concept.

> **Definition**
> **Mutually exclusive events** are events that cannot occur at the same time in the same experiment.

Two events are mutually exclusive if they do not have any outcomes in common. If one event happens, the other event cannot happen.

Examples

1. If a fair coin is tossed, there are two possible events (Heads and Tails) that can occur. One and only one of these events can occur at a time.

    If a Head is tossed, then a Tail cannot be tossed. Thus, the event of tossing a Head and the event of tossing a Tail are mutually exclusive events.

2. The two events of rolling an even number and rolling an odd number on one roll of a fair die are mutually exclusive events. The events cannot occur at the same time.

    It is impossible to roll a 4 and a 5 with just one roll of a die. If an even number is rolled, then an odd number cannot be rolled.

3. The two events of rolling an odd number, {1,3,5}, and a number less than 3, {1,2}, on one roll of a fair die are not mutually exclusive events. The two events have an outcome in common, 1.

Mutually exclusive events are events that have no outcomes in common. A related idea in set theory is that of disjoint sets. The Venn Diagram shows the situation that exists for two mutually exclusive events, A and B, in the sample space S. Observe that if A and B are mutually exclusive, then $A \cap B = \emptyset$.

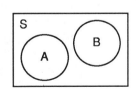

Problem Set 5.2

Determine in problems 1-5 whether the following events are mutually exclusive.

1. Rolling a 1 and a 2 on one roll of a fair die

    mutually exclusive

2. Rolling an even number and a number greater than 4 on one roll of a fair die

    not

3. Drawing an orange ball and a blue ball out of a bag that contains 3 orange balls and 3 blue balls if one ball is drawn at random

    mutually

4. Being President of the United States and being under 20 years of age

    mutually exclusive

5. Drawing a King and an Ace when drawing a single card from a deck of 52 playing cards

    mutually Exclusive

Review Competencies

6. A box contains one red, one white, and one blue marble. List the sample space for the experiment of drawing one marble at random from the box.

7. An urn contains three black balls (marked B_1, B_2, and B_3) and one white ball (W). If one ball is selected at random, what is the sample space associated with this experiment?

8. A commencement program has 5 speakers. In how many ways could their speeches be presented?

9. Using the digits 1, 2, 8, and 9, how many three digit numbers may be formed if any digit may be repeated and the number is greater than 800?

10. Express the fraction $\frac{45}{50}$ in lowest terms.

11. Add : $3\frac{1}{3} + 2\frac{1}{4}$

12. Express 450% as a decimal.

13. Express 3 as a percent.

14. Evaluate: $4 \cdot 3 \div 6 - 2 + 6 \cdot 2$

15. Find the next term in the following geometric progression.

    $$4, \ -2, \ 1, \frac{1}{2}, \ \underline{\hspace{1cm}}$$

16. Find the next term in the following harmonic progression. (The definition of a harmonic progression is in Chp. 6 Problem Set 15.2, #26.)

    $$\frac{3}{4}, \frac{3}{7}, \frac{3}{10}, \frac{3}{13}, \ \underline{\hspace{1cm}}$$

Section 6   Probability

If a fair coin is tossed, the **chances** of obtaining a head are 1 in 2 or 50%. This may be expressed as the ratio 1 to 2 or as the fraction $\frac{1}{2}$ (since ratios are often expressed as fractions) or the decimal equivalent 0.5.

If a fair die is rolled, the **chances** of obtaining a 4 on the face that is up are 1 in 6. This may be expressed as the ratio 1 to 6 or as the fraction $\frac{1}{6}$. The chances of obtaining an odd number (1,3,5) on a die are 3 in 6. Again this may be expressed by the ratio 3 to 6 or as the fraction $\frac{3}{6}$ which is, of course, $\frac{1}{2}$ or the decimal equivalent 0.5. Instead of referring to the "chances" of an event's occurring, we can proceed in a different way.

> ### Definition
> The **probability** of an event A, denoted by P(A), may be expressed as the ratio of the number of outcomes in event A to the total number of equally likely (equiprobable) outcomes in the sample space S of the experiment.
> $$P(A) = \frac{\text{the number of outcomes in A}}{\text{the number of outcomes in S}}$$

The probability of an event is expressed as a fraction in which the numerator is the number of outcomes in the event, and the denominator is the number of outcomes in the sample space.

Examples

1. If a fair coin is tossed, the chances of obtaining a Head is $\frac{1}{2}$, and also the probability of obtaining a Head is $\frac{1}{2}$ or 0.5. A probability of $\frac{1}{2}$ does not mean that if a coin is tossed two times, then exactly one Head will appear. The interpretation of $\frac{1}{2}$ is that it is the fraction of the time that Heads would appear if the coin were tossed an infinite number of times. This is a theoretical interpretation, but, in a more practical sense, Heads would appear about half the time in a very large number of tosses. Notice that we are not asserting that precisely half of the tosses would be Heads. The probability means that if the coin is repeatedly tossed, **approximately** half of the results will be Heads. The greater the number of tosses, the more likely the number of Heads will be close to half of the tosses.

2. The chances of rolling a 4 on a fair die are $\frac{1}{6}$, and the probability of rolling a 4 is $\frac{1}{6}$. The greater the number of times the die is rolled, the closer to one-sixth of the times a 4 will appear.

Chapter 7

Problem 1   If a fair die is rolled, what is the probability of rolling a number that is less than 3?

Solution   Let A be the event of rolling a number that is less than 3. Notice that the number of outcomes in event A is 2 since there are 2 numbers on the die that are less than 3, namely 1 and 2. The number of total outcomes in the sample space S is 6, since there are 6 possible outcomes on a die.

Therefore, $P(A) = \dfrac{\text{the number of outcomes in A}}{\text{the number of outcomes in S}} = \dfrac{2}{6} = \dfrac{1}{3}.$

If events in an experiment are mutually exclusive, the sum of the probabilities associated with each event always equals 1.

In tossing a fair coin, the probability of getting Heads is $\frac{1}{2}$, and the probability of getting Tails is $\frac{1}{2}$. Adding these probabilities together, we see that $\frac{1}{2} + \frac{1}{2} = 1$. The probability of rolling any of the six numbers on a fair die is $\frac{1}{6}$. Adding these probabilities, we obtain:

$$\begin{array}{cccccc} (1) & (2) & (3) & (4) & (5) & (6) \\ \frac{1}{6} + & \frac{1}{6} + & \frac{1}{6} + & \frac{1}{6} + & \frac{1}{6} + & \frac{1}{6} = \frac{6}{6} = 1 \end{array}$$

Problem 2   A mother wishes to take her children on a vacation. She can visit Disney World, Busch Gardens, or Everglades National Park. The probability of her visiting Disney World is four times the probability of her visiting Busch Gardens. The probability of her visiting Everglades National Park is five times the probability of her visiting Busch Gardens. (Assume the events are mutually exclusive.) Find the probability that she takes her children to Busch Gardens.

Solution   This type of problem is approached in the same way as are the verbal problems in algebra.

Let x be the probability that she visits Busch Gardens.
Then 4x is the probability that she visits Disney World and
5x is the probability that she visits Everglades National Park.

450

Since the sum of the probabilities should equal 1, we solve the following equation:

[handwritten: greater = mayor (NO IGUAL)]

$$x + 4x + 5x = 1$$

$$10x = 1$$

[handwritten: At least minimo # (para arriba)]

[handwritten: At most = máximo ≥]

$$x = \frac{1}{10} \text{ or } 0.1$$

[handwritten: less = menos (no igual)]

[handwritten right margin: <2 ≤ 2, At least = minimum, 2 or more = 2 or 3, at most = cant pass that. It could be less but not more.]

The probability that she visits Busch Gardens is 0.1.

Problem Set 6.1

1. Can the probability of an event ever be greater than 1? Why? *[handwritten: No. Bcoz the numerator can't b greater than the denominator]*

2. A fair die is tossed. Find the probability that:
   a. the number 2 is obtained. *[handwritten: 1/6]*
   b. a number greater than 2 is obtained. *[handwritten: 4/6, 2/3]*
   c. a number different from 6 is obtained. *[handwritten: 5/6]*
   d. a number obtained is at least 3. *[handwritten: 4/6 2/3]*
   e. a number obtained is at most 5. *[handwritten: 5/6]*
   f. a number obtained is less than 2. *[handwritten: 1/6]*

3. A bag contains 3 black balls and 6 red balls. One ball is drawn at random. What is the probability that the ball is black? *[handwritten: $\frac{3}{6} = \frac{3}{9}$]*

4. A spinner for a child's game may point to any of 7 numbers equally. The numbers are 1, 2, 3, 4, 5, 6, and 7. What is the probability that the spinner will point to an odd number? What is the probability that the spinner will point to a number less than 5? *[handwritten: 3/7, 4/7, 1/2]*

5. Twenty numbered slips of paper are placed in a bowl. The slips are numbered from 1 to 20. What is the probability of drawing a number that is exactly divisible by 5? What is the probability of drawing a number that is less than 9? *[handwritten: 4/20, 8/20 2/5]*

6. If a fair coin is tossed ten times and the first nine tosses are Heads, what is the probability that the tenth toss is a Head? *[handwritten: 1/2]*

7. A fair coin is tossed two times in succession. Recall that the sample space S associated with this experiment is S = {HH, HT, TH, TT}. Since there are 4 possible outcomes in the sample space, the probability of each outcome is $\frac{1}{4}$. *[handwritten: 1/2]*
   a. What is the sum of the probabilities associated with each outcome in the sample space? *[handwritten: 4/4 = 1]*
   b. What is the probability of obtaining two Heads? *[handwritten: 1/4]*
   c. What is the probability of obtaining a Head and a Tail? *[handwritten: 2/4]*
   d. What is the probability of obtaining a Head on the second toss? *[handwritten: 2/4]*
   e. What is the probability of obtaining at least one Head? *[handwritten: 3/4]*

451

8. A fair coin is tossed three times in succession. Recall that the sample space S associated with this experiment is:

S = {HHH, HHT, HTH, HTT, THH, THT, TTH, TTT}

Since there are 8 possible outcomes in the sample space, the probability of each outcome is $\frac{1}{8}$.

a. What is the sum of the probabilities in the sample space? $\frac{8}{8} = 1$
b. What is the probability of obtaining all Heads? $\frac{1}{8}$
c. What is the probability of obtaining exactly one Tail? $\frac{3}{8}$
d. What is the probability of obtaining at least two Heads? $\frac{4}{8}$

9. If a fair coin is tossed 4 times in succession, how many outcomes are obtained? How many outcomes are obtained if a fair coin is tossed n times in succession? $16 - 2^n$

10. Three hundred people attend a dance and are each given a number from 1 to 300 as they enter the front door. If one number is randomly selected, what is the probability that the number selected:
a. is 148? $\frac{1}{300}$
b. is greater than 75? $\frac{225}{300}$ $\frac{45}{60}$ $\frac{9}{12}$ $\frac{3}{4}$
c. has the same three digits? $\frac{200}{300}$ $\frac{2}{3}$

11. A college student must take an elective course. The available courses are: Music Appreciation, Florida History, Oceanography and Judo. The student knows that the probability of taking Florida History is twice the probability of taking Music Appreciation. The probability of taking Oceanography is three times the probability of taking Music Appreciation, and the probability of taking Judo is four times the probability of taking Music Appreciation. What is the probability of the student taking:
a. Music Appreciation? $\frac{1}{10}$
b. Florida History? $\frac{2}{10}$
c. Oceanography? $\frac{3}{10}$
d. Judo? $\frac{4}{10}$

12. If A, B, and C are the only mutually exclusive events for some experiment and P(A) = 0.4, P(B) = 0.3, and P(C) = 0.2, is this an accurate assignment of probabilities? Why? No, the sum of the probabilities is less than one

13. The sample space for the experiment involving rolling two dice is listed in example 5 of Section 5. Find the probability of each of the following events.
a. The sum of the numbers on the dice is 6. $\frac{5}{36}$
b. The numbers on the dice are both even. $\frac{9}{36}$
c. The sum of the numbers on the dice is divisible by 4.

Review Competencies

14. Which of the following pairs of events are mutually exclusive?
a. being 5 years old and having a valid license to drive a car ME
b. doing one's homework and listening to the radio No ME

15. A box contains two marbles (one blue and one green). Two marbles are drawn in succession with the first marble replaced in the box before the second is drawn. List the sample space associated with this experiment.

    (BB, BG, GB, GG)

16. How many groups of 3 books can be selected from a certain 9 books?

    $C = \frac{9}{3(9-3)} = \frac{9}{3(6)} = \frac{9 \times 8 \times 2}{3 \times 2} = \frac{504}{6} = \boxed{84}$

17. How many five digit telephone numbers may be formed using the decimal digits (0–9) if zero is not permitted as the first digit and any digit may be repeated?

18. Express $4\frac{5}{6}$ as an improper fraction. $\frac{26}{6}$

19. Multiply: $1\frac{1}{2} \cdot 1\frac{2}{5}$ $\frac{3}{2} \times \frac{7}{5} = \frac{21}{10}$

20. Divide: $0.123 \div 0.003$ *41*

21. What is 15% of 70? 10.5

$9\,\underline{10}\,\underline{10}\,\underline{10}\,\underline{10} = 90.000$

Let us now consider some special probability situations.

> **An event has a probability of 0 if it cannot possibly occur.**

Example

The probability of rolling a 7 on a die is 0. Let A be the event of rolling a 7 on a fair die. Since there is no 7 on a die, the number of outcomes in A is 0. The total number of outcomes in the sample space is 6. Using the definition of probability, we have:

$$P(A) = \frac{0}{6} = 0.$$

$(1\cdot5)(2+4)(4+2)(3+3)$

> **An event has a probability of 1 if it is absolutely certain to occur.**

Examples

1. The probability of obtaining fewer than 2 Heads in one toss of a fair coin is 1. The event is absolutely certain to occur. If A is the event of obtaining fewer than 2 Heads, the number of outcomes in A is 2 (obtaining 1 Head or 0 Heads). The total number of outcomes is 2. By the definition of probability:

$$P(A) = \frac{2}{2} = 1.$$

2. The probability of rolling a number on a fair die that is less than 7 is 1. This event is absolutely certain to occur. If A is this event, we use the definition of probability and obtain:

$$P(A) = \frac{6}{6} = 1.$$

> **The probability of an event A, P(A), must
> always be between zero and one inclusive.**
> $$0 \leq P(A) \leq 1$$

[handwritten: can take any value for 0% or 0, 0 nonnegative #.]

The probability of an event can never be a negative number.

If the probability of an event's occurring is P(A), then the probability that the event will not occur is represented by P(A'). Events A and A' are said to be **complementary events**.

Since an event must either occur or not occur, the sum of the probability that the event will occur and the probability the event will not occur is 1.

$$P(A) + P(A') = 1 \ \text{ or } \ P(A') = 1 - P(A)$$

Examples

1. The probability of rolling a 4 on a fair die, P(A), is $\frac{1}{6}$. The probability of not rolling a 4, P(A'), is $\frac{5}{6}$ since there are five numbers on the die that are not 4. Since $P(A') = 1 - P(A)$, then $P(A') = 1 - \frac{1}{6}$ or $\frac{5}{6}$.

2. If the probability of tossing a Head, P(A), on a fair coin is $\frac{1}{2}$, then the probability of not tossing a Head, P(A'), is $1 - \frac{1}{2}$ or $\frac{1}{2}$.

The idea of complementary events can be related to notions in set theory. In the Venn diagram on the right, the sample space S serves as the universal set. Event A and event A', the complementary event, are represented. Observe that

$$A \cup A' = S \ \text{ and } \ A \cap A' = \emptyset .$$

Problem 3

Two common sources of entertainment for Americans are movies and live theater. Sixty percent of Americans go to the movies but not live theater, while 15% go to both. What is the probability that a randomly selected American does not go to the movies?

Solution

The Venn diagram on the right shows 60% of Americans attending movies but not live theater and 15% attending both. Since $P(A') = 1 - P(A)$,

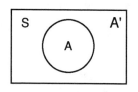

$$P(\text{not going to movies}) = 1 - P(\text{going to movies})$$
$$= 1 - (0.60 + 0.15)$$
$$= 1 - 0.75$$
$$= 0.25$$

The probability that a randomly selected American does not go to the movies is 0.25 (or equivalently 25% or $\frac{1}{4}$).

Problem Set 6.2

1. What is the probability of rolling a sum of 13 by tossing a single fair die two times in succession? *Not possible* 0

2. A bag contains 3 white balls and 3 black balls. If one ball is drawn from the bag, what is the probability that it is black or white? *1 out of 6*

3. The probability of obtaining all Tails if a fair coin is tossed three times in succession is $\frac{1}{8}$. What is the probability of not obtaining all Tails? *7/8*

4. If the probability that it will rain on a given day is 0.25, what is the probability that it will not rain? *0.75*

5. What is the probability that the next person you meet was not born on a Friday? *1 out of 7 ?*

6. A fair die is tossed. What is the probability that the number obtained is not 7? *True, 6/6 = 1*

7. A bag contains 4 green marbles, 3 red marbles, and 5 blue marbles. If one marble is drawn from the bag:
   a. What is the probability that it is not green? *8/12*
   b. What is the probability that it is purple? *0*
   c. What is the probability that it is not purple? *12/12*

8. Two popular kinds of television programs for American adults are comedies and dramas. Forty percent of American adults watch comedies but not dramas, while 15% watch both. What is the probability that a randomly selected American adult who watches television does not watch comedies? *45%*

Review Competencies

9. Ten numbered slips of paper are placed in a bowl. The slips are numbered from 1 to 10. What is the probability of drawing an even number that is exactly divisible by 3?

10. If A, B, and C are the only mutually exclusive events for some experiment and P(A) = 0.5, P(B) = 0.3, and P(C) = 0.3, is this an acceptable assignment of probabilities? Why?

11. Are the following events mutually exclusive? Drawing a green ball and a blue ball in one draw out of a box that contains 4 green balls and 5 blue balls.

12. A scholarship winner is to be selected from a group consisting of Pete, Ann and Steve. List the sample space associated with this experiment.

13. In how many ways can 6 people line up at a water fountain?

14. Using the digits 2, 3, 4, and 5, how many three digit numbers less than 400 can be formed if any digit cannot be repeated?

15. Divide: $4\frac{2}{5} \div 1\frac{1}{2}$

16. What percent of 40 is 15?

17. Evaluate: $6 \div (4 - 2) + 3 \cdot 2 - 5$

18. Express 0.04% as a decimal.

19. Subtract: $6.03 - 0.125$

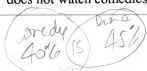

455

Chapter 7

A deck of 52 playing cards helps to illustrate the basic ideas of probability. If you are not familiar with playing cards the following information is useful:

The deck is composed of 4 suits. Each suit contains the following cards: 2, 3, 4, 5, 6, 7, 8, 9, 10, jack, queen, king, ace

$$
4 \text{ suits} \left\{ \begin{array}{l} 13 \text{ spades} \\ 13 \text{ clubs} \\ 13 \text{ hearts} \\ 13 \text{ diamonds} \end{array} \right. \quad \begin{array}{l} \{ 26 \text{ black cards} \\ \\ \{ 26 \text{ red cards} \end{array}
$$

There are 12 **face cards** (4 jacks, 4 queens, 4 kings). There are 4 aces (1 in each suit).

Problem 4

What is the probability of drawing a heart from a deck of 52 playing cards in a single draw?

Solution

Since there are 13 hearts in the deck, the probability of drawing a heart is $\frac{13}{52}$ or $\frac{1}{4}$.

Problem Set 6.3

1. A single card is drawn from a deck of 52 playing cards. Find the probability of:
   a. drawing a spade.
   b. drawing an ace.
   c. drawing a red card.
   d. drawing the ace of spades.
   e. drawing a face card.
   f. drawing a queen.
   g. drawing a black jack.
   h. not drawing an ace.
   i. not drawing a diamond.
   j. not drawing a face card.
   k. drawing a card that is greater than 5 and less than 8
   l. drawing an 11.
   m. drawing a red card or a black card.

2. A box contains 8 fuses of which 2 are defective. If one fuse is selected at random, what is the probability that it is nondefective?

3. What is the probability that the next person you meet was born on February 30?

4. A bag contains 3 red marbles and 2 white marbles. If one marble is drawn at random, what is the probability that it is not blue?

5. Two hundred people participate in a raffle in which each is given a number from 1 to 200. If one number is randomly selected, what is the probability that the number selected is divisible by 40?

456

6. If A, B, and C are the only mutually exclusive events for some experiment, and P(A) = 0.6, P(B) = 0.5, and P(C) = -0.1, is this an acceptable assignment of probabilities? Why?

7. A box contains 4 batteries (B$_1$, B$_2$, B$_3$, and B$_4$). If one battery is selected at random from the box, what is the sample space associated with this experiment?

8. How many sequential arrangements can be made using the letters of the word GATOR if any letter may not be repeated?

9. Express $5\frac{6}{7}$ as an improper fraction.

10. Express $3\frac{3}{4}\%$ as a decimal.

Section 7   The Addition Rules for Probability

In this section, we introduce some very important ideas in probability theory.

Consider the Venn diagram shown in which the numbers represent the number of elements contained in each region.

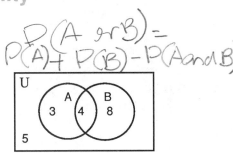

We know that the number of elements in A $\cup$ B, n(A $\cup$ B), is 3 + 4 + 8 or 15. The number of elements in set A, n(A), is 3 + 4 or 7. The number of elements in set B, n(B), is 4 + 8 or 12. Is it true that n(A $\cup$ B) = n(A) + n(B)? No. We can verify this by:

$$n(A \cup B) = n(A) + n(B)$$
$$15 \quad \neq \quad 7 + 12$$

The reason for this inequality is that the number of elements in **both** set A and set B (4) was counted **twice** in computing n(A) + n(B). We should have counted the 4 elements only **once**. Thus, we must subtract one of the quantities that we originally added twice. Note that this quantity lies within A $\cap$ B.

$$n(A \cup B) = n(A) + n(B) - n(A \cap B)$$
$$15 \quad = \quad 7 + 12 - 4$$
$$15 \quad = 15$$

This relationship in set theory will have a corresponding application in probability theory and we refer to it as the:

> **General Addition Rule for Probability:**
> If A and B are any two events for an experiment, then
> $$P(A \cup B) = P(A) + P(B) - P(A \cap B)$$

457

If A and B are two events in a sample space, the union of A and B, (A ∪ B), is the event composed of all the outcomes in event A **or** event B. It will be helpful to interpret P(A ∪ B) as P(A or B). The intersection of A and B, (A ∩ B), is the event composed of all the outcomes which are in **both** A and B. It will be helpful to interpret:

$$P(A \cap B) \text{ as } P(A \text{ and } B \text{ occur simultaneously}).$$

The General Addition Rule is used to find the probability of two events connected by the word "or." Thus, the rule is often restated in any of the following ways:

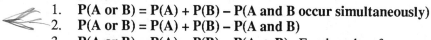

1. **P(A or B) = P(A) + P(B) – P(A and B occur simultaneously)**
2. **P(A or B) = P(A) + P(B) – P(A and B)**
3. **P(A or B) = P(A) + P(B) – P(A ∩ B)** For the sake of consistency, we will use this form in our solved problems.

Problem 1

What is the probability of rolling a 1 **or** a 5 on one roll of a fair die?

Solution

Let A to be the event of rolling a 1.

$$\text{Then, } P(A) = \frac{1}{6}$$

Let B to be the event of rolling a 5.

$$\text{Then, } P(B) = \frac{1}{6}$$

Notice that event A and event B are **mutually exclusive events.** They cannot occur at the same time. It is not possible to roll a 1 and a 5 on the same toss. Since it is impossible for event A and event B to occur at the same time, the probability of this happening is 0. In other words, P(A ∩ B) = P(A and B occur simultaneously) = 0.

$$P(A \text{ or } B) = P(A) + P(B) - P(A \cap B)$$
$$= \frac{1}{6} + \frac{1}{6} - 0$$
$$= \frac{2}{6} \text{ or } \frac{1}{3}$$

There are two numbers on the die that are involved in this problem, namely, 1 and 5. The probability of rolling 1 or 5 is $\frac{2}{6}$ or $\frac{1}{3}$.

Problem 2

A committee consists of 5 Democrats, 4 Republicans, and 2 Independents. If one person is selected at random, what is the probability the person is a Democrat **or** a Republican?

Solution If A is the event the person selected is a Democrat, then $P(A) = \frac{5}{11}$.

If B is the event the person selected is a Republican, then $P(B) = \frac{4}{11}$.

The events are mutually exclusive. One cannot be both a Democrat and Republican. Consequently,

$$P(A \cap B) = P(A \text{ and } B \text{ occur simultaneously}) = 0.$$
and
$$P(A \text{ or } B) = P(A) + P(B) - P(A \cap B)$$
$$= \frac{5}{11} + \frac{4}{11} - 0$$
$$= \frac{9}{11}$$

Thus, the probability that the person is a Democrat or a Republican is $\frac{9}{11}$.

The two preceding problems give rise to the:

> **Special Addition Rule for Probability**:
> If A and B are two mutually exclusive events, then
> $$P(A \cup B) = P(A) + P(B)$$

An equivalent form of the Special Addition Rule is:
$$P(A \text{ or } B) = P(A) + P(B)$$

If A, B, and C are three mutually exclusive events, then this Special Addition Rule can be extended as:
$$P(A \cup B \cup C) = P(A) + P(B) + P(C)$$

A convenient, equivalent form is:
$$P(A \text{ or } B \text{ or } C) = P(A) + P(B) + P(C)$$

Problem 3 A single card is drawn from a deck of 52 bridge cards. What is the probability that the card is a jack, a queen **or** a king?

Solution If A is the event of drawing a jack, then $P(A) = \frac{4}{52}$ or $\frac{1}{13}$.

If B is the event of drawing a queen, then $P(B) = \frac{4}{52}$ or $\frac{1}{13}$.

If C is the event of drawing a king, then $P(C) = \frac{4}{52}$ or $\frac{1}{13}$.

Since events A, B, and C are mutually exclusive, we obtain:

$$P(A \text{ or } B \text{ or } C) = P(A) + P(B) + P(C)$$
$$= \frac{1}{13} + \frac{1}{13} + \frac{1}{13}$$
$$= \frac{3}{13}$$

Chapter 7

Problem Set 7.1

1. An urn contains 6 white balls, 8 green balls, and 5 blue balls. If one ball is drawn at random, find the probability that the ball is:
   a. white or green.
   b. white, green, or blue.
   c. neither white nor blue.

2. A single card is drawn from a deck of 52 playing cards. Find the probability that the card selected is:
   a. a king or a queen.
   b. an ace or a jack.
   c. a red card or a black card.
   d. an ace or a 10.
   e. an 8, 9, or 10.
   f. a 2, 3, 4, 5, 6, or 7.

3. Joyce has an accounting test tomorrow. She feels there is a probability of 0.2 that she will get an A, a probability of 0.4 that she will get a B, a probability of 0.2 that she will get a C, a probability of 0.1 that she will get a D, and a probability of 0.1 that she will get an F. Find the probability that Joyce will:
   a. get an A or B.
   b. get a C or higher.
   c. get lower than a C.

4. The probability that a typist will make 0, 1, 2, 3, or 4 mistakes on a page are 0.1, 0.3, 0.3, 0.2, and 0.1 respectively. Find the probability that the typist will make:
   a. at least 3 mistakes ("at least 3" means 3 or more.)
   b. at most 2 mistakes.
   c. an odd number of mistakes.
   d. 1, 2, or 3 mistakes.
   e. more than 1 mistake.
   f. less than 3 mistakes.

Review Competencies

5. If the probability that it will snow on a given day is 0.4, what is the probability that it will not snow?

6. A box contains 3 orange balls and 3 blue balls.
   a. What is the probability that one ball drawn at random from the box will be orange or blue?
   b. What is the probability that the ball drawn will be red?

7. A spinner for a child's game may point to any of 5 numbers equally (1, 2, 3, 4, 5). What is the probability that the spinner will point to a number greater than or equal to 2?

8. An urn contains a red marble and a white marble (R and W). Two marbles are drawn in succession with the first marble replaced in the box before the second is drawn. List the sample space associated with this experiment.

9. A woman has 10 blouses and 6 skirts. How many possible outfits does she have?

10. Express $\frac{35}{60}$ in lowest terms.

11. Add: $4.08 + 0.006 + 28.1$

12. Evaluate: $5(6 \div 2) + 3(2 \cdot 3) - 5$

13. Solve: $2(x - 3) \geq 5(x + 2)$.

14. Express 2 as a percent.

15. $\frac{4}{5} = \frac{?}{30}$

Let us now consider the addition rule as it relates to events that are not mutually exclusive.

Problem 4   What is the probability of rolling a 5 **or** a number greater than 4 on one roll of a fair die?

Solution   Let us consider A to be the event of rolling a 5.

Therefore, $P(A) = \frac{1}{6}$.

Let us consider B to be the event of rolling a number greater than 4 (5 or 6).

Hence $P(B) = \frac{2}{6}$.

Observe here that event A and event B are **not** mutually exclusive events. They could both occur simultaneously. Certainly a 5 on the die is both an outcome in event A and event B. Since the probability of obtaining a 5 is $\frac{1}{6}$, we say that

$$P(A \cap B) = P \text{ (A and B occur simultaneously)} = \frac{1}{6}.$$

To compute the probability we have:

$$P(A \text{ or } B) = P(A) + P(B) - P(A \cap B)$$
$$= \frac{1}{6} + \frac{2}{6} - \frac{1}{6}$$
$$= \frac{2}{6} \text{ or } \frac{1}{3}$$

Problem 5   The probability that Doug will read a book on a Saturday night is 0.6, and the probability that he will listen to records is 0.4, and the probability that he will do both is 0.3. What is the probability that Doug will read a book **or** listen to records?

Solution   These events are **not** mutually exclusive. They could occur simultaneously.

If P(A) is the probability that Doug will read a book,
then P(A) = 0.6.

If P(B) is the probability that Doug will listen to records,
then P(B) = 0.4.

$P(A \cap B)$ is the probability that Doug will simultaneously read a book and listen to records.

$$P(A \cap B) = 0.3$$
$$P(A \text{ or } B) = P(A) + P(B) - P(A \cap B)$$
$$= 0.6 + 0.4 - 0.3$$
$$= 0.7$$

Thus, the probability that Doug will read a book or listen to records is 0.7.

Chapter 7

1. A single card is drawn from a deck of 52 playing cards. Find the probability that the card selected is:
   a. an ace or a spade.
   b. a diamond or a jack.
   c. an ace or a red card.
   d. a red card or a face card.
   e. a face card or a queen.

2. A fair die is rolled. What is the probability that the number obtained:
   a. will be an even number or a number greater than 2?
   b. will be an odd number or a number greater than 1?

3. Twenty numbered slips of paper are placed in a box. The slips are numbered from 1 to 20. What is the probability of drawing a number that is odd or exactly divisible by 5?

4. Suppose one letter of the English alphabet is selected at random. Find the probability that the letter selected:
   a. is a vowel.   b. precedes the letter m.
   c. precedes the letter m or is a vowel.

5. The probability that a tourist will visit Key West is 0.6. The probability that he/she will visit St. Augustine is 0.4, and the probability that he/she will visit both Key West and St. Augustine is 0.3. What is the probability that the tourist will visit Key West or St. Augustine?

6. At an international meeting, 30 percent of the delegates speak English, 40 percent of the delegates speak Spanish, and 20 percent speak both English and Spanish. If a delegate is randomly selected, what is the probability that the delegate speaks English or Spanish?

7. In a civic club, the probabilities that a person selected will have the following characteristics are: a politician: 0.75, corrupt: 0.15, a politician and corrupt: 0.10. Find the probability that the person:
   a. is a politician or corrupt.
   b. is neither a politician nor corrupt.

8. The probability that Theresa will mow her lawn is 0.5. The probability that she will wash her car is 0.3. The probability that she will trim her hedge is 0.2. Theresa plans to engage in only one activity. What is the probability that:
   a. she will wash her car or trim her hedge?
   b. she will mow her lawn or wash her car?

9. A box contains 6 red marbles, 5 white marbles, and 3 blue marbles. If one marble is drawn at random, find the probability that the marble is:
   a. white or blue.   b. neither red nor white.

10. A coin is tossed twice. What is the probability of obtaining either two Heads, one Head and one Tail, or two Tails?

11. A single card is drawn from a bridge deck of 52 playing cards. What is the probability that the card selected is a queen or an ace?

12. What is the probability that the next person you meet was not born in May?

13. In how many ways can a teacher line up 4 students by 4 front seats in a classroom?

14. How many six digit telephone numbers may be formed using the decimal digits (0 - 9) if zero is not permitted as the first digit, and no digit may be repeated?

Fort Lauderdale (handwritten)

15. Three slips of paper numbered 1, 2, and 3 are placed in a hat. List the sample space for the experiment in which one slip of paper is drawn at random from the hat.

16. Express $4\frac{3}{5}$ as an improper fraction.

17. Divide: $0.015 \div 0.5$

18. Express $2\frac{3}{10}\%$ as a decimal.

19. Multiply: $3\frac{1}{3} \cdot 1\frac{1}{2}$

A table describing the relative frequency of occurrence of all possible outcomes often forms the basis of solving real-world problems involving probability. The table below is based on the fact that between 1970 and 1977 approximately 12,000 people were killed on scheduled and non-scheduled airline flights. The table below provides the "when and how" of these air deaths. (Assume all events are mutually exclusive.)

When and How	Percent of All Air Deaths
During Take-Off and Climb	20
en Route	14
During Approach	31
During Landing	19
As a Result of Terrorism	7
As a Result of Collisions	9

Problem 6

What is the probability that an air death occurs during landing?

Solution

This information is provided directly by the table. We see that 19% of the air deaths occur during landing ($\frac{19}{100}$), so the probability can be expressed as 0.19.

Problem 7

What is the probability that an air death occurs as a result of terrorism or collisions?

Solution

Since the events in the table are mutually exclusive, if A is the event "terrorism," and B is the event "collisions,"
$$P(A \text{ or } B) = P(A) + P(B)$$
$$= 0.07 + 0.09$$
$$= 0.16$$

The probability that an air death occurs as a result of terrorism or collision is 0.16.

Problem 8

What is the probability that an air death is neither the result of terrorism nor collisions?

Solution

The outcome "neither terrorism nor collisions" is the complement of "terrorism or collisions." If C is the event "terrorism or collisions,"

$$
\begin{aligned}
\text{P(neither terrorism nor collisions)} \quad &= P(C') \\
&= 1 - P(C) \\
&= 1 - 0.16 \text{ (from problem 7)} \\
&= 0.84
\end{aligned}
$$

The probability that an air death is neither the result of terrorism nor collisions is 0.84.

Problem 9

How many of the next 400 air deaths would one expect to take place en route or during approach?

Solution

Since the events are mutually exclusive, the probability of an air death en route or during approach is $0.14 + 0.31 = 0.45$. We would expect 45% of the next 400 air deaths or $(0.45)(400) = 180$ deaths to take place en route or during approach.

Problem 10

If it is known that an air death was not a result of terrorism or collisions, find the probability that it took place during landing.

Solution

One way to approach this problem is to first exclude the last two rows from the given table. Thus,

$$
\text{P(landing)} \quad = \quad \frac{\text{percent of deaths during landing}}{\substack{\text{percent of deaths in all categories excluding} \\ \text{terrorism and collisions}}}
$$

$$
= \frac{19}{20 + 14 + 31 + 19} = \frac{19}{84}
$$

The probability that an air death took place during landing given that it was not a result of terrorism or collisions is $\frac{19}{84}$.

Problem Set 7.3

Use the table on pg. 463 to answer these questions.

1. What is the probability that an air death occurs during approach? *0.31*

2. What is the probability that an air death occurs en route or during landing? *0.33*

3. What is the probability that an air death occurs neither en route nor during landing? *0.67*

4. How many of the next 400 air deaths would one expect to occur during take-off and climb or en route? *136*

5. How many of the next 400 air deaths would one expect to occur neither en route nor during landing? *132 256*

6. If it is known that an air death did not occur during approach or landing, find the probability that it took place en route. *0.14 0.28*

7. If it is known that an air death did not occur en route, during approach, during landing, as a result of terrorism, or as a result of collisions, find the probability that it took place during take-off and climb.

The following is a description of students at a college by gender and student classification.

	Freshman	Sophomore	Junior	Senior
Male	13%	10%	29%	8%
Female	15%	8%	7%	10%

Use this table to answer problems 8 through 15.

8. Find the probability that a randomly selected college student is female. *40%*

9. Find the probability that a randomly selected college student is a freshman. *28%*

10. Find the probability that a randomly selected college student is a junior or a senior. *54%*

11. Find the probability that a randomly selected college student is a sophomore if it is known that the student is female.

12. Find the probability that a randomly selected college student is a freshman or a sophomore if it is known that the student is male. *23%/60*

13. Find the probability that a randomly selected college student is neither a freshman nor a junior if it is known that the student is not female. *18/60 0.3*

14. If it is known that a student is not a junior or a senior, find the probability of randomly selecting a sophomore. *18/46 = 9/23*

15. If it is known that a student is not a freshman or a sophomore, find the probability of randomly selecting a junior or a senior. *36/54 = 0.66*

Review Competencies

16. There are six finalists in an essay contest. First and second place prizes will be awarded to two of the finalists. In how many ways can first and second prizes be awarded?

17. The probability of having a fair coin land on heads five times in a row is $\frac{1}{32}$. What is the probability of not having the coin land on heads five times in a row?

18. The spinner shown is to be randomly spun twice. What is the sample space associated with this experiment?

   a. {(1, 2), (1, 3), (2, 1), (2, 3),(3, 1), (3, 2)}
   b. {1, 2, 3}
   c. {(1, 2), (1, 3),(2, 3)}
   d. {(1, 1), (1, 2), (1, 3), (2, 1), (2, 2), (2, 3), (3, 1), (3, 2), (3, 3)}

19. No matter which flight you choose to fly to London, you can travel first class, business class, or coach. From city A there are 4 flights to London, resulting in 12 travel options. How many travel options will you have from a city that has 15 flights into London?
   a. 18   b. 42   c. 45   d. 180

Section 8 The Multiplication Rule for Probability

In this section, we consider the probability that two or more events will occur in succession. Let us first begin with a definition.

Definition
Two events are said to be **independent** if the occurrence of one event does not affect the probability of the occurrence of the other event.

Examples

1. If a fair coin is tossed two times in succession, the outcome of the first toss does not affect the probability of the outcome of the second toss. These two events are independent events.

2. If a fair die is rolled two times in succession, the outcome of the first roll does not affect the probability of the outcome of the second roll. These two events are considered independent events.

3. If two cards are drawn in succession from a deck of 52 playing cards with the first card not replaced before the second card is drawn, the occurrence of the first draw affects the probability of the occurrence of the second draw. These two events are not independent events.

The notion of independent events gives rise to the:

General Multiplication Rule for Probability:
If A and B are any two events, the probability of A occurring followed by B occurring equals the probability of A occurring, P(A), multiplied by the probability of B occurring given that A has already occurred, denoted P(B|A).
P(A followed by B) = P(A) • P(B given that A has occurred)
$$= P(A) • P(B|A).$$

If two events A and B are independent events, then P(B|A) = P(B). This fact gives rise to:

> **Special Multiplication Rule for Probability**:
> If A and B are independent events, then
> **P(A followed by B) = P(A) • P(B)**

The Special Multiplication Rule can be extended to include three or more independent events. If A, B, and C are independent events, then
P(A followed by B followed by C) = P(A) • P(B) • P(C)

The multiplication rules are used to find the probability of events connected by "and." Thus, the General Multiplication Rule is often restated in one of the following ways: 1. **P(A and then B) = P(A) • P(B|A)**
   2. **P(A and B) = P(A) • P(B|A)** = $1 = given = (B given A)$
   3. **P(A ∩ B) = P(A) • P(B|A)**

In our solved problems, the relationships that will be most helpful to us are those that appear originally in the statements of the General and Special Multiplication Rules.

Problem 1

If a fair coin is tossed two times in succession, what is the probability of obtaining a Tail on the first toss **and** a Head on the second toss?

Solution

If A is the event of obtaining a Tail on the first toss, then $P(A) = \frac{1}{2}$. If B is the event of obtaining a Head on the second toss, then $P(B) = \frac{1}{2}$. Notice that the probability of event B is identical to the probability of event A although event A has already occurred. The occurrence of the first toss did not affect the probability of the occurrence of the second toss. Thus, events A and B are **independent events.** We obtain:

$$P(A \text{ followed by } B) = P(A) \cdot P(B)$$
$$= \frac{1}{2} \cdot \frac{1}{2}$$
$$= \frac{1}{4}$$

Therefore, the probability of obtaining a Tail and then a Head is $\frac{1}{4}$. Recall again the set representing the sample space when one coin is tossed two times: {HH, HT, TH, TT}. Notice that the probability of obtaining a Tail on the first toss and a Head on the second toss (TH) is indeed $\frac{1}{4}$. This agrees with the result obtained by using the multiplication rule.

467

Problem 2 Two cards are drawn in succession from a deck of 52 playing cards. The first card is replaced before the second card is drawn. What is the probability that the first card is a heart **and** the second card is an ace?

Solution If A is the event of drawing a heart on the first draw, then

$$P(A) = \frac{13}{52} \text{ or } \frac{1}{4}.$$

Let B be the event of drawing an ace on the second draw. Since event A has already occurred, and the card has been replaced in the deck,

$$P(B) = \frac{4}{52} \text{ or } \frac{1}{13}.$$

Events A and B are **independent events**.

$$P(A \text{ followed by } B) = P(A) \cdot P(B)$$
$$= \frac{1}{4} \cdot \frac{1}{13}$$
$$= \frac{1}{52}$$

The following problems illustrate situations in which events are not independent. If events are not independent, they are said to be **dependent**.

Problem 3 Two cards are drawn in succession from a deck of 52 playing cards. The first card is not replaced before the second card is drawn. Find the probability that **both** cards drawn are clubs.

Solution If A is the event of drawing a club on the first draw, then

$$P(A) = \frac{13}{52} \text{ or } \frac{1}{4}.$$

After the first card is drawn, but **not replaced**, there are only 51 cards left in the deck. We must assume that the first card drawn was a club. Thus, there are only 12 clubs left in the deck for the second draw. If P(B|A) is the probability of drawing a club on the second draw given that the first draw is a club, then

$$P(B|A) = \frac{12}{51} \text{ or } \frac{4}{17}.$$

Observe that the events in this problem are **not** independent events. The probability of the second event, P(B|A), is definitely affected by the outcome of the first event. Such events are **dependent** events. To compute the result, we must multiply the probability of A occurring, P(A), by the probability of B occurring given that A has already occurred, P(B|A).

$$P(A \text{ followed by } B) = P(A) \cdot P(B|A)$$
$$= \frac{1}{4} \cdot \frac{4}{17}$$
$$= \frac{1}{17}$$

$$P(ND \text{ and } ND) = \frac{5}{12} \times \frac{4}{11} =$$

Problem 4

A group consists of 7 Democrats, 3 Republicans, and 2 Independents. If two people are chosen in succession, and the first person is not returned to the group before the second is chosen, find the probability that **neither** person chosen is a Democrat.

or = no were
if you have

Solution

We must find the probability of a non-Democrat followed by a non-Democrat. If A is the event of selecting a non-Democrat on the first choice, then $P(A) = \frac{5}{12}$, since there are 5 non-Democrats and 12 people.

Again, the events in this problem are **dependent**. The probability of the second event, P(B|A), represents the probability of selecting a non-Democrat second given that we have already chosen a non-Democrat first. When we choose a second person, the group consists of 7 Democrats, 2 Republicans, and 2 Independents or 7 Democrats, 3 Republicans, and 1 Independent.

Thus, $P(B|A) = \frac{4}{11}$ since there are 4 non-Democrats and 11 people.

$$P(A \text{ followed by } B) = P(A) \cdot P(B|A)$$
$$= \frac{5}{12} \cdot \frac{4}{11}$$
$$= \frac{5}{33}$$

Problem 5

Twenty percent of the basketballs manufactured by a company are defective. If two balls are randomly selected with replacement, find the probability that at least one ball is defective.

Solution

Let D represent the event of selecting a defective ball and D' not selecting a defective ball. We are given that

$$P(D) = 20\% = \frac{1}{5}.$$
$$\text{Thus } P(D') = 1 - P(D) = 1 - \frac{1}{5} = \frac{4}{5}$$

(The probability of selecting a good ball, one that is not defective, is $\frac{4}{5}$.)

Now we must analyze the English expression "at least one ball is defective." This means that one or possibly both balls selected are defective. The only way that this would not happen is if we were to select two good balls in a row. These events are complements and we can write:

$$P(\text{at least one is defective}) = 1 - P(\text{both are good})$$

If both balls selected are good, we have:

$$P(D' \text{ followed by } D') = P(D') \cdot P(D')$$
$$= \frac{4}{5} \cdot \frac{4}{5} = \frac{16}{25}$$

Thus, $P(\text{at least one is defective}) = 1 - P(\text{both are good})$
$$= 1 - \frac{16}{25} = \frac{9}{25}$$

The probability that at least one ball is defective is $\frac{9}{25}$.

Chapter 7

Problem 6   A group consists of 10 women and 10 men. If two people are randomly selected without replacement, find the probability of selecting two women or two men.

Solution   We must first find the probability of two women, or a woman followed by a woman. If A is the event of selecting a woman on the first choice, $P(A) = \frac{10}{20} = \frac{1}{2}$, since there are 10 women and 20 people. The probability of the second event, $P(B|A)$, represents the probability of selecting a woman second given that we have already selected a woman. $P(B|A) = \frac{9}{19}$, since there are now only 9 women left and 19 people in the group. Thus,

$$
\begin{aligned}
P(A \text{ followed by } B) &= P(\text{two women}) \\
&= P(A) \cdot P(B|A) \\
&= \frac{1}{2} \cdot \frac{9}{19} \\
&= \frac{9}{38}
\end{aligned}
$$

In the same way, the probability of selecting two men is also $\frac{9}{38}$. Using the addition rule of probability:

P(Two Women **or** Two Men)

$$
\begin{aligned}
&= P(\text{Two Women}) + P(\text{Two Men}) - P(\text{Two Women} \cap \text{Two Men}) \\
&= P(\text{Two Women}) + P(\text{Two Men}) - P(\emptyset) \\
&= \frac{9}{38} + \frac{9}{38} - 0 = \frac{18}{38} = \frac{9}{19}
\end{aligned}
$$

The probability of selecting two women or two men is $\frac{9}{19}$. (Observe that selecting two women excludes selecting two men and vice-versa. Since these events are mutually exclusive, we subtracted 0, the probability of them both happening together.)

$$\frac{1}{10} \cdot \frac{1}{9} = \frac{12}{49}$$

Problem Set 8.1

$P(H \text{ and } H \text{ and } H \text{ and } H) = \frac{1}{2} \times \frac{1}{2} \times \frac{1}{2} \times \frac{1}{2} = \frac{1}{16}$

1. A fair coin is tossed 4 times in succession. What is the probability that all four tosses are Heads?

2. A box contains 5 red balls and 5 white balls. If two balls are drawn in succession from the box without replacement, what is the probability that they are both red?

3. Two cards are drawn from a deck of 52 bridge cards. The first card is replaced before the second card is drawn. Find the probability that:
   a. the first card is a club and the second card is a spade.
   b. both cards are aces of spades.
   c. both cards are face cards.
   d. both cards are numbers less than 5.

a) $P(\clubsuit \text{ and } \spadesuit) \; \frac{13}{52} \times \frac{13}{52} = \frac{169}{2704}$

c) (face and face) $\frac{12}{52} \times \frac{12}{52} = \frac{144}{2704}$

b) $P(\text{ and }) \; \frac{1}{52} \times \frac{1}{52} = \frac{1}{2704}$

d) $(5 \text{ and} < 5) \; \frac{13}{52} \times \frac{13}{51} = \frac{169}{2652}$

470

4. Repeat problem #3 if the first card is not replaced before the second card is drawn.

5. A fair die is rolled three times in succession. Find the probability that:
   a. an odd number is tossed on the first toss.
   b. a 1 is tossed on the first two tosses.
   c. a 1 is tossed on the first toss and not tossed on the second toss or the third toss.

6. The probability that x will be alive in 30 years is 0.8, and the probability that y will be alive in 30 years is 0.6. What is the probability that they will both be alive in 30 years? Assume independence of x and y.

7. On a given day the probability that it will rain in Miami is 0.7 and the probability that it will rain in Jacksonville is 0.4. Assuming independence, find the probability that:
   a. it will rain in both cities.
   b. it will rain in neither city.
   c. it will rain only in Miami.

8. The probability of a husband voting in an election is 0.3 and the probability of a wife voting is 0.5. What is the probability of their both not voting? Assume independence of the husband and wife.

9. A box contains 3 red balls, 2 white balls, and 5 blue balls. Two balls are drawn in succession from the box without replacement. What is the probability that neither ball is red?

10. 10% of the pens manufactured by a company are defective. Suppose that two pens are randomly selected with replacement.
    a. Find the probability that both of them are defective.
    b. Find the probability that at least one pen is defective.

11. A box contains 10 tennis balls, of which 2 are defective. If two balls are randomly selected without replacement:
    a. Find the probability that both of them are defective.
    b. Find the probability that at least one ball is defective.

12. A group contains 5 women and 5 men. If 2 people are randomly selected without replacement, find the probability that they are both women or both men.

Review Competencies

13. Suppose 6 slips of paper numbered 1, 2, 3, 4, 5, 6 are placed in a box and one slip is selected at random. What is the probability that:
    a. the number is an even number or a number greater than 4?
    b. the number is odd or even?
    c. the number is not less than 3?
    d. the number is 7?

14. How many groups of 3 gifts can be selected from a group of 6 gifts?

15. At a drive-in restaurant a person can order a hamburger rare, medium, or well-done; with or without ketchup; with or without onions; and with or without tomato. In how many ways can the person order the hamburger?

16. Add: $\frac{7}{8} + \frac{5}{6}$

17. Express 0.05% as a decimal.

18. What is 60% of 70?

19. Express $\frac{1}{8}$ as a decimal.

471

20. Express 70% as a fraction.

21. Evaluate: $\frac{1}{3}(\frac{1}{3} \div 3 + 2 \cdot \frac{1}{3})$

22. The **expected value** of an event is the amount that would be won multiplied by the probability of winning. For example, a lottery offers a prize of $10,000 and sells 5000 tickets.

Expected Value for 1 ticket =

$10,000 \cdot \frac{1}{5000} = \$2$.

Find the expected value if $2 can be won by tossing a fair coin and obtaining a Head.

23. Find the smallest positive multiple of 5 which leaves a remainder of 3 when divided by 4 and a remainder of 1 when divided by 7.

Section 9   Applications of Counting: Probability and Permutations

In this section we see how **The Fundamental Counting Principle** and permutations can be used to determine probabilities.

Problem 1   Consider the digits 3, 4, 5, and 8. If a three digit number is formed, what is the probability that the number is less than 500? Assume that no digit may be used more than once.

Solution   Our probability fraction is:

$$\frac{\text{The number of arrangements giving numbers less than 500}}{\text{The total number of arrangements using three digits}}$$

To obtain the numerator of the probability fraction, we must realize that to get a number less than 500, the first digit must be either 3 or 4. Thus, there are **two** possible ways to fill the first position. Once 3 or 4 is selected, and since the digits may not be repeated, there are **three** numbers remaining to choose from for the second position and three ways to fill the second position. Once the first two digits are used, there are two more digits to choose from for the third position or **two** possible ways to fill the third position. Using **The Fundamental Counting Principle**, the total number of arrangements giving numbers less than 500 is: 2 • 3 • 2 or 12. This number is the numerator of the probability fraction.

To obtain the denominator of the probability fraction, it is necessary to find the total number of arrangements (permutations) using the given digits. The first position can be filled in **four** ways, the second in **three** ways, the third in **two** ways. By **The Fundamental Counting Principle**, the total number of arrangements using three digits is 4 • 3 • 2 or 24. Thus, 24 is the denominator of the probability fraction. The probability that the number selected is less than 500 is $\frac{12}{24}$ or 0.5.

Problem Set 9.1

1. Consider the digits 2, 5, 6, 7, and 8. No digit is to be used more than once.
   a. If a three digit number is formed, what is the probability that the number is less than 800?
   b. If a three digit number is formed, what is the probability that the number is odd? (Hint: The number must end with 5 or 7.)
   c. If a three digit number is formed, what is the probability that it begins with 5?
   d. If a four digit number is formed, what is the probability that the number is even?

2. Using the digits 2, 5, 8, and 9 and the digits may be repeated, what is the probability of forming a three digit even number greater than 900?

3. Consider the letters S, T, A, and R. If a four letter arrangement is made repeating any letter, what is the probability that it will begin with A and end with S?

4. Martha, Lee, Marilyn, Paul, and Ann have all been invited to a dinner party. If they arrive randomly and at different times, what is the probability that:
   a. Martha will arrive second?
   b. Marilyn will arrive first and Lee last?
   c. they will arrive in the following order: Paul, Martha, Marilyn, Lee, and Ann?

Review Competencies

5. A box contains 2 blue balls, 3 white balls, and 4 green balls. If two balls are drawn from the box in succession without replacement, what is the probability that:
   a. the balls are both blue?
   b. the first ball is white and the second ball is green?
   c. neither ball is white?

6. Six slips of paper numbered from 1 to 6 are placed in a bowl. One slip is selected at random. What is the probability that the number obtained:
   a. will be an odd number or greater than 2?
   b. will be a number less than 3 or greater than 3?
   c. will neither be an even number nor an odd number?
   d. is not less than or equal to 2?

7. A dog gives birth to 2 puppies. Using m to represent a male and f for a female, list the sample space for this experiment.

8. Eight runners compete in a race for which there are first, second, and third prizes. In how many ways may the prizes be won?

9. Using the digits 2, 5, 6, and 7, how many 2 digit even numbers can be formed if no digit may be repeated?

10. Subtract: $6 - 1\frac{3}{8}$

11. Divide: $1 \div 0.0002$

12. Multiply: $\frac{3}{4} \cdot \frac{5}{6} \cdot \frac{8}{11}$

13. What percent of 50 is 28?

14. Evaluate: $10 \cdot 2 + 8 \div 4 - 5 \cdot 1 + 3$

15. A Fibonacci sequence is one in which the first two numbers are 1 and each term thereafter is obtained by adding the two preceding terms. Find the next term in the following Fibonacci sequence.

$$1, 1, 2, 3, 5, 8, 13, \underline{\quad}$$

Chapter 7

Section 10 Applications of Counting: Probability and Combinations

In this section we will see that there are times in which combinations must be used along with **The Fundamental Counting Principle.**

Problem 1   A shipment of 7 fuses contains 5 good fuses and 2 defective fuses. If 3 fuses are selected at random from the shipment:
  a.   how many ways are there of selecting one defective fuse and two good fuses?
  b.   how many ways are there of selecting three good fuses?

Solution   a.   Since order is not important, the number of ways of selecting 1 defective fuse is $\binom{2}{1}$ or 2. Corresponding to any one way of selecting one defective fuse, the number of ways of selecting two good fuses is $\binom{5}{2}$ or 10.

Using **The Fundamental Counting Principle** we obtain:
$$\binom{2}{1} \cdot \binom{5}{2} = 2 \cdot 10 = 20$$
Thus, the total number of ways of selecting one defective fuse and two good fuses is 20.

  b.   The number of ways of selecting 3 good fuses, $\binom{5}{3}$ or 10, and no defective fuses, $\binom{2}{0}$ or 1, is: $\binom{5}{3} \cdot \binom{2}{0}$ or $10 \cdot 1 = 10$.

Problem Set 10.1

1.   A committee of 8 people consists of 5 females and 3 males.
  a.   How many sub-committees consisting of 4 people can be formed at random if each is to consist of 2 females and 2 males?
  b.   How many sub-committees consisting of 4 people can be formed if each is to consist of 1 female and 3 males?

2.   A carton contains 10 apples of which 3 are rotten.
  a.   How many selections consisting of 5 apples each can be formed if each is to contain 3 good apples and 2 rotten apples?
  b.   How many selections consisting of 5 apples each can be formed if each is to contain 5 good apples?

3. A basketball team has 7 good foul shooters and 4 poor foul shooters.
   a. How many teams of 5 players each can be formed at random if there are 4 good foul shooters and 1 poor foul shooter?
   b. How many teams of 5 players each can be formed at random if there is 1 good foul shooter and 4 poor foul shooters?

4. A bag contains 5 orange, 4 blue, and 2 white marbles.
   a. How many selections consisting of 6 marbles each can be formed if each is to consist of 3 orange, 2 blue, and 1 white marble?
   b How many selections consisting of 4 marbles each can be formed if each is to consist of 3 blue and 1 white marble?

Review Competencies

5. Consider the letters M, A, T, and H. If a four letter arrangement is formed, what is the probability that it will begin with A? Any letter may be repeated.

6. Consider the digits 1, 3, 5, 6, and 8. If a three digit number is formed, what is the probability that the number begins with 3 and is even? Any digit may be repeated.

7. If A, B, and C are the only mutually exclusive events for some experiment and P(A) = 0.3, P(B) = 0.5, and P(C) = 0.1, is this an acceptable assignment of probabilities? Why?

8. How many arrangements can be made using all the letters of the word TURKEY if any letter may not be repeated?

9. A box contains three flash cubes marked C_1, C_2, C_3 for a camera. You are to select one flash cube. List the sample space for this experiment.

10. Subtract: $3\frac{1}{4} - 1\frac{1}{8}$

11. Divide: $1\frac{1}{5} \div 5$

12. Express $\frac{40}{65}$ in lowest terms.

13. Express 425% as a decimal.

14. Evaluate: $5(3 \cdot 2 - 6 \div 3) - 3(8 \div 4 - 1)$

There are times when combinations must be used along with **The Fundamental Counting Principle** to determine probabilities.

Problem 2   A committee contains 7 people of whom 4 are women. Three people are randomly selected to write the committee report. Find the probability that exactly two of the three people selected are women.

Solution   The phrase "exactly two of the three people selected are women" implies two women and one man. The number of ways of selecting two women is $\binom{4}{2}$ because there are four women on the committee. The order is not important. Corresponding to each way of selecting two women there are $\binom{3}{1}$ ways of selecting a man. **The Fundamental Counting Principle** states that the total number of ways of selecting two women and one man is $\binom{4}{2} \cdot \binom{3}{1} = 6 \cdot 3 = 18$.

The total number of ways of selecting three people from seven is $\binom{7}{3}$ or 35. The probability fraction is:

$$\frac{\text{the number of ways of selecting two women and one man}}{\text{the total number of ways of selecting three people from seven people}}$$

The probability that exactly two of the three people are women is:

$$\frac{\binom{4}{2} \cdot \binom{3}{1}}{\binom{7}{3}} = \frac{6 \cdot 3}{35} = \frac{18}{35}$$

Problem 3   An urn contains 8 balls. Five are black and three are white. Three balls are randomly selected from the urn. What is the probability that they are all black?

Solution   The order in which the balls are selected is not important. If three black balls are selected out of 5, $\binom{5}{3}$, and none of the white balls, $\binom{3}{0}$, the **Fundamental Counting Principle** gives $\binom{5}{3} \cdot \binom{3}{0} = 10 \cdot 1 = 10$

The total number of ways of selecting three balls out of eight is $\binom{8}{3}$ or 56. The probability fraction is:

$$\frac{\text{the number of selecting 3 black balls out of 5}}{\text{the total number of ways of selecting 3 balls out of 8}}$$

The probability that all 3 balls selected are black is:

$$\frac{\binom{5}{3} \cdot \binom{3}{0}}{\binom{8}{3}} = \frac{10 \cdot 1}{56} = \frac{10}{56} = \frac{5}{28}$$

Problem 3 may also be solved by using the multiplication rule as:
P (black ball followed by black ball followed by black ball)
$$= \frac{5}{8} \cdot \frac{4}{7} \cdot \frac{3}{6} = \frac{5}{28}$$

Problem Set 10.2

1. A box contains 6 beach balls consisting of 2 orange balls and 4 blue balls. If 4 balls are drawn at random from the box, find the probability that:
   a. one is orange and 3 are blue.
   b. all four are blue.
   c. 2 are orange and 2 are blue.

2. If 3 students are selected at random from a class of 5 men and 7 women, find the probability that:
   a. all 3 are men.
   b. all 3 are women.
   c. there is 1 man and 2 women.

3. A committee of 5 people is selected at random from a legislative delegation of 8 Democrats, 5 Republicans, and 2 Independents. Find the probability that the committee consists of:
   a. 3 Democrats and 2 Republicans.
   b. 1 Democrat, 3 Republicans, and 1 Independent.
   c. 5 Republicans.
   d. 2 Democrats, 1 Republican, and 2 Independents.

Review Competencies

4. A carton contains 8 good tomatoes and 2 rotten tomatoes. How many selections consisting of 4 tomatoes each can be formed if each is to contain 2 good tomatoes and 2 rotten tomatoes?

5. If a four digit number is formed using the digits 1, 3, 4, and 5, what is the probability that the number is greater than 4000? The digits are chosen randomly and no digit may be repeated.

6. On a certain day the probability that it will snow in Chicago is 0.6 and the probability that it will snow in Boston is 0.5. Assuming the events are independent, find the probability that:
   a. it will snow in both cities.
   b. it will snow in neither city.
   c. it will snow only in Boston.

7. If a letter is selected randomly from the English alphabet, what is the probability that the letter:
   a. precedes the letter G?
   b. is not a vowel?
   c. precedes the letter G or is a vowel?

8. List the sample space for the experiment of tossing a coin and selecting a marble from a box containing 1 green marble (G) and 1 white marble (W).

9. Evaluate: $\binom{6}{6}$

10. Evaluate: $_7P_3$

11. Multiply: $\frac{4}{5} \cdot 2\frac{1}{2} \cdot \frac{2}{3}$

12. Express $\frac{9}{4}$ as a decimal.

13. Express 28.5% as a decimal.

14. Express $2\frac{3}{5}\%$ as a decimal.

15. 20% of what number is 50?

Section 11   Calculating Odds   *against yer winning.*

Statements involving odds are encountered very frequently. One might be concerned with the odds of some baseball team winning the pennant or of some candidate winning an election. In this section we will consider methods for computing the **odds in favor** or the **odds against** some event.

> ### Definition
>
> The **odds in favor** of the occurence of some event is the ratio of the probability that the event will occur, P(A), to the probability that the event will not occur, P(A').
>
> The odds in favor of an event is:
>
> $$\frac{P(A)}{P(A')} = \frac{\text{the probability the event will occur}}{\text{the probability the event will not occur}}$$

The probability of an event's occurring is different from the odds in favor of the event.

odds in favor = $\dfrac{\text{# of the favorable outcomes}}{\text{# of unfavorable outcomes}}$

Problem 1   What are the odds in favor of obtaining an ace in one draw from a deck of 52 playing cards?

Solution   If A is the event of obtaining an ace, then $P(A) = \frac{4}{52}$ or $\frac{1}{13}$

The probability of not obtaining an ace, P(A'), is $1 - \frac{1}{13}$ or $\frac{12}{13}$.

$$\frac{P(A)}{P(A')} = \frac{\text{the probability the event will occur}}{\text{the probability the event will not occur}} = \frac{\frac{1}{13}}{\frac{12}{13}} = \frac{1}{13} \cdot \frac{13}{12} = \frac{1}{12}$$

Thus, the odds in favor of obtaining an ace are 1 to 12 which may also be written as 1:12. Notice that we do not express odds as a fraction.

> ### Definition
>
> The **odds against** the occurrence of some event A is the ratio of the probability that the event will not occur, P(A'), to the probability that the event will occur, P(A). The odds against an event is:
>
> $$\frac{P(A')}{P(A)} = \frac{\text{the probability the event will not occur}}{\text{the probability the event will occur.}}$$

odds against = $\dfrac{\text{# of the unfavorable outcomes}}{\text{# of favorable outcomes}}$

Notice that the odds against an event is the **reciprocal** of the odds in favor of the event. Recall that the reciprocal of $\frac{1}{2}$ is $\frac{2}{1}$ and the reciprocal of $\frac{4}{3}$ is $\frac{3}{4}$. In the previous problem, we determined that the odds in favor of drawing an ace from a deck of 52 playing cards are 1:12. The odds against drawing an ace are 12:1.

Problem 2   A fair die is tossed. What are the odds against obtaining a number that is less than 3?

Solution   If A is the event of obtaining a number that is less than 3, then $P(A) = \frac{2}{6}$ or $\frac{1}{3}$. The probability of not obtaining a number that is less than 3, $P(A')$, is $1 - \frac{1}{3}$ or $\frac{2}{3}$. The odds against obtaining a number that is less than 3 are:

$$\frac{P(A')}{P(A)} = \frac{\text{the probability the event will not occur}}{\text{the probability the event will occur}} = \frac{\frac{2}{3}}{\frac{1}{3}} = \frac{2}{3} \cdot \frac{3}{1} = \frac{2}{1}$$

The odds against obtaining a number that is less than 3 are 2 to 1 or 2:1.

The odds in favor of an event is the reciprocal of the odds against an event. Since the odds against obtaining a number that is less than 3 is 2 to 1, the odds **in favor** of obtaining a number that is less than 3 is 1 to 2.

We have seen that it is possible to convert probabilities to odds. We will now see that it is possible to convert odds to probabilities by using the following relationships:

> If the **odds in favor** of the occurrence of some event are a to b, then the probability that the event will occur is:
>
> $$\frac{a}{a + b}$$
>
> The probability that the event will not occur is:
>
> $$\frac{b}{a + b}$$

Problem 3   The odds in favor of a candidate's winning an election are 3 to 7. What is the probability that the candidate will win the election?

Solution   Since the odds in favor of the candidate's winning are 3 to 7, then a is 3 and b is 7. Thus, the probability that the candidate will win is:

$$\frac{a}{a + b} = \frac{3}{3 + 7} = \frac{3}{10}$$

Problem 4   The odds against a team's winning are 5 to 2. What is the probability that the team will lose the game?

Solution   Since the odds against a team's winning are 5 to 2, the odds in favor of the team's winning are 2 to 5. Thus, a is 2 and b is 5. The probability that the team will lose (**not win**) the game is:

$$\frac{b}{a + b} = \frac{5}{2 + 5} = \frac{5}{7}$$

Obviously, it is not necessary to first determine the odds in favor of the event when given the odds against the event. If the odds in favor of an event are a to b, then the odds against the event are b to a. Thus, if the odds against a team's winning a game are 5 to 2, then we can conclude directly that b is 5 and a is 2. The probability that the team will lose (not win) the game is still

$$\frac{b}{a + b} = \frac{5}{2 + 5} = \frac{5}{7}$$

Problem Set 11.1

1. What are the odds in favor of obtaining a diamond in one draw from a deck of 52 playing cards?

2. What are the odds against obtaining a spade in one draw from a deck of 52 playing cards?

3. What are the odds in favor of obtaining a face card in one draw from a deck of 52 playing cards?

4. Two decks of playing cards are shuffled together.
   a. What are the odds in favor of obtaining a face card in one draw?
   b. What are the odds against obtaining a queen in one draw?
   c. What are the odds against obtaining the ace of spades in one draw?

480

5. A fair die is tossed.
    a. What are the odds in favor of obtaining a number that is different from 2? $\frac{5}{1}$
    b. What are the odds against obtaining a number that is greater than 4? $\frac{2}{2} = 1$

6. Five balls numbered 1, 2, 3, 4, 5 are placed in a box. If a single ball is drawn:
    a. what are the odds in favor of obtaining an odd number? $\frac{3}{2}$
    b. what are the odds against obtaining an even number? $\frac{3}{2} =$

7. A department store is giving a free gift to any customer whose receipt has a star on it. The department store has printed a star on every tenth receipt. What are the odds in favor of a customer getting a free gift? $\frac{1}{9}$

8. The odds in favor of a horse's winning a race are 2 to 3. With these odds, what is the probability that the horse will win the race? What is the probability that the horse will not win the race? favor $\frac{2}{5}$ against $\frac{3}{5}$

9. The odds against the occurrence of an event are 7 to 5. With these odds, what is the probability that the event will occur? 5 out of 7 $\frac{5}{12}$

Review Competencies

10. A box contains 3 red balls and 5 white balls. If 4 balls are drawn from the box in succession, find the probability that:
    a. 2 balls are red and 2 are white.
    b. all the balls are white.

11. If a three digit number is formed randomly from the digits 1, 2, and 3, what is the probability that it is even and the digits are not repeated?

12. Two delegates to a convention are selected at random from a group of 10 people consisting of 5 women and 5 men. What is the probability that the two delegates are either two men or two women?

13. Add: 3.006 + 15.1 + 0.07

14. Express 65% as a fraction.

15. Express $5\frac{3}{4}$% as a decimal.

16. Multiply: $\frac{5}{6} \cdot \frac{3}{10} \cdot \frac{2}{5}$

17. Evaluate: $5(3 \cdot 2 - 1 + 8 \div 4) - 3(6 \div 2 - 1 \cdot 2)$

18. A television that originally cost $220 was sold for $140. What fractional part of the cost does one save by purchasing the television at the reduced price?

481

Chapter 7 Summary

1. The Fundamental Counting Principle

If task 1 can be performed in m ways, task 2 in n ways, task 3 in p ways, task 4 in q ways, etc., then all tasks can be performed in m • n • p • q ... ways.

2. Permutations

The number of permutations (arrangements) of n distinct objects taken r at a time, denoted by $_nP_r$, can be computed by:

a.   The "short-cut" formula: $n \cdot (n - 1) \cdot (n - 2) \cdot ... \cdot (n - r + 1)$

b.   The factorial formula: $\dfrac{n!}{(n - r)!}$

c.   The Fundamental Counting Principle. Order is important in permutation situations.

3. Combinations

The number of combinations (selections) of n distinct objects taken r at a time, denoted by $\binom{n}{r}$, is $\dfrac{n!}{r!\,(n - r)!}$. Order is not important in combination situations.

4. Sample Space

The set of all possible outcomes in an experiment is called the sample space, denoted by S. An event, A, is a subset of the sample space.

5. Probability

a. $P(A) = \dfrac{\text{the number of outcomes in A}}{\text{the number of outcomes in the sample space}}$

b.   If A cannot occur, $P(A) = 0$.

c.   If A is certain to occur, $P(A) = 1$.

d. $0 \le P(A) \le 1$

e.   P(A') = the probability that A will not occur = 1 − P(A)

f.   P(A or B) = P(A) + P(B) − P(A and B occur simultaneously)
        Equivalently: P(A or B) = P(A) + P(B) − P(A ∩ B)

g.   If A and B are mutually exclusive, P(A ∩ B) = 0.
        Thus, P(A or B) = P(A) + P(B).

h.   If A and B are independent events (the occurrence of A does not affect the occurrence of B), then: P(A followed by B) = P(A) • P(B).

i.   If A and B are dependent events (the occurrence of A affects the occurrence of B), then:

   P(A followed by B) = P(A) • P(B given that A has occurred).

   Equivalently: P(A followed by B) = P(A) • P(B|A)

j.   The probability of a permutation (arrangement) is

   $$\frac{\text{the number of ways of forming the permutation}}{\text{the total number of arrangements}}$$

   The numerator and denominator are computed using the Fundamental Counting Principle (see item 1).

k.   The probability of a selection (combination) is

   $$\frac{\text{the number of ways of making that selection}}{\text{the total number of selections.}}$$

   The numerator and denominator of the probability fraction are computed using the combination formula (see item 3). The numerator may also require use of The Fundamental Counting Principle (see item 1).

6. Odds

a.   The odds in favor of A = $\frac{P(A)}{P(A')}$

b.   The odds against A = $\frac{P(A')}{P(A)}$

c.   If the odds in favor of A are a : b (a to b),

   then P(A) = $\frac{a}{a + b}$ and P(A') = $\frac{b}{a + b}$.

Chapter 7 Self-Test

1. Two-digit numbers larger than 88 are written on separate slips of paper and placed in a hat. If the number on the slip selected is odd, which set represents an appropriate sample space associated with this experiment?
   a. $\{90, 92, 94, 96, 98\}$
   b. $\{89, 90, 91, 92, 93, 94, 95, 96, 97, 98\}$
   c. $\{x \mid 88 < x \le 99\}$
   d. $\{91, 93, 95, 97, 99\}$
   e. none of the above

2. A bowl contains three red balls (R_1, R_2, R_3). If two balls are drawn in succession without replacement, which set is an appropriate sample space for this experiment?
   a. $\{R_1, R_2, R_3\}$
   b. $\{(R_1,R_2), (R_1,R_3), (R_2,R_1), (R_3,R_1)\}$
   c. $\{(R_1,R_2), (R_1,R_3), (R_2, R_3), (R_2, R_1),$
      $(R_3, R_1), (R_3, R_2)\}$
   d. $\{(R_1,R_2), (R_1,R_3), (R_2, R_1), (R_2, R_3),$
      $(R_3, R_1)\}$
   e. none of the above

3. A bag contains 2 green balls, 3 red balls, and 4 white balls. One ball is chosen at random from the bag. What is the probability that the ball is white?
   a. $\frac{4}{9}$
   b. $\frac{4}{5}$
   c. $\frac{7}{9}$
   d. $\frac{2}{9}$
   e. none of these

4. If a fair coin is tossed 5 times in succession, the probability that all tosses are Heads is $\frac{1}{32}$. What is the probability that not all tosses are Heads?
   a. 32
   b. $\frac{1}{2}$
   c. $\frac{31}{32}$
   d. $\frac{15}{16}$
   e. none of these

5. What is the probability of obtaining a Head or a Tail if a fair coin is tossed one time?
   a. 0
   b. 1
   c. 0.5
   d. 0.25
   e. none of these

6. Five slips of paper numbered from 1 to 5 are placed in a hat. If one slip is drawn at random, what is the probability that the number is even or greater than 3?
   a. $\frac{2}{5}$
   b. $\frac{4}{5}$
   c. $\frac{3}{5}$
   d. $\frac{4}{25}$
   e. none of these

7. A student is selected at random from a group of 100 students in which 6 take math, 10 take English and 4 take both. What is the probability that the selected student takes math or English?
   a. 0.16
   b. 0.12
   c. 0.60
   d. 0.20
   e. none of these

8. The probability that Mary has a dog is 0.5. The probability that a dog has fleas is 0.8. What is the probability that Mary has a dog with fleas?
   a. 1
   b. 0.6
   c. 0.4
   d. 0.2
   e. none of these

9. A box contains 5 white balls, 4 green balls, and 3 blue balls. Two balls are drawn in succession without replacement. What is the probability that the first ball is green and the second ball is blue?
   a. $\frac{1}{11}$
   b. $\frac{1}{9}$
   c. $\frac{14}{33}$
   d. $\frac{2}{3}$
   e. none of these

10. If two balls are drawn in succession without replacement from a bag containing 3 red balls and 4 yellow balls, what is the probability that neither ball is yellow?
   a. $\frac{6}{49}$
   b. $\frac{2}{7}$
   c. $\frac{1}{7}$
   d. $\frac{16}{49}$
   e. none of these

11. How many 3 digit numbers may be formed using the digits 4, 6, 8, and 9 if no digit may be repeated?
    a. 12
    b. 24
    c. 48
    d. 64
    e. none of these

12. In how many different ways can five students sit in five chairs arranged in a row?
    a. 24
    b. 1
    c. 60
    d. 5
    e. 120

13. How many selections of 4 gifts may be chosen from a collection of 6 different gifts?
    a. 30
    b. 60
    c. 15
    d. 120
    e. none of these

14. A box contains 3 red balls and 5 yellow balls. If one ball is drawn at random, what are the odds in favor of obtaining a red ball?
    a. 3 to 8
    b. 3 to 5
    c. 5 to 3
    d. 5 to 8
    e. none of these

15. Six slips of paper numbered from 1 to 6 are placed in a container. One slip is selected at random. What are the odds against obtaining a number greater than 5?
    a. 1:6
    b. 6:1
    c. 1:3
    d. 5:1
    e. none of these

16. The odds against a candidate's election are 2 to 1. With these odds, what is the probability that the candidate will be elected?
    a. $\frac{2}{3}$
    b. $\frac{1}{3}$
    c. $\frac{3}{2}$
    d. $\frac{3}{1}$
    e. none of these

17. Two people are chosen from a group of six Europeans and six Africans. What is the probability that the two people selected are either both Europeans or both Africans if the random selection is done without replacement?
    a. $\frac{5}{11}$
    b. $\frac{5}{22}$
    c. $\frac{1}{2}$
    d. $\frac{2}{6}+\frac{2}{6}$
    e. none of these

18. The three letters A, B, and C form three two-member combinations: (A,B), (A,C), (B,C). In considering these pairs, the order in which the letters appear is not important. Consequently, four letters A, B, C, and D form six two-member combinations: (A,B), (A,C), (A,D), (B,C), (B,D), (C,D). Similarly, five letters A, B, C, D, and E form ten two-member combinations: (A,B), (A,C), (A,D), (A,E), (B,C), (B,D), (B,E), (C,D), (C,E), (D,E). How many two-member combinations can be formed using the six letters A, B, C, D, E, and F?
    a. 18
    b. 15
    c. 21
    d. 12
    e. none of these

19. Two cars enter a supermarket parking lot. If there are two parking spaces available, there will be two different ways to park the two cars. If there are three spaces available, there will be six different ways to park them. If six parking spaces are available, how many different ways will there be to park the two cars?
    a. 20
    b. 24
    c. 30
    d. 36
    e. none of these

20. A tourist wishes to visit Miami, Ft. Lauderdale, and Palm Beach in a random order. What is the probability that the tourist will visit Palm Beach first, Ft. Lauderdale second, and Miami third?
    a. $\frac{1}{3}$
    b. $\frac{1}{6}$
    c. $\frac{1}{9}$
    d. $\frac{1}{27}$
    e. none of these

21. Box A contains two cards numbered 2 and 4. Box B contains three cards numbered 1, 3, and 5. An experiment consists of selecting a card from each box, first from box A and then from box B. The sample space is:
    S = {(2,1), (2,3), (2,5), (4,1) (4,3), (4,5)}
    What is the probability of selecting one element from this set whose sum is divisible by 3 or greater than 6?
    a. $\frac{1}{3}$
    b. $\frac{2}{3}$
    c. $\frac{1}{6}$
    d. $\frac{5}{6}$
    e. none of these

22. A door prize at a dance is awarded to people who obtain a receipt stamped with the letter P. The management has stamped every 15th receipt with a P. Under these conditions, what are the odds against winning a door prize?
   a. 1:15   b. 15:1   c. 1:14
   d. 14:1   e. none of these

23. A box contains 6 orange marbles and 2 blue marbles. If 3 marbles are drawn at random from the box, find the probability that 2 are orange and 1 is blue.
   a. 30   b. $\frac{15}{56}$ c. $\frac{15}{28}$
   d. $\frac{5}{14}$ e. none of these

24. How many four digit odd numbers can be formed using the digits 1, 2, 4, 5, and 6 if no digit can be repeated?
   a. 12   b. 30   c. 48
   d. 120   e. none of these

25. The well-balanced spinner shown below is spun twice. What is the probability that the pointer will stop at red on the first spin and then on green on the second spin?

   Red | Green
   Green | Red
   Red | Yellow

   a. $\frac{1}{2} \cdot \frac{1}{3}$ b. $\frac{1}{2} \cdot \frac{2}{5}$ c. $\frac{1}{6} \cdot \frac{1}{6}$
   d. $\frac{1}{2} + \frac{2}{5}$ e. none of these

26. Consider all four-letter arrangements using the letters A, B, C, and D. Letters are not to be repeated. If all such arrangements are written on a card with one arrangement per card and these cards are placed into a box, what is the probability of randomly selecting an arrangement with A first and D last?
   a. $\frac{1}{6}$ b. $\frac{1}{8}$ c. $\frac{1}{12}$
   d. $\frac{1}{24}$ e. none of these

27. How many ways can 10 people be awarded first, second, and third prize in an art contest, assuming no person can win more than one prize?
   a. 120   b. 720   c. 1000
   d. 10!   e. none of these

28. On a two question multiple choice test, each question has four possible answers, one of which is correct. If you must guess at both questions, what is the probability that you will answer both questions correctly?
   a. $\frac{1}{2}$ b. $\frac{1}{4}$ c. $\frac{1}{8}$
   d. $\frac{1}{16}$ e. none of these

29. One group contains four men and two women. A second group contains three men and five women. Suppose that a person is selected from each group, first from group one and then from group two. What is the probability that both selections are men **or** both selections are women?
   a. $\frac{11}{24}$ b. $\frac{13}{24}$ c. $\frac{11}{21}$
   d. $\frac{13}{21}$ e. none of these

30. Two common sources for receiving the news are magazines and television. Forty percent of American adults get the news from television but not magazines, while 25% receive the news from both television and magazines. What is the probability that a randomly selected adult does not receive the news from television?
   a. 0.85   b. 0.65   c. 0.35
   d. 0.15   e. none of these

The table below shows the distribution of the cause of fires in a particular state.

Cause	Percent of All Fires
Electrical System	8
Heating System	6
Smoking	20
Cooking	37
Appliances	19
Other	10

Use the table to answer problems 31-33.

31. How many of the next 600 fires would one expect to be caused by heating systems or smoking?
a. 26   b. 154   c. 156
d. 444   e. none of these

32. What is the probability that a fire is caused by neither cooking nor appliances?
a. 0.56   b. 0.44   c. 0.34
d. 0.10   e. none of these

33. If it is known that a fire has not been caused by the electrical or heating systems, find the probability that it was caused by smoking.
a. $\frac{10}{43}$ b. $\frac{33}{43}$ c. $\frac{1}{5}$
d. $\frac{7}{10}$ e. none of these

34. A condominium community contains apartments in five different models, each offered in three different color schemes, either with or without a screened porch. How many buying options are available?
a. 3   b. 10   c. 15
d. 30   e. none of these

35. Six men and four women are competing to become members of a tennis team that will consist of three men and three women. How many different teams can be selected?
a. 80   b. 24   c. 10
d. 9   e. none of these

36. A particular automobile model can be purchased with or without the following options: radial tires, air conditioning, stereo, cruise control. How many different arrangements of these options are available?
a. 16   b. 8   c. 4
d. 1   e. none of these

37. Ten percent of the pens that are made by a ballpoint pen manufacturer leak. If two pens are randomly selected with replacement, find the probability that at least one of them leaks.
a. $\frac{1}{5}$ b. $\frac{1}{10}$ c. $\frac{19}{100}$
d. $\frac{81}{100}$ e. none of these

38. A box contains eight pens, of which two are defective. If two pens are randomly selected without replacement, find the probability that at least one pen is defective.
a. $\frac{1}{4}$ b. $\frac{1}{28}$ c. $\frac{13}{28}$
d. $\frac{15}{28}$ e. none of these

39. A recent survey indicated that ten percent of high school students participate in collegiate theater. Of these, two percent become professional actors. What is the probability that a randomly selected high school student will both participate in collegiate theater and become a professional actor?
a. 0.002   b. 0.003   c. 0.02
d. 0.03   e. none of these

40. If Earl is one of 20 students registered for college algebra, and two people are randomly selected to attend a convention, what is the probability that Earl will be one of the selected people?
a. $\frac{1}{10}$ b. $\frac{1}{20}$ c. $\frac{1}{40}$
d. $\frac{1}{400}$ e. none of these

eight

Statistics

Few of us are very far removed from statistics of one form or another. Statistical studies confront us daily in newspapers and magazines and on radio and television. Statistics are used to analyze the results of public opinion polls, to compute the cost of living, to determine levels of unemployment, and to calculate insurance mortality rates. They are also used in advertising, industry and educational testing.

Some people distrust statistics, believing they can be twisted to prove anything. To these people, statistics are mumbo-jumbo specifically designed to overwhelm and manipulate an unwary person. There are many abuses of statistics, but the blame should not rest with the statistical methods themselves. The problem lies with the individuals who use them. If there are indeed abuses, the reality is not that statistics lie, but that liars use statistics. It is important to detect these abuses of statistics and examine statistical results carefully before accepting them.

Statistical methods include all the methods used in the collection, presentation, description, and interpretation of data. Statistics are used to make decisions, inferences and predictions. This chapter will introduce you to some very basic statistical concepts and formulas so you will have some orientation should you encounter further work with them in the future. Statistics is a valuable tool of the sciences. A knowledge of statistics (and probability) is required of students in business, psychology, social science and education. These students must keep informed about the current research in their field and possess a familiarity with the terminology and concepts of statistics. Even though computers perform many of the tedious statistical calculations, statistics is more than just a bag of tools. It is a way of making decisions in the face of uncertainty.

Section 1 The Population, the Sample, and Sampling Techniques

It is important that the objectives of any statistical study be carefully defined first. Then, an experiment or procedure for collecting data must be carefully designed.

Data just do not magically appear. They must be collected. Thus, the first task facing the statistician is to define the population very carefully from which the data will be collected.

> ### Definition
> A **population** is the well-defined set containing all the objects, individuals or measurements whose properties are to be analyzed by the data collector.

Examples

1. The set of full-time students attending Florida universities is an example of a population.

2. The set of heights of all the royal palm trees in Florida is an example of a population.

3. The set of large community colleges in Florida is not an example of a population since this set is not well-defined.

We would like to analyze the entire population. It is not always easy, however, to collect every piece of information about a population because it is too time consuming, too expensive, or because the population is simply too large and inaccessible. When it is unrealistic to canvas the entire population, only a portion is analyzed. This portion is called a sample.

> ### Definition
> A **sample** is a subset of the population.

The objective of statistics is to make inferences, decisions or predictions about a population based upon data contained in a sample.

It is very important that the sample be **representative** of the population in that the sample has the same distribution of characteristics as the population. For example, if the religious composition of some community is 20% Catholic, 65% Protestant and 15% Jewish, the composition of any sample taken from this community should reflect similar percentages as closely as possible.

Problem 1   The labor force in a certain city is 30% federally employed, 60% factory workers and 10% self-employed. Which of the following samples is most representative of this labor force?

	federal	factory	self	total
sample A	7	63	0	70
sample B	50	220	30	300
sample C	27	58	5	90
sample D	59	122	19	200

Solution   In sample A, the percent of the 70 people who should be federally employed (30%) is 0.30(70) or 21. However, there are only 7 people in the sample who are federally employed. The percent of the 70 people who should be factory workers (60%) is 0.60(70) or 42. However, there are 63 people in the sample who are factory workers. The percent of the 70 people who should be self-employed (10%) is 0.10(70) or 7. In the sample, there are 0 people listed who are self-employed. For convenience, each ideal number should be listed in parentheses next to the corresponding number given in the sample as follows:

	federal	factory	self	total
sample A	7 (**21**)	63 (**42**)	0 (**7**)	70

The same procedure is followed for sample B, sample C, and sample D.

	federal	factory	self	total
sample B	50 (**90**)	220 (**180**)	30 (**30**)	300
	(0.30 •300)	(0.60 •300)	(0.10 •300)	
sample C	27 (**27**)	58 (**54**)	5 (**9**)	90
	(0.30 •90)	(0.60 •90)	(0.10 •90)	
sample D	59 (**60**)	122 (**120**)	19 (**20**)	200
	(0.30 •200)	(0.60 •200)	(0.10 •200)	

We must select the sample that differs the least in all categories from the actual percents in the population, especially when compared to the total number of items in the sample. Observe that samples A and B can be quickly eliminated.

Sample C appears representative. Sample D, however, is actually the most representative since the numbers in the sample differ the least from the ideal percents in the population.

When the population is large and diverse, a sample method must be designed to assure that the sample will be representative and random. This leads us to the following definition:

> ## Definition
> A **random sample** is a sample obtained in such a way that every element in the population has an equal chance to be selected for the sample.

It is important that the term "random" not be confused with "haphazard." The sample cannot be random if any element in the population is systematically excluded in any way.

There are a number of techniques for obtaining a random sample. These include:

1. Simple random sampling
2. Systematic sampling *A method of time how you go to select*
3. Stratified sampling
4. Cluster sampling

One technique for obtaining a random sample is called **simple random sampling**. This method requires that each element in the population be identified and assigned a number. For example, numbers corresponding to each element are written on pieces of paper. The papers are then placed in a box. The numbers drawn from the box become the random sample. Obviously, this task can become tedious when the population is large. In this situation, it is helpful to refer to a table of random numbers which contains series of numbers arranged in random order. The results from the table take the place of the numbers in the box. Unfortunately, the use of a table still does not eliminate the task of numerically identifying each element of the population.

Another type of sampling technique is called **systematic sampling**. In this method, every nth item is selected from the population for the sample. For example, every 50th name in the telephone book or every corner house on a city block might be selected for the sample. The disadvantage of this method is that some recurring characteristic may periodically manifest itself. This might produce a **bias** in the results. To illustrate, selecting every 50th name in the telephone book might result in underselecting certain ethnic groups because names that are associated with ethnic ties are often alphabetically clustered together.

The third type of sampling technique is called **stratified sampling**. In this technique, the population is divided into a number of smaller homogeneous subgroups or strata, and a sample is drawn from each stratum. For example, a population could be divided into age groups and a random sample taken from each group. These smaller strata are easier to work with than the large population. It is possible to stratify by income, sex, religion, occupation, ethnic background or virtually any characteristic. Stratification reduces the chances of a nonrepresentative sample by assuring that a specified number of elements will be selected from each stratum.

The final sampling technique that we will discuss is called **cluster sampling**. Similar to stratified sampling, this method divides the population into a number of subpopulations called clusters. However, samples are not drawn from each cluster. Rather, some of the clusters are randomly selected, and sampling is carried out only in those clusters. A community, for example, can be divided into city blocks as its clusters. Several blocks are then randomly selected. After this, residents on the selected blocks are randomly chosen, providing a sampling of the entire community.

The sampling plan used in any situation depends on such factors as the nature of the population, the difficulty of sampling and the cost of sampling. These factors must be carefully considered before any sampling technique is used.

97, 73, 61, 75, 59, 71

97 84 76 71 =

Problem Set 1.1

1. The following scores were obtained by 15 students on a test: 97, 56, 73, 82, 84, 61, 93, 94, 73, 76, 82, 59, 69, 70, 71
   Make a systematic sample of the test scores by selecting every third score on the list.

2. A university wishes to study the income level of its alumni. A questionnaire is mailed to each 20th alumnus on an alphabetical list. Is there a danger of bias? Explain. *Yes* .

3. If 45% of the students attending a university are males and 55% are females, how many males and females should be included in a sample of 200 students to insure a representative sample?

   55
   45
   100

   110 F
   90 M
   $200

4. A group of hotel owners in a large city are interested in doing an attitudinal questionnaire regarding opinions about legalized casino gambling. The owners decide to conduct a survey among citizens of the city to discover their opinions about casino gambling. Which of the following would be most appropriate for selecting an unbiased sample?
   a. randomly surveying people who live in oceanfront condominiums in the city
   b. surveying the first two hundred people whose names appear in the city telephone directory
   c. randomly selecting geographic clusters of the city and then randomly surveying people within the selected clusters
   d. randomly surveying people at six of the largest nightclubs in the city

493

5. Which of the following samples is most representative of a population comprised of 10% upper, 70% middle and 20% lower income families?

sample	upper	middle	lower	total
a.	30	180	70	280
b.	30	210	60	300
c.	40	320	40	400

6. The ethnic composition of a certain community is 50% White, 30% Hispanic and 20% Black. Which of the following samples is most representative of this ethnic composition?

sample	White	Hispanic	Black	total
a.	30	23	7	60
b.	103	59	38	200
c.	150	102	48	300
d.	15	12	13	40

Review Competencies

7. Add: $1\frac{1}{3} + 1\frac{1}{4}$

8. Subtract: $3\frac{1}{6} - 1\frac{5}{6}$

9. Multiply: $\frac{2}{3} \cdot \frac{4}{5} \cdot \frac{3}{8}$

10. Divide: $4\frac{1}{4} \div 4$

11. Subtract: $20.03 - 0.069$

12. Divide: $36 \div 0.009$

13. Express $\frac{2}{5}$ as a decimal.

14. Express 5% as a fraction.

15. Express 165% as a decimal.

16. Express 0.019 as a percent.

17. Evaluate: $5 \cdot 3 - 8 \div 2 + 6 - 5$

18. Express $4\frac{1}{2}\%$ as a decimal.

Section 2   Non-Graphic Presentation of Data

After data have been collected, the next task facing the statistician is to present the data in a condensed and manageable form. Then, the data can be more easily interpreted. In this section, we examine some non-graphic ways of presenting data.

Data collected from a random sample can be presented by using an **ungrouped frequency distribution**. Such a distribution consists of two columns of numbers. The possible data values are listed in one column (generally from the smallest to the largest). The adjacent column is labeled "frequency" or "f" and indicates the number of times each value occurs.

Problem 1   Consider the following random sample in which the data items are:

4, 3, 7, 5, 5, 1, 2, 2, 6, 4, 3, 1, 2, 7, 6, 6, 5, 4, 3, 5, 4, 4, 3

Construct an ungrouped frequency distribution.

Solution   Two columns are formed. One lists all possible data values (from smallest to largest). The other indicates the number of times the value occurs in the sample.

data value	f	
1	2	(The value 1 occurs twice.)
2	3	(The value 2 occurs 3 times.)
3	4	
4	5	
5	4	
6	3	
7	2	

An ungrouped frequency distribution can be cumbersome when there is a large number of items in the data. We will see that there is a more compact form for presenting such data.

Consider the following statistics test scores made by a group of 40 students.

82	47	75	64	57	82	63	93
76	68	84	54	88	77	79	80
94	92	94	80	94	66	81	67
75	73	66	87	76	45	43	56
57	74	50	78	71	84	59	76

If the grades are displayed in this manner, can you adequately answer the question, "How well did the group do?" The answer is no. Can you easily tell how the score 73 compared to the rest of the group? Again, the answer is no.

Since it is difficult to make any generalizations about the group's performance on the basis of the way in which the scores appear, it might be helpful to organize this information so the results can be more meaningful.

One way of doing this is to tally the scores into intervals. Obviously, the intervals should not overlap. Each score should belong to just one interval, and no score should be left out.

On the next page, a series of intervals and the tally of the number of scores in each interval is shown.

In the table below, each of the intervals has a width of 10 units. Certainly these are not the only possible intervals that could have been developed. We might have selected intervals in which the width is 5 units. Whatever width is selected, it is important that each interval possess the same width. We seldom use fewer than 5 or more than 15 intervals in a distribution.

Interval	Tally	Total Number in Each Interval
40 – 49	III	3
50 – 59	ЖІ	6
60 – 69	ЖІ	6
70 – 79	Ж Ж І	11
80 – 89	Ж ІІІІ	9
90 – 99	Ж	5

Observe that this data have been grouped into a table that indicates the way in which data are distributed among certain intervals. This table is called a **grouped frequency distribution**. The word "class" is often used instead of "interval." The **class frequency**, denoted by f, is the number of items that appear in the class.

The following is the standard form of a grouped frequency distribution (often called, simply, a "frequency distribution") constructed from the test scores:

class	f
40 – 49	3
50 – 59	6
60 – 69	6
70 – 79	11
80 – 89	9
90 – 99	5

The class 40 – 49 is called the **bottom class** of the frequency distribution even though it appears at the top. The bottom class contains the lowest scores in the frequency distribution.

Observe that the class 40 – 49 has a width of 10 units as do all the other classes. The number representing the width of a class is called the **class interval**. Thus, the class interval of the class 40 – 49 is 10. The class interval is the range of values that the class contains. Notice that the class interval is not found by taking the difference of 49 and 40, which is 9.

Unfortunately, a frequency distribution possesses some disadvantages that cannot be avoided. Notice that each test score has lost its individual identity in a class. It is no longer possible to identify the largest and smallest scores in each class or how often a particular score occurs in each class. We will see later that this sacrifice of information can slightly influence the accuracy of certain results. However, the advantages of the frequency distribution out-weigh the disadvantages.

Let us now mention the following important definitions.

> ### Definitions
> The **class limits** in a frequency distribution are the largest and smallest values that can appear in a class. The largest value is called the **upper class limit**, and the smallest value is called the **lower class limit**.

Example

Consider, again, our frequency distribution of test scores. In the class $40 - 49$, 40 is the lower class limit, and 49 is the upper class limit. In the class $50 - 59$, 50 is the lower class limit, and 59 is the upper class limit. Notice that the class interval (10) can be determined by subtracting a lower class limit (40) from the next lower class limit (50).

> ### Definition
> The **class boundaries** in a frequency distribution are dividing points between the classes and are determined by adding the upper class limit of one class to the lower class limit of the next class and dividing the sum by 2.

The class boundaries are numbers that are halfway between the upper class limit of one class and the lower class limit of the next class.

Example

Consider, again, the frequency distribution of test scores and especially the two lowest classes:

$$40 - 49$$
$$50 - 59$$

The class boundary between these two classes is determined by adding the upper class limit of the class $40 - 49$ (49) to the lower class limit of the class $50 - 59$ (50) and dividing the sum by 2. We obtain: $\frac{49 + 50}{2} = \frac{99}{2} = 49.5$

Observe that this value is both the upper class boundary of the class 40 – 49 and the lower class boundary of the class 50 – 59. Observe also that the lower class boundary of the class 40 – 49 is 39.5. The class boundaries are values that cannot occur in the sample data. For example, it is not possible for a student to receive a score of 49.5 on the test.

$\frac{A+D}{2}$

> ### Definition
> The **class mark** is the middle point of the class and is determined by adding the class limits (or boundaries) together and dividing the sum by 2.

Example

The class mark of the class 40 – 49 is determined by adding together the class limits (40 and 49) and dividing the sum by 2. This gives: $\frac{40 + 49}{2} = \frac{89}{2} = 44.5$

The class mark is the value that is the exact middle of the class.

The class mark of the class 40 – 49 is 44.5, and the class mark of the class 50 – 59 is 54.5. Notice that the distance between the class marks (10) is the same as the class interval.

Problem Set 2.1

1. Construct an ungrouped frequency distribution for the following data:
   {3, 5, 4, 3, 2, 1, 2, 3, 3, 4}

2. Consider a frequency distribution having the classes 15 – 19, 20 – 24, 25 – 29, 30 – 34 and 35 – 39.
   a. What is the class interval?
   b. What are the related class boundaries?
   c. What are the related class marks?

Review Competencies

3. The London Council needs to estimate the average income of the citizens of London, England. To do so, a survey is conducted. Which one of the following would be most appropriate for selecting an unbiased sample?

   a. randomly select geographic regions in all of England and then randomly survey persons within the regions.
   b. dividing the population of London into subpopulations and using cluster sampling.
   c. surveying all the residents of Chelsea, a large wealthy subdivision in London.
   d. randomly surveying persons at Harrods, London's largest department store.

4. Express $\frac{37}{8}$ as a mixed number.

5. Express $\frac{15}{40}$ as a fraction in simplest form.

6. Multiply: $2\frac{1}{3} \cdot \frac{3}{5} \cdot \frac{5}{7}$

7. Subtract: $5 - 0.0004$

8. Divide: $0.125 \div 0.05$

9. Express 0.03% as a decimal.

10. Express 1.5 as a fraction.

11. Express 65% as a fraction.

12. What is 15% of 40?

13. Evaluate: $5(3 - 1 + 2 \cdot 4 - 1)$

14. Convert 1101_2 to base ten.

Let us now discuss a common non-graphic method for presenting data that is derived from the frequency distribution.

> **Definition**
>
> In a frequency distribution, the **cumulative frequency**, denoted by c.f., associated with any given class, is the sum of the frequency in that class and the frequencies in all the classes that appear below the given class in the distribution.

Total frequency!!

Knowing the cumulative frequencies associated with each class enables us to construct a **cumulative frequency distribution**.

Example

Consider the following cumulative frequency distribution:

class	f	c.f. (cumulative frequency)
40 – 49	3	3 (3 + 0)
50 – 59	6	9 (6 + 3)
60 – 69	6	15 (6 + 9)
70 – 79	11	26 (11 + 15)
80 – 89	9	35 (9 + 26)
90 – 99	5	40 (5 + 35)

Problem Set 2.2

1. Use the frequency distribution at the right. Indicate the cumulative frequencies associated with each class in this distribution.

class	f	c.f.
50 – 69	8	8
70 – 89	13	21
90 – 109	19	40
110 – 129	25	65
130 – 149	16	81
150 – 169	12	93
170 – 189	7	100

Review Competencies

2. If a frequency distribution has classes 10 – 29, 30 – 49, 50 – 69, and 70 – 89:
    a. What is the class interval?
    b. What are the related class boundaries?
    c. What are the related class marks?

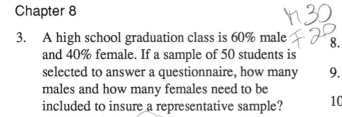

3. A high school graduation class is 60% male and 40% female. If a sample of 50 students is selected to answer a questionnaire, how many males and how many females need to be included to insure a representative sample?

4. Multiply: $5 \cdot 3\frac{1}{5}$ 5. Divide: $1 \div \frac{5}{6}$

6. Express $4\frac{1}{8}$ as an improper fraction.

7. Divide: $4.82 \div 0.2$

8. Express $\frac{3}{8}$ as a decimal.

9. Express 185% as a decimal.

10. 30% of what number is 60?

11. Evaluate: $6(\frac{1}{2} \div \frac{4}{5} - \frac{1}{3})$

12. Express $1\frac{1}{4}\%$ as a decimal.

13. Express 2,850,000 in scientific notation.

Section 3 Graphic Presentation of Data: Histograms and Frequency Polygons

It is helpful to present a frequency distribution in visual form. In this section, we consider two ways to present a frequency distribution graphically.

The first graph used to illustrate a frequency distribution is called a **histogram**. The following problem will outline the steps for constructing a histogram.

Problem 1 Construct a histogram for the frequency distribution shown at the right.

class	f
40 – 49	3
50 – 59	6
60 – 69	6
70 – 79	11
80 – 89	9
90 – 99	5

Solution

step 1
Draw two lines, one horizontal and one vertical, that intersect in the following way:

step 2
A scale must be constructed on the horizontal line. The numbers for the scale should be the **class boundaries** of each class in the frequency distribution. In this distribution, it is not necessary to begin this scale at 0.

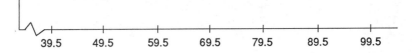

39.5   49.5   59.5   69.5   79.5   89.5   99.5

step 3

A scale must be constructed on the vertical line. This scale will measure the **class frequencies**. Since the largest frequency in the distribution is 11, the scale might be marked off in units of 2 beginning with 0, although there are other possible numbers (such as 0, 3, 6, 9, 12) that could be used. If the largest frequency were, say, 50, then neither of these schemes would really be appropriate. In this instance, a suitable scale might include: 0, 10, 20, 30, 40, 50.

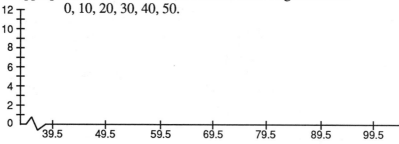

step 4

A histogram presents a frequency distribution using rectangles. The bases of the rectangles are equal to the widths of the classes, and the heights are equal to the frequencies of the classes. Thus, the histogram is constructed as follows:

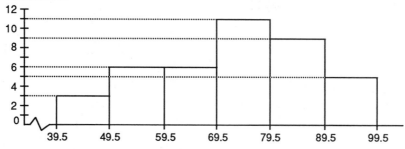

step 5

The completed histogram appears below.

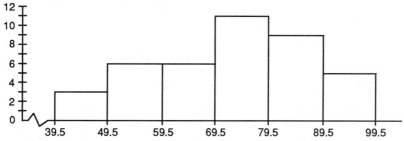

Notice that the bases of the rectangles begin and end at the class boundaries. Notice also that we are using heights of rectangles to indicate the frequency size of each class. The larger rectangles correspond to the larger frequencies in the distribution.

Chapter 8

Histograms are valuable because they quickly reveal the form of the distribution of data. There are a variety of ways in which data may be distributed. Thus, histograms can assume a variety of shapes. Consider the following shapes and the terms that describe them:

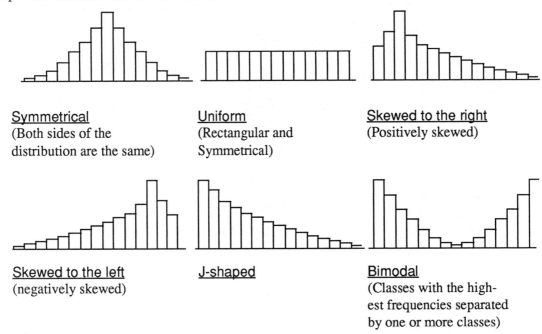

Symmetrical
(Both sides of the
distribution are the same)

Uniform
(Rectangular and
Symmetrical)

Skewed to the right
(Positively skewed)

Skewed to the left
(negatively skewed)

J-shaped

Bimodal
(Classes with the highest frequencies separated by one or more classes)

Another type of graphic presentation is based upon the histogram and emphasizes the continuous rise or fall of the frequencies.

> ### Definition
> A **frequency polygon** is a graph formed by connecting the midpoints of the upper bases of the rectangles in a histogram with straight lines.

Problem 2   Superimpose a frequency polygon on the following histogram.

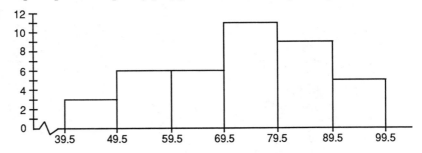

Solution

step 1

Determine the midpoints of each of the upper bases of the rectangles in the histogram.

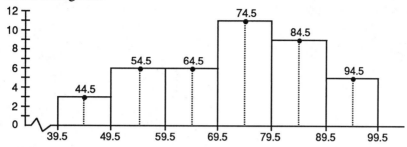

step 2

Connect each of these midpoints with straight lines.

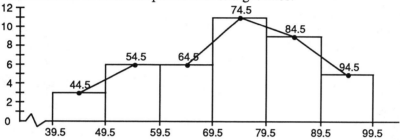

step 3

To complete the frequency polygon at both ends, the lines must be drawn down to touch the horizontal line. Certainly the classes on either side of the histogram would contain frequencies of 0.

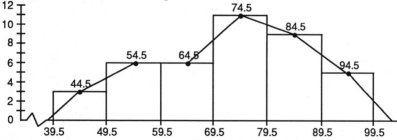

The frequency polygon appears below without the histogram.

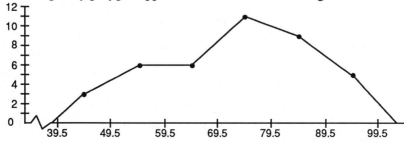

Problem Set 3.1

1. Use the frequency distribution at the right to answer the following:

    a. Construct a histogram for this distribution.

    b. Superimpose a frequency polygon on the histogram.

class	f
50 – 54	5
55 – 59	8
60 – 64	11
65 – 69	16
70 – 74	20
75 – 79	15
80 – 84	12
85 – 89	9
90 – 94	6

2. Use the frequency distribution at the right to answer the following:

    a. Construct a histogram for this frequency distribution.

    b. Superimpose a frequency polygon on the histogram.

class	f
10 – 29	27
30 – 49	39
50 – 69	38
70 – 89	26
90 – 109	20

Review Competencies

3. Use the frequency distribution at the right to answer the following:

    a. Indicate the cumulative frequencies associated with each class in the distribution.

    b. What is the class interval?

    c. What are the related class boundaries?

    d. What are the related class marks?

class	f
5 – 9	4
10 – 14	6
15 – 19	8
20 – 24	10
25 – 29	7
30 – 34	3

4. A college has 40% freshmen, 25% sopho-mores, 20% juniors and 15% seniors. Which of the following samples is most representative of the students at the college?

	A	B	C	D
freshmen	35	36	69	163
sophomores	21	15	83	98
juniors	10	23	41	82
seniors	4	16	7	57
Totals	70	90	200	400

5. Subtract: $5 - 3\frac{1}{4}$

6. Multiply: $\frac{4}{5} \cdot \frac{1}{2} \cdot 1\frac{2}{3}$

7. Divide: $1\frac{3}{4} \div 2\frac{2}{3}$

8. 12 is what percent of 48?

9. Express $\frac{1}{4}$ as a percent.

10. Express 0.08% as a decimal.

11. Express 35% as a fraction.

12. Evaluate: $5 \cdot 3 + 4 \div 2 - 8(4 - 2)$

Section 4   Additional Graphic Methods for Presenting Data

In this section, we will see that there are other methods for presenting data graphically. The first graph that we will consider is the **bar graph.** In the bar graph, information is presented by using rectangles, called bars. Such graphs provide a quick comparison of information. The bars can run either vertically or horizontally.

Problem 1   Use the following data on the average amount of money spent in 1960 for each school pupil in different areas of the country.

north	$500
midwest	$450
south	$300
west	$400

Construct a bar graph to show the data.

Solution   Let the vertical scale represent increments of $100 spent per pupil. The horizontal base line is labeled using areas of the country. The heights of the bars correspond to the data given in the problem. Notice that the rectangles (bars) in the bar graph do not touch each other; this is not the case with the rectangles in a histogram.

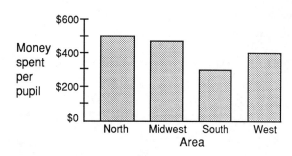

505

The **circle graph** shows how an entire quantity has been divided. The circle represents 100% of the quantity, and it can be divided into different sections or fractional parts.

Problem 2 Construct a circle graph to present the following data on the expenditure of each dollar of student activity fees at a university.

Athletics	$ .45
Newspaper	$ .25
Yearbook	$ .15
Entertainers	$ .10
Dances	$ .05

Solution The portion of the circle assigned to each activity needs to be determined. The size of the interior angle will be the activity's proportional share of 360° which is the number of degrees contained in a circle.

Since $.45 (45%) of each dollar goes to athletics, we obtain:
45% of 360° = 0.45(360°) = 162°
Thus, an angle of 162° is constructed and labeled "athletics." We can find the other portions in a similar way:

Newspaper	0.25(360°) = 90°
Yearbook	0.15(360°) = 54°
Entertainers	0.10(360°) = 36°
Dances	0.05(360°) = 18°

The completed circle graph is at the right.

A **line graph** is useful in showing trends of data.

Problem 3 Use the following data on the number of shares of a certain stock sold from June to December.

June	205	**July**	200	**August**	220
September	230	**October**	217	**November**	212
December	210				

Construct a line graph to present the data.

Solution Let the vertical scale represent increments of 10 shares of stock sold. The horizontal base line is labeled by the months. The completed graph is shown at the right.

Problem Set 4.1

1. Use the bar graph below showing crimes in a community:
   a. Approximately how many crimes were committed in 1974? *350*
   b. Between which two consecutive years occurred the greatest increase in crimes? *71 and 72.*
   c. Which two years had comparable numbers of crimes? *74 and 75'*
   d. About how many more crimes were committed in 1972 than 1973? *400*

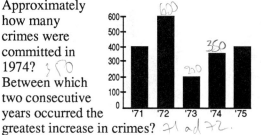

2. Use the graph below.
   a. Which is the largest state? *Alaska*
   b. Which two states have the greatest difference in area, Alaska and Texas or Arizona and Texas?

Largest States of the United States

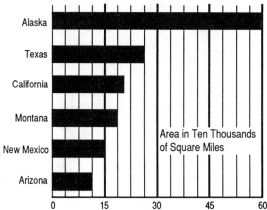

Area in Ten Thousands of Square Miles

3. In the circle graph shown, how much more money for every $10,000 is spent on education than on utilities? *200*

Allocations Per Dollar

4. Consider the accompanying line graph.
   a. From the graph tell the temperature each day of the week.
   b. Between which two successive days did the temperature rise the most? *Tues to Wed.*
   c. Between which two successive days did the biggest drop in temperature occur? *Thursday / Fri*

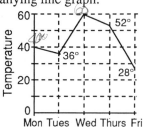

Review Competencies

5. Which of the samples shown below is most representative of a population composed of 5% upper, 80% middle and 15% lower income families?

	upper	middle	lower	total
sample A	*15* 8	121	21	150
sample B	*30* 15	*✓* 160	25 *50*	200
sample C	*✓* 15	250	35 *45*	300

20 *23* *240*

6. Use the frequency distribution at the right to answer:

class	f
10-19	4
20-29	7
30-39	9
40-49	10
50-59	8
60-69	3

   a. What is the class interval?
   b. What are the related class boundaries?
   c. What are the related class marks?
   d. Indicate the cumulative frequencies associated with each class.

7. Subtract: $10\frac{1}{3} - 3\frac{3}{4}$ 8. Multiply: $\frac{3}{5} \cdot \frac{7}{9} \cdot \frac{10}{13}$

9. Divide: $3\frac{1}{4} \div \frac{2}{3}$ 10. Divide: $15.5 \div 0.05$

11. Express $1\frac{1}{4}$ as a decimal.

12. Express 120% as a fraction in simplest form.

Chapter 8

Section 5   The Mean

One of the important tasks of the statistician is to describe a set of data quantitatively. Representative numbers are found that indicate where the data tend to cluster. Also found is the degree to which the data items differ from each other and are scattered or spread out.

In this section, we examine the first of the values around which the numbers in the data tend to cluster. Such values are called **measures of central tendency** or **measures of central location.** The word "average" is generally used to indicate these measures which are the most typical of a set of data.

> ### Definition
> The **mean**, denoted by $\bar{x}$ (read "x bar"), of a set of n numbers is the sum of those numbers divided by n.

The mean is the average with which most people are familiar.

The capital Greek letter sigma (Σ) is often used to represent the **sum** of numbers. Thus, the sum of the n values $x_1, x_2, ..., x_n$ may be written as:

$$\sum_{i=1}^{n} x_i = x_1 + x_2 + x_3 + ... + x_n.$$

The notation x_1 involves the use of 1 as a **subscript**. This subscript refers to the first addend. The second addend is x_2, etc. We shall use the notation Σx as a shorthand for $\sum_{i=1}^{n} x_i$.

> Using this summation notation, the formula for calculating the mean is
> $$\bar{x} = \frac{\Sigma x}{n}$$

Problem 1   Find the mean of the following set of data: $\{10, 13, 9, 4\}$

Solution   If we let each number in the data represent an x value, we have: $x_1 = 10$, $x_2 = 13$, $x_3 = 9$ and $x_4 = 4$. Since there are 4 numbers in the data, n = 4. Substituting in the formula we obtain:

$$\bar{x} = \frac{\Sigma x}{n} = \frac{10 + 13 + 9 + 4}{4} = \frac{36}{4} = 9$$

Thus, $\bar{x} = 9$

One and only one mean can always be determined for any set of numerical data, and the mean may or may not be one of the actual items in the data. The mean is dependent upon the value of every item in the data. A change in the value of any item will affect the mean. If the data in the previous problem had included 23 instead of 13, the mean would be 11.5 instead of 9. A single extreme value (very small or very large) can greatly affect the mean.

The fact that a single extreme value can greatly affect the mean can result in giving the extreme value too much influence. For example, the ages of five people are 3, 4, 5, 6 and 27. The mean age is $\frac{45}{5}$ or 9 years. However, to say that the average age is 9 years misrepresents the situation since 4 of the 5 people are younger than 9 years. In this case, the mean is not really the most typical number of the data it is supposed to describe. In the next section we will discuss another measure of central tendency which will provide a value that better describes this particular set of data.

Problem Set 5.1

1. What is the mean of the data in the following set? {9,14,10,12,8,10,3,6}

2. What is the mean of the data in the following set? {3.8,2.3,1.1,7.2,8.1}

3. An instructor's grade book showed that the grades of ten students on a test were 71, 75, 55, 81, 97, 41, 44, 77, 62, and 88. What was the mean of the grades by the students on the test?

4. Calculate the mean of the following set of data: 13, 18, 28, 31, 26, 35, 29. Suppose the 29 in the above set of data had been 49. What would happen to the value of the mean?

5. A small private school employs ten teachers with salaries ranging between $500 and $800 per week. Which of the following values could be a reasonable estimate of the semimonthly payroll for the teachers?
   a. $6,500
   b. $10,000
   c. $13,000
   d. $16,000

Review Competencies

6. Use the circle graph at the right. If Marion earns $2000 per month, how much more does she spend on food than on bills?

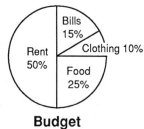

Budget

7. Construct a bar graph to represent the data below on the occupational composition of a community. Find the fraction of the community employed in clerical jobs.

	f
Clerical (C)	30
Managerial (M)	50
Professional (P)	40
Skilled Craftsmen (S)	65
Unskilled Labor (U)	85

8. Use the frequency distribution below to answer.
   a. What is the class interval?
   b. What are the related class boundaries?
   c. What are the related class marks?
   d. Indicate the cumulative frequencies associated with each class.

class	f
10 – 39	10
40 – 69	17
70 – 99	21
100 – 129	16
130 – 159	7

9. Express $\frac{42}{15}$ as a mixed number.

10. Multiply: $2\frac{1}{2} \cdot 3\frac{1}{3} \cdot 1\frac{1}{5}$

11. Divide: $\frac{5}{6} \div 2\frac{1}{3}$

12. Subtract: $6.07 - 2.089$

13. Express $\frac{1}{8}$ as a decimal.

14. Express 0.003 as a percent.

15. Evaluate: $2 \cdot 5 \div \frac{1}{2} - 5 + 3$

16. Express $2\frac{3}{4}\%$ as a decimal.

We will now see that it is possible to calculate the mean of a set of data that has been grouped into a frequency distribution.

Recall that each item in a frequency distribution has lost its individual identity. No knowledge is available relative to the actual distribution of the data within each class unless one has the individual data items at hand. It is therefore assumed that all the values in a class are evenly distributed. This means that the average value of all the data in the class coincides with the **class mark**. Naturally, it is not the case that the data will be evenly distributed in all classes. Also, the midpoint of any class will most likely differ from the mean of the actual items in that class. However, any errors would tend to balance out if the class interval is not too large and an adequate number of items is available in each class.

It should be noted that if the mean could possibly be calculated for the ungrouped data before being grouped into a frequency distribution, it would not precisely equal the mean of the grouped data.

> The **mean of a grouped frequency distribution** may be found by using the following formula:
> $$\bar{x} = \frac{\sum xf}{n}$$
> in which:
> x refers to the **class mark** of each class.
> f refers to the **frequency** of each class.
> $\sum xf$ refers to the **sum** of all the products obtained by multiplying each class mark by its frequency.
> n refers to the **total frequency** of the entire distribution.

Problem 2   Find the mean of the frequency distribution at the right.

class	f
40 – 49	3
50 – 59	6
60 – 69	6
70 – 79	11
80 – 89	9
90 – 99	5

$$\frac{\sum xf}{n}$$

✓

Solution

step 1

Determine the class marks (x values) for each class. For the first class,

$$\bar{x} = \frac{40 + 49}{2} = 44.5$$

Other class marks are similarly computed.

class mark

class	x	f
40 – 49	44.5	3
50 – 59	54.5	6
60 – 69	64.5	6
70 – 79	74.5	11
80 – 89	84.5	9
90 – 99	94.5	5

step 2

Construct another column in which each entry is the product of the class mark (x) and the frequency (f). This column will represent xf.

class mark

class	x	f	xf
40 – 49	44.5	3	133.5
50 – 59	54.5	6	327
60 – 69	64.5	6	387
70 – 79	74.5	11	819.5
80 – 89	84.5	9	760.5
90 – 99	94.5	5	472.5

step 3

Find the sum of the values ($\sum xf$) in this last column.

xf
133.5
327
387
819.5
760.5
472.5

$$\sum xf = 2{,}900.0$$

step 4

Substitute these values into the formula. Remember that n is the **total frequency** of the distribution or 40.

$$\bar{x} = \frac{\sum xf}{n} = \frac{2{,}900}{40} = 72.5$$

Note

Decimal numerals are frequently rounded to a
desired degree of accuracy. Answers in this
book have, when necessary, been rounded to
two decimal places.

Recall again the ungrouped statistics test scores (presented in Section 2) from
which the above frequency distribution was constructed. The sum of the
scores is actually 2,907, and the true mean is 72.68 which is extremely close
to the value obtained from the formula (72.5).

Problem Set 5.2

[handwritten: $\bar{x}$ 23.62]

1. Find the mean of the frequency distribution at the right.

class	f	xf
0 – 8	20	*80*
9 – 17	26	*338*
18 – 26	38	*836*
27 – 35	39	*1209*
36 – 44	27	*1080*

[handwritten: f 4, 13, 22, 31, 40; 152 ; 3543]

2. Calculate the mean of the frequency distribution at the right.

class	f	
45 – 49	3	*141*
50 – 54	12	*624*
55 – 59	19	*1083*
60 – 64	6	*372*
65 – 69	5	*335*
70 – 74	2	*144*
75 – 79	2	*154*
80 – 84	1	*82*

[handwritten: class values 47, 52, 57, 62, 67, 72, 77, 82; 50 ; 2936]

[handwritten: $\bar{x} = 58.7$]

3. Construct a circle graph that represents the following budget percents for a family:

food	30%
rent	25%
transportation	20%
utilities	15%
miscellaneous	10%

If the family income is $1000 per month, how
much money is spent for food and rent combined?

4. Use the bar graph to determine the truth or
falsity of the following statements.

a. More students earned B's than F's on the
exam.

b. The number of students earning the grade
of C was double the number earning the
grade of F.

c. 200 students took the exam.

d. If D is considered the lowest passing grade,
then 20% of the students failed.

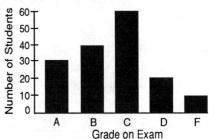

5. Use the frequency distribution at the right to construct a histogram and superimpose a frequency polygon.

class	f
5 – 9	10
10 – 14	6
15 – 19	3
20 – 24	7
25 – 29	8

6. Calculate the mean of: 12, 43, 19, 24, 32 and 26.

7. True or false? The mean of a set of data is always equal to one of the data items.

8. True or false? A set of data can contain more than one correct mean.

9. Express $4\frac{1}{7}$ as an improper fraction.

10. Divide: $1 \div 0.005$

11. Express 2.3% as a decimal.

12. Express $\frac{1}{5}$ as a percent.

13. 50 is what percent of 200?

14. Evaluate: $\frac{2}{3}(\frac{3}{4} \div \frac{1}{2} - \frac{1}{3})$

Section 6   The Median, the Percentile, the Quartile, and the Mode

To begin this section, we consider another measure of central tendency or central location that describes the middle of a set of data in a different way from the mean.

> **Definition**
>
> The **median**, denoted by x̃, (read "x tilde") of a set of ungrouped data arranged in order of size is the value of the middle item (when there is an odd number of items) or the mean of the two items nearest the middle (when there is an even number of items).

The median divides a set of data in such a way that half the items are above the median and half are below. The position of the median of a set of n numbers is $\frac{n+1}{2}$.

Examples

1. The median of the set of data {4,2,6,5,8} is not 6. The data must first be arranged in order of size. In increasing order, the data should appear as 2, 4, 5, 6, 8. Now the median is the value of the middle item. Thus, x̃ = 5. Notice that when there is an odd number of items, the median is the middle item. Since the data contains 5 numbers (n = 5), the position of the median is $\frac{n+1}{2}$ or $\frac{5+1}{2}$ or 3. This does not mean that the median is 3. It means that the median is the third number from either end of the data after the data are arranged in order of size.

2. The median of the set of data {2,5,7,8} is 6. Since there is no middle item, the median is the **mean** of the two items nearest the middle (5 and 7). Thus, $\bar{x}$ = 6. Since n = 4, the position of the median is $\frac{n+1}{2}$ or $\frac{4+1}{2}$ or $\frac{5}{2}$ or 2.5. The number 2.5 means that the median is halfway between the second and third number in the data after the data are arranged in order of size.

Like the mean, one and only one median can be determined for a set of data. Like the mean, the median may or may not be one of the data items. Unlike the mean, the median is not dependent upon every item in the data. The median is not affected by an extreme value. Rather, the median is affected by the position of each data item but not the value of each item. For the set of data {2,4,5,6,8}, $\bar{x}$ = 5. For the set of data {2,4,5,6,20}, $\bar{x}$ = 5.

In certain instances, the median better describes the most typical value than the mean. To illustrate, the mean of the set {3,4,5,6,27} is 9, which is greater than four of the five data items. The median is 5 and better describes the most typical value.

Problem Set 6.1

1. Find medians for the following sets of data.
   a. {1,8,3,6,9}
   b. {7,9,4,6,8,10}
   c. {6,7,7,8}
   d. {2,11,8,6,10,4,5,9,3}
   e. {1.1,2.3,3.9,7.2,}

2. True or false? For any distribution of data, the mean and median must coincide.

Review Competencies

3. Use the frequency distribution at the right.
   a. Calculate the mean.
   b. Indicate the cumulative frequencies associated with each class.

class	f
5 – 9	4
10 – 14	7
15 – 19	9
20 – 24	6
25 – 29	5

4. Construct a bar graph representing the following information on football ticket sales by classes:

freshmen	200
sophomores	350
juniors	375
seniors	425

5. Use the circle graph at the right to find the amount the government contributes for every $60 of income to private colleges.

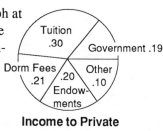

Income to Private Colleges

6. Calculate the mean of:
   116, 271, 94, 185, 202, 152, 138

7. True or false? The mean of a set of data can always be determined.

8. Add: 126.35 + 4.006 + 0.045

9. Express 0.126 as a percent.

10. Express 15% as a fraction in simplest form.

11. What percent of 60 is 15?

12. Evaluate: $10(6 - 4) + 3 \cdot 2 - 8 \div 4$

13. Express $5\frac{1}{10}$% as a decimal.

The process for finding the median of a set of grouped data is more involved since each item has lost its individual identity. It is possible, however, to determine the class in which the median is located.

If the grouped data are presented in the form of a histogram, the median would be the point that divides the histogram into two equal areas. If the grouped data are presented in a frequency distribution with n as the total number of items, then it is necessary to divide n in half and count $\frac{n}{2}$ items from the bottom class of the distribution.

> The **median of a grouped frequency distribution**, whose total frequency is n, may be determined by using the following formula:
>
> $$\bar{x} = L + \left(\frac{x}{f}\right) \cdot i$$
>
> in which: L refers to the **lower boundary** of the class that contains the median.
>
> x refers to the **number of items** still needed after L to reach the median.
>
> f refers to the **frequency** of the class containing the median.
>
> i refers to the **class interval.**

The following problem illustrates the application of this formula.

Problem 1   Determine the median of the distribution at the right.

class	f
40 – 49	3
50 – 59	6
60 – 69	6
70 – 79	11
80 – 89	9
90 – 99	5

Solution

step 1

Since n = 40, $\frac{n}{2}$ = 20. This means that we must count 20 items in from the bottom class of the distribution (the class 40 – 49). Observe that the twentieth item would occur in the class 70 – 79.

step 2

Since we know that this class contains the median, the lower boundary (L) of this class 70 – 79 is 69.5.

step 3

Observe that there are 15 items below the class 70 – 79 (3 + 6 + 6).
Therefore, we need 5 more items to reach the median (the twentieth item).
Thus, x = 5.

step 4

The frequency of the class 70 – 79 is 11. Thus, f equals 11 in our formula.

step 5

Since the class interval is 10, i = 10.

step 6

We must now substitute each of the values into the following formula to calculate the median.

$$\tilde{x} = L + \left(\tfrac{x}{f}\right) \cdot i \quad \text{when } L = 69.5,\ x = 5,\ f = 11,\ i = 10$$

$$= 69.5 + \left(\tfrac{5}{11}\right) \cdot 10$$

$$= 69.5 + 4.55 \ (\tfrac{50}{11} \text{ is approximately } 4.55)$$

$$\tilde{x} = 74.05 \ \text{(correct to two decimal places)}$$

Thus, 74.05 is the value below which 50 percent of the data falls.

Problem Set 6.2

Find the medians for the following frequency distributions:

1.

class	f
0 – 19	20
20 – 39	26
40 – 59	38
60 – 79	39
80 – 99	27

2.

class	f
45 – 49	4
50 – 54	11
55 – 59	19
60 – 64	6
65 – 69	5
70 – 74	2
75 – 79	2
80 – 84	1

Review Competencies

3. Find the median of the following data: {7, 2, 6, 8, 12}

4. Find the median of the following data: {5, 11, 13, 8}

5. True or false? The median of a set of data is always equal to one of the data items.

6. True or false? The median of a set of data can always be determined.

7. Use the frequency distribution at the right:
   a. Calculate the mean.
   b. Indicate the cumulative frequency associated with each class.

class	f
10 – 19	5
20 – 29	8
30 – 39	9
40 – 49	13
50 – 59	7
60 – 69	4

8. Construct a line graph to represent the monthly average number of inches of rainfall in St. Petersburg, Florida. Jan. 3.5; Feb. 5; March 5.3; April 6; May 6.6; June 7.5
Between which two months is there the greatest increase in rainfall?

9. Use the circle graph to determine the truth or falsity of each of the following statements.
   a. John spends $3,600 per year on mortgage payments.
   b. John spends $200 per month on his car.

Car 20% Utilities 10% Food 15% Misc. 10% Clothes 10% Medical 5% Mortgage 30%

How John Spent His $1000 Paycheck Each Month

c. John spends the same amount on utilities as on clothing.

d. John's medical expenses are $800 per year.

10. Subtract: $12.5 - 10.05$

11. Divide: $5.1 \div 0.17$

12. Express $\frac{7}{10}$ as a decimal.

13. Express 0.005% as a decimal.

14. Express 125% as a fraction.

15. 25% of what number is 75?

16. Express $6\frac{1}{2}\%$ as a decimal.

17. Simplify: $\sqrt{2} \cdot \sqrt{10}$

There are useful descriptive values other than measures of central tendency that can be determined from a set of data. These values are not called measures of central tendency. Rather, they are called **measures of position** since they describe the location of a specific data value in relation to the remaining data values. Let us consider the first of these measures of position.

Definition

A **percentile**, denoted by P_n, of a set of data is a value such that n percent of the data falls below P_n.

The formula for determining P_n is:

$$P_n = L + \left(\frac{x}{f}\right) \cdot i$$

in which:

L refers to the **lower boundary** of the class that contains P_n

x refers to the **number of items** still needed after L to reach P_n

f refers to the **frequency** of the class containing P_n

i refers to the **class interval**

Notice that the formula for determining a percentile is similar to the one used to determine the median.

Problem 2

Find the twenty-fifth percentile (P_{25}) of the data at the right

class	f
40 – 49	3
50 – 59	6
60 – 69	6
70 – 79	11
80 – 89	9
90 – 99	5

Solution

step 1

We are looking for the value below which 25 percent of the data falls. To find this value, we find 25 percent of the **total frequency** (40), which is 10. Count in 10 items from the bottom class of the distribution (the class 40 – 49). Notice that the tenth item occurs in the class 60 – 69.

step 2

P_{25} occurs in the class 60 – 69, and the lower boundary (L) of this class is 59.5.

step 3

Since we only need 1 more item from the class to reach the tenth item, we see that x = 1.

step 4

The frequency (f) of this class (60 – 69) is 6.

step 5

The length of the class interval (i) is 10.

step 6

What remains now is to substitute each of the values into the following formula to calculate P_{25}.

$$P_n = L + \left(\tfrac{x}{f}\right) \cdot i \quad \text{when } L = 59.5,\ x = 1,\ f = 6,\ i = 10$$

$$P_{25} = 59.5 + \left(\tfrac{1}{6}\right) \cdot 10$$

$$= 59.5 + 1.67 \ (\tfrac{10}{6} \text{ is approximately } 1.67)$$

$$P_{25} = 61.17 \ (\text{correct to two decimal places})$$

Thus, 61.17 is the value below which 25 percent of the data falls.

The twenty-fifth percentile (P_{25}) is also called the **lower quartile** and is represented by Q_1. The seventy-fifth percentile (P_{75}) is also called the **upper quartile** and is represented by Q_3. Q_3 is the value below which 75 percent of the data falls. The **middle quartile** (Q_2) is the same value as the fiftieth percentile (P_{50}) and is equal to the median. Thus, 25% of the data falls below Q_1, and 25% of the data falls above Q_3. The process for finding Q_1 or Q_3 is the same as for finding P_{25} or P_{75}.

Problem Set 6.3

Use the frequency distribution at the right to calculate:

class	f
10 – 19	2
20 – 29	7
30 – 39	8
40 – 49	17
50 – 59	23
60 – 69	17
70 – 79	15
80 – 89	7
90 – 99	4

1. P_{37}
2. P_{65}
3. P_{99}
4. P_{50}
5. Q_1
6. Q_3

Review Competencies

7. Use the frequency distribution at the right to:

class	f
10 – 29	7
30 – 49	10
50 – 69	15
70 – 89	11
90 – 109	5

    a. Find the median of the data.

    b. Find the mean of the data.

8. Calculate the median of the data: {12,9,13,16,10}.

9. Construct a bar graph representing the number of students enrolled in various mathematics courses at a college.

Algebra	450
Trigonometry	300
Analytic Geometry	150
Calculus	100

10. People in a community classified themselves as liberal, conservative, or moderate. If 25% are liberal, 40% conservative, and 35% moderate, how many of each should be included in a representative sample of 400 drawn from the community?

11. Express 45% as a decimal.

12. Add: $\frac{7}{8} + \frac{3}{5}$

13. Multiply: $2\frac{1}{6} \cdot 6$

14. Divide: $4 \div 2\frac{2}{3}$

15. Express $1\frac{1}{4}$ as a decimal.

16. Express $\frac{20}{45}$ in simplest form.

17. Express $\frac{5}{8}$ as a percent.

After deviating slightly and discussing some measures of position, let us now discuss the final measure of central tendency. It is defined below.

> ## Definition
> The **mode** of a set of data is the value or values which occur(s) the most often in that set. If there is no value that occurs most often, the set has no mode.

The mode is considered a measure of central tendency because the modal value(s) is/are often rather centrally located. Other data values occur less frequently in either direction.

Examples

1. The mode of the set of data {7,2,4,7,8,10} is 7 since this item occurs most often.
2. The set of data {5,2,8,5,9,10,9} has two modes (5 and 9). Such a set of data is said to be **bimodal**.
3. The set of data {2,1,4,5,3} has no mode at all.

Definition
The **modal class** of a grouped frequency distribution is the class with the largest frequency. The mode of this distribution is the class mark of the modal class.

Example

Consider the following frequency distribution:

class	f
40 – 49	3
50 – 59	6
60 – 69	6
70 – 79	11
80 – 89	9
90 – 99	5

The modal class is the class 70 – 79 since this class contains the largest frequency (11). The mode of this distribution is 74.5, which is the class mark of the modal class.

The mean, median, and mode represent different methods of describing the value around which the numbers in a set tend to cluster. Generally, these measures of central tendency will not result in identical values. For example, consider the set {6,7,8,9,9,10}. The mean $\bar{x}$ is 8.2, the median $\tilde{x}$ is 8.5, and the mode is 9.

Problem Set 6.4

1. Find the mode for each of the following sets of data, if it exists:
   a. $\{3, 5, 6, 7, 9, 5\}$ 5
   b. $\{1, 2, 3, 4, 5, 6\}$ none
   c. $\{3, 5, 7, 3, 4, 5, 9, 10\}$ 3 and 5

2. What is the modal class of the frequency distribution shown at the right?   69.5

class	f
0 – 19	20
20 – 39	26
40 – 59	38
60 – 79	39	69.5
80 – 99	27

Review Competencies

3. Use the frequency distribution at the right to:
   a. Calculate P_{62}
   b. Calculate Q_1
   c. Calculate $\bar{x}$

class	f
20 – 39	10
40 – 59	19
60 – 79	28
80 – 99	25
100 – 119	13
120 – 139	5

4. Find the median of the data:
   $\{12, 5, 19, 15, 13, 11\}$

5. True or false? The fiftieth percentile of a set of data is equal to the median of the set of data.

6. True or false? In any set of data, the value representing the fiftieth percentile is twice the value representing the twenty-fifth percentile.

7. Multiply: $\frac{6}{7} \cdot \frac{5}{9} \cdot \frac{3}{10}$

8. Subtract: $4\frac{1}{4} - 1\frac{1}{3}$

9. Add: $0.156 + 2.012 + 5.3$

10. Express 0.63 as a percent.

11. Express 0.35 as a fraction in simplest form.

12. 40 is what percent of 120?

13. Express $2\frac{4}{5}\%$ as a decimal.

Concepts of percentile, cumulative frequency, mean, and median enable us to interpret real-world data involving frequency and cumulative frequency tables.

Problem 3 The table below indicates scores on a 400-point exam and their corresponding percentile ranks.

Score	Percentile Rank
380	99
350	87
320	72
290	49
260	26
230	8
200	1

What percentage of examinees scored between 230 and 290?

Solution

The 49 to the right of 290 indicates that 49% of the scores fall below 290. The 8 to the right of 230 indicates that 8% of the scores fall below 230. Thus, the percentage of scores between 230 and 290 is 49% − 8% or 41%.

Problem 4

The table below shows the distribution of income levels in a community that appears to have a very small middle class.

Income Level	Percent of People
$0 - 9,999	7
10,000 - 14,999	24
15,000 - 19,999	3
20,000 - 24,999	2
25,000 - 34,999	7
35,000 - 49,999	1
50,000 - 79,999	25
80,000 - 119,999	11
120,000 and over	20

a. What percentage of people have incomes of at least $35,000?
b. Identify the amount below which 36% of the people in the community have lower incomes.

Solution

a. We must find the percentage of people whose incomes are $35,000 or more. If we focus on the last four classes, this percentage is 1% + 25% + 11% + 20%. The sum indicates that 57% of the people have incomes of at least $35,000.
b. To identify the amount below which 36% of the people in the community have lower incomes, we begin by adding the percents in the right column, starting with 7%, until we reach 36%.

$$7\% + 24\% + 3\% + 2\% = 36\%$$

We see that 36% of the people fall below the class $25,000 - 34,999. The lower class limit in this class is $25,000. Thus, $25,000 is the amount below which 36% of the people in the community have lower incomes.

Problem 5   The table below shows the distribution of the ideal number of children
based on a Gallup poll conducted in the spring of 1976 in America,
Australia, and Japan. (The poll question asked: "If a young married couple
could have as many or as few children as they wanted during their lifetime,
what number would you, yourself, suggest?")

Proportion of Respondents

Ideal Number of Children	USA	Australia	Japan
0	.06	.02	.00
1	.05	.06	.00
2	.52	.38	.41
3	.19	.25	.44
4	.18	.29	.15

a. Determine the median ideal number of children in the USA.
b. Determine the modal ideal number of children in Japan.
c. Determine the mean ideal number of children in Australia.

Solution   The problem becomes somewhat easier if we change the proportions to percents, and then drop the percent signs, thinking of the number of times that
each score occurs out of a total of 100 cases. We obtain the following
distribution:

Number of Respondents

Ideal Number of Children	USA f	Australia f	Japan f
0	6	2	0
1	5	6	0
2	52	38	41
3	19	25	44
4	18	29	15
	n = 100	n = 100	n = 100

a. The median ideal number of children in the USA is the score in
$\frac{n+1}{2}$ position. Since $\frac{n+1}{2} = \frac{100+1}{2} = 50.5$, we want the average of the scores
in 50th and 51st positions. The USA column indicates that the score 2 falls in
positions 12 through 63 inclusively, meaning that it certainly is the average
of scores in positions 50 and 51. The median ideal number of children in the
USA is 2.

b. The modal ideal number of children in Japan corresponds to the score with the greatest frequency. Focusing on the last column, we see that the greatest frequency is 44, corresponding to a score of 3. The modal ideal number of children in Japan is 3.

c. Even though the data are not grouped, we can compute the mean for the Australia column using the formula $\bar{x} = \dfrac{\Sigma x\, f}{n}$

x	f (Australia)	x f
0	2	0
1	6	6
2	38	76
3	25	75
4	29	116
		$\Sigma x\, f = 273$

$$\bar{x} = \frac{\Sigma x\, f}{n} = \frac{273}{100} = 2.73$$

The mean ideal number of children in Australia is 2.73.

Problem Set 6.5

1. Use the table in problem 3 to answer these questions.
   a. What percentage of examinees scored between 260 and 320?
   b. What percentage of examinees scored at least 350?

2. Use the table in problem 4 to answer these questions.
   a. What percentage of people have incomes of at least $25,000?
   b. What percentage of people have incomes of at most $14,999?
   c. Identify the amount below which 44% of the people in the community have lower incomes.
   d. Identify the amount above which 31% of the people in the community have higher incomes.

3. Use the table in problem 5 to answer these questions.
   a. Determine the median ideal number of children in Australia and Japan.
   b. Determine the modal ideal number of children in the USA and Australia.
   c. Determine the mean ideal number of children in the USA and Japan.

Review Competencies

4. For the following sets of numbers, the value x represents the same type of measure of central tendency.

$$\{1, 1, 2, 3, 3\} : x = 2$$
$$\{2, 3, 7, 9\} : x = 5$$

What is the value of x for the following set of numbers? $\{3, 4, 4, 6, 7, 8\}$

   a. 4   b. 5   c. $5\frac{1}{3}$ d. Both 4 and 6

5. The government of a large city needs to determine if the city's residents will support the construction of a new jail. The government decides to conduct a survey of a sample of the city's residents. Which one of the following procedures would be most appropriate for obtaining an unbiased sample of the city's residents?

   a. Survey a random sample of the employees and inmates at the old jail.
   b. Survey every fifth person who walks into City Hall on a given day.
   c. Survey a random sample of persons within each geographic region of the city.
   d. Survey the first 200 people listed in the city's telephone directory.

6. The graph below represents the yearly average snowfall for Colorado in inches for 1977-1982.

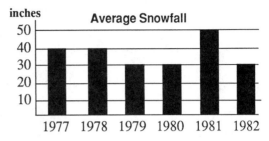

Which one of the following is true according to the graph?

   a. The percent increase in snowfall from 1980 to 1981 is 40%.
   b. The median amount of snowfall for the six years is 30 inches.
   c. Combined snowfall for 1977 and 1978 exceeded snowfall for 1979 by 50 inches.
   d. The mean amount of snowfall for the six years is 35 inches.

Section 7   Abuses of Statistics

The statement, "There are three kinds of lies: lies, damn lies, and statistics" is attributed to both Benjamin Disraeli and Mark Twain. It was mentioned at the beginning of this chapter that the real problem is not that statistics lie, but that liars use statistics. The following represent some common methods of distortion using statistics.

1. Distortion Due to Using the Mean, Median, or Mode to Represent "Average"

People will use the word "average" in a way that will yield the best results for them.

Example

A union represents all five workers in a certain factory. The union is distressed about the low yearly salaries of the workers and approaches the factory owner to negotiate a raise. The owner declines the union's request saying that the average (mean) salary of all individuals associated with the company is $14,792 which is certainly a "livable" salary.

Further research reveals that the yearly incomes of the five workers are $9,000, $9,500, $9,750, $10,000 and $10,500. However, the owner's yearly salary is $40,000.

While the mean salary of the five workers and the owner is indeed $14,792, this figure certainly misrepresents the true plight of the workers because it is significantly higher than five of the six people associated with the factory. Since the mean is affected by the extremely high salary of the owner, the median salary ($9,875) is actually a much more typical salary.

Example

Teacher A, teacher B, and teacher C try out their individual teaching methods on a group of 5 students. They then give a 10-point test to evaluate their teaching techniques.

Teacher A's scores are: 1, 2, 5, 8, 9
Teacher B's scores are: 1, 1, 6, 7, 8
Teacher C's scores are: 1, 2, 3, 7, 7

Teacher C quickly claims that her method is most effective because the modal "average" of her test scores is 7. Teacher A, on the other hand, says that the mean "average" of her test scores is 5 and teacher B says that the median "average" of her test scores is 6.

These claims are misleading since different measures of central tendency were used by each teacher to compute an "average." It is useful to determine the mean, median and mode for all three teachers. Then the results will be comparable.

	mean	median	mode
teacher A	5	5	none
teacher B	4.6	6	1
teacher C	4	3	7

In most cases, the mean of a set of data is the most reliable "average." Since teacher A's mean is the highest, she can claim that her teaching method is somewhat more effective than that of teacher B or teacher C.

2. Distortion Due to Samples that do not Represent the Population

If a sample is not representative of the population being described, distortion is inevitable. To avoid sampling error, the sample must be relatively large, representative and random. We certainly could not predict the outcome of an election by sampling only ten people. At times, however, even a massive sample can result in distortion.

Example

In 1936, the *Literary Digest* mailed out millions of ballots to voters throughout the country. The results poured in, and the *Literary Digest* predicted a landslide victory for Republican Alf Landon over Democrat Franklin Roosevelt.

The *Literary Digest* was wrong in their prediction because the mailing lists the editors used were from directories of automobile owners and telephone subscribers. People prosperous enough to own cars in the heart of the Depression tended to be somewhat more Republican than those who did not, so although the sample was massive, it was biased toward the affluent, and in 1936 many Americans voted along economic lines. (A victim of both the Depression and the 1936 fiasco, the *Literary Digest* folded in 1937.)

Advertisers distort the facts by using "recent independent surveys" rather than surveys based upon representative random samples. When you hear that "in an independent survey, it was found that 8 out of 10 doctors use Aspribuff for the relief of headache pain," you should wonder how many groups of 10 doctors were sampled. Perhaps 50 groups were sampled before the "independent survey" discovered a group in which 8 out of 10 recommended the product.

3. Distortion Due to Graphic Presentation of Data

The graphs shown below indicate the profits for a particular company from 1950 through 1954. Depending upon the scale used, one can greatly exaggerate the results.

the data

Year	1950	1951	1952	1953	1954
Profit (in millions)	30	30.2	31	31.4	31.6

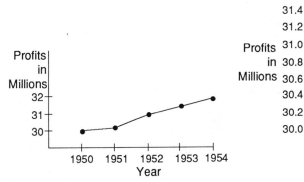

Profits From 1950 – 1954

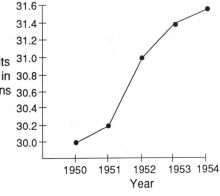

4. Distortion Due to Drawing Incorrect Conclusions from Data Presented in Tables and Graphs

Example

The number of salespersons and yearly sales revenue are given below.

Year	Number of Salespersons	Yearly Sales (in $100,000)
1986	45	3.70
1987	53	4.25
1988	57	5.36
1989	61	6.42
1990	70	7.30

We cannot conclude that increasing the number of salespersons caused increased sales, since increased yearly sales may be due to a number of other factors. Nor can we conclude that increased sales caused the company to expand its sales force. There certainly does seem to be a positive association between the number of salespersons and yearly sales—that is, as one increases the other increases, but this does not indicate a causal relationship in either direction. The number of salespersons might, however, provide executives of the company some useful information in terms of predicting yearly sales.

Problem Set 7.1

1. Explain how the following graphs are misleading.

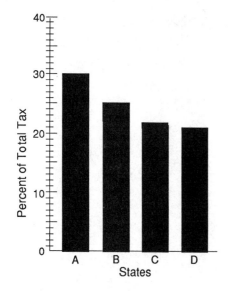
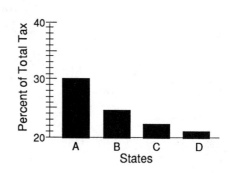

2. A student's parents promise to purchase a motorcycle for her if she has an A average. Her examination grades are 97%, 97%, 41%, 5% and 13%. The student tells her parents that her average is an A. How is this student lying with statistics?

3. The graph below shows the amount of sales generated by an advertising agency from 1983 through 1991.

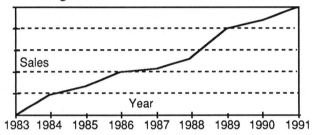

Which one of the following is true according to the graph?
a. From 1983 through 1991 sales increased by millions of dollars.
b. Sales began to stabilize from 1988 onward.
c. There is no trend in sales over time
d. Sales steadily increased over time.

4. The following plot depicts the number of seals sighted and the average temperature on Limantour Beach of the Point Reyes peninsula (Point Reyes, California) for 10 days.

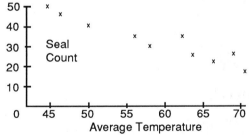

Which one of the following is true according to the graph?
a. There is no association between seal count and the average temerature.
b. Increased average temperature caused the decrease in seal count.

c. The seal count began to stabilize as the temperature rose above 55 degrees.
d. When the average temperature was 50 degrees, approximately 42 seals were sighted.

Review Competencies

5. For the following sets of numbers, the value x represents the same type of measure of central tendency.

$\{2, 2, 3, 4, 4\}: x = 3$
$\{1, 1, 1, 1\}: x = 1$
$\{2, 3, 5, 10\}: x = 5$

What is the value of x in the following set of numbers? $\{1, 1, 3, 7\}$
a. 1   b. 2   c. 3   d. 7

6. The graph below shows the temperature at five locations at noon on Dec. 25, 1970.

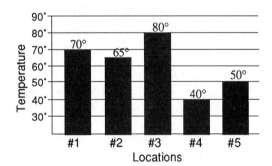

Which one of the following statements is true according to the graph?
a. The mean temperature for the five locations is 61°, and the median temperature for the five locations is 80°.
b. The mean temperature for the five locations is 60°, and the median temperature for the five locations is 65°.
c. The mean and the median temperatures for the five locations differ by 4°.
d. Location #4 had only $\frac{1}{3}$ the temperature as location #3 at noon on Dec. 25, 1970.

7. What is the median of the data in the following sample?

   26, 26, 26, 26, 1, 1, 13, 20, 28, 21

   a. 1　　b. 18.8　c. 23.5　　d. 26

8. The circle graph below shows the amount of money contributed to five charitable organizations. What percent of total money is contributed to organizations 2, 4, and 5 combined?

   a. 75%　　b. 37.5%

   c. 30%　　d. 15%

Section 8　The Standard Deviation

We have seen that it is important to describe a set of data by means of various measures of central tendency. It is also important to know how compactly the data items are distributed about these measures of central tendency or how they are scattered or spread out. For example, the two sets {80, 80, 80, 80} and {60, 60, 100, 100} both have means of 80, and yet the scores in the second set have far more variability than those in the first collection.

In this section, we will see that we can describe a set of data using numbers called **measures of dispersion** or **measures of variability.** These numbers describe the spread of the scores in the set. The particular measures of dispersion that we will discuss will tell us, informally speaking, the average of the distances that the data items deviate from the mean of the data.

The distance of an item (x) in a set of data from the mean ($\bar{x}$) of the data is called its **deviation**. The deviation may be expressed as $x - \bar{x}$. Thus, if $x = 15$ and $\bar{x} = 10$, then $x - \bar{x} = 15 - 10$ or 5. This means that the data item 15 is 5 units above the mean. If $x = 4$ and $\bar{x} = 10$, then $x - \bar{x}$ is $4 - 10$ or -6. Notice that a deviation can be negative, which means that the data item 4 is 6 units below the mean.

It might seem reasonable to calculate the average of these deviations as the measure of dispersion, i.e.

$$\frac{\Sigma(x - \bar{x})}{n}$$

Unfortunately, $\Sigma(x - \bar{x})$ is always 0 as we shall soon see. The sum of the negative deviations would cancel the sum of the positive deviations. To eliminate the problem caused by the negative signs in the negative deviations, we simply **square** each deviation and find the mean of the squared deviations.

To compensate for the fact that the deviations were squared, we take the **square root** of the mean of the squared deviations.

This discussion leads us to define the most important measure of dispersion.

Definition

The **standard deviation**, denoted by s, of a set of data is the square root of the mean of the squared deviations (distances) of the individual data items from the mean of the data.

A "small" standard deviation relative to the mean indicates that most items in the data tend to cluster about the mean. A "large" standard deviation relative to the mean indicates that the data items are more scattered away from the mean.

The standard deviation of a set of data may be calculated using the following formula:

$$s = \sqrt{\frac{\Sigma(x - \bar{x})^2}{n}}$$

in which:
  n is the **total number of items** in the set of data.
  $\bar{x}$ is the **mean** of the data.
  x represents each **individual item.**
  $(x - \bar{x})$ is the **difference** of each item from the mean.
  $(x - \bar{x})^2$ is the **square** of each of the differences.

  $\Sigma(x - \bar{x})^2$ is the **sum** of the squares of the differences.

Problem 1

Compute the standard deviation of the set of data at the right using the formula.

(A calculator is a helpful aid.)

x
7
4
9
8
2

531

Solution

step 1

First calculate the mean
of the data using n = 5.

$$\bar{x} = \frac{\Sigma x}{n} = \frac{30}{5} = 6$$

x
7
4
9
8
2

$\Sigma x = 30$

step 2

Now find the **difference** between each item in the data set and the mean.
In other words, complete a column for $(x - \bar{x})$:

	x	$(x - \bar{x})$	
$\bar{x} = 6$	7	1	$(7 - 6)$
	4	- 2	$(4 - 6)$
	9	3	$(9 - 6)$
	8	2	$(8 - 6)$
	2	-4	$(2 - 6)$
	30	0	

Notice that the sum of the $(x - \bar{x})$ column is 0. The sum of the negative
deviations cancel out the sum of the positive ones. This will always be the
case.

step 3

Now square each of the numbers in the $(x - \bar{x})$ column.
This new column will be labeled $(x - \bar{x})^2$.
The column will contain no negative numbers.

x	$(x - \bar{x})$	$(x - \bar{x})^2$	
7	1	1	$(1 \cdot 1)$
4	-2	4	$(-2 \cdot -2)$
9	3	9	$(3 \cdot 3)$
8	2	4	$(2 \cdot 2)$
2	-4	16	$(-4 \cdot -4)$

step 4
Now find the sum of the numbers in the $(x - \bar{x})^2$ column

$$(x - \bar{x})^2$$
$$1$$
$$4$$
$$9$$
$$4$$
$$\underline{16}$$
$$\Sigma(x - \bar{x})^2 = 34$$

step 5
Now calculate the standard deviation by using the formula.
Thus, we would have:

$$s = \sqrt{\frac{\Sigma(x - \bar{x})^2}{n}} = \sqrt{\frac{34}{5}} = \sqrt{6.8}$$

$\sqrt{6.8}$ is approximately equal to 2.6.
Informally speaking, we say that "on the average, the data items deviate 2.6 units from the mean."

Problem Set 8.1

1. Compute the standard deviation of the following set of data.

x
4
7
10
8
6
7

2. Compute the standard deviation of the following set of data:

x
5
8
11
4
6
2
4
8

3. Compute the standard deviation of the following set of data:

$$\{17, 8, 9, 9, 11, 18\}$$

4. Compute the standard deviation of the following set of data:

$$\{5, 5, 6, 8, 9, 9, 10, 10, 10\}$$

5. Given the data: 7, 7, 7, 7.
Without actually computing the standard deviation, which of the following best approximates the standard deviation?
   a. 7   b. 0   c. 28
   d. cannot be determined without actual computation

6. Given the data: 7, 9, 10, 11, 13.
Without actually computing the standard deviation, which of the following best approximates
the standard deviation?
a. 10   b. 6   c. 2   d. 20

7. Two classes have means of 70%. Group 1 has
a standard deviation of 5%. Group 2 has
a standard deviation of 15%. Which group
would be easier to teach?

Review Competencies

8. Find the mode of the data:
{17, 23, 5, 17, 6, 8, 9, 23, 17}

9. Use the frequency
distribution at the right to:
a. Calculate $\bar{x}$.
b. Calculate $\tilde{x}$.
c. Calculate P_{45}.
d. Calculate Q_3.
e. Indicate the cumulative
frequencies associated
with each class.

class	f
5 – 9	7
10 – 14	9
15 – 19	12
20 – 24	15
25 – 29	8
30 – 34	5
35 – 39	4

10. True or false? In any set of data the mean,
median and mode will always coincide (have
the same value).

11. Find the median of the data:
{11, 13, 7, 10, 6, 4, 12}

12. Construct a circle graph representing the
following family budget:

Shelter	30%
Food	25%
Automobile	15%
Utilities	15%
Savings	10%
Miscellaneous	5%

If the family's income is $800 per month, how
much is spent for shelter and utilities?

13. Consider the following graph indicating Bill's
monthly earnings from September through
May.

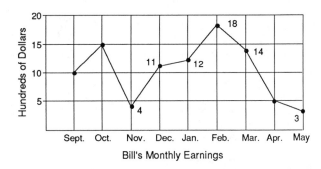

Bill's Monthly Earnings

a. Find the mean number of dollars Bill earns
from September through May.
b. Find the median number of dollars Bill
earns from September through May.
c. Find the modal number of dollars Bill
earns from September through May.

14. Divide: $4\frac{1}{2} \div 2$

15. Divide: $0.468 \div 0.2$

16. Evaluate: $10 \div (7 - 2) + 4(2 \cdot 3) - 6$

17. Rationalize the denominator: $\dfrac{4}{\sqrt{2}}$

Section 9   The Normal Curve

In Section 3 we observed that histograms can assume a variety of shapes depending on the nature of the distribution of data. The histogram below is of particular interest to us because its shape can be approximated by a continuous curve. The curve appears to be **bell-shaped** and extends indefinitely in both directions without touching the horizontal axis.

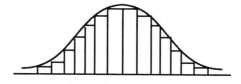

Such a distribution of data is called a **normal distribution** and the continuous curve is called a **normal curve.** We consider that the total area under the normal curve represents 100% of the data in the distribution. Stated another way, we say that the total area under the curve is equal to 1.

The normal distribution plays an important role in numerous situations where data are collected. Indeed, much of what comprises the formal study of elementary statistics involves the normal distribution. This is because measurements of data collected in nature often follow distributions that are approximately normal. The heights and weights of large populations of human beings are examples of distributions that are approximately normal. In these distributions, the data items tend to cluster around the mean and become more spread out as they differ from the mean. Theoretically, the normal curve represents a complicated mathematical formula rather than an actual set of data.

The mean and standard deviation play important roles in studying the normal curve. This is because there is one and only one normal distribution that has a particular mean and a particular standard deviation. Thus, there are many possible normal distributions. In order to deal conveniently with any normal distribution, we define a **standard normal distribution** which has a mean of 0 and a standard deviation (s) of 1.

In the figure at the right, a vertical axis is constructed at the number 0 on the horizontal axis. The peak of the curve lies above the point on the horizontal axis that represents the mean.

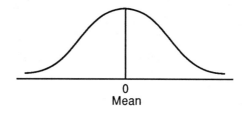

0
Mean

535

Values are assigned to points
on the horizontal axis. The unit
of distance from the mean 0 will
represent one standard deviation.
This standard deviation will order
the horizontal axis into several line
segments. In the figure shown,
notice the distances to the **right** of 0

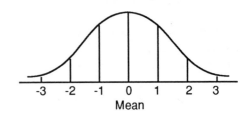

that represent 1, 2, and 3. There are corresponding distances to the **left** of 0
that represent -1, -2, and -3. The integers -3, -2, -1, 0, 1, 2, 3 are called
standard units. Notice that these standard units define particular areas under
the curve.

By looking at the shape of the normal curve, we can get an idea of six
important properties:

1. **In a normal curve, the mean,
    median and mode coincide
    (have identical values) and occur
    at the point 0 on the horizontal
    axis**. Notice that the vertical
    axis at this point (called the
    mean axis) divides the curve
    into two equal parts. This means
    that 50% of the data falls to the
    left of the mean (below the mean)
    and 50% of the data falls to the
    right of the mean (above the mean).

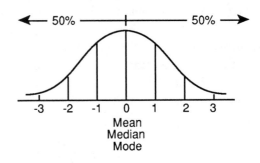

2. **The normal curve is symmetric about the mean.** The left side of the
    curve is the mirror image of the right side of the curve. If we fold the curve
    along the vertical axis, both halves would coincide.

3. **There is the same percent of area in any given region of the normal
    curve as in its corresponding region on the opposite side of the mean
    axis.** Because the curve is symmetric, the area between the mean and 1
    is the same as the area between the mean and -1. Finally, the area between
    2 and 3 is the same as the area between -2 and -3.

4. **The area between the mean and 1 standard deviation above the mean is approximately 34% of the total area under the curve.** This means that approximately 34% of the data will fall between the

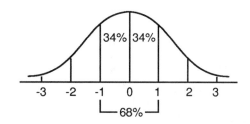

mean and 1 standard deviation above the mean. Similarly, the area between the mean and -1 is approximately 34% of the total area. Thus, approximately 68% of all the data will fall within one standard deviation of the mean.

5. **The area between 1 and 2 standard deviations above the mean is approximately 13.5% of the total area under the curve.** The area between -1 and -2 standard deviations is also approximately 13.5%. Thus, approximately 47.5% of the data falls between the mean and 2 standard deviations, and approximately 95% of the data will fall within two standard deviations of the mean.

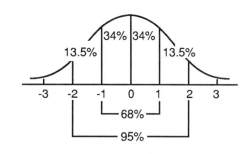

6. **The area between 2 and 3 standard deviations above the mean is approximately 2% of the total area under the curve.** The area between -2 and -3 standard deviations is also approximately 2%. Thus, approximately 49.5% of the data falls between the mean

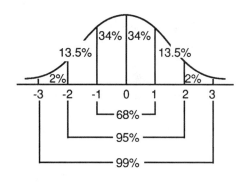

and 3 standard deviations, and approximately 99% of the data will fall within three standard deviations of the mean. Observe that approximately 1% of the data fall further than 3 standard deviations from the mean. Actually, some data fall further than 4 standard deviations from the mean.

Problem Set 9.1

1. Are these statements true or false for a set of test scores that is normally distributed?
   a. Approximately 68% of the scores are between -1 and 1 standard deviation units from the mean.
   b. Approximately 2% of the scores are greater than 3 standard deviations from the mean.
   c. The mean, median, and mode have different values.
   d. The curve is symmetric about the mean.
   e. There is only one mode.
   f. 50% of the scores lie below the mean.
   g. The percent of scores above any score is equal to the percent of scores below that score.

2. Which of the following statements is true for a set of test scores that is normally distributed?
   a. Approximately 5% of the scores will fall more than two standard deviations above the mean.
   b. The percent of scores between one and two standard deviations equals the percent of scores between two and three standard deviations.
   c. All but approximately 5% of the scores fall within two standard deviations of the mean.
   d. All of the scores must fall within four standard deviations of the mean.

Review Competencies

3. Consider the graph below indicating enrollment in Orange High School for grades 7 through 12.
   a. Find the mean number of pupils in grades 7 through 12.
   b. Find the median number of pupils in grades 7 through 12.
   c. Find the modal number of pupils in grades 7 through 12.

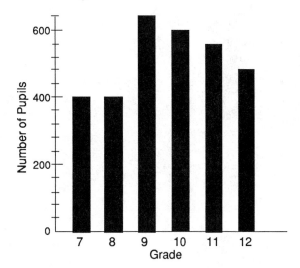

4. Three sets of data are shown at the right. In each case, the value of x represents the same type of measure of central tendency. What is the value of x for the following list of numbers? {5, 5, 10, 12}

8	3	2
9	5	2
9	5	3
10	7	9
x = 9	x = 5	x = 4

   a. 5   b. 7.5   c. 8   d. 10

Let us see the way in which our knowledge of the properties of the normal curve can be used to solve the following problem.

Problem 1   Consider a set of normally distributed test scores in which the mean is computed to be 70, and the standard deviation is 10. Approximately what percent of the students made a grade higher than 80?

Solution   **step 1**
Draw a normal curve
with a mean of 70.

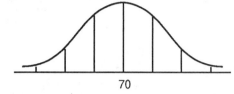

step 2
Since the mean is 70, this value corresponds to 0 in the standard normal distribution. The standard deviation is 10. Thus, the test score corresponding to 1 standard deviation above the mean is 80 (70 + 10). The test score corresponding to 2 standard deviations above the mean is 90, and the score corresponding to 3 standard deviations above the mean is 100. Corresponding to 1 standard deviation below the mean (-1) is a score of 60 (70 – 10), etc. This information is added to the normal curve diagram.

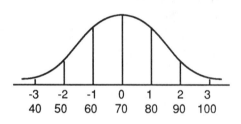

step 3
Shade the portion of the area under the normal curve that represents the percent of students whose grades are higher than 80.

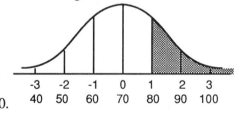

step 4
Since the normal curve is symmetric about the mean, 50% of the test scores fall above the mean. Recall that approximately 34% of the scores fall between the mean and 1 standard deviation above the mean. Therefore, the percent of students who scored between 70 and 80 is approximately 34%. The percent who scored higher than 80 on the test is 50% – 34% or approximately 16%. Thus, approximately 16% of the students made a grade higher than 80.

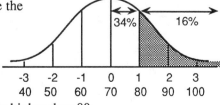

We can also determine the **probability** of selecting a value that falls in a certain region under the curve. There is a direct relationship between the percent of area under the curve and its corresponding probability. In the previous problem, approximately 16% of the students scored higher than 80 on the test. Thus, the probability that a student will make a grade higher than 80 is 0.16. We simply convert the percent to an equivalent decimal. Stated in another way, the probability of randomly selecting a student with a grade higher than 80 is 0.16.

Problem 2 Consider a set of normally distributed test scores in which the mean is computed to be 70, and the standard deviation is 10. What is the probability that a student will score below 80?

Solution Consider the normal curve at the right. Observe that the percent of students who scored below 80 is 84% (50% + 34%). The corresponding probability that a student scored below 80 is 0.84.

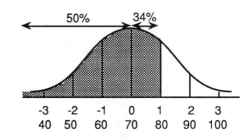

Problem Set 9.2

1. A normal distribution has a mean of 150 and a standard deviation of 15. Find:
    a. the number of standard deviations that 180 is above the mean.
    b. the number of standard deviations that 105 is below the mean.
    c. the number of standard deviations that 172.5 is above the mean.

2. In a large city the graduating scholastic averages of high school students are normally distributed with a mean of 78 and a standard deviation of 6. Find:
    a. the percent of students who have averages of more than 72.
    b. the probability that a student has an average of more than 72.
    c. the percent of students who have averages of less than 72.

    d. the probability that a student has an average of less than 72.
    e. the percent of students who have averages between 84 and 90.
    f. the probability that a student has an average between 84 and 90.
    g. the percent of students who have averages between 72 and 84.
    h. the probability that a student has an average between 72 and 84.

3. The mean weight of 1,000 male students at a certain college is 160 pounds, and the standard deviation is 12.5 pounds. Assuming that the weights are normally distributed, find:
    a. how many students weigh between 147.5 pounds and 172.5 pounds.
    b. how many students weigh less than 147.5 pounds.

4. Compute the standard deviation for the data:
   $\{6, 10, 11, 9, 7, 5\}$

5. Use the frequency dis-
   tribtion at the right to:

class	f
10 – 19	2
20 – 29	5
30 – 39	9
40 – 49	12
50 – 59	8
60 – 69	4

   a. Calculate $\bar{x}$.
   b. Calculate $\tilde{x}$.
   c. Calculate P_{80}.
   d. Calculate Q_1.
   e. Indicate the cumula-
      tive frequencies asso-
      ciated with each class.
   f. Find the modal class.

6. Calculate the mean of the data:
   $\{4.6, 3.8, 2.5, 4.6, 5.3\}$

7. Find the median of the data:
   $\{4, 10, 3, 1, 7, 5\}$

8. Find the mode of the data: $\{17, 21, 16, 19, 15\}$

9. Construct a line graph to represent the average monthly temperatures in Orlando, Florida.
   January 53, February 59,
   March 64, April 68, May 74
   Between which two months was there the smallest average increase in temperature?

10. Express $\frac{28}{35}$ as a fraction in simplest form.

11. Add: $\frac{3}{4} + 3\frac{2}{3}$

12. Divide: $\frac{7}{8} \div 1\frac{4}{7}$

13. Express 60% as a fraction in simplest form.

14. What is 20% of 90?

15. Evaluate: $\frac{3}{5} \div \frac{6}{5} - \frac{1}{3} \cdot \frac{1}{4} + \frac{1}{6}$

We know that the integers (-3, -2, -1, 0, 1, 2, 3) on the horizontal axis of the standard normal distribution represent standard deviations below and above the mean. Each of these integral values is called a **z** value, a **z** score, or a **standard score**. We will see later that z values can also represent non-integers.

In previous problems, we observed that certain data items were converted to their corresponding standard deviations above or below the mean (z values). Thus, a z value is the position of a data item in terms of the number of standard deviations it is located from the mean.

We will now see that any item in the data can be converted to its matching z value by the use of the following conversion formula:

$$z = \frac{\textbf{(data item's value) - mean}}{\textbf{standard deviation}}$$

If x represents the data item to be converted, $\bar{x}$ is the mean, and s is the standard deviation, then the conversion formula can be stated symbolically as:

$$z = \frac{x - \bar{x}}{s}$$

Example

A normal distribution has a mean of 70 and a standard deviation of 10.
To find the number of standard deviations (z) that 80 is away from the mean, we use the formula:

$$z = \frac{x - \bar{x}}{s}$$

$$z = \frac{80 - 70}{10} = \frac{10}{10} = 1$$

Thus, the value 80 is 1 standard deviation above the mean.

Example

A normal distribution has a mean of 70 and a standard deviation of 10.
To find the number of standard deviations (z) that 50 is away from the mean, we use the formula:

$$z = \frac{x - \bar{x}}{s}$$

$$z = \frac{50 - 70}{10} = \frac{-20}{10} = -2$$

Thus, the value 50 lies 2 standard deviations below the mean.

In the following problem, we see the way in which the conversion formula is applied.

Problem 3   Suppose the height of college students is normally distributed with a mean of 70.2 inches and a standard deviation of 5.3 inches. What percent of the students have a height of less than 59.6 inches?

Solution   **step 1**
Draw a normal curve in which the mean of 70.2 corresponds to the z value of 0 on the standard normal curve and represent 59.6 on the horizontal axis.

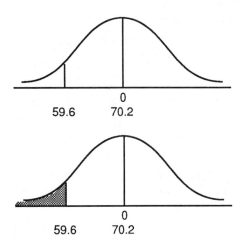

step 2
Shade the portion of the area under the curve that represents the percent of the students having a height of less than 59.6 inches.

step 3
Convert 59.6 to a z value.

$$z = \frac{x - \bar{x}}{s} = \frac{59.6 - 70.2}{5.3} = \frac{-10.6}{5.3} = -2$$

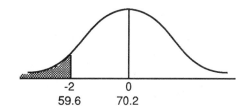

-2   0
59.6   70.2

Thus, the value 59.6 lies 2 standard deviations below the mean.

step 4
Recall that the area between
the mean and 2 standard
deviations below the mean is
approximately 47.5%
(34% + 13.5%). Thus, the
shaded area represents
50% – 47.5% or 2.5%.

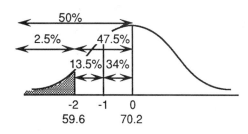

50%
2.5%   47.5%
13.5% 34%
-2 -1 0
59.6   70.2

Approximately 2.5% of the students have a height of less than 59.6 inches.

You should attempt to use the conversion formula whenever possible.

Problem Set 9.3

1. The grade point averages of a group of college students are normally distributed with a mean equal to 2.6 and a standard deviation of 0.4. Use the conversion formula to find the following:
  a. the percent of the students who obtain a grade point average of better than 3.0
  b. the percent of the students who obtain a grade point average between 2.2 and 3.0

2. The speeds of racing cars are normally distributed with a mean of 142.8 mph and a standard deviation of 10.7 mph. Use the conversion formula to find the percent of cars that travel between 142.8 mph and 164.2 mph.

3. The mean wage in a certain company is $5.25 per hour with a standard deviation of $0.75. Assume that the distribution of wages is approximately normally distributed. Using the conversion formula, find the percent of the employees who receive wages between $3.75 and $6.00.

Review Competencies

4. What is the value of the mean and the standard deviation in the standard normal distribution?

5. Approximately what percent of all the data in the standard normal distribution will fall within one standard deviation of the mean? Within two standard deviations? Within three standard deviations?

6. The mean I.Q. for a group of people is 100 with a standard deviation of 15. Assuming I.Q.'s are normally distributed, find:
   a. the percent of people who have an I.Q. greater than 85.
   b. the percent of people who have an I.Q. greater than 115.

7. Construct a circle graph representing the following distribution of grades in a statistics course:

A	15%	B	20%
C	35%	D	25%
F	5%		

8. Compute the standard deviation for the data:
   8, 12, 14, 13, 7, 6

9. Use the frequency distribution at the right to:
   a. Calculate $\bar{x}$.
   b. Calculate $\tilde{x}$.
   c. Calculate P_{75}.
   d. Calculate Q_1

class	f
30 – 49	12
50 – 69	15
70 – 89	21
90 – 109	14
110 – 129	10

10. What is 35% of 60?

11. Express 0.08% as a decimal.

12. Evaluate: $\frac{2}{3}(\frac{5}{6} \div \frac{2}{3}) - 2(\frac{1}{2} \cdot \frac{1}{2})$

We know that integral z values such as -3, -2, -1, 0, 1, 2, and 3 define particular areas under the normal curve between these values and the mean. If the conversion formula yields a non-integer z value such as 1.4, then a special table is required to find the corresponding area. Such a table, called the Z table, appears below. Its entries give the percent of the data between the mean (z = 0) and z values to one decimal place (z < 3).

Z Table

Each entry in the table under "Area" is a percent of the area under the normal curve that is between z = 0 and the table value of z. To find the actual percent, simply convert the three-place decimal number to its corresponding percent.

z	Area	z	Area	z	Area
0.0	0.000	1.0	0.341	2.0	0.477
0.1	0.044	1.1	0.364	2.1	0.482
0.2	0.079	1.2	0.385	2.2	0.486
0.3	0.118	1.3	0.403	2.3	0.489
0.4	0.155	1.4	0.419	2.4	0.492
0.5	0.192	1.5	0.433	2.5	0.494
0.6	0.226	1.6	0.445	2.6	0.495
0.7	0.258	1.7	0.455	2.7	0.496
0.8	0.288	1.8	0.464	2.8	0.497
0.9	0.316	1.9	0.471	2.9	0.498

In the following problems, we will see how the Z table is used to find the percent of the data that falls in various regions under the normal curve.

Problem 4 Find the percent of the data under the normal curve that falls between the mean and 1.6 standard deviations above the mean.

Solution In the accompanying diagram, the shaded area represents the area between the mean and 1.6 standard deviations. The table on pg. 544 is used, and the entry 0.455 is next to the z value 1.6. This means that 44.5% of the area is shaded.

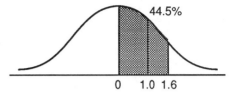

Since the normal curve is symmetric about the mean, the preceding problem also shows that the area between the mean and 1.6 standard deviations below the mean is also 44.5% of the area. In such a case the z value would be negative (-1.6), but the process of solving the problem would be the same.

Problem 5 Find the percent of the data under the normal curve that falls between z = -0.6 and z = 2.1.

Solution The accompanying figure shows the shaded area to be found as a percent of the whole. The problem is done by finding

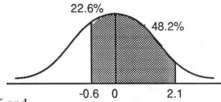

(1)   the area between z = 0 and z = -0.6 and
(2)   the area between z = 0 and z = 2.1.

The table gives 0.226 (22.6%) as the result for (1) and 0.482 (48.2%) as the result for (2). The total area is found by adding 0.226 and 0.482 for a sum of 0.708 or 70.8%. The areas of (1) and (2) do not overlap each other. Consequently, the areas are added to find the correct answer.

Problem 6 Find the percent of the data under the normal curve that falls between z = 1.4 and z = 2.2.

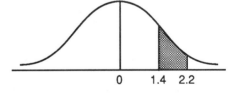

Solution The accompanying figure shows the shaded area under consideration.
Again, the problem is done by finding:

(1)   the area between z = 0 and z = 1.4 and
(2)   the area between z = 0 and z = 2.2.

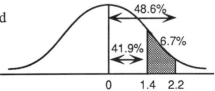

The area of (1) is 0.419 (41.9%), and the area of (2) is 0.486 (48.6%), but this time the numbers are subtracted to obtain 0.067 or 6.7%. The areas of (1) and (2) overlap each other. Consequently, the areas are subtracted to find the correct result.

Problem 7   The mean score of 100 students on a test is 75 with a standard deviation of
10. Assuming that the scores are approximately normally distributed, find
the percent and number of students who scored higher than 88.

Solution

step 1

Draw a normal curve in which
the mean of 75 corresponds to the
z value of 0 on the standard normal
curve and represent 88 on the
horizontal axis.

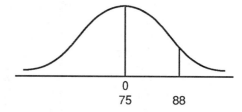

step 2

Shade the portion of the area under
the curve that represents the
percent of the students who scored
higher than 88.

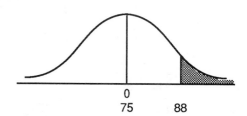

step 3

Convert 88 to a z value.

$$z = \frac{x - \bar{x}}{s} = \frac{88 - 75}{10} = \frac{13}{10} = 1.3$$

Thus, the value 88 lies 1.3 standard
deviations above the mean.

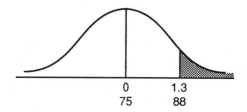

step 4

The area between z = 0 and z = 1.3
is 0.403 (40.3%). The total area
above the mean is 0.500 (50%).
The smaller area is subtracted from
the larger area to give the result 0.097
or 9.7%.

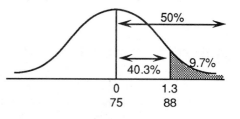

Thus, the percent of students who scored higher than 88 is 9.7%. The
probability of selecting a student who scored higher than 88 is 0.097. The
number of students scoring above 88 is 9.7% of 100 or 9.7. The number 9.7
is rounded to 10.

Problem Set 9.4

1. Find the percent of the area under the normal curve:
   a. between $z = -2.2$ and $z = 0.4$.
   b. greater than $z = -0.5$.
   c. less than $z = -1.9$.
   d. between $z = -2.1$ and $z = -1.6$.

2. The mean weight of 100 women is 115 with a standard deviation of 10. If the weights are normally distributed, find:
   a. the percent of women who weigh more than 130 pounds.
   b. the number of women who weigh less than 118 pounds.
   c. the number of women who weigh between 110 and 126 pounds.
   d. the percent of women who weigh between 90 and 99 pounds.

3. The mean age of 1,000 people living in a retirement community is 68 years with a standard deviation of 4 years. Assume that the ages are normally distributed. Find:
   a. the number of people with ages between 66 and 74 years.
   b. the percent of people whose ages exceed 70 years.
   c. the number of people whose ages are between 62 and 66 years.

Review Competencies

4. True or false? In a set of data, if most items are close in value but there are one or two extreme values, then the median is a better measure of central tendency than the mean.

5. Compute the standard deviation for the data: 10, 12, 14, 8, 10, 6, 10.

6. Use the frequency distribution at the right to:

class	f
5 - 9	11
10 - 14	13
15 - 19	17
20 - 24	9
25 - 29	4
30 - 34	2

   a. Calculate $\bar{x}$.
   b. Calculate $\bar{x}$.
   c. Calculate P_{25}.
   d. Calculate Q_3.
   e. Find the modal class.
   f. Indicate the cumulative frequencies associated with each class.

7. Construct a line graph representing the number of yearly victories of a professional football team:

1970 - 10	1971 - 12	1972 - 13
1973 - 9	1974 - 4	1975 - 7

8. The personnel of a factory is 65% women and 35% men. How many women and men should be included in a 500 person sample to make it as representative as possible?

9. Multiply: $\frac{2}{3} \cdot \frac{9}{14} \cdot \frac{7}{8}$

10. Evaluate: $10 \div 2 \cdot 3 - 4 + 6$

11. Express 330% as a decimal.

12. Express 45% as a fraction in simplest terms.

Recall that distributions of data can assume shapes other than the shape of the normal distribution. Histograms of such distributions appeared at the end of Section 3, Chapter 8.

The histograms below picture distributions that are said to be skewed. In figure (a), the distribution is said to be **skewed to the right** or **positively skewed**. In figure (b), the distribution is said to be **skewed to the left** or **negatively skewed**.

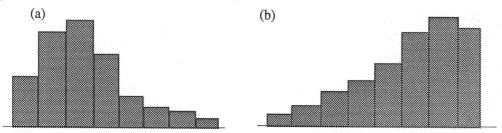

In figures (c) and (d) below, the shapes of the above histograms are approximated by continuous curves. Notice that each of the curves has more of a "tail" on one side than on the other side.

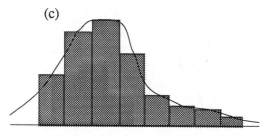
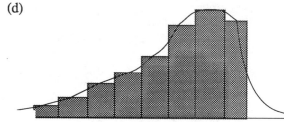

We know that the mean, median, and mode have the same value in the normal curve. Figure (e) illustrates that if a curve is positively skewed, the mean value will be greater than the median value. Thus, the mean value will lie to the right of the median value. This is because the mean of a distribution is affected by extreme values in the data and is pulled in the direction of the extreme values. Notice that the mode lies to the left of the median since it is the value that occurs most often.

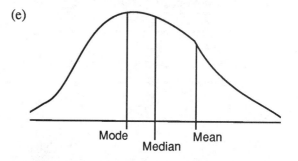

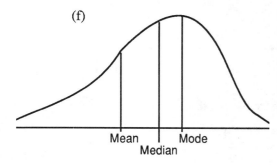

Figure (f) illustrates that if a curve is negatively skewed, then mean value will be less than the median value. Thus, the mean value will lie to the left of the median value. Again, this is because the mean of a distribution is affected by extreme values in the data. Notice that the mode lies to the right of the median.

Figure (g) displays a continuous curve superimposed on a histogram of a J-shaped distribution. This J-shaped distribution is also positively skewed.

(g)

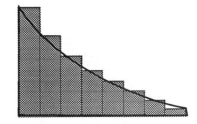

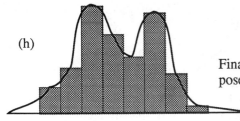

(h)

Finally, figure (h) shows a continuous curve superimposed on a histogram of a bimodal distribution.

Example

In a mathematics class, half the students scored 85 on an achievement test. Most of the remaining students scored 75 except for a few who scored 20. In this distribution of relatively high scores, we can certainly conclude that the mode is 85, since it is the score that occurs most frequently. Since the mean is pulled in the direction of the extreme scores, we see that the mean will be lowered by the few scores of 20. Consequently, the mode is greater than the mean. Depending on the number of scores, the median will either be 85 or the average of 85 and 75, although it will certainly have a greater value than the mean which, again, is pulled downward by the few 20s.

Example

On a statistics test, over half the students scored 75. Most of the remaining students scored 50. A few scored 100. In this distribution, we can immediately conclude that the mode is 75, since it is the score that occurs most frequently. The median is also 75 because if any score occurs more than half the time it must be the score in the middle when data are arranged in order. Thus, the mode and the median have the same value. Every score of 50 and every score of 100 will keep the mean balanced at 75, but since there are more 50s than 100s, the mean will be pulled below 75. Thus, the mean is less than both the mode and the median.

Chapter 8

1. True or false? In a skewed distribution, the mode value is closer to the median value than the mean value.

2. True or false? In a skewed distribution, the median value lies between the value of the mean and the mode.

3. True or false? In a bimodal distribution, the mean value and the median value lie between the two mode values.

4. The graph shown below depicts the distribution of scores for a final examination in mathematics. Select the statement that is true about the distribution of scores.

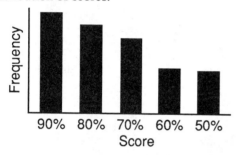

a. The median is greater than the mode.
b. The mean is less than the mode.
c. The median and the mode have identical values.
d. The mean and the mode have identical values.

5. This past year 400 waterfront homes were sold in the Upper Keys. Precisely 300 homes sold for $150,000. Some homes sold in the $150,000 to $250,000 range, and some sold in the $250,000 to $1,000,000 range. Six of the houses sold for more than $1,000,000. Which of the following is true about the distribution for the prices of the homes?
a. The mean, median, and mode have identical values.
b. The mean is less than the median.
c. The mean is greater than the median.
d. The mode is greater than the mean.

6. In a class of 100 students, 60 students scored 55% on an examination. The remaining students scored between 90% and 95%. Select the statement that is true about the distribution of scores.
a. The mode is greater than the mean.
b. The mode and the median have the same value.
c. The mode is greater than the median.
d. The median is greater than the mean.

7. The graph below shows a distribution of years of education for people in a particular community. Select the statement that is true about the distribution of years of education.

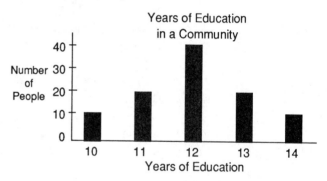

a. The mode is greater than the median.
b. The mean is less than the mode.
c. The mean, median, and mode are equal.
d. The mode is less than the median.

Review Competencies

8. For the following sets of numbers, the value x represents the same type of measure of central tendency.

$$\{4, 4, 4, 4, 4\} \qquad x = 4$$
$$\{2, 3, 3, 5, 8\} \qquad x = 3$$
$$\{5, 6, 8, 9\} \qquad x = 7$$

What is the value of x for:
$\{0.01, 0.01, 0.1, 0.4\}$?

a. 0.01   b. 0.105   c. 0.55   d. 0.055

9. Which one of the following statements is true for a set of data that is normally distributed?
   a. The percent of scores between the mean and 1 standard deviation equals the percent of scores between 2 and 3 standard deviations.
   b. 34% of the scores fall between the mean and the mode.
   c. The probability of randomly selecting a score that falls more than 1 standard deviation above the mean is approximately 0.16.
   d. All of the scores must fall within four standard deviations of the mean.

10. Select the correct formula for determining the surface area of the cube shown below.
    a. $SA = 6b^3$
    b. $SA = 8b^2$
    c. $SA = 6b^2$
    d. $SA = b^3$

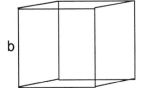

11. In 1980 Amy earned $20,000 and spent 15% of her income on car expenses. In 1981, Amy earned $25,000 and spent 20% of her income on car expenses. Find the percent of increase in the amount that Amy spent on her car from 1980 to 1981.

Chapter 8 Summary

1. Samples

   a. Samples should be representative of the population from which they are drawn.
   b. Random samples are obtained in such a way that each member of the population has an equal chance of being selected.

2. Presentation of Data

   a. Data can be presented non-graphically in ungrouped or grouped frequency distributions.
   b. Vocabulary Associated with a Grouped Frequency Distribution
      1. Class Frequency: Number of items that appear in the class.
      2. Class Interval: Number representing the width of the class.
      3. Class Limits: Largest and smallest values that can appear in a class.
      4. Class Boundaries: Numbers between the upper class limit of one class and the lower class limit of the next class; dividing point between the classes.
      5. Class Mark: Middle point of the class
      6. Cumulative Frequency (c.f.): Sum of the frequencies in a class and the frequencies in all classes below the given class.
   c. Data can be presented graphically in histograms, frequency polygons, bar graphs, circle graphs, and line graphs.

Chapter 8

3. Measures of Central Tendency

a. The Mean

   1. Ungrouped data: $\bar{x} = \dfrac{\Sigma x}{n}$

   2. Grouped data: $\bar{x} = \dfrac{\Sigma xf}{n}$

   (x = class mark, f = frequency of each class)

b. The Median: The middle item or the mean of the two items nearest the middle when the data are arranged in order of size.

   1. Ungrouped data: $\bar{x}$ = the score in the $\dfrac{n+1}{2}$ position when scores are in order

   2. Grouped data: $\bar{x} = L + \left(\dfrac{x}{f}\right) \cdot i$

      L = lower boundary of the class containing the median;
      x = number of items still needed after L to reach the median;
      f = frequency of the class containing the median;
      i = the class interval

c. The Mode: Value(s) occurring most often in a set of data
   1. Ungrouped data: Select the most frequent value or values.
   2. Grouped data: The mode is the class mark of the class with the largest frequency.

4. Percentile

a. Percentile, P_n, is a value such that n percent of the data falls below P_n.

b. $P_n = L + \left(\dfrac{x}{f}\right) \cdot i$

      L = lower boundary of the class containing P_n;
      x = number of items needed after L to reach P_n;
      f = frequency of the class containing P_n;
      i = the class interval

c. P_{25} (or Q_1) is called the lower quartile and is the value below which 25% of the data fall.

d. P_{75} (or Q_3) is called the upper quartile and is the value below which 75% of the data fall.

e. P_{50} (or Q_2) is the middle quartile, the fiftieth percentile, and the median.

5. Statistical distortion can occur using "average" in a way that best suits one's needs; by using samples that are not representative of the population; by presenting data graphically in misleading ways.

6. The Standard Deviation

a. Informally speaking, the standard deviation describes the average of the distances that the data items deviate from the mean.

b. $s = \sqrt{\dfrac{\Sigma(x - \bar{x})^2}{n}}$

7. The Normal Curve

a. Properties of the Normal Curve

1. The mean, median and mode have the same value.

2. The curve is symmetric with respect to the mean.

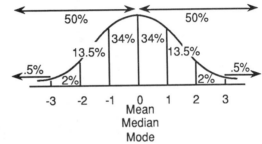

3. The horizontal axis is expressed in z values, standard scores or standard deviation units, where $z = \dfrac{x - \bar{x}}{s}$.

b. To solve a problem involving the normal curve:
Make a sketch and shade the desired region. Change all scores to z values. Use the z table or, if z is an integer, the percents that appear under the normal curve in item 7a above.

8. Skewed Distributions

a. These distributions lack symmetry.

b. Positively skewed distributions (skewed to the right) involve many low scores. The mean is pulled in the direction of the extreme scores.

c. Negatively skewed distributions (skewed to the left) involve many high scores. Again, the mean is pulled in the direction of the extreme scores.

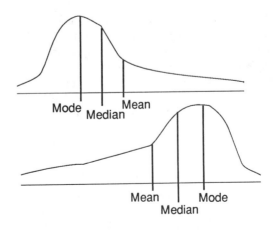

553

Chapter 8 Self-Test

1. Find the mean of the data:
   {7, 4, 8, 9, 5, 3, 6, 10, 7, 11}
   a. 5   b. 6   c. 7   d. 8

2. Find the median of the data:
   {7, 5, 9, 4, 3, 10, 8, 6, 2}
   a. 3   b. 6   c. 6.5   d. 7

3. Find the median of the data:
   {7, 3, 9, 7, 3, 8, 6, 4}
   a. 6   b. 7   c. 8   d. 6.5

4. Find the mode of the data:
   {3, 6, 9, 5, 3, 6, 5, 3, 8}
   a. 3   b. 5   c. 6   d. 3 and 5

5. Three sets of data are shown at the right. The x-numbers are the same measure of central tendency.

5	4	5
2	2	7
6	9	9
1	8	11
7	3	12
	10	
x = 5	x = 6	x = 9

   Find the value of x for the set of data shown at the right.

   | |
   | 5 |
   | 10 |
   | 6 |
   | 5 |
   | x = __ |

   a. 5   b. 5.5   c. 6   d. 6.5

6. To be eligible for the math club, a student must have a mean grade on three qualifying exams of at least 88% and a median grade of at least 90%. Mary has grades of 87%, 87% and 100%. Bob has grades of 80%, 95% and 100%. Susan has grades of 100%, 75% and 100%. Which student(s) is/are eligible for the math club?
   a. Bob and Mary   b. Mary and Susan
   c. Bob and Susan   d. Bob only

7. The city council of a large city needs to know if its residents will support the building of three new schools. The council decides to conduct a survey of a sample of the city's residents. Which procedure would be most appropriate for obtaining an unbiased sample of the city's residents?
   a. Survey a random sample of teachers who live in the city.
   b. Survey 100 individuals who are randomly selected from a list of all people living in the state in which the city in question is located.
   c. Survey a random sample of persons within each neighborhood of the city.
   d. Survey every tenth person who enters City Hall on a randomly selected day.

8. The bar graph below shows the market value of a home for each of five years. Which one of the following is true according to the graph?

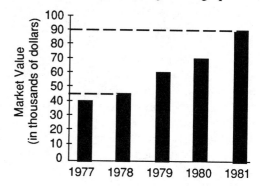

   a. The percent increase in value from 1977 to 1979 is $33\frac{1}{3}\%$.
   b. The value of the home in 1981 was $9,000.
   c. The mean value of the home for the five years exceeds the median value by $1,000.
   d. The mean value of the home for the five years is $60,000.

9. The graph below represents the monthly temperature for 6 months of the year. The readings for March and June fall exactly midway between the horizontal lines. Which one of the following is true according to the graph?

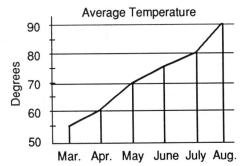

Average Temperature

a. The percent increase in temperature from April to August is $33\frac{1}{3}\%$.

b. The mean and median temperatures for the six months differ by less than one degree.

c. The modal temperature for the six months is 90 degrees.

d. The value of the mean temperature for the six months is exactly 71 degrees.

10. The circle graph below shows money contributed to five charities (numbered from 1 to 5). What percent of the money is contributed to charities 1 and 2 combined?

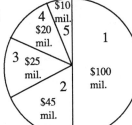

a. 145%
b. 95%
c. 72.5%
d. 70.5%

11. Which of the following statements is true for a set of scores that is normally distributed?

a. The percent of scores between the mean and one standard deviation equals the percent of scores between one and two standard deviations.

b. The percent of scores above any score is equal to the percent of scores below that score.

c. All of the scores must fall within four standard deviations of the mean.

d. The mean, median and mode have the same value.

12. Study the three examples below in which each score for a normal distribution is converted to the z value at the right.

score	standard z value
100	-1
105	0
115	2

What is the z value for a score of 117?

a. 2.25   b. 2.4   c. 2.5   d. 3

13. Use the table shown below to answer the question.

Standard deviations above mean	Proportion of area between mean and indicated standard deviation above mean
0.00	.000
0.25	.099
0.50	.192
0.75	.273
1.00	.341
1.25	.394
1.50	.433
1.75	.460
2.00	.477
2.25	.488
2.50	.494
2.75	.497
3.00	.499

One hundred high school students took an aptitude test which has a mean of 50 and a standard deviation of 10. If their scores are normally distributed, approximately what proportion of the students scored between 55 and 65?

a. 0.9   b. 0.24   c. 0.43   d. 0.63

14. On a recent examination over half the students had a score of 85%. Of the remaining students, most scored between 60% and 80% with the exception of 5% of the group that scored below 20%. Which one of the following is true about the distribution of scores in this situation?
    a. The mode and the mean have the same value.
    b. The mode and the median have the same value.
    c. The mean is greater than the mode.
    d. At least one student scored 10%.

15. The graph below represents the distribution of ages of students at a day camp. Select the statement that is true about the distribution of ages.

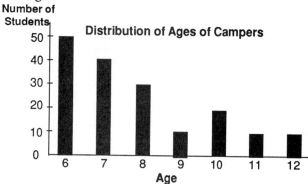

a. The mode is equal to the median.
b. The mode is equal to the mean.
c. The mode is greater than the mean.
d. The mode is less than the median.

16. In a class of 100 students, 45 students scored 80% on an examination. Of the remaining 55 students, 20 scored between 60% and 70%, and 35 scored under 50%. Select the statement that is true about this distribution of scores.
    a. The mode is greater than the mean.
    b. The mode and the median have the same value.
    c. The mean, median, and mode have the same value.
    d. The exact value for all three statistics of central tendency is unknown.

17. In a set of scores that is normally distributed, 140 has a z-score of 0 and 132 has a z-score of -1. What score has a z-score of 1.25?
    a. 150   b. 152   c. 152.25   d. 152.5

18. Compute the standard deviation for the following set of data: {8, 10, 12, 16, 19}
    a. 4   b. 13   c. 16
    d. None of these

Refer to the following frequency distribution to answer problems 19-22:

class	f
10-19	2
20-29	6
30-39	9
40-49	15
50-59	25
60-69	18
70-79	14
80-89	7
90-99	4
	100

19. Calculate the mean of this frequency distribution.
    a. 56.0   b. 56.9   c. 57.1
    d. None of these

20. Calculate the mean of this frequency distribution.
    a. 56.0   b. 57.2   c. 56.7
    d. None of these

21. Calculate P_{42}.
    a. 52.0   b. 53.5   c. 54.3
    d. None of these

22. Calculate Q_1.
    a. 44.8   b. 44.0   c. 45.3
    d. None of these

23. The table below indicates scores on a 400-point exam and their corresponding percentile ranks.

Score	Percentile Rank
380	99
350	87
320	72
290	49
260	26
230	8
200	1

What percentage of examinees scored between 230 and 290?

a. 41%   b. 49%   c. 72%   d. 98%

The table below shows the distribution of income levels in a community. Refer to this table to answer problems 24 and 25.

Income Level	Percent of People
$0-4,999	7
5,000-9,999	9
10,000-14,999	12
15,000-19,999	18
20,000-24,999	24
25,000-29,999	20
30,000-34,999	7
35,000-39,999	2
40,000 and over	1

24. What percentage of people have incomes of at least $25,000?
a. 10%   b. 30%   c. 54%   d. 70%

25. Identify the income level which has 28% of the people in the community with lower incomes.
a. $10,000   b. $14,999
c. $15,000   d. $19,999

The table at the top of the next column shows the distribution of the number of children in families of a community. Refer to this table to answer problems 26 and 27.

Number of Children	Proportion of Families
0	.45
1	.25
2	.20
3	.05
4	.04
5	.01

26. What is the median number of children per family?
a. 0   b. 0.5   c. 1   d. 1.5

27. What is the mean number of children per family?
a. 1   b. 1.01   c. 1.03   d. 1.5

28. The table below indicates the fastest running times (for 26.5 miles) from 1980 through 1989.

Year	Fastest Running Time (in minutes)
1980	155
1981	148
1982	140
1983	134
1984	130
1985	127
1986	126
1987	126
1988	126
1989	126

Which one of the following is true?
a. The fastest running times began to stabilize in 1985.
b. There is no trend in running times for the years shown in the table.
c. With each year in the 1980s, running time steadily decreased.
d. By the mid-1980s, running times only slightly decreased or stabilized.

29. The graph below depicts the number of whales sighted (o) and the number of seals sighted (x) at Point Reyes National Seashore in California.

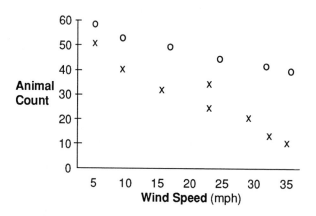

Which one of the following is true?
a. Increased wind speed was directly responsible for the decrease in the number of whales and seals that were sighted.
b. There is an association between wind speed and animal count only for the seals.
c. Approximately 45 seals were sighted when the wind speed was 15 mph.
d. Fewer whales and seals tend to be sighted as the wind speed increases; a stronger relationship exists for the seals than for the whales.

Appendix: Review Competencies Covering
the Entire Text

Answers with brief explanations begin on page 576.

1.   Sets A, B, C, and U are related as shown in the
diagram below. Which of the following statements is true assuming none of the regions
is empty?

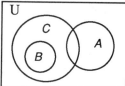

a.   All elements of set U are elements of set C.
b.   Some elements belong to all four sets A, B,
C, and U.
c.   Some elements belong to A ∩ C but not
to set B.
d.   Any element which is a member of set C
must also be a member of set B.

2.   If $3(t-1) = 2[t - 3(t-2)]$, then:
a. $t = \frac{7}{15}$ b. $t = \frac{-9}{7}$ c. $t = \frac{15}{7}$ d. $t = \frac{-7}{9}$

3.   A box contains ten tennis balls, of which four
are defective. If two balls are randomly selected without replacement, find the probability
that at least one ball is defective.
a. $\frac{2}{15}$ b. $\frac{1}{3}$ c. $\frac{2}{3}$ d. $\frac{13}{15}$

4.   All of the following arguments have true
conclusions, but one of the arguments is not
valid. Select the argument that is not valid.
a.   All people who are outstanding athletes
possess excellent eye-hand coordination.
Only outstanding athletes are eligible for
athletic scholarships. Therefore, to be
eligible for an athletic scholarship, one
must possess excellent eye-hand
coordination.
b.   Prolonged periods of rain will damage
desert plants. The cactus is a desert plant.
Therefore, the cactus will be damaged by
prolonged periods of rain.

c.   All multiples of 15 are multiples of 5.
Twenty-one is not a multiple of 5. Therefore, 21 is not a multiple of 15.
d.   Clocks are designed to inform people of
the time. Time-measuring devices are also
designed to inform people of the time.
Therefore, clocks are time-measuring
devices.

5.   For each of the statements below, determine
which one(s) has/have -3 as a solution.
i. $5x + 1 > -13$
ii. $2y^2 + 7y = -3$
iii. $\frac{4}{3}x - 4 = 0$
a.   i only   b.   ii only
c.   iii only   d.   ii and iii only

6.   What is the area of the shaded region?
The diameter of each small semi-circle is
the radius of the larger semicircle.
a. 25π sq. ft
b. 50π sq. ft
c. 75π sq. ft
d. 100π sq. ft

7.   Choose the inequality equivalent to the
following: $-7x < -70$
a. $x < -10$ b. $x < 10$ c. $x > -10$ d. $x > 10$

8.   The roots of the equation $x^2 = 4 - 4x$ are:
a. $-2 + 4\sqrt{2}$ and $-2 - 4\sqrt{2}$
b. $-2 + 2\sqrt{2}$ and $-2 - 2\sqrt{2}$
c. $2 + 4\sqrt{2}$ and $2 - 4\sqrt{2}$
d. $2 + 2\sqrt{2}$ and $2 - 2\sqrt{2}$

9.   A college student can compute his or her grade
point average by converting A to 4 points, B to
3 points, C to 2 points, D to 1 point, and F to 0
points. Bernie has so far accumulated 10 A's, 7
B's, 5 C's, and no D's or F's. What is his grade
point average?
a.   2.2   b.   3.1   c.   3.2   d.   22

10. To qualify for a rental lease at Lakeview-at-the-Hammocks, one must have a yearly income of at least $20,000, no more than 2 children, only 1 pet not in excess of 20 pounds, and at least $5,000 in some kind of savings. Read the qualifications of the three applicants listed below. Then identify who qualifies for the rental lease.

Frank Martinez, single, earns $60,000 yearly, has no children, one German shepherd weighing 45 pounds, with $40,000 in a savings account.

Dr. Selma Rubin, divorced, earns $200,000 yearly as an internist. Her three children live with her. They have no pets, and Dr. Rubin has nearly one million dollars in U.S. Govenment bonds.

Bill and Roberta Raboid are husband and wife. Bill earns $14,000 yearly and Roberta earns $6,500 yearly. They have two cats, each weighing 11 pounds, with $6,000 in a savings account.

a. Frank Martinez   b. Dr. Selma Rubin
c. Bill and Roberta Raboid
d. No one qualifies for the rental lease.

11. Select the numeral for $(2 \cdot \frac{1}{10}) + (2 \cdot \frac{1}{10^6})$.
   a. 0.020002   b. 0.200020
   c. 0.200002   d. 2.000002

12. $2[\frac{1}{2} - \frac{1}{2}(10 - 2 \cdot 3)] =$
   a. $\frac{-23}{2}$ b. -2   c. -3   d. $\frac{-3}{2}$

13. $\frac{1}{4} \cdot 12c + 6c \div \frac{1}{3} - c^2 + 4c =$
   a. $25c - c^2$ b. $9c - c^2$
   c. $25c$ d. $17c - c^2$

14. Find the median of the numbers in the list:
   5, 6, 7, 3, 7, 7, 4, 1, 3, 6

   a. 6   b. 7   c. 5.5   d. 5

15. Select the statement that is not logically equivalent to: If Sarah is in San Francisco, then she is in California.
   a. If Sarah is not in California, then she is not in San Francisco.
   b. If Sarah is in California, then she is in San Francisco.
   c. Sarah is in California when she is in San Francisco.
   d. Sarah is not in San Francisco or she is in California.

16. Select the statement that is the negation of: A Supreme Court nominee receives majority approval by the Senate or the nominee is not a judge on the Supreme Court.
   a. A Supreme Court nominee does not receive majority approval by the Senate or the nominee is a judge on the Supreme Court.
   b. A Supreme Court nominee does not receive majority approval by the Senate and the nominee is a judge on the Supreme Court.
   c. A Supreme Court nominee receives majority approval by the Senate and the nominee is a judge on the Supreme Court.
   d. If a Supreme Court nominee receives majority approval by the Senate, then the nominee is a judge on the Supreme Court.

17. The table below shows the percent of all car sales in the United States in 1979.

Manufacturer	Percent of All Car Sales
GM	46
Ford	20
Chrysler	9
AMC	2
Imports	23

If it is known that the manufacturer was not Ford or Chrysler, find the probability that the car was an import.

a. $\frac{23}{71}$ b. $\frac{29}{71}$ c. $\frac{23}{100}$ d. $\frac{29}{100}$

18. Which statement listed below is true?
    a. Opposite angles of a parallelogram are not necessarily congruent.
    b. Diagonals of a trapezoid are perpendicular.
    c. Consecutive angles of a rectangle are complementary.
    d. Adjacent sides of a rhombus have equal measure.

19. Simplify $\sqrt{36} + \sqrt{6}$
    a. $\sqrt{42}$ b. $\sqrt{6} + 6$ c. $6\sqrt{6}$ d. $7\sqrt{6}$

20. Simplify $\dfrac{12}{\sqrt{6}}$
    a. $2\sqrt{6}$ b. $\dfrac{\sqrt{6}}{2}$ c. $\sqrt{2}$ d. $\dfrac{\sqrt{72}}{6}$

21. For the following lists of numbers, the value of x represents the same type of measure of central tendency.

    $$6,7,7,8:\quad x = 7$$
    $$3,5,8,12:\quad x = 7$$
    $$2,2,3,8:\quad x = 4$$

    What is the value for x in the list 1.1, 2.3, 3.8, 7.2, 8.1?
    a. 3.5   b. 3.8   c. 4.5   d. 7.0

22. Which shaded region below identifies the portion of the plane in which $x \le -1$ and $y \ge 0$?

    a.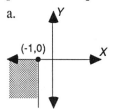

    b.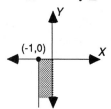

    c.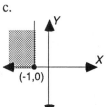

    d.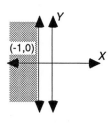

23. Which option gives the condition(s) that correspond(s) to the shaded region of the plane shown at the right?
    a. $y - x < 2$
    b. $y - x > 2$
    c. $x > -2$ and $y < 2$
    d. $y > -2$ and $x < 2$

24. If $y > y - (16 - 2y)$, then:
    a. $y > 8$ b. $y < 8$ c. $y > -8$ d. $y < -8$

25. Select the statement that is the negation of the statement: If it is hot then at least one municipal pool is open.
    a. If it is not hot, then at least one municipal pool is not open.
    b. If all municipal pools are not open, then it is not hot.
    c. It is hot and no municipal pool is open.
    d. It is not hot and at least one municipal pool is not open.

26. $(-0.006) \div (-0.12) =$

    a. 0.0005%   b. 0.05%
    c. 5%   d. 50%

27. 108.7%
    a. $108\frac{7}{10}$ b. $10\frac{87}{100}$ c. $1\frac{87}{100}$ d. $1\frac{87}{1000}$

28. What is the length of AB in meters?
    a. 0.13 m
    b. 1300 m
    c. 0.17 m
    d. 1700 m

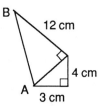

29. Study the examples:
$$x^9 + x^6 - x^3 = x^{18}$$
$$x^4 + x^6 - x^2 = x^{12}$$
$$x^5 + x^4 - x^2 = x^{10}$$
Select the equation which is compatible with the data.
a. $x^a + x^b - x^c = x^{a+b+c}$
b. $x^a + x^b - x^c = x^{bc}$
c. $x^a + x^b - x^c = x^{\frac{ac}{b}}$
d. $x^a + x^b - x^c = x^{\frac{ab}{c}}$

30. The formula for the sales price (P) of an article with cost (C) and mark-up (M) is $P = C + MC$. What is the sales price of an object costing $12 at a 40% mark-up?
a. $52   b. $492   c. $12.48   d. $16.80

31. Choose the equation that is not true for all real numbers.
a. $7x + 14y = 7(2y + x)$
b. $6a^2b(3x + y) = (y + 3x)6ba^2$
c. $(x - y)(x + y) = (x + y)(y - x)$
d. $(6x^3)y^2 = 6(x^3y^2)$

32. Which whole number is divisible by 5 and is also a factor of 45?
a. 9   b. 15   c. 20   d. 90

33. In 1980, Joe earned $20,000 and spent 15% of his income on car expenses. In 1981, Joe earned $25,000 and spent 20% of his income on car expenses. Find the percent increase in the amount that Joe spent on his car from 1980 to 1981.
a. 5%   b. $66\frac{2}{3}\%$ c. 40%   d. 60%

34. Two classes each have an enrollment of 40 students. On a certain day, $\frac{4}{5}$ of one class and $\frac{7}{8}$ of the other are present. How many students are absent from the two classes?
a. 8   b. 13   c. 26   d. 67

35. Given the function defined by $f(x) = x^3 - 6x^2 - x + 2$, find f(-2).
a. -32   b. -28   c. 16   d. 20

36. On a very difficult examination, half the students scored 45%. Most of the remaining students scored 75%. However, a few students had scores of 95% or better. Which of the following is true about this distribution?
a. The mean is less than the median.
b. The mean is greater than the median.
c. The mean and the median have identical values.
d. The mode is greater than the median.

37. Which one of the following is false for the accompanying figure?

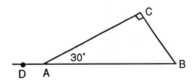

a. $(AC)^2 + (BC)^2 = (AB)^2$
b. $m \angle DAC = m \angle ABC + 90°$
c. $\triangle ABC$ is right and scalene.
d. $m \angle ABC \neq 60°$

38. A linear factor of the expression $8x^2 - 14x + 3$ is:
a. $2x - 3$ b. $2x + 3$ c. $4x + 1$ d. $8x + 3$

39. Use the table below to answer this question.

Standard deviations above mean	Proportion of area between mean and indicated standard deviation above mean
.00	.000
.25	.099
.50	.192
.75	.273
1.00	.341
1.25	.394
1.50	.433
1.75	.460
2.00	.477
2.25	.488
2.50	.494
2.75	.497
3.00	.499

Two hundred students took an examination having a mean of 60 and a standard deviation of 20. If the scores are normally distributed, approximately what proportion of the students scored between 30 and 85?

a. 0.827   b. 0.866   c. 0.173   d. 0.039

40. Use scientific notation to find the answer:
$0.00073 \cdot 8,200,000 =$

a. $5.986 \cdot 10^3$ b. $5.986 \cdot 10^2$
c. $5.986 \cdot 10$ d. $1.55 \cdot 10^2$

41. In November, 40 people each purchased $30 worth of merchandise. In December, there were 10% fewer customers, but each customer increased their purchase by 20%. What is the change in sales from November to December?

a. $96 increase   b. $98 increase
c. $44 decrease   d. None of these

42. A dinner at a restaurant costs $60 before 5% sales tax was added to the bill. A customer left a tip equal to 25% of the final bill (the bill after the sales tax was added to the dinner bill). Afterwards, the customer decided that the tip should have been 25% of the dinner bill, not the final bill. What would have been saved by leaving a tip only on the dinner bill?

a. $0.75   b. $1.25   c. $3.00   d. $5.00

43. Which statement is true for a set of scores that is normally distributed?

a. The percent of scores above any score is equal to the percent of scores below that score.
b. The probability of randomly selecting a score that falls between one standard deviation above or below the mean is approximately 0.68.
c. All of the scores must fall within four standard deviations from the mean.
d. The mean, median, and standard deviation all have the same value.

44. Study the premises below. Select a logical conclusion if one is warranted.

All physicists are scientists. All biologists are scientists. All scientists are college graduates. Florence is a scientist.

a. Florence is a physicist or a biologist.
b. All biologists are college graduates.
c. Florence is not a college graduate.
d. None of the above is warranted.

45. All students who talk during lecture are rude. No lovable people are rude. Eubie is a lovable person. Which one of the following cannot be logically concluded from these premises?

a. All students who talk during lecture are not lovable people.
b. Eubie is not a student who talks during lecture.
c. Some lovable people are students who talk during lecture.
d. Eubie is not rude.

46. In a mathematics class consisting of 40 students, 15 students are older than 30, 11 students are majoring in business, and 5 students are business majors older than 30. If one student is selected at random, what is the probability that the selected student is older than 30 or a business major?

    a. $\frac{15}{40} + \frac{11}{39} - \frac{5}{38}$ b. $\frac{15}{40} + \frac{11}{40} - \frac{5}{40}$

    c. $\frac{15}{40} \cdot \frac{11}{39} - \frac{5}{38}$ d. $\frac{5}{40} + \frac{5}{40}$

47. A rectangular roof measuring 40 feet by 30 feet has a square porch 10 feet long on a side extending from it. The roof material cost is $70 for bundles covering 100 square feet and the labor cost for the roof is $275. What is the total cost of the labor and material to roof the building?

    a. $345   b. $1,115   c. $1,185   d. $275

48. Five men and four women are trying out for a vocal group that will consist of two men and two women. How many different vocal groups can be selected?

    a. 60   b. 20   c. 9   d. 4

49. Which one of the following is not valid?

    a. If a person earns an A in finite mathematics, then that person does not miss more than two classroom lectures. Jacques missed five classroom lectures in finite mathematics. Thus, Jacques did not earn an A in finite mathematics.

    b. Maggie does not have a good time when she listens to music whose lyrics are racist. At this moment Maggie is listening to a new rock album that contains racist lyrics throughout. Consequently, at this moment Maggie is not having a good time.

    c. A Supreme Court nominee must receive majority approval by the Senate or the nominee is not a judge on the Supreme Court. Bork, a Reagan nominee, did not receive majority approval by the Senate. Therefore, Bork is not a judge on the Supreme Court.

    d. If one is to answer question 28 correctly, then one must first find the correct length of the hypotenuse. Ana correctly found the length of the hypotenuse. Therefore, Ana got the right answer to question 28.

50. If there is a mathematics and English requirement, all college bookstores do a great business. The college bookstore at Tate College is not doing a great business. Tate College has an English requirement. What can be logically concluded?

    a. Tate College has no mathematics requirement.

    b. Tate College has a mathematics requirement.

    c. If there is no mathematics requirement or no English requirement, some college bookstores do a great business.

    d. If college bookstores do a great business, there is a mathematics and English requirement.

51. An executive decides to visit cities A, B, C, D, E, and F in random order. What is the probability that city E will be visited first, city D second, and city A last?

    a. $\frac{1}{30}$ b. $\frac{1}{120}$ c. $\frac{1}{360}$ d. $\frac{1}{720}$

52. How much molten steel is needed to construct a solid spherical statue having a diameter of 6 feet?

    a. 9π cubic ft   b. 36π cubic ft

    c. 72π cubic ft   d. 288π cubic ft

53. Suppose that a consumer protection agency wants to determine whether the citizens of Dade County think that Florida Power and Light company charges them fair rates for electricity. Which of the following is the most appropriate procedure to select an unbiased sample of Florida Power and Light company customers?
    a. Survey all the customers who pay their electric bills at Government Center in Miami on the third day of the month.
    b. Survey all Florida Power and Light customers who live in thirty randomly selected blocks within Dade County.
    c. Survey 100 individuals who are randomly selected from telephone directories of Dade, Broward and Monroe Counties.
    d. Survey a random sample of employees who work for Florida Power and Light in Dade County.

54. What is the volume of the combined cylinder and cone shown below if both have a diameter of 6 feet and a height of 3 feet?

    a. 27π ft^3
    b. 36π ft^3
    c. 54π ft^3
    d. 144π ft^3

55. A building contractor is to dig a foundation 90 feet long, 60 feet wide, and 6 feet deep for an apartment's foundation. The contractor pays $12 per load for trucks to remove the dirt. Each truck holds 10 cubic yards. What is the cost to the contractor to have all the dirt hauled away?
    a. $1,440   b. $12,960
    c. $14,400   d. $129,600

56. Which one of the following is true for the accompanying figure showing four lines in the same plane?

    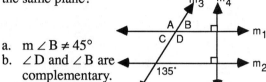

    a. m $\angle$ B $\neq 45°$
    b. $\angle$ D and $\angle$ B are complementary.
    c. m $\angle$ A $= 135°$
    d. None of the above is necessarily true.

57. Which one of the following statements is true for the triangles in the accompanying figure?
    a. m $\angle$ BAC = m $\angle$ DEC
    b. $\dfrac{AC}{DC} = \dfrac{DE}{AB}$
    c. If CE = 5 ft, BE = 4 ft, and AB = 10 ft, the measure of DE is $5\frac{5}{9}$ ft.
    d. $\dfrac{CE}{DC} = \dfrac{AC}{BC}$

    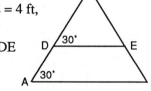

58. Which one of the following is false for the accompanying figure?

    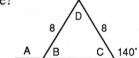

    a. m $\angle$ D $= 100°$
    b. m $\angle$ B $+$ m $\angle$ C $+$ m $\angle$ D $= 180°$
    c. m $\angle$ A $= 140°$ d. $8^2 + 8^2 = (BC)^2$

59. $\frac{1}{10}$ of a person's salary is spent for clothing, $\frac{1}{3}$ for food, and $\frac{1}{5}$ for rent. What percent of the salary is left?
    a. $46\frac{2}{3}\%$ b. $42\frac{1}{3}\%$ c. $37\frac{2}{3}\%$ d. $36\frac{2}{3}\%$

60. $2 - 5(a - 3b) =$
    a. $-3a - 9b$ b. $-3a + 9b$
    c. $2 - 5a + 15b$ d. $2 - 5a - 15b$

61. Select the symbol that shoud be place in the box. $-\frac{5}{6}$ [ ] -0.84

    a. =   b. <   c. >

62. Study the figures shown below.

    9 triangles in the interior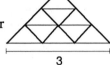

    16 triangles in the interior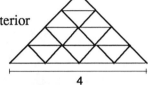

    How many triangles are contained within the interior when the base measures 12?

    a. 36   b. 78   c. 124   d. 144

63. How many prime numbers less than 29 and greater than 17 will yield a remainder of 3 when divided by 4?

    a. one   b. two   c. three   d. four

64. An object that weighs 7 pounds on planet A weighs 11.27 pounds on planet B. Find the weight on planet B of an object that weighs 6 pounds on planet A.

    a. 3.73 lbs.   b. 9.46 lbs.
    c. 9.66 lbs.   d. 10.16 lbs.

65. $5\frac{1}{2} - 2\frac{3}{4} =$

    a. $3\frac{3}{4}$ b. $2\frac{3}{4}$ c. $2\frac{1}{4}$ d. $-2\frac{1}{4}$

66. If 420 is decreased by 13% of itself, what is the result?

    a. 126   b. 355.4   c. 365.4   d. 366.4

67. What is the area of a square, in square meters, with a side measuring 10 centimeters?

    a. $1m^2$ b. $0.4m^2$ c. $0.1m^2$ d. $0.01m^2$

68. Two common sources for finding out about new movies are television reviews and newspaper reviews. Thirty percent of American adults find out about new movies from television reviews but not newspaper reviews, while 5% find out about new movies from both television and newspaper reviews. What is the probability that a randomly selected American adult did not find out about a new movie from television?

    a. 0.25   b. 0.65   c. 0.7   d. 0.75

69. Select the statement that is the negation of the statement: All autumn days are cool.

    a. Some autumn days are not cool.
    b. Some autumn days are cool.
    c. No autumn days are cool.
    d. If it is not an autumn day, then it is not cool.

70. Select the statement that is not logically equivalent to: It is not true that both Othello and Walden Pond were written by Shakespeare.

    a. If Shakespeare wrote Othello, he did not write Walden Pond.
    b. Shakespeare did not write Othello or he did not write Walden Pond.
    c. If Shakespeare wrote Walden Pond, he did not write Othello.
    d. Shakespeare did not write Othello and he did not write Walden Pond.

71. The circle graph below shows how a student's weekday is spent. What category contains 25% of the weekday?

    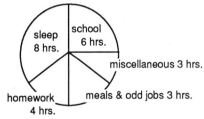

    a. School   b. Sleep
    c. Homework   d. Miscellaneous

72. The accompanying line graph indicates the temperature from Monday through Friday. Which of the following is true?

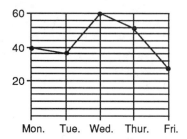

   a. There is exactly one modal temperature for the five days.
   b. The mean temperature for the five days is 42.2 degrees.
   c. The median temperature for the five days is 60 degrees.
   d. The percent increase in temperature from Tuesday to Wednesday is approximately 67%.

73. Select the rule of logical equivalence which directly (in one step) transforms statement i into statement ii.
   i. If a person studies, then that person does not fail.
   ii. If a person fails, then that person does not study.
   a. $(p \rightarrow \sim q) \equiv (q \rightarrow \sim p)$
   b. $(p \rightarrow \sim q) \equiv (\sim p \rightarrow q)$
   c. $\sim(p \rightarrow \sim q) \equiv (p \wedge q)$
   d. $(p \rightarrow \sim q) \equiv (\sim p \vee \sim q)$

74. $\frac{7}{14} =$
   a. 0.50% b. 20%   c. $33.\overline{3}\%$ d. 50%

75. A community offers homes in five different models, each available in four color schemes, each with or without a screened porch, and each with or without a pool. How many buying options are available?
   a. 13   b. 80   c. 120   d. 160

76. Select the conlusion that will make the following argument valid.
   If all people take exams honestly, then no supervision is needed. Some supervision is needed.
   a. Some people take exams honestly.
   b. If no supervision is needed, then all people take exams honestly.
   c. No people take exams honestly.
   d. Some people do not take exams honestly.

77. Five cans of juice cost $3. Let y be the cost of 11 cans of the same juice. Select the correct proportion reflecting the given conditions.
   a. $\frac{5}{3} = \frac{y}{11}$ b. $\frac{5}{3} = \frac{11}{y}$ c. $\frac{3}{5} = \frac{11}{y}$ d. $\frac{5}{11} = \frac{y}{3}$

78. Choose the correct solution set for the system of linear equations.
   $$6x + 5y = 9$$
   $$5x + 2y = 14$$
   a. $\{(-3, 4)\}$ b. $\{(-4, \frac{33}{5})\}$
   c. $\{(4, -3)\}$ d. $\emptyset$

79. After several tryouts, 20% of a football squad was discharged. The coach then had 32 players. How many players were on the squad at first?
   a. 80   b. 40   c. 39.   d. 26

80. Which one of the following would not be used to describe the amount of water in an aquarium?
   a. cubic meters   b. gallons
   c. liters   d. square feet

81. Round 27.349 cm^3 to the nearest tenth of a cubic centimeter.
   a. 20 cm^3 b. 27.3 cm^3
   c. 27.4 cm^3 d. 30 cm^3

82. In the accompanying figure, round the measure of AB to the nearest $\frac{1}{4}$ inch.

   a. $1\frac{1}{4}$ inch
   b. $1\frac{1}{2}$ inch
   c. $1\frac{7}{8}$ inch
   d. 1 inch

83. Select the statement below which is logically equivalent to: Gene is an actor or a musician.
   a. If Gene is an actor, then he is not a musician.
   b. If Gene is not an actor, then he is a musician.
   c. It is false that Gene is not an actor or not a musician.
   d. If Gene is an actor, then he is a musician.

84. Ten percent of the pens that are made by a ballpoint pen manufacturer leak. If two pens are randomly selected with replacement, find the probability that neither pen leaks.
   a. $\frac{1}{100}$
   b. $\frac{19}{100}$
   c. $\frac{81}{100}$
   d. $\frac{99}{100}$

85. The table below indicates scores on a 100-point exam and corresponding percentile ranks.

Score	Percentile Rank
90	99
80	87
70	72
60	49
50	26
40	8
30	1

What percentage of examinees scored between 60 and 80?
   a. 72%
   b. 51%
   c. 38%
   d. 13%

86. The table below shows the distribution of the number of children in families of a community.

Number of Children	Proportion of Families
0	.35
1	.20
2	.25
3	.10
4	.08
5	.02

Which one of the following is true?
   a. The modal number of children per family is 2.
   b. The median number of children per family is 0.5.
   c. The mean number of children per family is 1.22.
   d. None of the above is true.

87. A room measures 12 feet by 15 feet. The entire room is to be covered with tiles that measure 3 inches by 2 inches. If the tiles are sold at 10 for 30¢, what will it cost to tile the room?
   a. $10.80 b. $108 c. $129.60 d. $1,296

88. If a person reads extensively, that person has an excellent vocabulary. If a person studies mathematics, that person is logical. Trisha has an excellent vocabulary, but is not logical. What can be validly concluded about Trisha?
   i. Trisha reads extensively.
   ii. Trisha does not study mathematics.
   a. i only
   b. ii only
   c. Both i and ii
   d. Neither i nor ii

89. Two representatives are chosen without replacement from a group of 40 people consisting of 20 men and 20 women. What is the probability that the two representatives selected are either both men or both women?
   a. $\frac{20}{40} \cdot \frac{19}{39}$
   b. $\frac{19}{39}$
   c. $2 \cdot \frac{20}{40} \cdot \frac{19}{40}$
   d. $\frac{2}{40} + \frac{2}{40}$

Appendix

90. Select the statement below which is not logically equivalent to: Cassandra studies physics
or calculus.
    a. If Cassandra studies physics, then she does
       not study calculus.
    b. If Cassandra does not study calculus, then
       she studies physics.
    c. If Cassandra does not study physics, then
       she studies calculus.
    d. It is false that Cassandra does not study
       physics and does not study calculus.

91. The tens digit of a two-digit number is 3 more
than the units digit. If the number itself is 12
times the units digit, what equation can be used
to find t, the tens digit?
    a. $10t + (t - 3) = 12$
    b. $10(t + 3) + (t - 3) = 12$
    c. $10t + (t - 3) = 12(t - 3)$
    d. $10(t + 3) + (t - 3) = 12t$

92. The lengths in feet of two sides of a rectangle
are consecutive even integers. If ten more than
half the length of the shorter side is increased
by twice the length of the longer side, the result
is 59 feet. If x represents the length of the
shorter side, select the equation that can be
used to find the area of the rectangle.
    a. $\frac{1}{2}x + 10 + 2(x + 1) = 59$
    b. $\frac{1}{2}x + 10 + 2x + 1 = 59$
    c. $x(x + 2) = 59$
    d. $10 + \frac{1}{2}x + 2(x + 2) = 59$

93. Because of the lake shown below, a road cannot directly connect B to C, but must instead
connect B to A and then A to C. How long is
the road from B to A
and then from A to C?

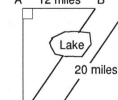

    a. 28 miles
    b. 29 miles
    c. 30 miles
    d. 32 miles

94. Find the area of the shaded region, the region
outside the isosceles right triangle whose legs
measure 1 foot each and inside the quadrilateral whose parallel sides measure 10 feet and 6
feet and whose height is 4 feet.

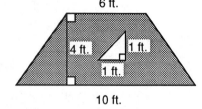

    a. 63.5 ft^2
    b. 63 ft^2
    c. 31.5 ft^2
    d. 31 ft^2

95. Let A = {1, 2}, B = {2, 3, 4} and
U = {1, 2, 3, 4, 5}. Select the choice
corresponding to (A ∩ B)'.
    a. {2}   b. {5}
    c. {3, 4}   d. {1, 3, 4, 5}

96. Which option gives the conditions which
correspond to the shaded region of the graph
shown below?

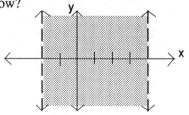

    a. x > -2 and x > 4   b. -2 > x > 4
    c. -2 < x < 4   d. y > -2 and y < 4

97. A person purchases eight gifts. The least
expensive gift costs $6 and the most expensive
gift costs $14. Which of the following is a
reasonable estimate of the total amount spent
on the eight gifts?
    a. $50   b. $92   c. $112   d. $126

98. Construct the truth table for (p→ q) ∧~q if the
statements p and q occur in the following
order:

p	q
T	T
T	F
F	T
F	F

Which one of the following represents the final column of truth values in the table?

a. F   b. F   c. F   d. T
   T   F   F   T
   F   F   T   T
   T   T   T   T

99. To be eligible for a scholarship, a student must have a grade point average of at least 3.6 and an activity index that exceeds .825. Marsha has a grade point average of 3.67 and an activity index of .83. Ethel has a grade point average of 3.7 and an activity index of .825. Paul has a grade point average of 3.8 and an activity index of .83. Who is not eligible for the scholarship?

a. Marsha only   b. Ethel only
c. Paul only
d. Both Marsha and Paul

100. The box shown below, with length A, width B and height C, is constructed with no top. What is its surface area?

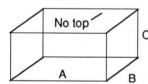

a. 5ABC   b. AC + 2BC + 2AB
c. 2AC + 2BC + 2AB
d. 2AC + 2BC + AB

101. The accompanying figure consists of a square with an equilateral triangle and a semicircle attached. One side of the square measures 4B linear units. What is the distance around this figure?

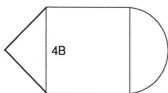

a. $16B + 4\pi B$ b. $16B + 2\pi B$
c. $24B + 4\pi B$ d. $24B + 2\pi B$

102. In how many ways can first, second and third prize be awarded to 5 people if no person can be given more than one prize?

a. 125   b. 120   c. 60   d. 10

103. Seven cards, numbered 1 through 7, are put into a box. If one card is randomly selected, what is the probability of selecting an even number or a number greater than 5?

a. $\frac{2}{7}$ b. $\frac{3}{7}$ c. $\frac{4}{7}$ d. $\frac{5}{7}$

104. The graph below depicts the number of fish caught (o) and the number of lobsters caught (x) by a group of six people on a number of outings in the Florida Keys.

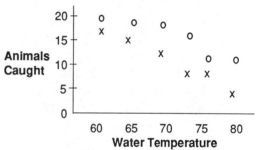

Which one of the following is true?

a. Increased water temperature was responsible for the decrease in the number of fish and lobsters that were caught.

b. There is an association between water temperature and animals caught only for the lobsters.

c. Approximately 10 lobsters were caught when the water temperature was 60 degrees.

d. Fewer fish and lobsters tend to be caught as the water temperature increases; a stronger relationship exists for the lobsters than for the fish.

105. Select the statement that is the negation of the statement: If it is cold, we will not go swimming.
    a. If it is not cold, then we will go swimming.
    b. It is cold and we do not go swimming.
    c. If we go swimming, then it is not cold.
    d. It is cold and we will go swimming.

106. Select the conclusion that will make the following argument valid: If vitamin C is effective, then I will not catch a cold. If I do not catch a cold, then I will not miss exercising.
    a. If vitamin C is not effective, then I will miss some exercising.
    b. If vitamin C is effective, then I will not miss exercising.
    c. If I do not catch a cold, then vitamin C was effective.
    d. If vitamin C is effective, then I will miss some exercising.

107. The table below shows the distribution of incomes in a community.

Income Level	Percent of People
$0 - 9,999	2
10,000 - 14,999	12
15,000 - 19,999	21
20,000 - 24,999	39
25,000 - 34,999	20
35,000 - 49,999	4
50,000 and over	2

What percentage of people in the community have incomes of at least $35,000?
a. 6%   b. 26%   c. 74%   d. 94%

108. The graph below represents the distribution of ratings (1 = poor, 2 = fair, 3 = average, 4 = good, 5 = excellent) for a movie that recently opened.

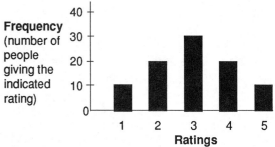

Select the statement that is true about the distribution of ratings.
    a. The mean, median, and mode have the same value.
    b. The mode is greater than the mean.
    c. The median is greater than the mode.
    d. The mode is less than the mean.

109. A recent survey indicated that 40% of the residents of Key Largo own boats. Of these, 5% use their boat four or more times per week. What is the probability that a randomly selected Key Largo resident will have a boat that is used four or more times per week?
    a. 0.45   b. 0.35   c. 0.2   d. 0.02

110. What is the volume in centiliters of a 3.25 liter bottle?
    a. 3.25 cl b. 32.5 cl c. 325 cl   d. 3250 cl

111. In the accompanying diagram $\overline{DA} \parallel \overline{CB}$ and $\overline{CD} \parallel \overline{BA}$. Lower-case letters indicate the measures of the various angles. (x = m < BCD, etc.)

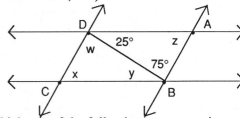

Which one of the following statements is true?
    a. z = 75° b. w = z   c. w = 25° d. x = 80°

112. 105 is 125% of what number?
 a. 8.4   b. 84   c. 86   d. 131.25

113. $(-0.73) + (0.296) =$
 a. 1.026 b. 0.434 c. -1.026 d. -0.434

114. What is the perimeter of a rectangle with dimensions of 50 centimeters and 6 meters?
 a. 13 cm b. 112 cm c. 300 cm d. 13 m

115. If Jason drives between 40 mph and 55 mph for 4 to 5 hours, estimate the distance traveled.
 a. 160 miles   b. 165 miles
 c. 217 miles   d. 265 miles

116. The electrical resistance of a wire varies directly as the length of the wire and inversely as the square of the wire's diameter. An 8-foot wire whose diameter is 1 inch has a resistance of 8 ohms. What is the resistance of a wire whose length is 6 feet and whose diameter is 2 inches?
 a. 1.5 ohms   b. 2.5 ohms
 c. 5 ohms   d. 12 ohms

117. The amount of floor surface that can be covered by the contents of a gallon of wood finishing liquid is given by which measure?
 a. square feet   b. cubic feet
 c. quarts   d. liters

118. A lending library charges a fee of 60¢ for the first day a book is borrowed and 30¢ for each additional day. A borrower paid $4.80 for loaning a book. For how many days was the book borrowed?
 a. 13   b. 14   c. 15   d. 16

119. Which one of the following correctly names a right triangle with one of its angles measuring 45°?
 a. scalene   b. isosceles
 c. equilateral   d. obtuse

120. $5\frac{4}{7} \div 1\frac{2}{7} =$
 a. $\frac{351}{49}$ b. $4\frac{1}{3}$ c. $4\frac{3}{7}$ d. $4\frac{1}{2}$

121. Which pairs of angles in the accompanying diagram are complementary?

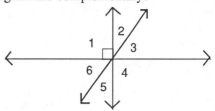

 a. 3 and 5   b. 1 and 4
 c. 2 and 5   d. 1 and 6

122. $\frac{6(3x-5y)}{2} - \frac{2(4x)^2}{2x} - (4x - 3y) =$
 a. $-11x - 12y$ b. $-11x - 18y$
 c. $x - 12y$ d. none of these

123. $\frac{6^6}{2^3} =$
 a. 3^2 b. 3^3
 c. $\frac{6+6+6+6+6+6}{2+2+2}$ d. $\frac{6 \cdot 6 \cdot 6 \cdot 6 \cdot 6 \cdot 6}{2 \cdot 2 \cdot 2}$

124. Identify the missing term in the following geometric progression.
$$-16, 4, -1, \frac{1}{4}, \text{---}, \frac{1}{64}$$
 a. $-\frac{1}{4}$ b. $-\frac{1}{8}$ c. $-\frac{1}{12}$ d. $-\frac{1}{16}$

125. Select the statement that is logically equivalent to: It is not true that both Jacksonville and Australia are cities.
 a. Jacksonville is not a city and Australia is not a city.
 b. If Australia is a city, then Jacksonville is a city.
 c. If Jacksonville is a city, then Australia is not a city.
 d. Jacksonville is a city or Australia is not a city.

126. Given that $\overline{BC} \parallel \overline{DE}$ and $\overline{AD} \perp \overline{DE}$, which one of the following statements is true for the figure shown? ($m\angle BAC = x$, etc.)

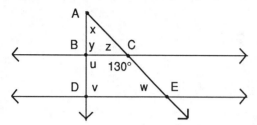

   a. $x = z$ b. $x = v$
   c. $\angle CBD$ and $\angle CED$ are supplementary angles.
   d. $w = 50°$

127. Look for a common relationship between the numbers in each pair. Then identify the missing term.

   $(25, 5), (1, 1), (\frac{4}{9}, \frac{2}{3}), (0.16, 0.4), (4, \underline{})$

   a. 4   b. 2   c. $\frac{4}{3}$ d. $\frac{1}{2}$

128. Select the statement that is logically equivalent to: If it is July, it does not snow.
   a. If it snows, it is not July.
   b. If it is not July, it snows.
   c. If it does not snow, it is July.
   d. If it does not snow, it is not July.

129. If $b < 0$, then $b^2 < ab + b$ is equivalent to which of the following?
   a. $b < a + 1$ b. $b > -a - 1$
   c. $b < -a - 1$ d. $b > a + 1$

130. Choose the equation equivalent to the following: $4x - 7 = 3x + 4$
   a. $4x - 3 = 3x$ b. $7x - 7 = 4$
   c. $4x - 11 = 3x + 4$ d. $x - 7 = 4$

Appendix Test Answers with Brief Explanations

1. $A \cap C$ is represented by the region where A and C intersect. This region is separate from circle B, so that elements in $A \cap C$ are not in set B.   [c]

2. $3(t - 1) = 2[t - 3(t - 2)]$
   $3t - 3 = 2[t - 3t + 6]$
   $3t - 3 = 2[-2t + 6]$
   $3t - 3 = -4t + 12$
   $7t - 3 = 12$
   $7t = 15$
   $t = \frac{15}{7}$ [c]

3. P(both are good) = P(first is good) • P(second is good)
   $= \frac{6}{10} \cdot \frac{5}{9} = \frac{1}{3}$

   P(at least defective) = 1 – P(both are good)
   $= 1 - \frac{1}{3} = \frac{2}{3}$ [c]

4. In option d, just because "clocks" and "time-measuring devices" are drawn inside the circle representing "things designed to inform people of the time," this does not mean that the "clocks" circle must be inside the "time-measuring devices" circle.   [d]

5. i. $5(-3) + 1 > -13$ ii. $2(-3)^2 + 7(-3) = -3$
      $-14 > -13$, false $18 + (-21) = -3$
                                       $-3 = -3$, true

   iii. $\frac{4}{3}(-3) - 4 = 0$
      $-4 - 4 = 0$
      $-8 = 0$, false   -3 is the solution to ii only   [b]

6. Area of shaded region = Area large semicircle – Area two smaller semicircles
   $= \frac{1}{2}\pi \cdot 10^2 - (\frac{1}{2}\pi \cdot 5^2 + \frac{1}{2}\pi \cdot 5^2)$
   $= 50\pi - (\frac{25\pi}{2} + \frac{25\pi}{2}) = 50\pi - 25\pi = 25\pi$ [a]

7. $-7x < -70$
   $\frac{-7x}{-7} > \frac{-70}{-7}$
   $x > 10$ [d]

8. $x^2 + 4x - 4 = 0$ $(a = 1, b = 4, c = -4)$ for $x = \frac{-b \pm \sqrt{b^2 - 4ac}}{2a}$

   $x = \frac{-4 \pm \sqrt{16 - 4(1)(-4)}}{2(1)} = \frac{-4 \pm \sqrt{32}}{2} = \frac{-4 \pm \sqrt{16 \cdot 2}}{2} = \frac{-4 \pm 4\sqrt{2}}{2} = -2 \pm 2\sqrt{2}$ [b]

9. $\text{GPA} = \frac{(10)(4) + (7)(3) + (5)(2) + (0)(1) + (0)(0)}{10 + 7 + 5} = \frac{71}{22} \approx 3.2$ [c]

10. Frank Martinez: not eligible because of the weight of his dog.
    Dr. Selma Rubin: not eligible because of 3 children.
    Raboids: not eligible because of 2 pets. [d]

11. $(2 \cdot \frac{1}{10}) + (2 \cdot \frac{1}{10^6}) = 0.200002$ [c]

12. $2[\frac{1}{2} - \frac{1}{2}(10 - 2 \cdot 3)] = 2[\frac{1}{2} - \frac{1}{2}(10 - 6)] = 2[\frac{1}{2} - \frac{1}{2}(4)] = 2[\frac{1}{2} - \frac{4}{2}] = 2(-\frac{3}{2}) = -3$
    [c]

13. $\frac{1}{4} \cdot 12c + 6c \div \frac{1}{3} - c^2 + 4c = 3c + 18c - c^2 + 4c = 25c - c^2$ [a]

14. Scores in order: 1, 3, 3, 4, 5, 6, 6, 7, 7, 7
    Position of median $= \frac{n+1}{2} = \frac{10+1}{2} = \frac{11}{2} = 5.5$

    median = average of scores of positions five and six $= \frac{5+6}{2} = 5.5$ [c]

15. p: Sarah is in San Francisco.
    q: Sarah is in California.
    Both $p \rightarrow q$ and $\sim p \rightarrow \sim q$ (converse and inverse) are not equivalent
    to $p \rightarrow q$. The converse appears in option b. [b]

16. p: A Supreme Court nominee receives majority approval.
    q: The nominee is a judge on the Supreme Court.
    $\sim(p \vee \sim q) \equiv \sim p \wedge q$ [b]

17. $\frac{\text{number of imports}}{\text{total number excluding Ford and Chrysler}} = \frac{23}{46 + 2 + 23} = \frac{23}{71}$ [a]

18. All sides of a rhombus are equal. [d]

19. $\sqrt{36} + \sqrt{6} = 6 + \sqrt{6} = \sqrt{6} + 6$ [b]

20. $\frac{12}{\sqrt{6}} = \frac{12}{\sqrt{6}} \cdot \frac{\sqrt{6}}{\sqrt{6}} = \frac{12\sqrt{6}}{6} = 2\sqrt{6}$ [a]

21. Group 1: Mean = 7, median = 7, mode = 7, so x can be the mean or
    median or mode.
    Group 2: Mean = 7, median = 6.5, no mode; since x = 7, it must be the
    mean, and there is no need to consider group 3.
    x = the mean $= \frac{1.1 + 2.3 + 3.8 + 7.2 + 8.1}{5} = \frac{22.5}{5} = 4.5$ [c]

22. Graph of $x \leq -1$: The line of $x = -1$ (parallel to the y-axis) and the portion
    of the plane to the left of this vertical line. Graph of $y \geq 0$: The x-axis
    $(y = 0)$ and the half-plane above the x-axis. [c]

23. Eliminate c and d since the graphs of $x = -2$ and $y = 2$ are lines parallel to the y-axis and x-axis respectively, and no such dashed lines appear in the figure. The line of $y - x = 2$ has an x-intercept (set $y = 0$) of -2 and a y-intercept (set $x = 0$) of 2, which is verified by the diagram. The test point $(0, 0)$ must satisfy the correct inequality since $(0, 0)$ lies in the shaded region. $y - x < 2$ becomes $0 - 0 < 2$, which is true.   [a]

24. $y > y - (16 - 2y)$ or $y > y - 16 + 2y$ or $y > 3y - 16$ or $-2y > -16$
    Therefore, $y < 8$ [b]

25. p: It is hot
    q: Some municipal pool is open.
    $\sim(p \rightarrow q)$ is $p \wedge \sim q$. Since the negation of "some" is "no," the negation is:
    It is hot and no municipal pool is open.   [c]

26. $.0060 \div .12 = .60 \div 12 = .05$ and $.05 = 5\%$ [c]

27. $108.7\% = 1.087 = 1\frac{87}{1000}$ [d]

28. Let C = hypotenuse of smaller triangle.
    $C^2 = 3^2 + 4^2$; $C^2 = 25$; $C = 5$
    Pythagorean Theorem in larger triangle:
    $5^2 + 12^2 = (AB)^2$; $169 = (AB)^2$; $AB = 13$
    13 cm $= 13(\frac{1}{100}$ m$) = 0.13$ m   [a]

29. The pattern in all three examples is to multiply the first two numbers and divide this product by the third number.   [d]

30. $P = C + MC = 12 + (.4)(12) = 12 + 4.8 = 16.8$ [d]

31. Since $x - y$ is not equal to $y - x$ (although $x - y = -y + x$), option c is not true for all real numbers.   [c]

32. 15 is divisible by 5 and $45 = (15)(3)$, so 15 is also a factor of 45.   [b]

33. 1980: $(.15)(20,000) = 3000$
    1981: $(.2)(25,000) = 5000$
    percent increase: $\dfrac{\text{increase}}{\text{original amount}} = \dfrac{2000}{3000} = \dfrac{2}{3} = .66\frac{2}{3} = 66\frac{2}{3}\%$ [b]

34. Class 1: $\frac{1}{5}$ absent $\frac{1}{5} \cdot \frac{40}{1} = 8$
    Class 2: $\frac{1}{8}$ absent $\frac{1}{8} \cdot \frac{40}{1} = 5$
    $8 + 5 = 13$ students absent   [b]

35. $f(-2) = (-2)^3 - 6(-2)^2 - (-2) + 2$
    $= -8 - 6(4) + 2 + 2$
    $= -8 - 24 + 2 + 2$
    $= -28$ [b]

36. The mode is 45%. The few scores of 95% or better pull the mean up in this (extreme) direction. The median is either 45% or the average of 45% and 75% (60%). Mode < Mean, Median < Mean   [b]

37. $m\angle ABC = 180° - (90° + 30°) = 180° - 120° = 60°$ Option d is false. [d]

38. $8x^2 - 14x + 3 = (2x - 3)(4x - 1)$ [a]

39. $z(30) = \dfrac{\text{Score} - \text{Mean}}{\text{Standard Deviation}} = \dfrac{30 - 60}{20} = -\dfrac{30}{20} = -1.5$

   $z(85) = \dfrac{85 - 60}{20} = \dfrac{25}{20} = 1.25$ From the table: $.433 + .394 = .827$ [a]

40. $(7.3 \cdot 10^{-4}) \cdot (8.2 \cdot 10^6) = 59.86 \cdot 10^2 = 5.986 \cdot 10 \cdot 10^2 = 5.986 \cdot 10^3$
                                                         [a]

41. November: $(40)(\$30) = \$1,200$
   December: $(40 - .1 \cdot 40)(30 + .2 \cdot 30) = (36)(\$36) = \$1,296$
   $\$96$ increase   [a]

42. Tax: $(.05)(\$60) = \$3.00$
   Save 25% of $3 = $(.25)(3) = 0.75$ [a]

43. With approximately 34% of the scores between the mean and 1 standard deviation above the mean, and 34% of the scores between the mean and 1 standard deviation below the mean, option b is true.   [b]

44. Using Euler circles, physicists and biologists are shown inside the scientists' circle, and the scientists' circle is inside the college graduates' circle. Thus, the biologists' circle falls inside the college graduates' circle.
                                                         [b]

45. Using Euler circles, talkers during lecture are inside the rude circle. The lovable people circle (containing Eubie) is separate from both these circles. We conclude that no lovable people talk during lecture. We cannot conclude that some lovable people talk during lecture.   [c]

46. P(over 30 or business) = P(over 30) + P(business) – P(over 30 $\cap$ business)
   $= \dfrac{15}{40} + \dfrac{11}{40} - \dfrac{5}{40}$ [b]

47. Area $= (40)(30) + (10)(10) = 1300 \text{ ft}^2$
   Bundles $= \dfrac{1300}{100} = 13$
   Cost $= (13)(\$70) + \$275 = \$1,185$ [c]

48. $\binom{5}{2} \cdot \binom{4}{2} = \dfrac{5!}{2!\,3!} \cdot \dfrac{4!}{2!\,2!} = 10 \cdot 6 = 60$ [a]

49. Consider option d. The form is:
   $p \rightarrow q$
   $\underline{q\qquad}$
   $\therefore p$, which is invalid (fallacy of the converse)   [d]

50. p: There is a mathematics requirement.
q: There is an English requirement.
r: College bookstores do a great business.

$(p \wedge q) \rightarrow r$ (given) $\sim p \vee \sim q$

$\underline{\sim r}$ (given) $\underline{q}$ (given)

$\therefore \sim(p \wedge q)$ $\therefore \sim p$ (Tate has no math requirement.) [a]

51. $\dfrac{\text{Number of ways of visiting E first, D second, A last}}{\text{Total number of arrangements}} = \dfrac{1 \cdot 1 \cdot 3 \cdot 2 \cdot 1 \cdot 1}{6 \cdot 5 \cdot 4 \cdot 3 \cdot 2 \cdot 1} = \dfrac{1}{120}$
[b]

52. $V = \dfrac{4}{3}\pi \, r^3 = \dfrac{4}{3}\pi(3)^3 = \dfrac{4}{3}\pi \cdot \dfrac{27}{1} = 36\pi$ [b]

53. see answer to the right   [b]

54. $V(\text{cone}) = \dfrac{1}{3}\pi \, r^2 h = \dfrac{1}{3}\pi \, (3)^2(3) = 9\pi \text{ ft.}^3$

$V(\text{cylinder}) = \pi \, r^2 h = \pi \, (3)^2(3) = 27\pi \text{ ft.}^3$

$9\pi + 27\pi = 36\pi \text{ ft.}^3$ [b]

55. $V = LWH = (30\text{yd})(20 \text{ yd})(2 \text{ yd}) = 1200 \text{ yd}^3$

$\text{Trips} = \dfrac{1200}{10} = 120$

$\text{Cost} = (120)(\$12) = \$1,440$ [a]

56. $m\angle A = 135°$ (congruent alternate exterior angles)   [c]

57. Corresponding sides of similar triangles are proportional.

$\dfrac{CE}{CB} = \dfrac{DE}{AB}$ or $\dfrac{5}{5+4} = \dfrac{DE}{10}$ or $9DE = 50$ and $DE = \dfrac{50}{9} = 5\dfrac{5}{9}$ [c]

58. $m\angle C = 180° - 140° = 40°$
$m\angle B = 40°$ (isosceles triangle)
$m\angle D = 180° - 80° = 100°$
The triangle is not a right triangle, so the
Pythagorean Theorem does not apply.   [d]

59. $\dfrac{1}{10} + \dfrac{1}{3} + \dfrac{1}{5} = \dfrac{3 + 10 + 6}{30} = \dfrac{19}{30}$

Fraction left $= 1 - \dfrac{19}{30} = \dfrac{11}{30} = .36\dfrac{2}{3} = 36\dfrac{2}{3}\%$ [d]

60. $2 - 5(a - 3b) = 2 - 5a + 15b$ [c]

61. $-\dfrac{5}{6} = -0.8\overline{3}$ Since $-0.8\overline{3}$ is farther to the right on the number line than
-0.84, $-\dfrac{5}{6} > -0.84$ [c]

62. The pattern is that the number of triangles in the interior is the square
of the base.
$12^2 = 144$ [d]

63. Prime numbers greater than 17 and less than 29: 19, 23
    Both yield a remainder of 3 when divided by 4. [b]

64. $\dfrac{\text{Planet A}}{\text{Planet B}}$: $\dfrac{7}{11.27} = \dfrac{6}{x}$ or $7x = (11.27)(6)$ or $7x = 67.62$ or $x = 9.66$ [c]

65. $5\frac{1}{2} = 5\frac{2}{4} = 4\frac{6}{4}$
    $-2\frac{3}{4} = -2\frac{3}{4} = -2\frac{3}{4}$
    $\phantom{-2\frac{3}{4} = -2\frac{3}{4} = }2\frac{3}{4}$ [b]

66. $420 - (.13)(420) = 420 - 54.6 = 365.4$ [c]

67. $10 \text{ cm} = 10\left(\frac{1}{100}\text{ m}\right) = \frac{1}{10}\text{ m}$ Area $= \left(\frac{1}{10}\right)\left(\frac{1}{10}\right) = \frac{1}{100}\text{ m}^2$ [d]

68. P(television) $= 0.3 + 0.05 = 0.35$ p(not-television) $= 1 - 0.35 = 0.65$ [b]

69. The negation of "all" is "some...not." [a]

70. p: Shakespeare wrote *Othello*.
    q: Shakespeare wrote *Walden Pond*.
    $\sim(p \wedge q)$ is not equivalent to $\sim p \wedge \sim q$. [d]

71. 25% of the total $= (.25)(24) = 6$, school [a]

72. M(40), T(36), W(60), Th(52), F(28)
    No temperature occurs most frequently, so there is no mode.
    Mean $= \dfrac{40 + 36 + 60 + 52 + 28}{5} = 43.2$
    Scores in order: 28, 36, 40, 52, 60   median $= 40$
    T to W: $\dfrac{\text{increase}}{\text{original}} = \dfrac{24}{36} = \dfrac{2}{3} \approx .67 = 67\%$ [d]

73. p: A person studies.   q: A person fails. $(p \rightarrow \sim q) \equiv (q \rightarrow \sim p)$ [a]

74. $\dfrac{7}{14} = \dfrac{1}{2} = 0.5 = 50\%$ [d]

75. $5 \cdot 4 \cdot 2 \cdot 2 = 80$ [b]

76. p: All people take exams honestly. q: No supervision is needed.
    $p \rightarrow q$
    $\underline{\sim q}$ (The negation of "no" is "some.")
    $\therefore \sim q$ (Some people do not take exams honestly)
     (The negation of "all" is "some...not.") [d]

77. $\dfrac{\text{cans}}{\text{cost}}$: $\dfrac{5}{3} = \dfrac{11}{y}$ [b]

78. $6x + 5y = 9$ Eliminate y by multiplying the $-12x - 10y = -18$
    $5x + 2y = 14$ equations by -2 and 5 respectively. $\underline{25x + 10y = 70}$

    $$\text{Add:} \quad 13x \qquad = 52$$
    $$x \quad = \quad 4$$

    $6x + 5y = 9$ becomes $6(4) + 5y = 9$ or $5y = -15$ or $y = -3$ $\{(4, -3)\}$ [c]

79. x: players originally on the team

    $x - .2x = 32$ simplifies to $.8x = 32$ or $x = \frac{32}{8} = 40$ [b]

80. Volume cannot be measured in square feet. Square units are used to measure area.   [d]

81. $27.349 \text{ cm}^3 \approx 27.3 \text{ cm}^3$ [b]
    $\smile$ less than 5

82. The answer is either $1\frac{1}{4}$ or $1\frac{1}{2}$, and since the point B is more than midway between these options, the answer is $1\frac{1}{2}$.   [b]

83. p: Gene is an actor.   q: Gene is a musician.

    $p \vee q$ is logically equivalent to $\sim p \rightarrow q$.   [b]

84. P(does not leak) $= 1 - 0.1 = 0.9$
    P(neither leaks) $=$ P(first does not leak) $\bullet$ P(second does not leak)
    $$= (0.9)(0.9) = 0.81 = \frac{81}{100}$$ [c]

85. $87\% - 49\% = 38\%$ [c]

86. Mode $= 0$ (score with greatest proportion)
    Median $= 1$
    Mean $= \frac{(0)(35) + (1)(20) + (2)(25) + (3)(10) + (4)(8) + (5)(2)}{100} = \frac{142}{100} = 1.42$ [d]

87. Area of room$= (12 \bullet 12 \text{ in.})(15 \bullet 12 \text{ in.}) = 25{,}920 \text{ in}^2$
    Area of tile $= (3 \text{ in.})(2 \text{ in.}) = 6 \text{ in}^2$
    Number of tiles $= \frac{25{,}920}{6} = 4{,}320$
    Number of groups of 10 tiles $= \frac{4320}{10} = 432$
    Cost $= (432)(.3) = \$129.60$ [c]

88. p: Person reads extensively.
    q: Person has an excellent vocabulary.
    r: Person studies math.
    s: Person is logical.

    $p \rightarrow q$ $r \rightarrow s$
    $\underline{q \qquad}$ $\underline{\sim s \qquad}$
    $\therefore p$ (invalid) $\therefore \sim r$ (valid)

    (Trisha does not study math.)   [b]

89. P(both men) = P(man first) • P(man second) = $\frac{20}{40} \cdot \frac{19}{39} = \frac{19}{78}$

   Similarly, P(both women) = $\frac{19}{78}$

   P(both men or both women) = $\frac{19}{78} + \frac{19}{78} = \frac{38}{78} = \frac{19}{39}$ [b]

90. p: Cassandra studies physics.   q: Cassandra studies calculus.
   $p \vee q$ is not equivalent to $p \rightarrow \sim q$.
   Cassandra can be studying both physics and calculus.   [a]

91. t: tens digit. Then t – 3: units digit. And, 10t + (t – 3): the number.
   "Number is 12 times units digit" translates as:
        $10t + (t - 3) = 12(t - 3)$ [c]

92. Sides: x and x + 2
   (10 more than half the shorter side) + (twice the longer side)
        $(\tfrac{1}{2}x + 10) + 2(x + 2) = 59$ [d]

93. By the Pythagorean Theorem:
   $(AC)^2 + 12^2 = 20^2$ or $(AC)^2 + 144 = 400$ or $(AC)^2 = 256$ or AC = 16
   Thus, AB + AC = 12 + 16 = 28   [a]

94. Area of trapezoid – area of triangle = $\frac{1}{2} \cdot 4(10 + 6) - \frac{1}{2} \cdot 1 \cdot 1$

        $= 32 - \frac{1}{2} = 31.5$ [c]

95. $(A \cap B)' = \{2\}' = \{1, 3, 4, 5\}$ [d]

96. Equations of the dashed lines: x = -2 and x = 4
   Thus, x > -2 and x < 4. Equivalently: -2 < x < 4   [c]

97. Least amount spent: 8 • $6 = $48
   Most amount spent: 8 • $14 = $112
   Options a, c, and d are either too close to $48 and $112, or
   outside this range.   [b]

98.

p	q	$p \rightarrow q$	$\sim q$	$(p \rightarrow q) \wedge \sim q$
T	T	T	F	F
T	F	F	T	F
F	T	T	F	F
F	F	T	T	T

     [b]

99. Ethel's activity index (.825) does not exceed .825, so
   she is not eligible.   [b]

100. Surface area = area bottom + area two sides + area front and back
        = AB + 2BC + 2AC   [d]

101. Perimeter = Length of four line segments that measure 4B each plus semicircle's circumference

$$= 4(4B) + \frac{1}{2}\pi\,(4B) = 16B + 2\pi B \qquad [b]$$

102. $5 \cdot 4 \cdot 3 = 60$ [c]

103. P(even or greater than 5)

$$= P(even) + P(greater\ than\ 5) - P(even \cap greater\ than\ 5)$$
$$= P(even) + P(greater\ than\ 5) - P(\{6\})$$
$$= \frac{3}{7} + \frac{2}{7} - \frac{1}{7} = \frac{4}{7} \qquad [c]$$

104. As one variable increases (water temperature), the other variables decrease, but the degree of association between variables does not mean that changes in one cause changes in the other.   [d]

105. p: It is cold.   q: We will not go swimming.

$\sim(p \to q)$ is $p \wedge \sim q$, which translates as: It is cold and we will go swimming.   [d]

106. p: Vitamin C is effective.   q: I will not catch a cold.
r: I will not miss exercising.

$$p \to q$$
$$\underline{q \to r}$$

$\therefore p \to r$ (If vitamin C is effective, then I will not miss exercising.)   [b]

107. Percent of people with incomes of \$35,000 or more = 4% + 2% = 6%   [a]

108. Mode = most frequent rating = 3   Median = middlemost rating = 3

Each rating of 4 is counterbalanced by a rating of 2, giving a rating of

$\frac{4+2}{2} = 3$. Similarly each 5 is counterbalanced by a 1. Thus, the mean

stays at 3. Mode = median = mean = 3.   [a]

109. $(0.4)(0.05) = 0.02$ [d]

110. Just as a meter contains 100 centimeters, a liter contains 100 centiliters.
3.25 liters = 3.25(100 cl) = 325 cl   [c]

111. $y = 25°$ and $w = 75°$ (congruent alternate interior angles)
$x = 180° - (25° + 75°) = 80°$ [d]

112. Let x = the number.

"125% of x = 105" translates as: $1.25x = 105$ or $x = \frac{105}{1.25} = 84$ [b]

113. The difference between 0.296 and 0.73 is:
$$\begin{array}{r} 0.730 \\ -0.296 \\ \hline 0.434 \end{array}$$

$(-0.73) + (0.296) = -0.434$ [d]

114. $P = 2L + 2W = 100 \text{ cm} + 12 \text{ m} = 1 \text{ m} + 12 \text{ m} = 13 \text{ m}$ [d]

115. Distance = rate • time
Minimum distance = $40 \cdot 4 = 160$ miles
Maximum distance = $55 \cdot 5 = 275$ miles [c]

116. R: resistance   L: length   D: diameter
$R = \dfrac{KL}{D^2}$ or $8 = \dfrac{K(8)}{1^2}$ or $8 = 8K$ and $1 = K$.
Thus, $R = \dfrac{(1)L}{D^2} = \dfrac{6}{2^2} = \dfrac{6}{4} = 1.5$ [a]

117. Floor surface refers to area, measured in square units. [a]

118. x: number of days book was borrowed
$.6 + .3(x - 1) = 4.8$
$.6 + .3x - .3 = 4.8$
$.3 + .3x = 4.8$
$.3x = 4.5$
$x = \dfrac{4.5}{.3} = 15$ [c]

119. Angles measure 90°, 45°, and $90° - 45° = 45°$. With two congruent angles, the triangle has two sides of equal measure and is isosceles. [b]

120. $5\frac{4}{7} \div 1\frac{2}{7} = \frac{39}{7} \div \frac{9}{7} = \frac{39}{7} \cdot \frac{7}{9} = \frac{39}{9} = 4\frac{3}{9} = 4\frac{1}{3}$ [b]

121. 2 and 3 are complementary. Since 2 and 5 have equal measure (congruent vertical angles), 5 and 3 are complementary. [a]

122. $\dfrac{6(3x - 5y)}{2} - \dfrac{2(4x)^2}{2x} - (4x - 3y) = 3(3x - 5y) - \dfrac{16x^2}{x} - (4x - 3y)$
$= 9x - 15y - 16x - 4x + 3y$
$= -11x - 12y$ [a]

123. Exponents in numerator and denominator indicate repeated multiplication. [d]

124. Since a geometric progression indicates that each number (after the first) is obtained from the previous number by multiplying by a constant number, in this progression the constant is $-\frac{1}{4}$. This can be seen most clearly from 4 to -1. The missing number is $(\frac{1}{4})(-\frac{1}{4}) = -\frac{1}{16}$. [d]

125. p: Jacksonville is a city.
q: Australia is a city.
$\sim(p \wedge q)$ is equivalent to $\sim p \vee \sim q$.
$\sim(p \wedge q)$ is equivalent to $p \rightarrow \sim q$, which becomes:
If Jacksonville is a city, then Australia is not a city. [c]

126. $u = 90°$ and $v = 90°$. Since the sum of the measures of the interior angles of a quadrilateral is 360°,
$$w = 360° - (90° + 90° + 130°) = 50°.$$
Equivalently: $z = 50°$ and $z = w$ (congruent corresponding angles), so $w = 50°$.   [d]

127. The second number in each pair is the principal square root of the first number. $\sqrt{4} = 2$ [b]

128. p: It is July.
q: It does not snow.
$(p \rightarrow q) \equiv (\sim q \rightarrow \sim p)$
$\sim q \rightarrow \sim p$: If it snows, it is not July.   [a]

129. $b^2 < ab + b$
$\dfrac{b^2}{b} > \dfrac{ab + b}{b}$
$b > \dfrac{ab}{b} + \dfrac{b}{b}$
$b > a + 1$ [d]

130.
$$4x - 7 = 3x + 4$$
$$4x - 7 + (-3x) = 3x + 4 + (-3x)$$
$$x - 7 = 4$$
                                                    [d]

Answers for Problem Sets and Chapter Tests

Chapter 1

1. yes
2. no
3. yes
4. no
5. no
6. 12
7. $\frac{38}{7}$

1. a. No braces enclose the elements.
   b. No braces enclose the elements. In addition, at least the first three ele-ments of the set should be listed in order to determine the pattern for obtaining the remaining elements.
   c. Parentheses are not used to enclose the elements.
   d. Both braces must be used to enclose the elements.
   e. The elements in a set must be separated by commas.
2. a. finite set   b. infinite set
   c. infinite set
3. a. not well-defined
   b. well-defined
4. $3\frac{5}{6}$

1. a. true   b. true
2. a. true   b. true
   c. false
3. a. true   b. false
   c. true
4. yes
5. infinite
6. $\frac{13}{12} = 1\frac{1}{12}$

1. equal sets
2. equal sets
3. unequal sets
4. true
5. finite set
6. false
7. $\frac{8}{45}$

1. The set of days of the week.
2. The set of southern states that border the Gulf of Mexico.
3. The set of counting numbers that are greater than 20.
4. The set of multiples of 4
5. The set of counting numbers from 1 to 50.
6. no
7. true
8. $8\frac{7}{12}$

1. a. {11,12,13,...}
   b. {5,6,7,...,19}
2. infinite set
3. no
4. yes
5. no
6. The set of multiples of 7
7. true
8. $3\frac{4}{7}$

1. U = {January,June,July}
2. U = {a,e,i,o,u}
3. The set of all the multiples of 5.
4. {3,4,5,6,7}
5. no
6. $\frac{73}{20} = 3\frac{13}{20}$

1. a. {1,3,5,7} b. {1,2}
   c. {2,4,6,8}
2. a. false   b. true
   c. true   d. false
3. yes
4. infinite
5. false
6. $\frac{2}{21}$

1. a. yes   b. yes
   c. yes   d. no
2. ∅

587

3. a. $\{x\};\{y\};\{x,y\};\emptyset$
   b. $\{x\};\{y\};\{z\};\{x,y\};\{y,z\};\{x,z\};\{x,y,z\};\emptyset$
   c. $\{0\};\emptyset$
   d. $\{x\};\{y\};\{\emptyset\};\{x,y\};\{y,\emptyset\};\{x,\emptyset\};\{x,y,\emptyset\};\emptyset$
   e. $\emptyset$

4. a. Number of elements in a set. Number of subsets that can be formed.

0	1	2	3
1	2	4	8

   b. $2^4 = 16$; $2^5 = 32$

5. a. no proper subset b. one c. three
   d. seven e. fifteen
   f. thirty-one

6. $\emptyset$ represents the set that contains no elements (the null set); $\{\emptyset\}$ represents a set that contains one element enclosed in braces.

7. a. false b. true c. true d. false
   e. true 8. no

9. no. $\emptyset$ contains no elements at all.

10. This set is well-defined since the set contains no elements.

11. no

12. infinite

13. true

14. $\frac{91}{6} = 15\frac{1}{6}$

Chp. 1, Section 2, Problem Set 2.4

1. a. $A' = \{K\}$ b. $B' = \{F,N\}$
   c. $X' = \emptyset$ d. $C' = \{I\}$
   e. $Y' = \{F,I,N,K\} = U$ f. $(D')' = \{I,N\} = D$

2. $A' = \{2,4,6,\ldots\}$

3. The complement of the universal set is the null set. The complement of the null set is the universal set.

4. It is not possible for any set to contain fewer elements than the null set. The null set contains no elements at all.

5. The set of all humans not in Florida universities.

6. 4 subsets, 3 proper subsets

7. false

8. true

9. $\frac{36}{35} = 1\frac{1}{35}$

Chp. 1, Section 3, Problem Set 3.1

1. a. no. For every father, there does not necessarily exist one and only one son.
   b. no c. no d. yes e. no

2. One could attempt to establish a one-to-one correspondence between the two sets and see which set contains elements that cannot be matched.

3. yes $\{0, 1, 2, \ldots\}$
   $\{1, 2, 3, \ldots\}$

4. $A' = \{F,O,R\}$

5. The universal set

6. $(A')' = \{1,3\}$ Thus, $(A')' = A$

7. $\{4,5,6,\ldots\}$

8. $\frac{85}{52} = 1\frac{33}{52}$

Chp. 1, Section 3, Problem Set 3.2

1. no. If set $A = \{1,2\}$ and set $B = \{a,b\}$, then set A is equivalent to set B but $A \neq B$.

2. yes

3. yes

4. yes. They both contain three distinct elements.

5. no

6. $\{2\},\emptyset$

7. the null set

8. no

9. $\frac{5}{12}$

Chp. 1, Section 3, Problem Set 3.3

1. a. 5 b. 3 c. 1 d. 4
   e. 2 f. 20 g. 0 h. 1

2. The cardinal number refers to the number of elements in a set and not the elements themselves. Also, the elements of a set need not always be numbers.

3. Yes. If set A is equivalent to set B, then the number of elements in set A is equal to the number of elements in set B.

4. $n(A) = 50$

5. $n(A) = 0$

6. a. Elements in set A can be placed into a one-to-one correspondence with elements in set B.

    b. Every element in set B is in set A.

    c. 2n

7. yes

8. $\emptyset = A$

9. no

10. 0.625

Chp. 1, Section 4, Problem Set 4.1

1. a. {2}   b. {3}   c. $\emptyset$ d. $\emptyset$
    e. $\emptyset$ f. 2   g. 0

2. a. {6}   b. $\emptyset$ c. $\emptyset$

3. yes

4. $\emptyset$

5. a. The set of all Florida citizens who are employed.

    b. The set of all employed Florida citizens who are not registered Democrats.

    c. The set of all Florida citizens who are employed and registered Democrats.

6. 2

7. no

8. 5

9. {1}, {1,$\emptyset$}, {$\emptyset$},$\emptyset$

10. 5.437

Chp. 1, Section 4, Problem Set 4.2

1. a. {1,2,3,4,6}   b. {1,2,3,4,6}
    c. {1,2,3,4,6}   d. {1,2,3,4,6}
    e. 4   f. 3

2. a. {1,2,4,5,6,7,8}   b. {1,2,3,...}
    c. {3,6,9, ...}

3. a. true   b. true

4. a. The set of all fans of the Miami Dolphins or the Tampa Bay Buccaneers.

    b. The set of all fans of the Miami Dolphins or the Tampa Bay Buccaneers or the Dallas Cowboys.

5. yes

6. $\emptyset$

7. 24

8. no

9. $A \cap B = \{5,9\}; n(A \cap B) = 2$

10. no

11. 600

Chp. 1, Section 4, Problem Set 4.3

1. a. $X' = \{1,2,4,6,8,9,10\}$ $Y' = \{2,3,4,5,8,9,10\}$
       $X' \cap Y = \{1,6\}$
       $X \cup Y' = \{2,3,4,5,7,8,9,10\}$
       $X' \cap Y' = \{2,4,8,9,10\}$
       $X' \cup Y' = \{1,2,3,4,5,6,8,9,10\}$

    b. $X' = \{1,3,5,7,9,10\}$ $Y' = \{5,6,7,8,9,10\}$
       $X' \cap Y = \{1,3\}$
       $X \cup Y' = \{2,4,5,6,7,8,9,10\}$
       $X' \cap Y' = \{5,7,9,10\}$
       $X' \cup Y' = \{1,3,5,6,7,8,9,10\}$

2. a. $\emptyset$ b. U   c. $\emptyset$ d. U
    e. A   f. U   g. $\emptyset$ h. A

3. $A \cup B = \{1,2,3,...,10\}; n(A \cup B) = 10$

4. {1,3,4}

5. $\frac{9}{20}$

1. a. {14,16}
   b. {2,4,6,8,10,12,18,20}
   c. {2,6,8,12,14,16,20}
   d. {2,6,8,12,14,16,20}
   e. {2,8,20}   f. {6,12}
   g. {2,4,8,10,14,16,18,20}
   h. {4,6,10,12,14,16,18}
   i. {4,10,18}   j. 5   k. 3
2. a. $A \cup B = A$ or $A \cup B = B$;
      $A \cap B = A$ or $A \cap B = B$
   b. $A \cup B$ is the set containing the elements in
      set A together with the elements in set B;
      $A \cap B = \emptyset$
3. a. true   b. false   c. false   d. true
4. a. $(A \cup B)' = \{3\}$ and $A' \cap B' = \{3\}$
   b. $(A \cap B)' = \{2,3,4,5,6\}$ and
      $A' \cup B' = \{2,3,4,5,6\}$
5. a. true   b. true   c. false
6. a. $A \cup B = \{1,2,3,4,5,6,7,8\}$;
      $B \cup A = \{1,2,3,4,5,6,7,8\}$. Thus,
      $A \cup B = B \cup A$. The order in which the
      sets are combined under set union does not
      affect the result.
   b. $A \cap B = \{2,4,5,6\}$; $B \cap A = \{2,4,5,6\}$.
      Thus, $A \cap B = B \cap A$. The order in which
      the sets are combined under set intersection
      does not affect the result.
   c. $(A \cup B) \cup C = \{1,2,3,4,5,6,7,8\}$;
      $A \cup (B \cup C) = \{1,2,3,4,5,6,7,8\}$.
      Thus, $(A \cup B) \cup C = A \cup (B \cup C)$. Under
      set union the manner of grouping does not
      affect the result.
   d. $(A \cap B) \cap C = \{5,6\}$; $A \cap (B \cap C) = \{5,6\}$.
      Thus, $(A \cap B) \cap C = A \cap (B \cap C)$. Under
      set intersection the manner of grouping does
      not affect the result.
   e. $A \cup (B \cap C) = \{1,2,3,4,5,6,7,8\}$;
      $(A \cup B) \cap (A \cup C) = \{1,2,3,4,5,6,7,8\}$.
      Thus, $A \cup (B \cap C) = (A \cup B) \cap (A \cup C)$.
      This statement is a form of the distributive

property. Note that both set union and set
intersection are involved in this property.
   f. $A \cap (B \cup C) = \{1,2,3,4,5,6\}$;
      $(A \cap B) \cup (A \cap C) = \{1,2,3,4,5,6\}$.
      Thus, $A \cap (B \cup C) = (A \cap B) \cup (A \cap C)$.
      This statement is another form of the
      distributive property.
7. The set of all visitors to both Disney World and
   Everglades National Park and not Busch Gardens.
8. {6,8}
9. 44%

1
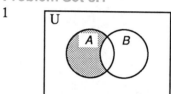

2.
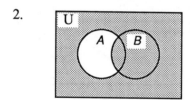

3.
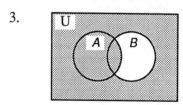

4.
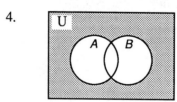

5. region 4
6. region 2
7. {4,10}

8. $(A \cup B)' = A' \cap B' = \{4\}$
9. yes
10. $A \cap (B \cup C) = (A \cap B) \cup (A \cap C) = \{1,2,3,4,5,6,8\}$
11. 0.176

Chp. 1, Section 5, Problem Set 5.2

1.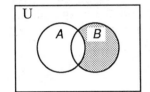

2.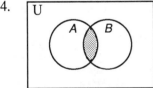

3.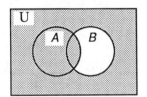

4.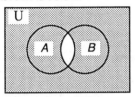

5. $(A \cap B)' = A' \cup B' = \{1,3,6,9\}$
6. $(A' \cap B)' = A \cup B'$
7. $A \cup (B \cap C) = (A \cup B) \cap (A \cup C) = \{1,2,3,5\}$
8. 36.5%

Chp. 1, Section 5, Problem Set 5.3

1.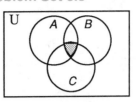

2.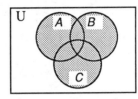

3.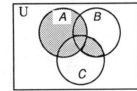

4.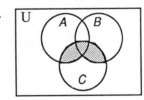

5. $(A' \cup B')' = A \cap B.$
6. region 1 and region 4
7. region 1, region 3, and region 4

Chp. 1, Section 5, Problem Set 5.4

1.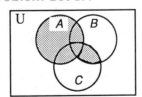

2.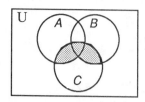

3. region 7
4. region 1 and region 8
5. region 3
6. region 5
7. regions 1,2,4,5, and 6
8. region 1
9. region 4 and region 8
10. region 3
11. regions 1, 9 and 12
12. region 7
13. $(A \cup B)' = A' \cap B'$; $(A \cap B)' = A' \cup B'$
14. $(A' \cap B')' = A \cup B$
15. 0.875
16. 60%

Chp. 1, Section 5, Problem Set 5.5

1. b
2. a
3. c
4. c
5. d
6. b
7. a
8. d
9. $\frac{1}{9}$
10. The revenue decreased by $48.
11. $-3\frac{17}{18}$

Chp. 1, Section 5, Problem Set 5.6

1.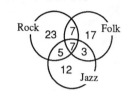

   a. 23 people liked rock music.
   b. 3 people liked both folk music and jazz but not rock music.

2.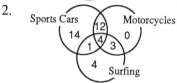

   a. 38 people were interviewed.
   b. 1 student

3.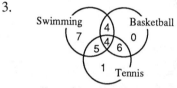

   a. 7 students are only on the swimming team.
   b. 4 students are on both the swimming and basketball teams but not the tennis team.
4. 8 subsets and 7 proper subsets
5. a. {1,2,3,4,6,7}   b. {1,2,3,4,7}
6. $(A' \cap B') \neq A \cap B'$
7. 300%
8. 30%
9. 20
10. 160

Chapter 1 Self-Test

1. d
2. d
3. c
4. b
5. b
6. d
7. d
8. d
9. b
10. c
11. c
12. a
13. c
14. b
15. d
16. d
17. b
18. d
19. d
20. b
21. c
22. a

Chapter 2

Chp. 2, Section 1, Problem Set 1.1

1. simple
2. not simple
3. not simple
4. not simple
5. simple
6. simple
7. simple
8. simple
9. not simple
10. not simple
11. 300
12. 200%
13. 68%

Chp. 2, Section 1, Problem Set 1.2

1. The simple statements are: (a) It is windy. (b) It is turning cold. The connective is "and."
2. The simple statements are: (a) The Yankees won their games. (b) The Mets won their games. The connective is "and."
3. The simple statements are: (a) Roses are red. (b) Violets are blue. The connective is "or."
4. The simple statements are: (a) Heathcliffe's friends have bad breath. (b) Heathcliffe's friends need a bath. The connective is "or."
5. The simple statements are: (a) It is beautiful today. (b) I am going to school. The connective is "and." However, it is expressed with the equivalent word "but."
6. The simple statements are: (a) I will take Mike to the basketball game. (b) I will take Carol to the basketball game. The connective is "or."
7. The simple statements are: (a) Sandy is a secretary. (b) Larry is a secretary. The connective is "and."
8. The simple statements are: (a) Tom reads rapidly. (b) Karen reads rapidly. The connective is "and."

9. a. not simple   b. simple   c. simple
10. 11
11. no

Chp. 2, Section 1, Problem Set 1.3

1. a. Mary has it or Alice has not had it.
   b. Mary does not have it or Alice has had it.
   c. Mary does not have it and Alice has not had it.
   d. It is not true that both Mary has it or Alice has had it.
   e. It is not true that both Mary has it and Alice has not had it.
   f. It is false that Alice has not had it. This is equivalent to: Alice has had it.
2. a. $\sim p \vee q$ b. $\sim(p \wedge \sim q)$
   c. $\sim p \wedge \sim q$ d. $\sim p \wedge \sim q$
3. a. The simple statements are: "John is an efficient secretary," "Sandy is an efficient secretary." The connective is "and."
   b. The simple statements are: "The children are noisy," "The children are rude." The connective is "or."
4. 18

Chp. 2, Section 2, Problem Set 2.1

1. a. Tampa is not in Florida.
   b. The negation is a false statement.
2. a. This statement is false.
   b. The negation is a true statement.
3. No people like baseball.
4. Some Floridians are not nice people.
5. Some parakeets weigh fifty pounds.
6. Some books do not have titles.
7. All teachers are interesting.
8. At least one person likes baseball. There are no Floridians who are not nice people. All parakeets do not weigh fifty pounds. There are no books that do not have titles. Not all teachers are interesting.

9. a. It is not windy or I am not going sailing.
    b. It is not both true that it is not windy today and I am not going sailing.
10. a. ~p ∨q   b. ~(p ∧~q)
11. $\frac{19}{28}$
12. 6.095
13. 0.48

Chp. 2, Section 2, Problem Set 2.2

1. false
2. true
3. true
4. true
5. true
6. false
7. true
8. false
9. a
10. a. Some books are not well written.
    b. No people are litterbugs.
    c. All mathematics courses are fun.
    d. Some people are fifteen feet tall.
11. Not all cats are cute.
12. 1.4%
13. 0.3
14. 6.795

Chp. 2, Section 2, Problem Set 2.3

(Only the final column of the truth table is listed.)

1.	2.	3.	4.
T	F	F	F
F	T	T	F
T	F	T	F
T	F	T	T

5. a. true   b. false
6. a. Some teachers are dull.
    b. Some cars are not lemons.
7. $\frac{6}{17}$
8. 60%
9. 4.5
10. 0.035

Chp. 2, Section 2, Problem Set 2.4

(Only the final column of the truth table is listed.)

1.	2.	3.	4.
F	F	T	T
F	F	F	F
F	T	T	F
T	F	T	F

5. a.	b.	c.
F	F	F
T	F	F
T	F	F
T	F	T

6.	7.	8.
F	F	T
F	F	T
F	F	T
F	F	T
T	F	F
T	F	T
F	T	T
T	F	T

9. a. No men are overweight.
    b. All adults are arrogant.
10. All triangles are not squares.
11. $\frac{7}{20}$
12. 20%
13. 0.0225

Chp. 2, Section 2, Problem Set 2.5

1. You live in America and you are not studying math.
2.
p	q	~q	p ∧~q
T	T	F	F
T	F	T	T
F	T	F	F
F	F	T	F

3. a. no   b. yes
4. a. yes   b. yes
5. a. no   b. yes
6. a. no   b. yes
7. a. All cars are expensive.
    b. Some persons are above the law.

8. $\frac{12}{5} = 2\frac{2}{5}$

9. $\frac{15}{4} = 3\frac{3}{4}$

10. 70

11. 0.025

Chp. 2, Section 2,
Problem Set 2.6

1. Jim is not tall or John is not fat.
2. Mary does not live in Florida or Nancy does not live in Georgia.
3. Estelle is not loud or Estelle is not abrasive.
4. The Dolphins are not in the playoffs or the Buccaneers are not in the playoffs.
5. Clearwater is not in Georgia or Atlanta is in Florida.
6. The Strikers are Jacksonville's team or the Bandits are not Tampa's team.
7. All movies are exciting or some jokes are funny.
8. Vince is tall or Mike is fat.
9. Some dogs are not faithful or no cats are cute.
10. Frank does not appreciate rock music or Judy does not appreciate rock music.
11. b
12. a. All people are dependable.
    b. Some babies are self-sufficient.
13. a. false   b. true
14. 30
15. $\frac{22}{5}$
16. 28%
17. 0.0006

Chp. 2, Section 2,
Problem Set 2.7

1. London is not in England and Paris is not in France.
2. The bill does not pass and it is a law.
3. Dave does not visit San Francisco and he does not visit London.
4. It is hot and it is humid.

5. Some mathematics books do not have sample tests and they do get published.
6. Antonio is Prospero's brother and Romeo is not Juliet's lover.
7. Some people do not carry umbrellas and no people get wet.
8. a. University of Florida is not a college or Sears is not a college.
   b. Some musicians do not write lyrics or no Broadway musicals lose money.
9. a
10. $6\frac{1}{12}$
11. 7000
12. 250%
13. 96

Chp. 2, Section 3,
Problem Set 3.1

1. a. Hypothesis: I work hard.
      Conclusion: I will pass the course.
   b. Hypothesis: A polygon is a quadrilateral.
      Conclusion: A polygon has four sides.
   c. Hypothesis: The sky is not overcast.
      Conclusion: It is not raining. Notice the hypothesis appears at the end of the statement.
2. a. If you have long hair, then you will get dandruff.
   b. If you get dandruff, then you have long hair.
   c. If you do not have long hair, then you will not get dandruff.
   d. If you do not get dandruff, then you do not have long hair.
3. a. If George buys a television, then it is a Sony.
   b. If one is a soldier, then one is not afraid.
   c. If the bell rings, then the class is over.
   d. If one is a student, then one will pass this test.
   e. If it is a Corvette, then it will not enter this race.

    f. If one is a criminal, then one will be punished.

    g. If a person is your teacher, then the person is not dull.

    h. If a person is tall, then the person should play basketball.

    i. If the National Anthem is not being played, then no one remains standing.

    j. If a member is not excused, then he is required to attend the meeting.

4. c

5. a. If you have avoided a ticket, then you have observed the speed limit.

    b. If you are making the flowers grow, then you have watered them.

    c. If you attend class regularly, then you will pass the course.

    d. If I live in the North, then I can endure the winters.

    e. If you start the car, then you have turned on the ignition.

    f. If you miss the bus, then you will be late for class.

6. a. Linda is a good skater and Larry is not a fast runner.

    b. Some dogs are not animals.

    c. Jim smokes or Laura drinks.

    d. No men are bad drivers.

7. $6\frac{9}{20}$

8. 14.999

9. $\frac{13}{20}$

10. 0.625

Chp. 2, Section 3, Problem Set 3.2

1. false

2. true

3. true

4. true

5. a. true   b. false   c. true   d. true

6. a. If one is a student, then one will graduate.

    b. If one is under seventeen, then one will not be allowed in the theatre.

7. a. If you work hard, then you will succeed.

    b. If you will succeed, then you work hard.

    c. If you do not work hard, then you will not succeed.

    d. If you will not succeed, then you do not work hard.

8. a. She is not a good wife or she is not a good mother.

    b. He does not like steak and she likes roast beef.

9. $\frac{7}{9}$

10. 25

11. 1.25; $\frac{5}{4}$

12. 40%

13. 0.0475

14. $700

Chp. 2, Section 3, Problem Set 3.3

(Only the final column is listed.)

1. T	2. T	3. T	4. T
T	T	T	F
T	F	F	T
F	T	T	T

5. The final column in each truth table, T, F, T, T is identical. These statements are logically equivalent.

6. a. If I do not pass the course, then I have not worked hard.

    b. If Bob doesn't live in Florida, then he doesn't live in Clearwater.

    c. If x is not an even number, then x^2 is not an even number.

7. The final column in each truth table, T, T, F, T is identical. These statements are logically equivalent.

8. a. false   b. true   c. true

9. a. He is a good administrator or she is a good teacher.
    b. Tom is the father and he is not the legal guardian.
    c. No rivers are polluted.
    d. All bridges are safe.
10. $1\frac{2}{3}$
11. $\frac{1}{4}$
12. 2.7
13. 27

Chp. 2, Section 3, Problem Set 3.4

(Only the final column is listed.)

1. T	2. T	3. T	4. T
T	T	T	T
T	T	F	T
T	F	T	F

5. T	6. T	7. T	8. F
T	T	T	F
F	F	T	F
T	T	T	T
		F	T
		T	T
		T	T
		T	T

9. a. true   b. true
10. a. Some television programs are not educational.
    b. Some cars are designed to last forever.
    c. Bill is not a good husband or he is a good politician.
    d. Miami is in Alabama and Florida is not a fast growing state.
11 If I will not go to the beach, then the sun does not shine.
12. 12
13. $\frac{32}{21} = 1\frac{11}{21}$
14. 15

Chp. 2, Section 4 Problem Set 4.1

1. Converse: If Nigel is in England, then he is in London. Inverse: If Nigel is not in London, then he is not in England. Contrapositive: If Nigel is not in England, then he is not in London.
2. Converse: If it is yellow, then it is a banana. Inverse: If it is not a banana, then it is not yellow. Contrapositive: If it is not yellow, then it is not a banana.
3. Converse: If Joan is having a good time, then she is at the movies. Inverse: If Joan is not at the movies, then she is not having a good time. Contrapositive: If Joan is not having a good time, then she is not at the movies.
4. Converse: If Erin does not wear red clothing, then it is St. Patrick's Day. Inverse: If it is not St. Patrick's Day, then Erin wears red clothing. Contrapositive: If Erin wears red clothing, then it is not St. Patrick's Day.
5. Converse: If one does not advocate freedom, then one advocates censorship. Inverse: If one does not advocate censorship, then one advocates freedom. Contrapositive: If one advocates freedom, then one does not advocate censorship.
6   Converse: $\sim q \rightarrow p$; Inverse: $\sim p \rightarrow q$; Contrapositive: $q \rightarrow \sim p$
7. Converse: $\sim q \rightarrow \sim p$; Inverse: $p \rightarrow q$; Contrapositive: $q \rightarrow p$
8. Converse: If it is not allowed in the restaurant, then it is an animal. Inverse: If it is not an animal, then it is allowed in the restaurant. Contrapositive: If it is allowed in the restaurant, then it is not an animal.
9. Converse: If one is a scientist, then one is a chemist. Inverse: If one is not a chemist, then one is not a scientist. Contrapositive: If one is not a scientist, then one is not a chemist.
10. If lines share a common point, then they are not parallel.

11. If the team does not win, then Jose does not play or Marsha does not play.

12. If one is not in the South, then one is not in Atlanta and one is not in New Orleans.

13. $(q \wedge \sim r) \rightarrow p$

14. $r \rightarrow (\sim p \vee q)$

15. If some schools are not closed, then there is not a hurricane.

16. If some students fail the test, then the review session is not successful.

17. a. I do not win the game and I do not collect the prize.
    b. It is not true that both I win the game or I do not collect the prize.

18. a. No drivers are reckless.
    b. Irma is not a fast typist or Theresa is not a good cook.

19. $\frac{5}{7}$

20. $2\frac{1}{5}$

21. 527.057

22. 1.75

23. 2

24. $270.50

Chp. 2, Section 4, Problem Set 4.2

1. a. If Sherry lives in Palm Beach, then Sherry lives in Florida. (True)
   b. If Sherry lives in Florida, then Sherry lives in Palm Beach. (Not necessarily true)
   c. If Sherry does not live in Palm Beach, then Sherry does not live in Florida. (Not necessarily true)
   d. If Sherry does not live in Florida, then Sherry does not live in Palm Beach. (True)

2. Converse: If I am happy, then I pass the test. Inverse: If I do not pass the test, then I will not be happy.

3. Converse: If Erin wears green, then it is St. Patrick's Day. Inverse: If it is not St. Patrick's Day, then Erin does not wear green.

4. Converse: If on is a writer, then one is a poet. Inverse: If one is not a poet, then one is not a writer.

5. Converse: If a number is divisible by 3, then it is divisible by 6. Inverse: If a number is not divisible by 6, then it is not divisible by 3.

6. Converse: If if is not blue, then it is a banana. Inverse: If it is not a banana, then it is blue.

7. Converse: If the musical is a success, then Mary and Ethel star. Inverse: If Mary does not star or Ethel does not star, then the musical is not a success.

8. Converse: If one is in the South, then one is in Atlanta or New Orleans. Inverse: If one is not in Atlanta and not in New Orleans, then one is not in the South.

9. Converse: If we celebrate and purchase new equipment, then the team wins. Inverse: If the team does not win, then we do celebrate or we do not purchase new equipment.

10. If one fails, then one does not work hard.

11. If the book is published, then the book has a test bank.

12. If Maggie is not having a good time, then she is not attending the opera.

13. If no people suffer, then some corporations do not place profit above human need.

14. $(\sim p \rightarrow q) \equiv (\sim q \rightarrow p)$

15. $[p \rightarrow (q \vee r)] \equiv [(\sim q \wedge \sim r) \rightarrow \sim p]$

16. a. If I will be happy, then I pass the test.
    b. If I do not pass the test, then I will not be happy.
    c. If I will not be happy, then I do not pass the test.

17. Converse: $q \rightarrow \sim p$; Inverse: $p \rightarrow \sim q$; Contrapositive: $\sim q \rightarrow p$

18. a. Some elephants are ballet dancers.
    b. Jane is not a good dancer and Martha is not a good singer.

19. $4\frac{5}{12}$

20. 48%

21. $3\frac{1}{8}$

22. $\frac{3}{4}$

23. 45 cents

**Chp. 2, Section 4,
Problem Set 4.3**

1. a. If a number is a multiple of 2, then the number is even, and if the number is even, then the number is a multiple of 2.
    b. A number being even is necessary and sufficient for a number being a multiple of 2.
    c. A number being a multiple of 2 is necessary and sufficient for a number being even.
    d. A number is a multiple of 2 if and only if (iff) a number is even.
    e. A number is even if and only if (iff) the number is a multiple of 2.
    f. A number is a multiple of 2 is equivalent to a number is even.
2. The statement if false. "If you are a brilliant physicist, then you are Albert Einstein" is not true.
3. The statement is true since the conditional works in both directions.
4. true
5. true
6. false
7. true
8. false
9. a. If you go to the fair, then you clean up your room.
    b. If you hand in the homework, then you will receive bonus points.
    c. If you are watching the soap opera, then you have turned on the television set.
10. If I stay home, then it rains.
11. If Martha is not afraid, then Martha is not alone.
12. a. All students pay attention.
    b. Some movies are not exciting.
    c. Paul does not have the answer and Dan does not have the question.
    d. Myra is cooperative or Lois is dependable.
13. $\frac{2}{5}$
14. 25%

15. $\frac{1}{7}$

**Chp. 2, Section 4,
Problem Set 4.4**

1. a. If Miami does not win the division, then Buffalo will win the division.
    b. If Kay does not go to the convention, then Joyce will go.
    c. If Professor Gill teaches this course, then I am withdrawing.
    d. If John is not selected for the committee, then I am not resigning.
    e. If Ron will win the award, then Doug will not win the award.
2. a. Steve does not go surfing or he will miss class.
    b. I do not try to stand up or my back will hurt.
    c. She appears or we can conduct the meeting in her absence.
    d. Michelle does not discover the flat tire or she will not be happy.
    e. Mark does work or he cannot pay the rent.
3. a. If Joan is nervous, then she is not tense.
    b. If Pete is going to the movies, then he is not going to the beach.
    c. If Richard is rich, then he is not famous.
    d. If Joe is our driver, then he is not our guide.
4. b
5. c
6. a
7. d
8. c
9. c
10. b
11. d
12. a. If I will play tennis, then the sun is shining.
    b. If the sun is not shining, then I will not play tennis.
    c. If I will not play tennis, then the sun is not shining.

13. a. Joan is not a good mathematician and Ruth is a computer programmer.

   b. Sally is obnoxious or Alice is friendly.

14. $\frac{77}{6} = 12\frac{5}{6}$

15. 300

16. 20

17. 3

18. 280

19. $1050

Chp. 2, Section 4, Problem Set 4.5

1. a. Barbara wins the election and I will not be sorry.

   b. You speak loudly and I cannot hear you.

   c. It is a nice day and I will not go to the beach.

   d. John is a criminal and he will not be punished.

   e. Bill is 18 and he is not eligible to vote.

2. a. A person studies Latin and that person does not have an excellent vocabulary.

   b. It is hot and no city pool will be open.

   c. Both Mary and Ethel star and the musical is not a success.

   d. There is a lottery and some schools will not receive increased funding.

3. a. Joe arrives in time or he will miss the show.

   b. Scott gets married or he will be very lonely.

4. a. If Liz is not the nominee, then Millie will resign from the committee.

   b. If Jack is going on vacation this year, then Jim will travel to Europe.

5. If everyone will listen, then she speaks.

6. a. Some mathematics books are not exciting.

   b. Don is not a stable person or Ron is even-tempered.

7. 24

8. $\frac{33}{8}$

9. 8

10. 0.05

11. $6160

12. $62.50

Chp. 2, Section 5, Problem Set 5.1

1. a. no   b. yes   c. yes   d. yes
   e. yes   f. yes

2. c

3. b

4. a. Some foods are harmful to the body.

   b. The sun is shining or I am not going to the beach.

   c. She is feeling well and she will not go to the party.

5. 81

6. $97.20

Chp. 2, Section 5, Problem Set 5.2

1. a. Arguments (a), (c) and (d) are tautologies. Thus, arguments in this form are valid arguments.

2. All of the arguments (a - e) are invalid.

3. a. valid argument   b. valid argument
   c. invalid argument   d. valid argument
   e. invalid argument   f. invalid argument
   g. valid argument   h. invalid argument
   i. invalid argument   j. invalid argument
   k. valid argument   (The conclusion is in the form of the contrapositive of the statement. This would make the argument valid. The contrapositive has the same truth value and is said to be logically equivalent.)
   l. invalid argument

4. a. valid   b. valid

5. a. If one is an outstanding mathematician, one should excel in engineering.

   b. If a course is meaningful, I find it enjoyable.

   c. If one is socially intelligent, then one is mature.

6. a. This argument does not yield the conclusion that "this administration is not the party of fiscal responsibility."
    b. This argument does not yield a valid conclusion.
7. a. tautology   b. tautology
8. a. valid   b. invalid   c. valid
    d. invalid   e. valid
9. a. I will work.
    b. The Dolphins finished in first place in their division.
    c. $b = 0$
10 a. not a tautology   b. not a tautology
11. a. invalid   b. invalid
12. ii
13. c
14. b
15. a, c
16. c
17. c
18. c
19. d
20. c
21. 40
22. $\frac{11}{30}$

Chp. 2, Section 5, Problem Set 5.3

1. Some people do not obey the rules
2. Some people smoke cigarettes.
3. Joel does not study mathematics.
4. Kathy does not have a knowledge of theatre.
5. Angie appreciates language.
6. We watched television on Friday.
7. Marco does not study biology.
8. Susan is not studying.
9. Fran did no homework problems.
10. If it is Tuesday, then Javier contemplates the meaning of existence.
11. If the medication is effective, then I will not fall behind.

12. If one lives in Tampa, then one lives in North America.
13. South Africa has no peace and must spend a great deal of money building jails.
14. An outstanding baseball player can hit and throw a ball.
15. It did not rain and it did not snow on Friday.
16. Samir does not study mathematics and he appreciates language.
17. We will watch television and we will not pass the exam.
18. Some winter days are not cold.
19. It is not cold or it is humid.
20. Tim will not do the homework and will pass the course.
21. It rains and we will use the pool.
22. c
23. $3\frac{17}{18}$
24. c
25. 20%

Chp. 2, Section 6, Problem Set 6.1

1. valid argument
2. invalid argument
3. invalid argument
4. valid argument
5. invalid argument
6. invalid argument
7. valid argument
8. d
9. d
10. d
11. d
12. $350
13. 80
14. -15

Competency in College Mathematics

1. invalid argument
2. valid argument
3. valid argument
4. invalid argument
5. valid argument
6. invalid argument
7. valid argument
8. valid argument
9. invalid argument
10. invalid argument
11. valid argument
12. b
13. c
14. c
15. c
16. c
17. a
18. b
19. 0.04
20. 2:45 P.M.

1. invalid argument
2. valid argument
3. invalid argument
4. invalid argument
5. invalid argument
6. valid argument
7. valid argument
8. invalid argument
9. valid argument
10. d
11. a
12. a. valid   b. valid   c. invalid
13. Some politicians are not dishonest.
14. It is not cold or it is snowing.
15. It is cloudy and we will go to the beach.
16. All people follow rules and some prisons are necessary.

17. 60 cents
18. 24

1. c
2. d
3. d
4. He first takes the goat across. He then returns and picks up the wolf. He leaves the wolf off and takes the goat back. He then leaves the goat at the starting place and takes the cabbage over to where the wolf is. He then returns and picks up the goat and goes where the wolf and the cabbage are waiting.
5. The man furthest from the wall either sees 2 tan hats or a black and a tan hat. If he saw 2 black hats, he would have known he had to have a tan hat. The middle man sees a tan hat because if he saw a black hat, he would know he must be wearing a tan hat from the first response. Therefore, the man facing the wall concludes he can only be wearing the tan hat the middle man sees.
6. Two letters (A,B) are left and one (C) is removed; three letters (D,E,F) are left and two (G,H) are removed, etc.
7. 276. The middle digit is not the sum of the other two digits.
8. Joan cannot be the art critic or architect. The aviator must be a man, so Joan must be the acrobat. Hence, Jane, the only other woman, must be the art critic. John cannot be the aviator, who is happily married, so he must be the architect.
9. b
10. d
11. $2588.93

Chp. 2, Section 8, Problem Set 8.1

1. a. Fallacy of Hasty Generalization
   b. Fallacy of Ambiguity
   c. Fallacy of Composition
   d. Fallacy of False Authority
   e. Fallacy of False Cause
   f. Fallacy of Division
   g. Fallacy of Emotion
   h. Fallacy of the Complex Question
2. c
3. d
4. b
5. A bill does not receive majority approval and it does become law.
6. d
7. a
8. a
9. 0
10. 26
11. $\frac{1}{8}$
12. $4770
13. 16

Chapter 2 Self-Test

1. c	11. c	21. b	31. a
2. b	12. e	22. c	32. b
3. d	13. d	23. d	33. b
4. c	14. d	24. d	34. a
5. c	15. b	25. a	35. c
6. b	16. d	26. b	36. a
7. d	17. d	27. c	37. c
8. b	18. d	28. d	38. b
9. b	19. c	29. b	39. c
10. d	20. b	30. d	40. c

Chapter 3

Chp. 3, Section 1, Problem Set 1.1

1. One
2. Yes
3. No. There is no counting number between 1 and 2 for example.
4. No. There is no counting number smaller than 1.
5. No. Dividing one counting number by another counting number does not always produce another counting number.
6. No; no
7. No; yes; yes
8. Yes; yes; yes
9. No. The product of, say 3 and 4, is not a one-digit counting number.
10. $\frac{41}{12} = 3\frac{5}{12}$
11. $\frac{31}{30} = 1\frac{1}{30}$
12. $\frac{91}{12} = 7\frac{7}{12}$
13. $\frac{15}{32}$
14. 5.677
15. 6.525
16. 50
17. 3.5%
18. $\frac{7}{11}$

Chp. 3, Section 2, Problem Set 2.1

1. a. r   b. x; z   c. b; c
2. no
3. no
4. yes
5. no
6. yes
7. no
8. 400
9. 1.25

10. 0.7

11. $\frac{7}{12}$

Chp. 3, Section 2, Problem Set 2.2

1. a. x; z
   b. b; a
   c. a
2. yes
3. no
4. $a + b = b + a$
5. no
6. yes
7. 87.5%
8. 0.0525
9. $\frac{47}{20} = 2\frac{7}{20}$
10. $3\frac{5}{8}$

Chp. 3, Section 2, Problem Set 2.3

1. p; r
2. no
3. a. yes   b. yes
   c. yes
4. $(10 - 6) - 4 = 0$ and $10 - (6 - 4) = 8$
5. $a + b = b + a$; $a \cdot b = b \cdot a$
6. no; yes
7. 28.8
8. 2.85
9. 12,000 students

Chp. 3, Section 2, Problem Set 2.4

1. 4; 3
2. no
3. yes
4. $(12 \div 6) \div 2 = 1$ and $12 \div (6 \div 2) = 4$

5. $(a + b) + c = a + (b + c)$
6. $(x + y) + z = x + (y + z)$
7. no
8. 0.0053
9. 10%

Chp. 3, Section 2, Problem Set 2.5

1. associative property; commutative property; commutative property
2. commutative property; associative property; commutative property
3. 5 and 20
4. $(a \cdot b) \cdot c = a \cdot (b \cdot c)$
5. $a + b = b + a$; $a \cdot b = b \cdot a$
6. no
7. no
8. no
9. 200

Chp. 3, Section 2, Problem Set 2.6

1. no
2. yes
3. $(a + b) + c = a + (b + c)$; $(a \cdot b) \cdot c = a \cdot (b \cdot c)$
4. $x \cdot (y \cdot z) = (x \cdot y) \cdot z$
5. no, no
6. 400
7. 3.1%

8. 50%

Chp. 3, Section 2, Problem Set 2.7

1. a. true
   b. false A one-to-one correspondence like the one shown could be established.
   {0, 1, 2, 3,...}
   | | | |
   {1, 2, 3, 4,...}
2. no
3. 1
4. yes
5. yes
6. 300
7. 9.433
8. $2\frac{1}{9}$

Chp. 3, Section 2, Problem Set 2.8

1. yes. The whole number is 0.
2. yes
3. true
4. 1
5. no
6. no
7. no
8. 300
9. $\frac{11}{5}$ or $2\frac{1}{5}$
10. 24%

Chp. 3, Section 2, Problem Set 2.9

1. a. $a = 4$
   b. $a = 2$
   c. $a = 2$

d. a can represent any whole number.
   e. $a = 4$
2. no
3. no
4. no
5. Multiplication is distributive over subtraction.
6. no
7. yes
8. 0
9. yes
10. $(x + y) + (y + z) = (y + z) + (x + y)$
11. no
12. 1
13. 15
14. 12
15. $2\frac{11}{12}$

Chp. 3, Section 3, Problem Set 3.1

1. yes
2. yes
3. -6
4. $a(b + c) = ab + ac$
5. $(2 + 3) \cdot 4 = (2 \cdot 4) + (3 \cdot 4)$
6. $a + b = b + a$; $a \cdot b = b \cdot a$
7. $(a + b) + c = a + (b + c)$; $(a \cdot b) \cdot c = a \cdot (b \cdot c)$
8. 0; 1
9. no
10. 12
11. 2.5 ft.
12. 3.1
13. $\frac{17}{20}$

Chp. 3, Section 3, Problem Set 3.2

1. Yes. For every counting number there is a corresponding negative counting number.
2. no
3. the null set.
4. -5
5. $2 \cdot (3 + 4) = (2 \cdot 3) + (2 \cdot 4)$
6. yes
7. yes
8. $(a + b) + c = (b + a) + c$
9. yes
10. 9
11. $2\frac{7}{10}$
12. $117\frac{1}{3}$ lbs.

Chp. 3, Section 3, Problem Set 3.3

1. no. They are both names for the same set of numbers.
2. no. The union would not include the integer 0 which is neither a positive nor a negative integer.
3. yes
4. yes
5. 10
6. true
7. 2
8. x can be any number
9. false
10. 9
11. $\frac{1}{6}$
12. 28%

Chp. 3, Section 3, Problem Set 3.4

1. a. 2   b. 0   c. -2
   d. 5   e. -4

2. a. 3   b. -4   c. -4
   d. -3   e. -9
3. a. 7y   b. -2x   c. 3a
   d. 0   e. -7b   f. 9π
   g. -2ab   h. 4x + 3a
   i. 6y + a + b
   j. $4x^2 - x$
4. yes
5. yes
6. yes
7. false
8. 1; 0
9. yes
10. yes
11. 170%
12. 84
13. 1000
14. $4\frac{1}{4}$
15. $\frac{13}{20}$
16. 1.75

Chp. 3, Section 3, Problem Set 3.5

1. a. 3   b. -3   c. -3
   d. -5
2. a. -5   b. -7
   c. -13   d. 6
   e. -2   f. 2x   g. 2y
   h. -a   i. -5b   j. -2y
   k. -x   l. -13a
   m. -7x - a
   n. -bc - 4y
   o. 2xy - 4x - 2y
3. yes
4. no
5. a. -4   b. -2   c. 3
6. a. 7x   b. -y   c. 0
   d. -3c   e. π
   f. 2xy + y
7. false
8. 6.997

9. 0.03
10. 0.125

Chp. 3, Section 3, Problem Set 3.6

1. -15
2. -20
3. 0
4. -18
5. -30
6. -216
7. 10
8. 56
9. 4
10. 6
11. 1
12. 3
13. 3
14. 7
15. 0
16. -5
17. -7
18. -7
19. -15x
20. -6yz
21. 20t
22. -8x + 12y
23. 6x - 8y
24. -2x - y + 3z
25. 3x + 6y - 9z
26. a. -2   b. -5   c. -5
    d. 3y   e. -2x   f. -7a
27. a. 4   b. -1   c. -5
    d. 3a   e. -3π   f. -3x
28. -8
29. 6
30. 28
31. $3\frac{23}{30}$
32. $\frac{16}{27}$

33. $\frac{6}{11}$

34. 0.0046

35. $\frac{7}{20}$

36. $\frac{41}{10}$

37. 0.085

Chp. 3, Section 3
Problem Set 3.7

1. The multiplicative inverse of 0 would be $\frac{1}{0}$. This expression is undefined.

2. a. $\frac{1}{5}$ b. 4 c. $\frac{4}{5}$

3. no

4. yes, if $b \neq 0$

5. a. -10 b. 24 c. 27
   d. 8 e. 7 f. 0
   g. 2 h. -4 i. -3
   j. -1 k. -9x
   l. -6x m. -10

6. 44%

7. 117

8. 300

9. $3\frac{7}{8}$

10. 2.4

Chp. 3, Section 3
Problem Set 3.8

1. yes; no

2. no

3. a. $\frac{1}{10}$ b. 5 c. $\frac{3}{2}$

4. a. -16 b. -36 c. 10
   d. 7 e. -6 f. -4
   g. -6a h. -2a i. 3π
   j. -8

5. 500%

6. 40%

7. 0.175

8. 0.61

9. $\frac{9}{4}$ or $2\frac{1}{4}$

10. $\frac{3}{16}$

Chp. 3, Section 3
Problem Set 3.9

1. a. -4 b. -2 c. -2
   d. 3 e. -1 f. $\frac{1}{2}$
   g. $-\frac{1}{3}$ h. $-\frac{1}{3}$

2. a. -24 b. $-\frac{28}{15}$ c. $\frac{21}{32}$

3. a. -2x b. -2x c. x
   d. x e. 2x
   f. -2x g. -4xy
   h. 2ab i. -2abc

4. yes; yes

5. yes

6. yes

7. yes

8. a. $\frac{1}{4}$ b. 7 c. $\frac{6}{5}$

9. a. -20 b. 40 c. 12
   d. 8 e. 2 f. 1
   g. 0 h. -6
   i. -13 j. -a
   k. -12x

10. $\frac{11}{20}$

11. 10%

12. 24%

13. 6.996

14. $2\frac{13}{21}$

Chp. 3, Section 4
Problem Set 4.1

1. a. 5
   b. cannot be determined exactly
   c. 6 d. $\frac{2}{5}$
   e. cannot be determined exactly
   f. 0

2. a. rational
   b. not rational
   c. rational
   d. not rational

3. no

4. a. -2 b. -1 c. 1
   d. $\frac{1}{3}$ e. $\frac{1}{4}$
   f. -25 g. $-\frac{3}{4}$ h. 16
   i. -2x j. 3x
   k. -4x l. 42
   m. -11 n. -10

5. $-\frac{3}{4}, \frac{-3}{4}, \frac{3}{-4}$

6. $\frac{5}{24}$

7. 6

8. 5000

9. $\frac{35}{9}$

10. 0.047

Chp. 3, Section 4
Problem Set 4.2

1. a. $2\sqrt{3}$ b. $5\sqrt{6}$
   c. $3\sqrt{2}$ d. $2\sqrt{5}$
   e. $3\sqrt{7}$
   f. cannot be simplified
   g. $6\sqrt{2}$

2. a. 8 b. $\frac{4}{5}$ c. 0

3. a. -4 b. $\frac{7}{6} = 1\frac{1}{6}$
   c. -10 d. -12

4. 2.5; $\frac{5}{2}$

5. 5.8%

Chp. 3, Section 4
Problem Set 4.3

1. a. $19\sqrt{5}$ b. $-\sqrt{6}$
   c. cannot be combined
   d. $6\sqrt{3}$ e. $-6\sqrt{2}$
   f. $4\sqrt{7}$ g. $-\sqrt{3}-\sqrt{2}$

2. a. $4\sqrt{2}$ b. $6\sqrt{2}$

3. b and c are rational

4. $\frac{3}{4}$

5. 6

Chp. 3, Section 4
Problem Set 4.4

1. a. $\sqrt{21}$ b. $\sqrt{5}$ c. 48
   d. $-72\sqrt{6}$ e. $30\sqrt{2}$ f. -16
   g. $5\sqrt{5}$
2. a. $5\sqrt{8} = 10\sqrt{2}$ b. $-\sqrt{6}$
3. a. 8   b. 3   c. 7
   d. -2   e. -1   f. -10
   g. -8b h. -8
4. $4\frac{5}{8}$
5. 60%
6. $5\frac{1}{3}$

Chp. 3, Section 4
Problem Set 4.5

1. a. $\sqrt{5}/5$ b. $\sqrt{3}/12$
   c. $3\sqrt{7}/7$ d. $\sqrt{6}/3$
   e. $-7\sqrt{6}/6$ f. $-4\sqrt{15}/15$
   g. $5\sqrt{3}/3$
2. a. $4\sqrt{3}$ b. -36
   c. -15
3. $-\sqrt{3}$

Chp. 3, Section 4
Problem Set 4.6

1. a. irrational b. rational
   c. irrational d. rational
   e. rational
2. the null set
3. no
4. a. $\frac{3\sqrt{2}}{4}$ b. $\frac{-\sqrt{2}}{4}$

Chp. 3, Section 4
Problem Set 4.7

1. yes
2. the set of rational numbers
3. yes

4. yes
5. yes
6. yes
7. yes
8. b
9. a
10. d
11. b
12. a irrational b. rational
    c. irrational
13. $\frac{\sqrt{5}}{2}$
14. $3\sqrt{3}$
15. -20
16. $\frac{4}{15}$
17. $1,250
18. 20

Chp. 3, Section 5
Problem Set 5.1

1. 1,2,3,4,6,8,12,24
2. 7,14,21,28,35
3. d
4. one
5. 28
6. two
7. b
8. d
9. 13 and 17
10. $6\sqrt{2}$
11. $5\sqrt{5}$
12. a decrease of $6
13. 0.04
14. c
15. d

Chp. 3, Section 5
Problem Set 5.2

1. $2 \cdot 3^2 \cdot 5 \cdot 7$
2. $2^3 \cdot 3^2 \cdot 11$
3. one (7)
4. two (2 and 3)

5. none
6. 35
7. $6 + \sqrt{5}$
8. no
9. 80
10. -0.434
11. $-3\frac{17}{18}$
12. c

Chp. 3, Section 6
Problem Set 6.1

1. 14
2. 23
3. 4
4. 19
5. $3\frac{7}{8}$
6. 12
7. 2
8. 12
9. 23
10. $5\frac{1}{3}$
11. 70
12. 20%
13. 120
14. 12%
15. 250

Chp. 3, Section 6
Problem Set 6.2

1. 4
2. 31
3. 0
4. 2
5. 2
6. 36
7. 1
8. 0
9. $9\frac{3}{4}$
10. 15
11. -22

607

12 a. 8 b. 13

13. $1\frac{19}{20}$

14. 6.798

15. 3.25

16. 0.0425

17. 35 games

Chp. 3, Section 6
Problem Set 6.3

1. $14,250

2. 10^0

3. $16.80

4. 123 miles per hour

5. -12

6. 12

7. 23

8. -42

9. a. 23 b. $-\frac{5}{4}$ c. $-\frac{1}{12}$

10. $25.20

Chp. 3, Section 6
Problem Set 6.4

1. -4x

2. 1

3. $3b^2$

4. $2x + 4x^2$

5. $10c + c^2$

6. 0

7. 5x

8. -2b

9. -11p

10. 14x

11. 1.25

12. 0.0001

13. 2000

14. 6

Chp. 3, Section 6
Problem Set 6.5

1. -6

2. -12

3. -14a

4. -3x

5. 60x

6. $9x - 12$

7. $5x - 10$

8. $2x^2y - 6xy^2$

9. $-24x^2 - 20x$

10. 0

11. $\frac{3}{5}$

12. 0.047

13. $-\frac{1}{9}$

14. $\frac{1}{9}$

15. b

Chapter 3 Self-Test

1. a   2. a   3. c
4. e   5. e   6. d
7. b   8. e   9. d
10. d   11. c   12. c
13. a   14. d   15. b
16. c   17. c   18. a
19. c   20. a   21. c
22. b   23. d   24. d
25. b   26. c   27. b
28. b   29. c   30. b
31. a   32. a   33. d
34. b   35. a

Chapter 4

Chp. 4, Section 1
Problem Set 1.1

1. 8

2. 64

3. 32

4. 10

5. -216

6. 100

7. 1,000

8. 0.25

9. 18

10. 64

11. 20

12. 300

13. 4,000

14. 36

15. 75

16. 36

17. 32

18. 13

19. 64

20. 64

21. 11

22. 4

23. yes

24. no

25. d

26. d

27. b

28. 24

29. $\frac{65}{8} = 8\frac{1}{8}$

30. $5\frac{7}{12}$

31. $\frac{9}{20}$

32. 52%

33. 37.5%

34. 300

35. 0.025

Chp. 4, Section 1
Problem Set 1.2

1. 1

2. 1

3. 1

4. 5

5. 8

6. 8

7. 0

8. a. 27   b. 250   c. 27
    d. 200   e. 72   f. 64
    g. 43   h. 23

9. $\frac{5}{18}$

10. 5.997

11. 17

12. 16

Chp. 4, Section 1
Problem Set 1.3

1. a. $(6 \cdot 10^1) + (9 \cdot 10^0)$

    b. $(5 \cdot 10^2) + (8 \cdot 10^1) + (7 \cdot 10^0)$

    c. $(3 \cdot 10^3) + (2 \cdot 10^2) + (5 \cdot 10^1) + (6 \cdot 10^0)$

    d. $(4 \cdot 10^2) + (9 \cdot 10^0)$

2. a. ones or 10^0

    b. tens or 10^1

    c. hundreds or 10^2

    d. thousands or 10^3

3. a. 1   b. 1   c. 6
    d. 0   e. 16   f. 75
    g. 300   h. 81

4. 3

5. 1.35

6. 20%

7. 4

Chp. 4, Section 1
Problem Set 1.4

1. $\frac{1}{16} = 0.0625$

2. $\frac{1}{8} = 0.125$

3. $\frac{1}{16} = 0.0625$

4. $\frac{1}{10} = 0.1$

5. $\frac{1}{10000} = 0.0001$

6. $\frac{3}{4} = 0.75$

7. $\frac{5}{100} = 0.05$

8. $\frac{7}{1000} = 0.007$

9. $\frac{3}{10} = 0.3$

10. 1

11. $(4 \cdot 10^3) + (3 \cdot 10^2) + (5 \cdot 10^1) + (7 \cdot 10^0)$

12. a. 1   b. 6   c. 12
    d. 216

13. $\frac{21}{10} = 2\frac{1}{10}$

14. $\frac{93}{14} = 6\frac{9}{14}$

15. $\frac{18}{5} = 3\frac{3}{5}$

16. 0.09

17. 700

18. 28

19. thousands or 10^3

Chp. 4, Section 1
Problem Set 1.5

1. a. $(4 \cdot 10^0) + (3 \cdot 10^{-1}) + (9 \cdot 10^{-2})$

    b. $(2 \cdot 10^1) + (3 \cdot 10^0) + (2 \cdot 10^{-1}) + (9 \cdot 10^{-2}) + (6 \cdot 10^{-3})$

    c. $(9 \cdot 10^{-3})$

    d. $(6 \cdot 10^3) + (8 \cdot 10^{-2})$

2. a. hundredths or 10^{-2} or $\frac{1}{10^2}$

    b. thousandths or 10^{-3}

       or $\frac{1}{10^3}$

3. a. $\frac{1}{9}$ b. $\frac{1}{10} = 0.1$

    c. $\frac{1}{1000} = 0.001$ d. $\frac{2}{9}$

    e. $\frac{7}{100} = 0.07$ f. 1

    g. 3   h. 40   i. 54

4. $(8 \cdot 10^2) + (6 \cdot 10^0)$

5. 12

6. $\frac{35}{18} = 1\frac{17}{18}$

7. $\frac{1}{6}$

8. 2

9. ones or 10^0

10. c

Chp. 4, Section 1
Problem Set 1.6

1. 532.36

2. 804.03

3. 0.367

4. 0.0354

5. d

6. a

7. $(1 \cdot 10^1) + (2 \cdot 10^0) + (3 \cdot 10^{-2}) + (1 \cdot 10^{-3})$

8. a. $\frac{1}{16}$ b. $\frac{1}{1000} = 0.001$
    c. 0   d. 32
    e. 300   f. 18

9. 0

10. 30%

11. $\frac{7}{20}$

12. 0.41

13. thousandths or 10^{-3} or $\frac{1}{10^3}$

14. $635.50

15. $1,862

16. $300

Chp. 4, Section 2
Problem Set 2.1

1. a. a^{11} b. x^6 c. $2x^9$
    d. $6x^7$ e. $\frac{1}{y^3}$ f. $1/x^7$
    g. x^3 h. 1   i. $1/x^7$
    j. $-3x^4$ k. $1/x^{18}$ l. $16a^4$
    m. $\frac{x^3}{y^3}$

2. no, $a^x \cdot a^y = a^{x+y}$

3. no, $2^2 \cdot 2^3 = 2^5$

4. a. $x^{20}y^{12}$
    b. x^6/y^{14}
    c. $64y^6$ d. y^6
    e. $40y^{11}$
    f. $27b^2$

5. d

6. d

7. c

8. b

9. 2.03

10. $(2 \cdot 10^2)+(1 \cdot 10^0)+(4 \cdot 10^{-3})$

11. a. $\frac{1}{4}$ b. $\frac{1}{10} = 0.1$

    c. 1   d. 81

    e. 4000   f. $\frac{3}{4}$

    g. 64   h. 7

12. 24

13. $1\frac{6}{11}$

14. 10

15. $\frac{11}{3} = 3\frac{2}{3}$

16. hundredths or 10^{-2} or $\frac{1}{10^2}$

17. $1\frac{1}{4}$

18. a

19. $326.25

Chp. 4, Section 3
Problem Set 3.1

1. a. $3.25 \cdot 10^5$

   b. $4.85 \cdot 10^8$

   c. $7.12 \cdot 10^1$

   d. $3.86 \cdot 10^0$

   e. $6 \cdot 10^{-4}$

   f. $1.75 \cdot 10^{-1}$

   g. $7.96 \cdot 10^{-7}$

   h. $1 \cdot 10^{-1}$

2. a. 29,000 b. 0.0053

   c. 6.9

3. a. x^7 b. $8x^9$ c. $1/a^7$

   d. y^4 e. a^3 f. $5x^3$

   g. x^9 h. $27y^3$ i. $\frac{9}{16}$

4. 6020.03

5. $7 \cdot 10^{-3}$

6. a. $\frac{1}{64}$ b. 6000   c. 4

   d. 32   e. 1

---

7. 4.8

8. e

Chp. 4, Section 3
Problem Set 3.2

1. a. $6 \cdot 10^1 = 60$

   b. $8 \cdot 10^7 = 80{,}000{,}000$

   c. $2 \cdot 10^3 = 2000$

   d. $1 \cdot 10^4 = 10{,}000$

   e. $1 \cdot 10^8 = 100{,}000{,}000$

   f. $2 \cdot 10^{-10} = 0.0000000002$

   g. $9 \cdot 10^{-9} = 0.000000009$

2. a. 0.52   b. 0.0345

   c. 0.007   d. 0.0002

3. a. $9.25 \cdot 10^8$

   b. $6.2 \cdot 10^{-5}$

4. 75,000

5. 2.3

6. a. a^5 b. $-15a^7$ c. $5x^5$

   d. $\frac{1}{50}$ e. 75

7. 7.1%

8. 5

9. 38%

10. thousandths or 10^{-3}

11. d

Chp. 4, Section 3
Problem Set 3.3

1. a. 1,922,000

   b. 1,531,500

   c. 251,500

   d. 4,496,500

2. 320,000,000

3. 0.0041

4. a. x^3 b. $1/x^7$ c. $-5x^6$

5. 0.0135

6. $(5 \cdot 10^3)+(2 \cdot 10^2)+$

   $(9 \cdot 10^1)+(7 \cdot 10^0)+$

   $(3 \cdot 10^{-1})+(1 \cdot 10^{-2})$

7. a. $\frac{1}{25}$ b. $\frac{3}{500}$

   c. 6   d. 144   e. 48

---

8. $\frac{47}{6}$

9. $2\frac{5}{18}$

10. $\frac{11}{10} = 1\frac{1}{10}$

11. 11.975

12. $\frac{8}{5}$, 1.6

13. 150

Chp. 4, Section 4
Problem Set 4.1

1. 24_7

2. 20_9

3. 22_8

4. 30_6

5. $2.635 \cdot 10^7$

6. 3

7. 7.5

8. a. $1/a^5$ b. x^4 c. $10a^2$

   d. $\frac{1}{10000} = 0.0001$ e. $\frac{1}{27}$

   f. 64   g. $\frac{1}{16}$

   h. $8b^3$ i. $\frac{16}{25}$

9. 40

10. $2\frac{2}{3}$

11. $6\frac{4}{5}$

12. 200%

13. -16

14. b

Chp. 4, Section 4
Problem Set 4.2

1. a. 23   b. 25   c. 14

   d. 38   e. 7   f. 32

   g. 35   h. 3   i. 43

   j. 18   k. 4   l. 2

   m. 6

2. Yes. $20_4 = 8$ and $2_4 = 2$.

3. Yes. Each numeral is equal to 2.

4. a. 15_9 b. 24_5 c. 32_4

5. a. $1.28 \cdot 10^5$
   b. $2.6 \cdot 10^{-3}$

6. 100,000

7. a. x^4 b. $20a^8$
   c. $1/x^6$ d. $\frac{1}{32}$ e. 1
   f. $\frac{1}{1000}$ g. $\frac{1}{25}$ h. 16
   i. 16

8. $(2 \cdot 10^1)+(3 \cdot 10^0)+$
   $(1 \cdot 10^{-1})+(2 \cdot 10^{-2})+$
   $(5 \cdot 10^{-3})$

9. 0.103

10. $\frac{11}{20}$

11. $\frac{7}{32}$

12. 52%

13. 54

14. 3

15. hundredths or 10^{-2} or $\frac{1}{10^2}$

16. d

Chp. 4, Section 5
Problem Set 5.1

1. a. $(3 \cdot 6^2)+(5 \cdot 6^1)+(4 \cdot 6^0)$
   b. $(2 \cdot 8^3)+(7 \cdot 8^2)+$
      $(5 \cdot 8^1)+(4 \cdot 8^0)$

2. base 3

3. c

4. a. 2^2 b. 3^3 c. 5^1
   d. 8^4 e. 4^{-2}

5. a. 36 b. 20 c. 5

6. a. 22_7 b. 31_5 c. 17_9

7. $6.5 \cdot 10^{-6}$

8. 1,500

9. a. a^2 b. $1/x^6$ c. x^5
   d. $-3x^3$ e. 1 f. 4
   g. 50 h. 1 i. a^{21}
   j. a^6b^6 k. x^7/y^7

10. $(2 \cdot 10^2)+(1 \cdot 10^1)+(2 \cdot 10^0)$
    $+(1 \cdot 10^{-2})+(5 \cdot 10^{-3})$

11. 40%

12. 500

13. 9

14. 0.0475

15. $\frac{1}{4}$

Chp. 4, Section 5
Problem Set 5.2

1. a. 113 b. 423 c. 375
   d. 11 e. 150 f. 246

2. a. 14_8 b. 15_7 c. 30_4

3. 6,000,000

4. a. a^2 b. $-30x^6$ c. $1/y^3$
   d. 1 e. 4 f. 1
   g. 100 h. 27 i. $\frac{3}{4}$

5. $(1 \cdot 10^{-4})$

6. 12

7. 0.07

8. $\frac{17}{20}$

9. 2

10. 8^3

11. $\frac{1}{25}$

12. -16

13. -54

14. 21

Chapter 4 Self-Test

1. c
2. d
3. b
4. c
5. a
6. d
7. b
8. c
9. a
10. a
11. d
12. b
13. d
14. a
15. d
16. b
17. c
18. b
19. c
20. d
21. b
22. d
23. c
24. a
25. a

Chapter 5

Chp. 5, Section 1
Problem Set 1.1

1. a, c, d and e are identities
2. yes, x can be equal to 5
3. yes, y can be equal to -6
4. yes, z can equal any real number
5. no
6. a. yes b. no
7. a. 27 b. 1 c. $\frac{1}{16}$
   d. 18
8. $a + b = b + a$
9. $1\frac{3}{7}$
10. 0.15
11. 18
12. 20 miles

Chp. 5, Section 1
Problem Set 1.2

1. yes; {3}
2. no
3. yes; {-4}
4. yes; {-4}
5. yes
6. yes
7. yes
8. yes
9. yes
10. yes
11. no
12. yes
13. a. 36 b. 6 c. $\frac{1}{8}$
    d. 36
14. $(a \cdot b) \cdot c = a \cdot (b \cdot c)$
15. $5\frac{3}{5}$
16. 6.094
17. 250%
18. 28

Competency in College Mathematics

1. {2}
2. {5}
3. {-5}
4. {4}
5. $\{-\frac{19}{3}\}$
6. {-25}
7. {24}
8. {3}
9. $\{-\frac{15}{4}\}$
10. $\{\frac{21}{5}\}$
11. c
12. c
13. no
14. yes
15. all real numbers
16. a. 18   b. 2   c. $\frac{1}{9}$
    d. -7
17. $a(b + c) = ab + ac$
18. $\frac{10}{7} = 1\frac{3}{7}$
19. 4
20. $\frac{9}{20}$
21. 20%

1. {5}
2. {10}
3. $\{\frac{5}{3}\}$
4. {-7}
5. {-5}
6. {-4}
7. {-9}
8. {13}
9. {2}
10. a. {10}   b. {-4}   c. {6}
    d. $\{-\frac{12}{5}\}$

11. a. $-6x^6$ b. $-12x^2$ c. 7
12. -5
13. 69
14. $\frac{53}{15} = 3\frac{8}{15}$
15. 12
16. $\frac{6}{7}$
17. 0.4
18. 75%
19. 14
20. 0.075

1. {1}
2. {-2}
3. {3}
4. {-1}
5. {-3}
6. {10}
7. {6}
8. {-5}
9. {1}
10. a. {3}   b. {-5}   c. {3}
    d. {3}   e. $\{-\frac{9}{4}\}$
11. a. $-12/x^7$ b. $-2x^2$ c. 0
    d. -7
12. yes
13. 9
14. 13
15. 18
16. 0.0005
17. 40%
18. $\frac{1}{2}$
19. 5^3

1. {4}
2. $\{-\frac{13}{2}\}$

3. $\{-\frac{8}{3}\}$
4. {-2}
5. {2}
6. {3}
7. {1}
8. a. {-4}   b. {-4}   c. {1}
   d. $\{-\frac{24}{9}\}$
9. a. x^2 b. $2x^2$ c. -2
10. $\frac{1}{3}$
11. 1
12. 116
13. $\frac{11}{6} = 1\frac{5}{6}$
14. $\frac{13}{20}$
15. 8
16. 0.0375
17. 0.625
18. 30

1. $35x^2 + 26x + 3$
2. $27x^2 + 33x - 20$
3. $40x^2 - 31x + 6$
4. $15x^2 - 17x - 4$
5. $36x^2 + 43x - 35$
6. $4x^2 - 49$
7. $16x^2 + 38x + 21$
8. $16x^2 + 78x - 55$
9. $49x^2 - 16$
10. $25x^2 + 70x + 49$
11. 40%
12. 115
13. 70
14. It snows and some people do not ski.

Chp. 5, Section 1
Problem Set 1.8

1. $(3x + 5)(x + 1)$
2. $(2x + 7)(x + 1)$
3. $(5x + 1)(x + 11)$
4. $(4x + 1)(x + 2)$
5. $(2x + 1)(4x + 3)$
6. $(6x - 5)(x - 3)$
7. $(8x + 9)(2x - 3)$
8. $(4x - 3)(2x - 3)$
9. $(4x - 3)(x - 6)$
10. $(4x + 3)(3x - 7)$
11. $(4x - 9)(x + 2)$
12. $(3x + 1)(2x + 3)$
13. $(2x - 3)^2$
14. $(4x - 7)(2x - 3)$
15. $(3x - 1)(2x + 7)$
16. 20%
17. 28
18. $4.16 \cdot 10^3$
19. 52
20. b

Chp. 5, Section 1
Problem Set 1.9

1. $\frac{2}{3}$ or -4
2. $-\frac{1}{2}$ or 3
3. $\frac{3}{5}$ or 1
4. $\frac{2}{7}$ or 4
5. 2 or -1
6. -5 or -3
7. $\frac{2}{3}$ or 5
8. $\frac{3}{4}$ or 2
9. 9 or -6
10. $\frac{3}{2}$ or 1
11. $-\frac{3}{2}$ or 1
12. 8 or -2

13. 5 or 3
14. 1 or $-\frac{5}{7}$
15. 25%
16. -10
17. inverse property of addition
18. 2^4 or 16

Chp. 5, Section 1
Problem Set 1.10

1. -1 or $-\frac{4}{5}$
2. $\frac{1}{4}$ or $-\frac{1}{3}$
3. $-1 + \sqrt{5}$ or $-1 - \sqrt{5}$
4. $\frac{1+\sqrt{7}}{3}$ or $\frac{1-\sqrt{7}}{3}$
5. $\frac{3+\sqrt{41}}{8}$ or $\frac{3-\sqrt{41}}{8}$
6. $\frac{-7+\sqrt{53}}{2}$ or $\frac{-7-\sqrt{53}}{2}$
7. $\frac{-1+\sqrt{5}}{2}$ or $\frac{-1-\sqrt{5}}{2}$
8. 7 or -1
9. 9 or -5
10. $-\frac{1}{5}$
11. 80
12. If $x > 5$ and $x < -5$, then $|x| > 5$.
13. $13\sqrt{2}$
14. 14
15. $2.49 \cdot 10^{12}$

Chp. 5, Section 2
Problem Set 2.1

1. yes
2. a, b and e are true
3. a. $4 > -1 > -12$
   b. $-2 > -4 > -5$
   c. $-\frac{1}{3} > -\frac{1}{2} > -\frac{2}{3}$
4. $\frac{1}{a} > \frac{1}{b}$

5. $\frac{1}{a} > \frac{1}{b}$
6. a. $\frac{3}{4} > \frac{2}{3}$
   b. $-\frac{4}{8} < -\frac{3}{10}$
   c. $-\frac{5}{8} > -\frac{2}{3}$
   d. $-\frac{5}{2} < -2$
   e. $.70 < \frac{3}{4}$
   f. $4.306 < 4.31$
   g. $5.3 > 5.31$
7. d
8. a. $\{-1\}$ b. $\{-\frac{9}{4}\}$
9. irrational number
10. 218
11. a. 50 b. $\frac{1}{2}$ c. 1
12. 4000
13. 11
14. 56%
15. 1.25

Chp. 5, Section 2
Problem Set 2.2

1. yes
2. yes
3. no, the inequality reverses
4. yes
5. a. $4 > 3$ b. $2 < 3$
   c. $3 < 4$ d. $3 < 6$
   e. $-1 > -2$
6. d
7. b
8. c
9. c
10. a. false b. true
    c. false d. true
11. $-3 < -1 < 3$
12. $\{2\}$
13. $\frac{7}{20}$
14. 10.5

15. 9

16. $2900

17. 280

18. $\frac{21}{32}$

Chp. 5, Section 2
Problem Set 2.3

1. x < -4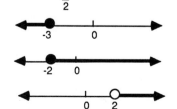

2. y ≥ 5

3. x > -2

4. a ≤ -2

5. a. true   b. false   c. true   d. true
6. true
7. no
8. yes
9. yes
10. a 3 > 2   b. 2 < 6   c. -2 > -6
11. a. true   b. false   c. true   d. false
    e. true   f. true
12. a. {-2}   b. {-21}   c. {-1}   d. {-8}
    e. {-$\frac{2}{5}$}   f. {-$\frac{4}{3}$}
13. 4.5 • 10⁻⁵

14. 86
15. 37.5%
16. -1
17. two (5 and 35)

Chp. 5, Section 2
Problem Set 2.4

1. x > $\frac{1}{2}$

2. x ≤ -3

3. x ≥ -2

4. 2 < a or a > 2

5. d
6. c
7. 1 ≤ x ≤ 4
8. a. x > 3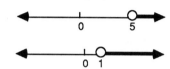

    b. y ≤ 1

    c. x > 5

    d. y > 1

9. yes
10. yes
11. false
12. yes
13. a. {1}   b. {1}   c. {-2}   d. {-21}
    e. {-$\frac{20}{3}$}
14. a. 3x   b. -3   c. -1
15. yes
16. 9.25 • 10⁸
17. $\frac{21}{5} = 4\frac{1}{5}$
18. 0.72
19. 300
20. $\frac{11}{2} = 5\frac{1}{2}$
21. true
22. -8

Chp. 5, Section 2
Problem Set 2.5

1. x ≥ -2

2. x > 4

3. x > 2

4. a > -2

5. x ≤ -3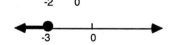
6. a
7. b

8. b

9. c

10. a

11. d

12. d

13. a. x < 2

b. x ≥ -1

c. a ≥ 3

d. x < 4

e. x ≥ -2

14. no

15. yes

16. no

17. false

18. a. {4}   b. {$-\frac{3}{2}$}

19. a. 3x   b. -3   c. 3000   d. 81

20. b

21. d

22. $\frac{35}{4}$ or $8\frac{3}{4}$

Chp. 5, Section 3
Problem Set 3.1

1. x + 5

2. x + 12

3. x – 10

4. x – 6

5. 2x – 4

6. 3x

7. 4x

8. $\frac{x}{5}$

9. b

10. a

11. c

12. no

13. x > -1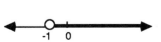

14. 0.000042

15. $\frac{-3}{2}$ > -2

16. 15

17. three (2,3 and 5)

Chp. 5, Section 3
Problem Set 3.2

1. 2x + 10 = 32

2. $\frac{x}{6}$ = 4

3. 4x + 3 = 2x + 11

4. x + 10 = 3x – 16

5. 3x

6. x – 10

7. 20 – x

8. $\frac{1}{2}$ • x

9. a. 2x + 3   b. 3 • $\frac{x}{4}$

10. a

11. -7

12. a. $-6x^2$ b. -3   c. -4

13. x > -2

14. no

Chp. 5, Section 3
Problem Set 3.3

1. 8

2. 45

3. 17

4. 9

5. 8

6. 12 and 24

7. 10 and 13

8. 500

9. 40%

10. 3; 12

11. $2050

12. a

13. {2}

14. b > 4

15. $13,625

1. 9, 10, 11
2. 77, 78, 79
3. 64, 66, 68
4. 15, 17, 19
5. 51, 53, 55, 57
6. 42, 44, 46, 48
7. No integers satisfy these conditions.
8. No odd integers satisfy these conditions.
9. 10, 12, 14
10. 17, 19, 21
11. 8, 10, 12
12. 17, 19, 21
13. The conditions are true for any three consecutive even integers.
14. No three consecutive odd integers satisfy these conditions.
15. d
16. 9
17. 13

1. 72
2. 59
3. 64
4. 93
5. c
6. d
7. 186
8. b
9. 1
10. d

1. a. $\frac{1}{2}$ b. $\frac{5}{7}$ c. $\frac{2}{7}$
   d. $\frac{2}{3}$ e. $\frac{5}{4}$ f. $\frac{4}{3}$
2. $\frac{3}{4}$
3. $\frac{7}{4}$; $\frac{7}{11}$
4. $\frac{3}{1}$
5. 7, 8, 9
6. $18,000
7. a
8. 150%
9. $\frac{2}{3}$ or $\frac{-1}{2}$

1. a. true proportion
   b. not a true proportion
   c. true proportion
2. 42
3. 12
4. 8
5. 144 in.
6. 430
7. 42
8. 40
9. $5\frac{5}{7}$ in.
10. 630
11. 20.8 lbs.
12. a
13. b
14. a. 4   b. 15
15. $x \le -6$

16. no
17. 5^2 or 25
18. 0.03
19. $\frac{2}{3}$
20. 24

21. yes (23)

1. $17\frac{1}{2}$
2. $5\frac{2}{3}$
3. 400
4. 108
5. $10\frac{6}{25}$
6. a. $-6x^6$ b. $1/a^5$
   c. $\frac{3}{25}$ d. 81
   e. 5   f. 8
7. 24
8. $\frac{3}{11}$
9. 1
10. false
11. 45.6%
12. 15

1. a. $A = kr^2$
   b. $A = kbh$
   c. $F = k/s^2$
   d. $I = kW/D^2$
   e. $V = kr^2h$
   f. $kx^2\sqrt{z} / R^2$
   g. $P = kT/V$
2. $1\frac{2}{3}$
3. 150
4. 384
5. $\frac{5}{6}$
6. 12
7. 2
8. 16 amps
9. 15 days
10. 10

11. 40 yrs.

12. 36

13. 7

14. two (2 and 3)

Chp. 5, Section 6
Problem Set 6.1

1. a. {(0,4),(1,2),(2,0)} b. {(0,0),(1,0),(2,0)}
   c. {(0,0),(0,1),(1,0),(2,0),(0,2),(1,1)}

2. a. {(0,1),(1,2),(2,3)} b. {(2,0),(0,2),(1,1)}
   c. {(1,1),(3,0)}
   d. {(2,0),(0,2),(1,0),(0,1),(0,0),(1,1)}
   e. {(0,2),(1,0),(0,1),(0,0),(0,3),(1,1)}

3. a. {(-2,-2),(-1,-1),(0,0),(1,1),(2,2)}
   b. {(0,-1),(2,1),(1,0),(-1,-2)}
   c. {(2,0),(0,2),(1,1)} d. {(2,2),(1,0),(0,-2)}

4. a. {(2,1),(2,0),(2,-1),(1,0),(1,-1),(0,-1)}
   b. {(2,1),(2,0),(2,-1),(1,0),(1,-1),(0,-1)}
   c. {(1,2),(0,1),(0,2),(-1,0),(-1,1),(-1,2)}

5. 24

6. 9

7. {-$\frac{6}{5}$}

8. 50

9. no

10. 350,000

11. x ≥ 1

12. -7

Chp. 5, Section 6
Problem Set 6.2

1–9

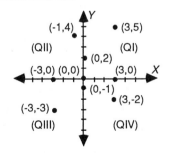

10. {(0,0),(0,1),(1,0),(1,1),(2,0),(0,2)}

11. 26

12. {2}

13. a. 16   b. 0   c. $\frac{3}{4}$
    d. x^2 e. $\frac{1}{a}$

14. a(b + c) = ab + ac

15. $\frac{7}{17}$

16. 600

17. 240

Chp. 5, Section 6
Problem Set 6.3

1.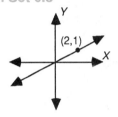

2.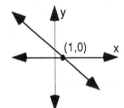

3.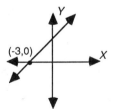

4.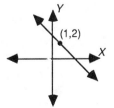

5.

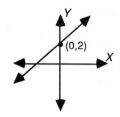

2.

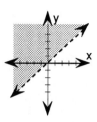

6.

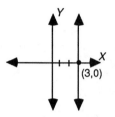

3.

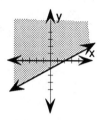

7.

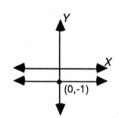

4.

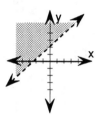

8. $2\frac{1}{3}$

9. 40

10. 16

11. $x \leq 2$

5.

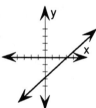

12. yes

13. 3.998

14. 16%

15. true

Chp. 5, Section 6
Problem Set 6.4

1.

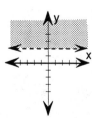

6. $\frac{2}{3}$

7. 210 pounds

8. 4

9. no

10. $-\frac{3}{4} < -\frac{2}{3}$

11. $a \cdot b = b \cdot a$

12. 10

13. $\{-\frac{6}{5}\}$

14. 1.75

15. $1\frac{3}{5}$

1.

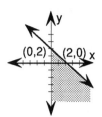

2.

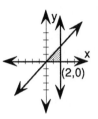

3.

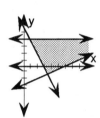

4.

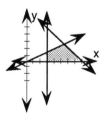

5. a. $x < -1$ b. $y < -2$
    c. $x \geq 0$ and $y \leq 0$
    d. $x \geq 0$, $y \geq 0$ and $y \leq 2$
    e. $y > x + 1$
6. a
7. d

8.

9. $y = 4$
10. no
11. 129
12. 3200
13. $(a + b) + c = a + (b + c)$
14. false

**Chp. 5, Section 7
Problem Set 7.1**
1. $\{(-31, -13)\}$
2. $\{(3, \frac{5}{2})\}$
3. $\{(2, -3)\}$
4. $\{(4, 0)\}$
5. $\{(5, -5)\}$
6. $\{(-2, -5)\}$
7. $\{(5, 4)\}$
8. $x < 12$
9. a
10. 17, 19, 21, 23
11. 42

**Chp. 5, Section 7
Problem Set 7.2**
1. $\{(5, 2)\}$
2. $\{(2, -1)\}$
3. $\{(2, -3)\}$
4. $\{(-4, 11)\}$
5. $\{(-1, -3)\}$
6. $\{(3, -1)\}$
7. $\{(4, \frac{1}{3})\}$
8. $\{(4, 2)\}$
9. $\{(-2, -1)\}$
10. $\{(7, 4)\}$
11. $\{(1, -2)\}$

12. $\{(0, 4)\}$
13. 85
14. a
15. c
16. b
17. b

**Chp. 5, Section 7
Problem Set 7.3**
1. dependent
2. inconsistent
3. inconsistent
4. inconsistent
5. inconsistent
6. inconsistent
7. dependent
8. -2, -1, 0
9. three
10. b

Chapter 5 Self-Test

1. d	14. b	27. c
2. b	15. c	28. d
3. d	16. d	29. c
4. a	17. c	30. b
5. a	18. a	31. c
6. d	19. a	32. a
7. c	20. c	33. d
8. c	21. c	34. b
9. a	22. b	35. c
10. c	23. d	36. c
11. b	24. c	37. d
12. b	25. d	
13. b	26. d	

Appendix

Chapter 6

Chp. 6, Section 1
Problem Set 1.1

1. If two points were not distinct, one point would co-incide with the other. In other words, there would be two representations for the same location. Thus, an infinite number of lines could be drawn through that location.

2. a. yes   b. yes; yes
   c. six
   $(\overleftrightarrow{AB}, \overleftrightarrow{BA}, \overleftrightarrow{BC}, \overleftrightarrow{CB}, \overleftrightarrow{AC}, \overleftrightarrow{CA})$
   d. no
   e. no. Point B is not included.

3. three

4. six

5. no. A line has no endpoints.

6. yes. If two lines coincide they both would occupy the same position. Any point on one line would also be a point on the other.

7. $\frac{25}{36}$

8. $\frac{1}{2}$

9. $\frac{5}{9}$

10. 38

Chp. 6, Section 2
Probem Set 2.1

1. If the three distinct points all lie on the same line, then an infinite number of planes could pass through those points.

2. no

3. yes
4. yes
5. no
6. no
7. infinitely many
8. infinitely many
9. yes
10. 10
11. 3.997
12. 30
13. $\frac{7}{20}$
14. 0.045
15. $\frac{3}{5}$

Chp. 6, Section 3
Problem Set 3.1

1. a. $\overleftrightarrow{NK}$ b. $\overrightarrow{FK}$
   c. $\overleftrightarrow{NF}$ d. $\overrightarrow{FI}$
   e. point N

2. no

3. a. false. Two half-lines result.
   b. false

4. An infinite number of planes

5. no

6. no

7. 6

8. 9%

9. $1\frac{11}{12}$

10. 2.1

11. $\frac{3}{25}$

12. $3\sqrt{2}$

13. $5\sqrt{3}$

Chp. 6, Section 3
Problem Set 3.2

1. yes
2. a. true   b. true
   c. false   d. true
   e. true

3. yes
4. a. $\overleftrightarrow{BC}$
   b. Point C
5. yes
6. no
7. 18
8. 16
9. 18

Chp. 6, Section 3
Problem Set 3.3

1. a. point Z and point Y
   b. $\angle ZXY$
   c. $\overrightarrow{ZY}$ and point X
   d. point Z

2. a. Exterior $\angle HAM$
   b. $\overrightarrow{AT}$

3. a. point P and point U
   b. Interior $\angle DCJ$

4. no. It represents the entire plane except for the set of points of the angle itself.

5. a. $\varnothing$ b. $\overleftrightarrow{AD}$
6. no
7. 3
8. 1.2
9. 12
10. 4000
11. 0.0275
12. $\frac{1}{40}$

Chp.6, Section 4
Problem Set 4.1

1. no. A curve forms one continuous path on a plane. One must lift the pencil from the paper to draw this particular figure.

2. no

3. no

4. It is not a simple curve because the curve passes through some point more than once.

5. It is neither a simple nor a closed curve because it passes through some points more than once, and the endpoints do not coincide.

6. a. Point C and Point D
    b. Interior ∠ABC

7. a. $\overset{\circ}{Q}\overset{\circ}{R}$ b. $\overset{\leftrightarrow}{QR}$

8. A straight line

9. 25%

10. $3\frac{4}{5}$

11. $4\frac{1}{6}$

12. $\frac{1}{8}$

13. $4.1 \cdot 10^{-4}$

1. yes

2. A polygon is a simple closed curve. Any curve formed by only two line segments joined together could not be closed.

3.

4. This curve is not a polygon because some portions are not line segments.

5. a. quadrilateral
    b. quadrilateral, parallelogram
    c. quadrilateral, parallelogram, rectangle
    d. quadrilateral, parallelogram, rectangle, rhombus, square
    e. quadrilateral

---

6. c
7. d
8. d
9. a
10. no
11. yes
12. a. $\overset{\leftrightarrow}{BA}$
    b. ∠CBD
13. infinitely many planes
14. 3
15. $\frac{25}{6}$
16. 20
17. $\frac{13}{20}$
18. $a \cdot b = b \cdot a$

1. a. yes   b. no
    c. no
2. a. not convex
    b. convex
    c. not convex
3. no
4. a. Interior ABD
    b. $\overset{\leftrightarrow}{AC}$
    c. The entire plane
5. a. Interior ACD
    b. $\overset{\leftrightarrow}{CD}$ and point A
    c. Interior ABF
6. Pentagon
7. no
8. Interior ABC
9. lines in the same plane that never intersect.
10. 50
11. $1\frac{3}{5}$
12. -21
13. $(a + b) + c = a + (b + c)$

---

1. a. 4; 3   b. 6
    c. 4, triangular region
    d. $4 - 6 + 4 = 2$
2. a. 5   b. 8
    c. 5
    d. $5 - 8 + 5 = 2$
3. a. 6   b. 9
    c. 5
    d. $6 - 9 + 5 = 2$
4. a. 6   b. 12
    c. 8
    d. $6 - 12 + 8 = 2$
5. 4
6. 6
7. 4
8. no
9. yes
10. Interior ∠ABC
11. 6
12. yes
13. no
14. 150%
15. $3\frac{1}{5} = \frac{16}{5}$
16. $\frac{7}{10}$
17. 0.0575
18. $\frac{9}{5}$

1. a. 9 yd.   b. 5 yd.
    c. 4 ft.   d. 4 yd.
    e. 7 in.   f. 38 cm.
    g. 6 km.   h. 2,501 km.
    i. 2.7 km.   j. 24.27 cm.
    k. 62.302 m.   l. 5,500 ft.
2. a. 4 inches   b. $3\frac{1}{2}$ inches
    c. $3\frac{3}{4}$ inches   d. $3\frac{5}{8}$ inches

e. 4 inches   f. 4 inches

g. 4 inches   h. $4\frac{1}{8}$ inches

3. five
4. no
5. yes
6. $\overrightarrow{BC}$
7. yes
8. eight
9. 20
10. 0.08
11. $7\frac{2}{5}$
12. 0.5
13. $25.30

Chp. 6, Section 7
Problem Set 7.1

1. 42.8 yd.
2. 30 in.
3. 7 in.
4. 5 ft.
5. 180 ft.
6. $133.33
7. 96
8. W = 10, L = 20
9. W = 4, L = 11
10. 31.4 ft.
11. 44 in.
12. r = 5 units, d = 10 units
13. 31.4 in.
14. 31.4 yd.
15. 6 in.
16. 11
17. a. 6 ft.   b. 4.3 cm.

c. 6 m

18. 4
19. no
20. no
21. yes
22. $\frac{31}{6}$
23. 15

24. 30
25. 6%

Chp. 6, Section 8
Problem Set 8.1

1. 20°, acute
2. 60°, acute
3. 30°, acute
4. 130°, obtuse
5. 70°, acute
6. 120°, obtuse
7. 90°, right
8. 180°, straight
9. yes
10. four
11. 20 units
12. 88 in.
13. 13.2 ft.
14. 1 m
15. $32\frac{1}{3}$
16. yes
17. $2\frac{2}{9}$
18. $\frac{3}{4}$
19. 24

Chp. 6, Section 8
Problem Set 8.2

1. a. ∠3 and ∠5

b. ∠3

2. a. 17°   b. 66°
3. a. 63°   b. 91°
4. 90°
5. 45°
6. 60°
7. The angles are
    supplementary
8. 90°
9. supplement; 90°
10. a. 50°, 130° b. 140°, 40°

c. 46°, 44°   d. 60°, 30°

11. a. yes   b. yes
12. a. straight   b. obtuse

c. acute

13. 62.8 ft.
14. 80 in.
15. 4.08 km.
16. yes
17. 200
18. 0.046
19. $\frac{-1}{8}$
20. $600

Chp. 6, Section 9
Problem Set 9.1

1. ∠1, ∠8, ∠6
2. ∠6, ∠8, ∠3, ∠1
3. ∠4, ∠5, ∠7
4. ∠6, ∠8, ∠1, ∠3
5. ∠2 =12°, ∠5 = 12°,
    ∠6 = 12°, ∠1 = 168°,
    ∠3 = 168°, ∠4 = 168°,
    ∠7 = 168°
6. 140°
7. 45°
8. x = 25; The corresponding
    angles are each 125°.
9. b
10. b and c
11. a. ∠2, ∠4   b. ∠4

c. ∠2, ∠4

12. x = 10; 30°, 60°
13. x = 7; 25°
14. 220 ft.
15. $16\frac{2}{3}$ ft.

Chp. 6, Section 10
Problem Set 10.1

1. yes
2. no
3. yes
4. yes
5. yes

6. a. isosceles; right
   b. isosceles; acute
   c. scalene; acute
   d. equilateral; acute
   e. scalene; obtuse
7. 118°
8. 61°, 43°, 76°
9. 79°
10. 130°
11. 53°
12. $180° - 2x°$
13. a. true   b. true
    c. true   d. false
    e. true
14. c
15. a. $\angle 5; \angle 4; \angle 1$
    b. $\angle 8; \angle 5; \angle 4; \angle 1$
    c. $\angle 5; \angle 8; \angle 1$
    d. $\angle 2; \angle 3; \angle 7; \angle 6$
16. x = 3; 20°
17. x = 40; 80°, 10°
18. 16,000 lb.
19. 5 ft.
20. $\frac{3}{5}$
21. $\frac{19}{12}$ or $1\frac{7}{12}$
22. 2.985
23. $\frac{4}{5}$
24. 4.5
25. 15
26. $\frac{21}{10}$

1. 720°, 1,080°
2. 18°
3. 140°
4. 1,800°
5. 125°; 55°
6. a. 70°   b. 48°
   c. 18°   d. 110°

7. 145°
8. 135°
9. 12 ft.
10. 10π in.
11. 13.018 cm.
12. $3\frac{7}{8}$
13. $1\frac{3}{4}$
14. 13.026
15. 0.375
16. 200
17. 12
18. $\frac{107}{500}$
19. {-4}
20. ii and iii

1. a. 17 in.   b. 7 ft.
   c. 8 yds.   d. 2 cm.
   e. 7 m.   f. 1 in.
2. a. 15 miles b. 24 ft.
   c. 25 ft.   d. 20 ft.
   e. 25 ft.   f. yes
3. $2,640,000
4. b
5. 540°
6. 150°
7. 6a + 7b
8. m $\angle 1 = 130°$ m $\angle 2 = 50°$
   m $\angle 3 = 130°$ m $\angle 4 = 130°$
   m $\angle 5 = 50°$ m $\angle 6 = 50°$
   m $\angle 7 = 130°$
9. m $\angle x = 170°$
10. 55°
11. c
12. 6300 ft.
13. 25
14. $\frac{34}{15} = 2\frac{4}{15}$
15. 46.035
16. $\frac{10}{7} = 1\frac{3}{7}$

17. 0.275
18. 200
19. 12
20. $\frac{13}{20}$
21. x < -3

1. DF = 6 and DE = 8
2. AC = 7.5 and BC = 15
3. DE = 4
4. AB = 10
5. BC = 16 and EC = 6
6. i and iii
7. i and iv
8. i, ii and iii
9. 8, 10 and 12
10. 10 in. and 15 in.
11. 21 ft.
12. 125 ft.
13. 48 in.
14. 5 ft.
15. x = 24; 72°, 108°
16. 37 cm.
17. 40%
18. 0.0001
19. 0.022
20. $8\frac{3}{7}$
21. 20
22. 10

1. a. 18 ft^2 or 2592 in^2
   b. 6 yd^2 or 54 ft^2
   c. 2 yd^2 or 2592 in^2
2. $272.58
3. 900 ft^2
4. 230 ft^2
5. $84

6. $30
7. 376 ft^2
8. 96 in^2
9. false
10. 30 ft.
11. m ∠1 = 125°
12. 3 ft; 3 ft; 8 ft
13. g
14. 6.8 m
15. $\frac{2}{3}$
16. 0.625
17. $\frac{11}{25}$
18. 0.0225
19. 16
20. $\frac{2007}{1000}$
21. $\frac{11}{24}$

Chp. 6, Section 14
Problem Set 14.2

1. 81 m^2
2. $1\frac{1}{4}$ ft^2; 180 in^2
3. $1\frac{1}{3}$ yd^2; 12 ft^2
4. 14 in^2
5. 69 ft^2
6. 44 in^2
7. 800 ft^2
8. 70.5 cm^2
9. 18 ft^2
10. four
11. $144
12. yes
13. Two angles of ΔABC are congruent, respectively, to two angles of ΔCDE.
14. 5 ft.
15. 52°

16. x = 11; 35°, 55°
17. f
18. 47,300 ft.
19. $5\frac{13}{20}$
20. $2\frac{7}{12}$
21. 0.0484
22. 5
23. 21
24. 0.0004
25. $\frac{123}{100}$

Chp. 6, Section 14
Problem Set 14.3

1. 2826 ft^2
2. $38\frac{1}{2}$ in^2
3. 757 ft^2
4. 307 in^2
5. 16π cm^2
6. 4 π yd^2
7. 92.52 cm^2
8. 42.14 ft^2
9. The area of the new circle is 4 times the area of the original circle.
10. 314 mi^2
11. four
12. iv
13. v
14. Two angles of ΔABC are congruent, respectively, to two angles of ΔDEF.
15. 24
16. i and iii
17. e
18. ii
19. 8 m

20. 1.75
21. $\frac{11}{20}$
22. 1
23. $\frac{18}{7} = 2\frac{4}{7}$
24. 200%
25. 4
26. 2.7 • 10^5

Chp. 6, Section 15
Problem Set 15.1

1. a. 9   b. 3
   c. 27 in^3
2. a. 32 ft^3 b. 24 yd^3
   c. 6 ft^3
3. yes
4. three 4-inch cubes
5. 64 yd^3
6. 600
7. $\frac{35}{48}$ ft^3
8. 4 cm.
9. 9:16
10. 4 ft.
11. vi
12. 30 m^2
13. $\frac{43}{8}$
14. 27.0445
15. $\frac{1}{4}$
16. 0.0525
17. >

Chp. 6, Section 15
Problem Set 15.2

1. a. 32 in^3 b. 144 ft^3
2. a. 360 yd^3 b. 2592 in^3
3. a. 125.6 in^3 b. 1570 ft^3
4. a. 157 m^3 b. 47.1 ft^3
5. 4186.7 ft^3

6. 33.5 ft^3

7. $750

8. $128

9. $72

10. a. linear (L) b. square (S)
    c. cubic (C)

11. 84 cm^2

12. 26 m^2

13. $13.50

14. 21 ft.

15. width = 12 yd.,
    length = 48 yd.

16. $3\frac{2}{3}$ sq. ft.

17. 20 in.

18. a. $\angle 1, \angle 4, \angle 5, \angle 8$
    b. 80°

19. $\frac{37}{12} = 3\frac{1}{12}$

20. 1.23

21. 0.375

22. 0.875

23. 20

24. $\frac{111}{250}$

25. -6

26. $\frac{1}{12}$

27. c

Chp. 6, Section 16
Problem Set 16.1

1. a. 3,700   b. 520
   c. 91   d. 0.0048

2. 1,670 m. = 1.67 km.

3. $2,000

4. a. 0.01   b. 200,000

5. 1,370 cubic centimeters

6. a. 2 cm.   b. $1\frac{1}{2}$ cm.

7. 30 grams

8. a. cm.   b. cm^2

   c. 1 liter

Chapter 6 Self-Test

1. a	18. a	35. c
2. d	19. d	36. a
3. a	20. d	37. d
4. a	21. c	38. b
5. d	22. d	39. c
6. b	23. c	40. c
7. b	24. d	41. b
8. a	25. c	42. b
9. d	26. a	43. a
10. c	27. a	44. c
11. b	28. c	45. c
12. c	29. d	46. c
13. c	30. c	47. a
14. b	31. c	48. b
15. a	32. c	49. c
16. a	33. a	50. a
17. c	34. c	

Chapter 7

Chp. 7, Section 1
Problem Set 1.1

1. 720

2. 120

3. 625

4. 96

5. 72

6. 9,000,000

7. a. 125

b. 25

8. 3,276,000

9. a. 18   b. 15

10. 12

11. $1\frac{8}{15}$

12. 0.35

13. 0.0225

14. 2

Chp. 7, Section 2
Problem Set 2.1

1. No. $_5P_6$ cannot be evaluated
   because r cannot be greater
   than n. It is meaningless to
   consider 5 objects taken 6 at
   a time.

2. a. 4   b. 42
   c. 720   d. 360

3. no

4. 210

5. 120

6. 6

7. 48

8. $\frac{23}{5}$

9. 4

10. 7.348

11. $\frac{13}{20}$

Chp. 7, Section 2
Problem Set 2.2

1. 5,040

2. 30

3. 720

4. 1

5. 120

6. 144

7. 120

8. 2

9. 24

10. 1

11. 22

12. 696

13. 20

14. 360

15. 10

16. 15

17. 1

18. 360

19. 24

20. 105

21. 27
22. 32
23. 200
24. 0.2
25. $1\frac{1}{20}$
26. 45

Chp. 7, Section 2
Problem Set 2.3

1. 120
2. 42
3. 720
4. 1
5. yes
6. a. 210   b. 1
   c. 720
7. 360
8. 12
9. 62.5%
10. 0.0001
11. $1\frac{1}{35}$
12. 30.246

Chp. 7, Section 2
Problem Set 2.4

1. 720
2. 120
3. 720
4. 24
5. 120
6. 210
7. 360
8. 3,024
9. 360
10. 1
11. 120
12. 400
13. 2.7
14. $6\frac{3}{4}$
15. $\frac{2}{3}$

16. 200
17. $7 \cdot 10^{-3}$

Chp. 7, Section 3
Problem Set 3.1

1. No. It is meaningless to consider 3 objects 4 at a time.
2. a. 15   b. 56
   c. 210   d. 1
   e. 1
3. a. 21   b. 56
   c. 1   d. 1
   e. 5
4. 24
5. 1
6. 3,024
7. 6
8. $3\frac{1}{9}$
9. 1.5
10. 1.25
11. 0.0525
12. 7
13. 27

Chp. 7, Section 3
Problem Set 3.2

1. 70
2. 56
3. 35
4. 10
5. 35
6. 56
7. 1,4,6,4,1
8. 21
9. 720
10. 60
11. 60

Chp. 7, Section 4
Problem Set 4.1

1. $_7P_7$

2. $\binom{8}{3}$
3. $\binom{8}{4}$
4. $\binom{5}{2}$
5. $_{15}P_2$
6. $_5P_5$
7. $\binom{10}{3}$
8. $_{10}P_{10}$
9. $_4P_4$
10. $_6P_6$
11. $\binom{8}{3}$
12. $\binom{10}{5}$
13. $_7P_2$
14. 126
15. 120
16. $\frac{1}{3}$
17. $\frac{16}{35}$
18. 0.056
19. 0.875
20. 0.00012
21. $\frac{31}{60}$
22. $\left\{\frac{7}{3}\right\}$

Chp. 7, Section 5
Problem Set 5.1

1. {a,e,i,o,u}
2. {bb,bg,gb,gg}
3. {OO,OG,OW,GW,GO,WO, WG,GG,WW}; {OO,GG,WW}
4. {1R, 1W, 2Y, 2G}
5. d
6. a. $\binom{10}{6} = 210$ b. $_8P_2 = 56$
7. 24
8. $4\frac{2}{5}$
9. $\frac{9}{20}$

10. $2\sqrt{3}$
11. 60%
12. 0.047
13. 457

Chp. 7,Section 5
Problem Set 5.2
1. Mutually exclusive
2. Not mutually exclusive
3. Mutually exclusive
4. Mutually exclusive
5. Mutually exclusive
6. {red marble, white marble, blue marble}
7. $\{B_1, B_2, B_3, W\}$
8. 120
9. 32
10. $\frac{9}{10}$
11. $5\frac{7}{12}$
12. 4.5
13. 300%
14. 12
15. $\frac{1}{4}$
16. $\frac{3}{16}$

Chp. 7, Section 6
Problem Set 6.1
1. No.There can never be more outcomes in an event than there are total outcomes in the experiment. The numerator in a probability fraction can never be greater than the denominator, so that a probability is never greater than 1.
2. a. $\frac{1}{6}$ b. $\frac{2}{3}$
   c. $\frac{5}{6}$ d. $\frac{2}{3}$
   e. $\frac{5}{6}$ f. $\frac{1}{6}$

3. $\frac{1}{3}$
4. $\frac{4}{7}$; $\frac{4}{7}$
5. $\frac{1}{5}$; $\frac{2}{5}$
6. $\frac{1}{2}$
7. a. 1   b. $\frac{1}{4}$
   c. $\frac{1}{2}$ d. $\frac{1}{2}$
   e. $\frac{3}{4}$
8. a. 1   b. $\frac{1}{8}$
   c. $\frac{3}{8}$ d. $\frac{1}{2}$
9. $16; 2^n$
10. a. $\frac{1}{300}$ b. $\frac{3}{4}$
   c. $\frac{1}{150}$
11. a. 0.1   b. 0.2
   c. 0.3   d. 0.4
12. No. The sum of the probabilities is less than one.
13. a. $\frac{5}{36}$ b. $\frac{1}{4}$
   c. $\frac{1}{4}$
14. a. Mutually exclusive
   b. Not mutually exclusive
15. {BB,BG,GB,GG}
16. 84
17. 90,000
18. $\frac{29}{6}$
19. $2\frac{1}{10}$
20. 41
21. 10.5

Chp. 7, Section 6
Problem Set 6.2
1. 0
2. 1
3. $\frac{7}{8}$
4. 0.75

5. $\frac{6}{7}$
6. 1
7. a. $\frac{2}{3}$ b. 0
   c. 1
8. 0.45
9. $\frac{1}{10}$
10. No. The sum of the probabilities is greater than one.
11. yes
12. {Pete,Ann,Steve}
13. 720
14. 12
15. $2\frac{14}{15}$
16. 37.5%
17. 4
18. 0.0004
19. 5.905

Chp. 7, Section 6
Problem Set 6.3
1. a. $\frac{1}{4}$ b. $\frac{1}{13}$
   c. $\frac{1}{2}$ d. $\frac{1}{52}$
   e. $\frac{3}{13}$ f. $\frac{1}{13}$
   g. $\frac{1}{26}$ h. $\frac{12}{13}$
   i. $\frac{3}{4}$ j. $\frac{10}{13}$
   k. $\frac{2}{13}$ l. 0
   m. 1
2. $\frac{3}{4}$
3. 0
4. 1
5. $\frac{1}{40}$
6. No. A probability value cannot be negative.
7. $\{B_1,B_2,B_3,B_4\}$
8. 120
9. $\frac{41}{7}$

10. 0.0375

1. a. $\frac{14}{19}$ b. 1

   c. $\frac{8}{19}$

2. a. $\frac{2}{13}$ b. $\frac{2}{13}$

   c. 1   d. $\frac{2}{13}$

   e. $\frac{3}{13}$ f. $\frac{6}{13}$

3. a. 0.6   b. 0.8

   c. 0.2

4. a. 0.3   b. 0.7

   c. 0.5   d. 0.8

   e. 0.6   f. 0.7

5. 0.6

6. a. 1   b. 0

7. $\frac{4}{5}$

8. {RR,RW,WR,WW}

9. 60

10. $\frac{7}{12}$

11. 32.186

12. 28

13. $x \le -\frac{16}{3}$

14. 200%

15. 24

1. a. $\frac{4}{13}$ b. $\frac{4}{13}$

   c. $\frac{7}{13}$ d. $\frac{8}{13}$

   e. $\frac{3}{13}$

2. a. $\frac{5}{6}$ b. 1

3. 0.6

4. a. $\frac{5}{26}$ b. $\frac{6}{13}$

   c. $\frac{7}{13}$

5. 0.7

6. 0.5

7. a. 0.8   b. 0.2

8. a. 0.5   b. 0.8

9. a. $\frac{4}{7}$ b. $\frac{3}{14}$

10. 1

11. $\frac{2}{13}$

12. $\frac{11}{12}$

13. 24

14. 136,080

15. {1,2,3}

16. $\frac{23}{5}$

17. 0.03

18. 0.023

19. 5

1. $\frac{31}{100}$ or 0.31

2. $\frac{33}{100}$ or 0.33

3. $\frac{67}{100}$ or 0.67

4. 136

5. 268

6. $\frac{14}{50}$ or 0.28

7. 1

8. 0.40

9. 0.28

10. 0.54

11. 0.20

12. $\frac{23}{60}$

13. 0.30

14. $\frac{9}{23}$

15. 1

16. 30

17. $\frac{31}{32}$

18. d

19. c

1. $\frac{1}{16}$

2. $\frac{2}{9}$

3. a. $\frac{1}{16}$ b. $\frac{1}{2704}$

   c. $\frac{9}{169}$ d. $\frac{9}{169}$

4. a. $\frac{13}{204}$ b. 0

   c. $\frac{11}{221}$ d. $\frac{11}{221}$

5. a. $\frac{1}{2}$ b. $\frac{1}{36}$

   c. $\frac{25}{216}$

6. 0.48

7. a. 0.28   b. 0.18

   c. 0.42

8. 0.35

9. $\frac{7}{15}$

10. a. $\frac{1}{100}$ b. $\frac{19}{100}$

11. a. $\frac{1}{45}$ b. $\frac{17}{45}$

12. $\frac{4}{9}$

13. a. $\frac{2}{3}$ b. 1

   c. $\frac{2}{3}$ d. 0

14. 20

15. 24

16. $1\frac{17}{24}$

17. 0.0005

18. 42

19. 0.125

20. $\frac{7}{10}$

21. $\frac{7}{27}$

22. $1

23. 15

Chp. 7, Section 9
Problem Set 9.1

1. a. $\frac{4}{5}$ b. $\frac{2}{5}$

    c. $\frac{1}{5}$ d. $\frac{3}{5}$

2. $\frac{1}{8}$

3. $\frac{1}{16}$

4. a. $\frac{1}{5}$ b. $\frac{1}{20}$

    c. $\frac{1}{120}$

5. a. $\frac{1}{36}$ b. $\frac{1}{6}$

    c. $\frac{5}{12}$

6. a. $\frac{5}{6}$ b. $\frac{5}{6}$

    c. 0   d. $\frac{2}{3}$

7. {mm,mf,fm,ff}

8. 336

9. 6

10. $4\frac{5}{8}$

11. 5,000

12. $\frac{5}{11}$

13. 56%

14. 20

15. 21

Chp. 7, Section 10
Problem Set 10.1

1. a. 30   b. 5
2. a. 105   b. 21
3. a. 140   b. 7
4. a. 120   b. 8

5. $\frac{1}{4}$

6. $\frac{2}{25}$

7. No. The sum of the probabilities is less than one.

8. 720

9. {C_1,C_2,C_3}

10. $2\frac{1}{8}$

11. $\frac{6}{25}$

12. $\frac{8}{13}$

13. 4.25

14. 17

Chp. 7, Section 10
Problem Set 10.2

1. a. $\frac{8}{15}$ b. $\frac{1}{15}$

    c. $\frac{2}{5}$

2. a. $\frac{1}{22}$ b. $\frac{7}{44}$

    c. $\frac{21}{44}$

3. a. $\frac{80}{429}$ b. $\frac{160}{3003}$

    c. $\frac{1}{3003}$ d. $\frac{20}{429}$

4. 28

5. $\frac{1}{2}$

6. a. 0.3   b. 0.2

    c. 0.2

7. a. $\frac{3}{13}$ b. $\frac{21}{26}$

    c. $\frac{9}{26}$

8. {HG,HW,TG,TW}

9. 1

10. 210

11. $1\frac{1}{3}$

12. 2.25

13. 0.285

14. 0.026

15. 250

Chp. 7, Section 11
Problem Set 11.1

1. 1 to 3
2. 3 to 1
3. 3 to 10
4. a. 3 to 10   b. 12 to 1

    c. 51 to 1

5. a. 5 to 1   b. 2 to 1
6. a. 3 to 2   b. 3 to 2

7. 1:9

8. $\frac{2}{5};\frac{3}{5}$

9. $\frac{5}{12}$

10. a. $\frac{3}{7}$ b. $\frac{1}{14}$

11. $\frac{1}{3}$

12. $\frac{4}{9}$

13. 18.176

14. $\frac{13}{20}$

15. 0.0575

16. $\frac{1}{10}$

17. 32

18. $\frac{4}{11}$

Chapter 7 Self-Test

1. e		21. b
2. c		22. d
3. a		23. c
4. c		24. c
5. b		25. a
6. c		26. c
7. b		27. b
8. c		28. d
9. a		29. a
10. c		30. c
11. b		31. c
12. e		32. b
13. c		33. a
14. b		34. d
15. d		35. a
16. b		36. a
17. a		37. c
18. b		38. c
19. c		39. a
20. b		40. a

Appendix

Chapter 8

Chp. 8, Section 1
Problem Set 1.1

1. 73, 61, 75, 59 and 71
2. Every 20th alumnus could be unemployed, and this might distort the true picture of the income level of all alumni.
3. 90 males and 110 females.
4. c
5. sample b
6. sample b
7. $2\frac{7}{12}$
8. $1\frac{1}{3}$
9. $\frac{1}{5}$
10. $\frac{17}{16} = 1\frac{1}{16}$
11. 19.961
12. 4000
13. 0.4
14. $\frac{1}{20}$
15. 1.65
16. 1.9%
17. 12
18. 0.045

Chp. 8, Section 2
Problem Set 2.1

1.
	f
1	1
2	2
3	4
4	2
5	1

2. a. class interval 5
   b. boundaries 14.5, 19.5, 24.5, 29.5, 34.5, 39.5
   c. class marks 17, 22, 27, 32, and 37
3. b
4. $4\frac{5}{8}$

5. $\frac{3}{8}$
6. 1
7. 4.9996
8. 2.5
9. 0.0003
10. $\frac{3}{2}$
11. $\frac{13}{20}$
12. 6
13. 45
14. 13

Chp. 8, Section 2
Problem Set 2.2

1.
class	c.f
50- 69	8
70- 89	21
90-109	40
110-129	65
130-149	81
150-169	93
170-189	100

2. a. 20
   b. 9.5, 29.5, 49.5, 69.5, 89.5
   c. 19.5, 39.5, 59.5, 79.5
3. 30 males, 20 females
4. 16
5. $\frac{6}{5} = 1\frac{1}{5}$
6. $\frac{33}{8}$
7. 24.1
8. 0.375
9. 1.85
10. 200
11. $\frac{7}{4} = 1\frac{3}{4}$
12. 0.0125
13. $2.85 \cdot 10^6$

Chp. 8,Section 3
Problem Set 3.1

1.
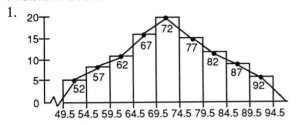

2.
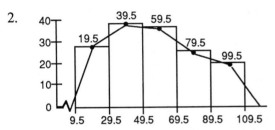

3. a. 4, 10, 18, 28, 35, 38   b. 5
   c. 4.5, 9.5, 14.5, 19.5, 24.5, 29.5, 34.5
   d. 7, 12, 17, 22, 27, 32
4. sample D
5. $1\frac{3}{4}$
6. $\frac{2}{3}$
7. $\frac{21}{32}$
8. 25%
9. 25%
10. 0.0008
11. $\frac{7}{20}$
12. 1

Chp. 8, Section 4
Problem Set 4.1

1. a. 350   b. 1971 and 1972
   c. 1971 and 1975   d. 400
2. a. Alaska   b. Alaska and Texas
3. $2,000
4. a. Monday: 40°, Tuesday: 36°, Wednesday:
      60°, Thursday: 52°, Friday: 28°
   b. Tuesday and Wednesday
   c. Thursday and Friday
5. sample A

6. a. 10
   b. 9.5, 19.5, 29.5, 39.5, 49.5, 59.5, 69.5
   c. 14.5, 24.5, 34.5, 44.5, 54.5, 64.5
   d. 4, 11, 20, 30, 38, 41
7. $6\frac{7}{12}$
8. $\frac{14}{39}$
9. $4\frac{7}{8}$
10. 310
11. 1.25
12. $\frac{6}{5}$

Chp. 8, Section 5
Problem Set 5.1

1. 9
2. 4.5
3. 69.1
4. The mean is 25.71, but would become 28.57
   since it is affected by a change in any data
   value.
5. c
6. $200
7. $\frac{1}{9}$
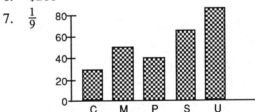
8. a 30
   b. 9.5, 39.5, 69.5, 99.5, 129.5, 159.5
   c. 24.5, 54.5, 84.5, 114.5, 144.5
   d. 10, 27, 48, 64, 71
9. $2\frac{4}{5}$
10. 10
11. $\frac{5}{14}$
12. 3.981
13. 0.125
14. 0.3%
15. 18
16. 0.0275

Appendix

Chp. 8, Section 5
Problem Set 5.2

1. $\bar{x} = 23.62$
2. $\bar{x} = 58.7$
3. $550

4. only a is true
5.

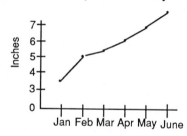

6. 26
7. false
8. false
9. $\frac{29}{7}$
10. 200
11. 0.023
12. 20%
13. 25%
14. $\frac{7}{9}$

Chp. 8, Section 6
Problem Set 6.1

1. a. 6   b. 7.5   c. 7   d. 6
   e. 3.1
2. false
3. a. 17.16   b. 4, 11, 20, 26, 31
4.

5. $11.40
6. 165.43
7. true
8. 130.401
9. 12.6%
10. $\frac{3}{20}$
11. 25%
12. 24
13. 0.051

Chp. 8, Section 6
Problem Set 6.2

1. $\bar{x} = 54.76$
2. $\bar{x} = 57.13$
3. 7
4. 9.5
5. false
6. true
7. a. 39.07   b. 5, 13, 22, 35, 42, 46
8. Between January and February

9. a. True   b. True   c. True   d. False
10. 2.45
11. 30
12. 0.7
13. 0.00005
14. $\frac{5}{4} = 1\frac{1}{4}$
15. 300
16. 0.065
17. $2\sqrt{5}$

Chp. 8, Section 6
Problem Set 6.3

1. $P_{37} = 50.8$
2. $P_{65} = 64.21$
3. $P_{99} = 97$
4. $P_{50} = 56.46$
5. $Q_1 = 44.21$
6. $Q_3 = 70.17$
7. a. $\bar{x} = 58.83$ b. $\bar{x} = 58.25$
8. 12
9.

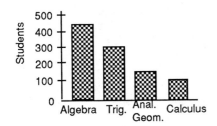

10. 100 liberals, 160 conservatives, and 140 moderates
11. 0.45
12. $\frac{59}{40} = 1\frac{19}{40}$
13. 13
14. $\frac{3}{2} = 1\frac{1}{2}$
15. 1.25
16. $\frac{4}{9}$
17. 62.5%

Chp. 8, Section 6
Problem Set 6.4

1. a. 5   b. The mode does not exist.
   c. 3 and 5
2. The modal class is 60-79.
3. a. 83.5   b. 55.29   c. 74.5
4. 12.5
5. true
6. false
7. $\frac{1}{7}$

8. $2\frac{11}{12}$
9. 7.468
10. 63%
11. $\frac{7}{20}$
12. $33\frac{1}{3}\%$
13. 0.028

Chp. 8, Section 6
Problem Set 6.5

1. a. 46%   b. 13%
2. a. 64%   b. 31%
   c. $50,000   d. $79,999
3. a. Australia-3, Japan-3
   b. USA-2, Australia-2
   c. USA-2.38, Japan-2.74
4. b
5. c
6. c

Chp. 8, Section 7
Problem Set 7.1

1. The numerical scale in the first graph exaggerates the results.
2. She reports the mode which in this case is a poor reflection of central tendency.
3. d
4. d
5. c
6. c
7. c
8. b

Chp. 8, Section 8
Problem Set 8.1

1. 1.82 which approximates $\sqrt{3.33}$
2. 2.69 which approximates $\sqrt{7.25}$
3. 4
4. 2
5. b
6. c

Appendix

7. Group 1
8. 17
9. a. 20.25   b. 20.17   c. 19.08
   d. 25.75   e. 7,16,28,43,51,56,60
10. False
11. 10
12. $360

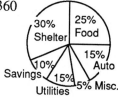

13. a. $\bar{x} = 10.22$ (or $1,022)
$$\frac{10 + 15 + 4 + 11 + 12 + 18 + 14 + 5 + 3}{9}$$
   b. $\bar{x} = 11$ (or $1,100)
   c. No mode is present.
14. $\frac{9}{4} = 2\frac{1}{4}$
15. 2.34
16. 20
17. $2\sqrt{2}$

Chp. 8, Section 9
Problem Set 9.1

1. a. True   b. False   c. False   d. True
   e. True   f. True   g. False
2. c
3. a. 513.3   b. 520   c. 400
4. c

Chp. 8, Section 9
Problem Set 9.2

1. a. 2   b. 3   c. 1.5
2. a. 84%   b. 0.84   c. 16%   d. 0.16
   e. 13.5%   f. approximately 0.14
   g. 68%   h. 0.68
3. a. 680   b. 160
4. 2.16 which approximates $\sqrt{4.67}$
5. a. 42.25   b. 42.83   c. 54.5   d. 32.83
   e. 2,7,16,28,36,40   f. 40-49
6. 4.16
7. 4.5

8. There is no mode.
9. Between March and April

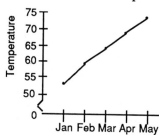

10. $\frac{4}{5}$
11. $4\frac{5}{12}$
12. $\frac{49}{88}$
13. $\frac{3}{5}$
14. 18
15. $\frac{7}{12}$

Chp. 8, Section 9
Problem Set 9.3

1. a. 16%   b. 68%
2. 47.5%(34% + 13.5%)
3. 81.5%(47.5% + 34%)
4. $\bar{x} = 0$, $s = 1$
5. 68%, 95%, 99%
6. a. 84%   b. 16%
7.

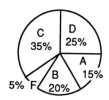

8. 3.11 which approximates $\sqrt{9.67}$
9. a. 78.11   b. 78.07
   c. 98.07   d. 57.5
10. 21
11. 0.0008
12. $\frac{1}{3}$

Chp. 8, Section 9
Problem Set 9.4

1. a. 64.1%   b. 69.2%
    c. 2.9%   d. 3.7%
2. a. 6.7%   b. 62   c. 56   d. 4.9%
3. a. 625   b. 30.8%   c. 241
4. True
5. 2.39 which approximates $\sqrt{5.71}$
6. a. 15.93   b. 15.68
    c. 10.65   d. 20.06
    e. 15-19   f. 11,24,41,50,54,56
7.

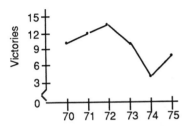

8. 325 women, 175 men
9. $\frac{3}{8}$
10. 17
11. 3.3
12. $\frac{9}{20}$

Chp. 8, Section 9
Problem Set 9.5

1. True
2. True
3. True
4. b
5. c
6. b
7. c
8. d
9. c
10. c
11. $66\frac{2}{3}\%$

Chapter 8 Self-Test

1. c
2. b
3. d
4. a
5. b
6. c
7. c
8. c
9. b
10. c
11. d
12. b
13. b
14. b
15. d
16. a
17. a
18. a
19. b
20. c
21. b
22. a
23. a
24. b
25. c
26. c
27. b
28. d
29. d

Index